GROL + RADAR

FCC COMMERCIAL RADIO LICENSE PREPARATION STUDY MANUAL

Element 1 – Radio Law
Element 3 – General Radiotelephone Operator
Element 8 – RADAR Endorsement

Gordon West, WB6NOA

with

Pete Trotter, KB9SMG
and
Eric P. Nichols, KL7AJ , KL7AJ, Technical Editor

This book was developed and published by:
Master Publishing, Inc.
Arlington, Texas

Author:
Gordon West, WB6NOA

Editing by:
Pete Trotter, KB9SMG
Eric Nichols, KL7AJ

Book Design and Typesetting by:
Kymm! Bang

Master Publishing, Inc.
2000 E. Randol Mill Road
Suite 608-A
Arlington, TX 76011
817-860-3800
817-548-9594 (fax)
e-mail: sales@w5yi.org
www.w5yi.org

QUESTION POOL RELEASE
The Element 1, Element 3, and Element 8 question pools contained in this book were adopted by the Federal Communications Commission on June 26, 2009, and must be used for all Commercial License examinations given on and after December 26, 2009. The question pools included in this book are the complete, current pools used to create examinations as of the date of publication of this book.

Revised First Edition

10 9

Table of Contents

Acknowledging Five Years of Contributing Experts!

In the Fall of 1993, the Federal Communications Commission entered into an agreement with private-sector organizations for the management of its commercial radio operator licensing program. The Commercial Operator License Examination Managers (COLEMs) replaced FCC testing, and privatized the examination program. Within a scant six months, a new set of question pools for Elements 1, 3, and 8 were developed by industry experts.

This original set of 1993 pools underwent a very minor revision in 1996, and has not been updated since. Needless to say, communications technology has changed greatly in the 16 years since the original question pools were developed. Communications industry experts all agreed that elements 1, 3, and 8 question pools all needed a big revision. Beginning in 2004, my publisher, Pete Trotter of Master Publishing, and I, became the leaders in this extensive question pool upgrade project. We had lots of help, and we wish to thank each and every one of the more than 100 communications experts who kept us focused and on track. Their contributions have helped to create the new question pools that are the subject of this new ***GROL + RADAR*** examination study guide.

We wish to publicly acknowledge the many electronics experts who dedicated a great deal of time, talent, and deep knowledge to the project:

Richard Glass, founder and former Executive Director of the Electronics Technician Association International (ETA-I), organized a "Blue Ribbon Subject Matter Experts Committee" that established subcategories to better organize the question pool, and developed hundreds of new questions for the Element 3 pool.

Ira Wiesenfeld, P.E., Senior CET, WA5GXP, was responsible for many of the revisions used to update the Element 3 examination, identifying obsolete questions that were deleted from the previous version of the FCC exam, adding many new questions for the current FCC exam, and reviewing the entire question pool for the FCC for accurate answers on the questions.

In addition to Mr. Glass and Mr. Wiesenfeld, the following ETA-I Subject Area Experts made significant contributions to the development of the new Element 3 Question Pool: Jim Arcaro, CETsr, Wireless Communications Technician; Richard Agard, CET, Fiber Optics; John Baldwin, CETsr, South East Technical College; Doug Ferguson, CET, Communications Technician; Dave Golieri, CETma, Tennessee Technology Center; Tom Janca, CETsr, Communications Technician, Power Engineers; Mike Powers, ITT Technical Institute; Don Thomas, CETsr, retired Avionics Educator; Tom Walker, CETsr, Pellissippi State Technical College; William Woodward, FOT, PE-Communications Engineer, Ursa Navigation.

At the same time, the Element 1 pool on FCC rules and regulations and operating procedures underwent extensive review and revision from experts with the Radio Technical Commission for Maritime Service (RTCM), and its national GMDSS (Global Marine Distress Safety System) implementation task force. Captain Bob Markle, Director of RTCM, and Captain Jack Fuechsel, U.S. Coast Guard (Retired), reviewed this pool and provided valuable information and input for Element 1.

The Element 8 question pool for the Radar Endorsement was reviewed and updated by Captain Charles Pascal, a well-known West Coast maritime and amateur radio instructor. His work was enhanced by airline pilot George Tamayo, WD6EJO, who added valuable aviation information.

All three question pools underwent several years of question-and-answer scrutiny, with some questions being eliminated, others updated, and many new technology questions added.

With this work accomplished, it became time to bring an organizational concept to the three pools. This vision was provided by Owen Anderson, Chairman of the national GMDSS Task Force Training Group. The GMDSS Task Force is charged by the FCC and the U.S. Coast Guard with maintaining the GMDSS test pools. Mr. Anderson is a member of the faculty working with the GMDSS training program at Pacific Maritime Institute, in Seattle.

His plan for the three pools resulted in the development of Subelement Topic Areas with Key Topics within each Subelement. Each Key Topic contains six questions, with one question to be selected at random on an applicant's examination. This approach ensures that a question from each and every Key Topic will appear on your exam. This two year effort on Mr. Anderson's part insures an even balance of subject matter and test questions on the elements 1, 3, and 8 exams.

Marine and aviation electronics experts and instructors also provided valuable input to these question pools – a true team effort!

Ghassan Khalek, Federal Communications Commission
Captain Dermanelian, Chief of USCG Communications
Steve Spitzer, Technical Director, National Marine Electronics Association, (NMEA)
Walter Fields, SN, United States Power Squadron telecommunications expert
William H. Scholz, US Coast Guard Auxiliary, Chief, Telecommunications Division, Response Department
Jorge Arroyo, US Coast Guard AIS Regulations
Joe Hersey, US Coast Guard MF-DSC Communications
Ralph Sponar, RTCM Communications Technical expert
Chuck Husick, RTCM Recreational Vessel Task Force, Boat U.S.
Captain Lisa Festa, US Coast Guard Rescue 21
Captains Carolyn and Jack West, Radar for Safer Boating
Captain Y. Portne, on-board RADAR photos
Russ Levin, GMDSS Question pools
Jim Dodez, KVH Industries, satellite communications
Chris Wahler, EPIRB expert, ACR Electronics
David Klein, Metal and Cable Corporation, copper foil grounding techniques
Spence Porter, Comm-Spec, rescue tags
Fred Maia, W5YI, FCC regulation updates
Ed Brady, US Coast Guard, High Frequency Communications
Frank Pfeiffer, math and formulas
Brian Mullan, INMARSAT
Captain Drew Rambo, US Coast Guard Coastal Network Upgrade to DSC
Chris O'Connors, NOAA NESDIS, COSPAS-SARSAT
Jack Hudson, W9MU, technical formula research
Larry Pollock, NB5X, National Radio Examiners COLEM, FCC Commission's Rules
Leroy Higgins, KI6MES, EIMAC, high-voltage technologies
Ken Sheehan, KD0AGV, US Navy, Retired, high math expert
Warren Moxley, K5WGM, Consultant to Elkins International COLEM, Antennas
Jim Shaw, AL7BA, parabolic antennas
Jerry Black, RF Engineering, aeronautical TACAS antennas
Alan Martin, KB3HIP, Retired FAA, aeronautical and mathematics
Don Leeds, KI6VRZ, antenna mathematics
Marc Goldman, WB6DCE, for his math and proofing expertise
James E. Hebert, KC1LU, math expert
Ian Keith, N8IK
Larry Pollock, NB5X

Our go-to RADAR and mathematics engineer is Eric Nichols, KL7AJ, in North Pole, AK. Eric is a fellow instructor for the GROL. He spent more than a year assisting with answer explanation rewrites. We appreciate his informative yet light-hearted style!

Many ham radio operators played an essential part in the final reviews of the material in our new book. The American Radio Relay League HANDBOOK is a valuable resource for communications electronics information. The U.S. Navy and its Point Magu RADAR courses were excellent additional resources.

I must also thank my wife Suzy, N6GLF, for her skill and patience typing and re-typing new questions, answers, and explanations.

Final credit goes to Mr. Pete Trotter, KB9SMG, this book's editor and publisher, who dedicated thousands of hours over the course of 5-plus years to coordinating the efforts of the hundreds of outside experts who created the test question pools for a well balanced examination preparation study manual.

Experts like YOU, in both training and teaching aviation and marine electronics, can help improve subsequent printings of this book. (We will gladly give you credit, too!) Please email us your suggestions, comments, and corrections! I personally see and review every comment. E-mail them to MasterPubl@aol.com, or directly to me at:WB6NOA@arrl.net

Thanks to everyone for years of hard work!

Gordon West, WB6NOA
Master Publishing, Inc.

Recommended Organizations and Reading

The United States Power Squadron
The USPS is the publisher of a guide, ***The Boat Owners' Guide to GMDSS and Marine Radio***, which may be ordered directly from USPS by calling 1-888-367-8777. Walter Fields is the author and the book includes a CD for Digital Selective Calling simulation.

National Marine Electronics Association
The NMEA offers books, and classes for training and certification as a marine electronics installer. Their Installation Standards for Marine Electronic Equipment Used on Moderate-Sized Vessels also includes American Boat and Yacht Council (ABYC) recommendations. The National Marine Electronics Association (NMEA) supports technical marine electronics dealers and YOU, the electronics technician, through training and certification courses and programs. Every major marine electronics dealer and manufacturer in the USA is a member of this non-profit organization. I am one of its original members, and I encourage all marine electronics technicians to look at their website, and consider joining NMEA. www.NMEA.org

Radio Technical Commission for Maritime Services
If you plan to work aboard large yachts and passenger ships, there is a good chance that RTCM standards and RTCM activities had a lot to do with the radio communication and navigation equipment aboard those vessels. RTCM is a non-profit scientific and educational organization focused on maritime radio communications. Their GMDSS task force committee is always looking for marine electronics technicians to join in on the recommendations they share with the US Coast Guard, the Federal Communications Commission, the general public, and the marine electronics industry. I have been a member of RTCM for years, and encourage you to learn more at www.RTCM.org/overview

Thanks to all of these organizations and all of these individuals who helped shape the new GROL and Radar Endorsement Question Pool and this one-of-a-kind book!

There is a great deal of complicated electronic theory and math covered in the questions included in the Element 3 question pool. While we have attempted to provide answer explanations that properly convey the understanding you need to know why the correct answer is indeed correct for individual questions, there is a great deal of background knowledge assumed in these explanations. You may find it helpful to round out your study of electronics theory and math with additional resources. Three such resources are the following books, which have been used by the author to provide illustrations and explanatory material in this edition of ***GROL + RADAR***:

Basic Electronics by Gene McWhorter and Alvis J. Evans
Basic Communications Electronics by Jack Hudson and Jerry Luecke
Basic Digital Electronics by Alvis J. Evans

These three books are written at an introductory level and can help provide you with an enhanced knowledge of the electronics associated with advanced radio communications used in today's marine, aviation, and RADAR fields. They are available from The W5YI Group on-line at www.w5yi.org, or by calling 800-669-9594.

How to Use This Book

Our ***GROL*** + ***RADAR*** FCC Commercial Radio License preparation manual is intended as a first and last reference in your preparation for the Commercial FCC radio examinations. Every question and multiple choice answer on your upcoming written examination for Elements 1, 3, and 8 appears word-for-word in this book. Examinations must be constructed from these official FCC question pools.

The only thing you will not see on your actual FCC exam is my spirited explanation of the correct answer! These explanations are written to help you understand the concepts behind the questions and to learn important material about the question topic. But they are not the only educational resource you will need to successfully learn the material covered by the question pools.

Each of the three question pools contained in this book are organized into Key Topics. Each Key Topic contains 6 questions, and one question from each Key Topic will appear on your written examination. Here is a summary of each pool:

- Element 1, on Radio Law, contains 24 Key Topics with a total of 144 questions in the pool. You must correctly answer a minimum of 18 of the 24 questions to pass your examination.
- Element 3, on electronics, contains 100 Key Topics with a total of 600 questions in the pool. You must correctly answer a minimum of 75 of the 100 questions to pass your examination.
- Element 8, on RADAR, contains 50 Key Topics with a total of 300 questions in the pool. You must correctly answer a minimum of 38 of the 50 questions to pass your examination.

The Element 1 Marine and Aviation Radio Law, Element 3 Electronic Fundamentals and Techniques, and Element 8 RADAR Endorsement question pools contained in this book were adopted by the Federal Communications Commission on June 26, 2009, and must be used for all Commercial License examinations given on and after December 26, 2009. The question pools included in this book are the complete, current pools used to create examinations as of the date of publication of this book.

Simply passing an FCC written examination, as hard as that may be, is not an indicator of your knowledge or skill. I encourage all of our FCC Commercial Radio Licensing readers to use this book in conjunction with classroom sessions, and with professional home-study schools via computer. Many of the COLEMs, found listed in the back of the book, offer world-class training programs. They have on-staff electronics professors who teach classes and offer online electronics courses. I encourage you to enroll in one soon!

Preface

FCC Commercial Radio Operator licenses are important confirmation of your knowledge and skill in the realm of aviation, marine, and land-mobile communications. These documents are key credentials that you will need if you are the Captain of a small passenger vessel, an electronics technician on a large ocean-going ship, or a technician repairing and maintaining ship, aircraft, or land-based radio equipment or marine RADARs.

This book is your guide to obtaining one or several of these valuable Federal Communications Commission Commercial Radio Licenses.

This book makes studying for your examination easy – I explain every exact question and its correct answer in detail, and apply this examination study to the real world of marine and aviation radio.

Chapter A is your history lesson. I tell where wireless began, the reasons for radio regulation and operator licensing, and the opportunities that await you in the wireless telecommunications field.

Chapter B explains FCC marine, land, and aviation radio operator licensing. If you dream about sailing the seven seas as a cruise ship radio operator, this chapter details how this book is the prerequisite to the higher grade cruise ship license requirements. As a marine and avionics radio and RADAR installer, this book will prepare you for the written exams.

Chapter C explains the Global Maritime Distress and Safety System – the international radio lifesaving system, and why aviation and distress radio channels are the same all around the world! I bring you up-to-date on satellite signaling EPIRBs, and even preview ship collision avoidance systems and search and rescue transponders.

The Element 1 Question Pool covering FCC radio law, rules and regulations is found in Chapter D. Every question gets my detailed explanation, along with fact charts illustrating why the right answer is indeed the correct one. Passing the Element 1 examination earns you your Marine Radio Operator Permit.

The Element 3 Question Pool is the FCC's technical exam and is found in Chapter E. Passing this exam earns you your General Radiotelephone Operator License (GROL). I have some fun shortcuts to help you understand the math, and to double-check complicated calculations.

The Element 8 Question Pool is in Chapter F. Passing this exam earns you your Ship RADAR Endorsement. This exam is required only if you plan to service air, land, and marine RADARs.

Passing one, two, or all three of these FCC Commercial License exams is an indication of your knowledge. But the foundation of any competent radio operator or maintainer is their training and education. Knowing the answers to the questions in this book is one thing – understanding what the questions mean in the real world is what is truly important. So I encourage you to purse your education well beyond the examinations in this book.

There are numerous ways to obtain classroom training; training that will bring these book Q & As into sharp focus in the marine and aviation electronics fields. Community colleges offer electronics training. Marine institutes offer "hands-on" training with live GMDSS consoles, ARPA RADAR screens, and more. There are study lesson plans on marine and aviation electronics with interactive, on-line computer electronic schools!

The NMEA offers marine electronic technician certification, advanced marine electronics installation training, and NMEA 2000 wiring classes throughout the country.

Flight schools may offer airframe electronics classes, and the Navy offers electronics and RADAR books over the internet. Most important, use this book with professional marine, land mobile, and aircraft "live gear" schools!

I look forward to meeting up with you down at the docks, out on a ship, or at a local aircraft hangar!

Gordon West, WB6NOA

CHAPTER A

A Brief History of Radio Regulation

Radio communication is more than 110 years old. It began with spark-gap dots and dashes, grew into a worldwide form of voice communications – thanks largely to amplitude modulation (AM) and shortwave frequencies – and today encompasses a wide array of modern technologies that can transmit data, voice, and pictures, employing everything from simple wire antennas to satellites orbiting the Earth.

From its earliest days as a viable form of communication, wireless has been seen as an invaluable tool to assure public safety at sea, in the air, and on land. *Table 1-1* lists some significant events and important actions between 1835 and 1910 that furthered electronic communications, which began with the invention of the telegraph.

Table 1-1. Electronic Communications 1835-1910

Year	Event and Action
1835	Electronic communications begin with the invention of the telegraph by Samuel F. B. Morse, a professor at New York University. Morse code is named after him, and was the international CW code used for many decades especially in the Maritime Service. With the advent of GMDSS, the primary use of this code today is in the Amateur Radio Service.
1849	Two European countries are linked by telegraph, causing the development of initial international agreements on rules and regulations governing the sharing of information.
1865	25 European nations meet in Paris to form the International Telegraphic Convention, which later becomes the International Telecommunications Union (ITU).
1865	Italian inventor Guglielmo Marconi proves the feasibility of radio communications, for which he receives a patent in 1897.
1899	Marconi makes the first successful transatlantic radio transmission from England to Newfoundland.
1901	Marine radio is born when the U.S. Navy adopts a wireless system.
1903	First international conference on governing radio communications is held in Berlin. Nine nations agree that public safety takes precedence over squabbles between commercial ventures.
1906	International conferees meeting in Berlin agree to require that ships be properly equipped with wireless transmitters and receivers, and to set the first international distress frequency as 500 kHz for ships to use to call for help. International regulation of wireless radio is added to the ITU's responsibilities.
1906	Reginald Aubrey Fessenden makes the world's first voice radio broadcast on Christmas Eve and again on New Year's Eve, one week later. Other Fessenden firsts besides voice over radio was the first two-way trans-Atlantic radio communications, the first voice heard across an ocean.
1909	Steamships Republic and Florida collide off the coast of New York and 1500 lives are saved by a distress call sent by radio operator Jack Binns. Later in the year, the S.S. Arapahoe calls for help using "SOS," which is adopted this year as the international radiotelegraph distress call. "Mayday" is adopted in 1927 as the international distress call for radiotelephony.

The Titanic Disaster

During the early years of the 20th century, the use of radio communications remained confined primarily to ships at sea. Few ships operated radio equipment, and those that did saw no reason to staff their stations around the clock. Doing so was considered an unnecessary luxury. All of that changed when the Titanic was ripped open by an iceberg in the North Atlantic and sent to the bottom of the sea just three hours later on the night of April 15, 1912.

Sinking of the Titanic led to the adoption of important radio regulations to help assure life safety of all ships at sea.

Source: www.titanic-experience.com

While the Titanic was sinking, her radio operator frantically called for help over the wireless. The Carpathia, 58 miles away, heard the call, responded, and managed to rescue 700 survivors. The Californian was just 20 miles away from the Titanic and could have rushed to the scene much faster. But the Californian radio operator had gone to bed, there was no one to relieve him, and the call went unheard. Some 1,500 passengers and crew perished!

The U.S. Congress had passed the Wireless Ship Act of 1910, which might have helped prevent the loss of so many lives, but the law's requirements were not comprehensive enough. Following the Titanic disaster, the Act was amended to require a minimum of two radio operators on board ships, a constant 24-hour watch, and the installation of backup power supplies to assure radio communications in emergency situations.

The Advent of Commercial Aviation

While shipping was a mature industry hundreds of years old, aviation was in its infancy. Wilbur and Orville Wright made their historic first flight just 9 years before the Titanic disaster. Aviation grew quickly, and the first U.S. airline was operating – however briefly – in Florida. It was the St. Petersburg-Tampa Airboat Line, which ran the world's first regularly scheduled airline service using heavier-than-air craft from January through March 1914.

Early Radio Legislation

Congress enacted the first law providing domestic control of general radio communications in 1912, partially in response to the Titanic. The Radio Act of 1912 was the beginning of government radio licensing. The Act placed the control of all wireless stations under the jurisdiction of the Department of Commerce and made access to the electromagnetic spectrum a privilege granted only by government approval. The law regulated the type of emissions, the transmission of distress calls, and set aside certain frequencies for government use.

At the time the Radio Act of 1912 was passed, the radio spectrum was so unoccupied that no one thought that frequencies would ever have to be assigned, much less shared. If you wanted to operate a transmitter, there was plenty of room in which to do it and all you had to do was apply for a license.

Radio was so new that very few understood its potential. Some barely knew what "wireless" meant. Radio pioneer Lee de Forest, the inventor of the Audion three-element vacuum tube amplifier (a very essential ingredient in the advancement of both wired and wireless communications) was prosecuted for mail fraud. The prosecutor accused de Forest of "...willfully and deliberately misleading the public by stating that soon it will be possible to transmit the human voice across the Atlantic Ocean."

A Pause in Development for World War I

The growth in commercial and public use of radio was placed on hold when the U.S. entered World War I in 1917. The Woodrow Wilson administration was concerned about possible misuse of radio by German spies. So the Federal government took over control of all commercial radio stations. And all amateur radio operators were required to cease operations and dismantle their stations under penalty of imprisonment. Radio manufacturers pooled their knowledge and expertise and turned their attention to putting this new technology to work winning the War.

World War I also provided the impetus for major advances in aircraft. New designs emerged for small fighter planes, as well as large, heavy bombers. These advances would lead to viable aircraft designs for commercial aviation after the War.

When World War I ended in November, 1918, the massive military market for radio transmitters and receivers disappeared.

The Growth of Commercial Broadcast Radio

In 1919, the American Marconi company was purchased by the Radio Corporation of America (RCA). David Sarnoff became its manager, and he led RCA into the radio business. Sarnoff envisioned a "radio music box" that would receive programs broadcast for public information and entertainment. Around the same time, General Electric and Westinghouse also began making radio receivers. Public demand for the new radio receivers was small at first because there were few stations to listen to. At the end of 1920, only 30 radio stations in the U.S. offered regular broadcasts. That changed rapidly as licensed stations went on the air and began regular broadcasting.

In Pittsburgh, Westinghouse engineer Dr. Frank Conrad set up an amateur radiotelephone station, 8XK, in 1916. Four years later, 8XK became KDKA, the nation's first commercial broadcast station transmitting on a wavelength of 360 meters (833 kHz). It signed on the air on election night, November 2, 1920, and today is the nation's oldest commercial radio station still in operation.

Radio history was made at 6:00 p.m. Tuesday, November 2, 1920 when four men gathered in a shack atop the Westinghouse Electric Building in Pittsburgh to broadcast election result reports to the public. KDKA, the world's first commercial broadcast radio station, was on the air!

Source: News Radio 1020, KDKA, Pittsburgh, PA

The 1920s saw a virtual explosion of growth in radio broadcasting. Licensing of broadcast stations on a regular basis began in 1921 when WBZ, Springfield, MA, became the first licensed station. By the end of 1922, there were 382 licensed stations, a number that nearly doubled to 733 stations by 1927! Most were operated by radio manufacturers, dealers and department stores selling receivers.

In today's connected world, it is difficult to imagine a time when the vast majority of citizens lived their entire lives without ever hearing the voices of the leaders of their own country. Radio changed that, and a great deal more. Radio receivers brought the world right into the home. People eagerly snapped up receivers to hear what was going on in the world around them. Music lovers could hear operas, symphonies, and more on their radios almost every night. Radio dramas and comedies became popular forms of in-home entertainment. Getting election results in hours instead of days was astonishing. Hearing the news faster than newspapers could print it was truly exciting.

The explosion of radio stations joining the airwaves proved overwhelming. From 1923 to 1926, Secretary of Commerce Herbert Hoover repeatedly submitted bills to Congress to straighten out the regulatory process. He convened conferences each year from 1922 to 1925 in an effort to entice voluntary cooperation from the radio industry. As the Federal government's man in charge, he used his power to grant or deny licenses, assign frequencies, and dictate the time of day when a station could operate. But Secretary Hoover understood that the Radio Act of 1912 no longer provided adequate regulation of the burgeoning radio industry.

Source: Herbert Hoover Presidential Library-Museum

Commerce Secretary Herbert Hoover listened to radio broadcasts on a special receiver installed in his home so he could better understand complaints received by his department and the needs of the listening public.

Although radio stations held licenses, they began bending the rules. Stations lengthened their broadcasting hours, changed frequencies, and increased power output without authorization. When one of these station's owners, Zenith, was taken to court, Zenith charged that the Secretary of Commerce had no legal authority to tell them when or where to transmit a radio signal. On July 8, 1926, the acting Attorney General of the United States agreed with Zenith, and decreed that the Secretary of Commerce had no legal authority to assign wavelengths, power levels, or hours of operation, or to restrict the length of a station's license. Officially, no government agency controlled radio for the next six months.

The predictable happened. Without regulations, hundreds of stations had a field day. They cranked up power levels and changed frequencies whenever they wished. New, unlicensed stations went on the air. The result was chaos. There was so much interference among stations that millions of listeners all across America switched off their receivers in disgust, and the sale of new radios slowed to a trickle. When radio receiver sales collapsed, the radio industry finally agreed that some form of government control was necessary.

The Federal Radio Commission

Secretary Hoover met with radio executives to discuss legislation and rules. With passage of the Radio Act of 1927, Congress created the Federal Radio Commission (FRC).

Led by five commissioners, Congress gave the FRC the authority to decide how much of the radio spectrum each service would be granted, and to change those grants if necessary. The FRC could legally refuse to grant license applications or renew them. Engineers were placed in charge of radio stations to keep up with the latest scientific developments, and such developments were incorporated into the FRC's rules and regulations as necessary. These changes were so sweeping and so well received that the Radio Act of 1927 has been called radio's "Magna Carta."

The International Radiotelegraph Convention also took place in Washington, DC, in 1927. Nearly every nation in the world joined in, deciding as a whole who would use what portions of the shortwave bands for different purposes.

FAA Origins

At the same time Secretary Hoover was dealing with radio regulation issues, he also was concerned about the emerging commercial aviation industry. The Air Commerce Act of May 20, 1926, was the cornerstone of the Federal government's regulation of civil aviation. This landmark legislation was passed at the urging of the aviation industry, whose leaders believed the airplane could not reach its full commercial potential without Federal action to improve and maintain safety standards.

Source: Wyoming State Museum

The Ford Tri-Motor, with a passenger capacity of 11, made its first flight on June 11, 1926. This one, operated by Standard Oil, drew a crowd at an air show in Casper, Wyoming, in 1928.

The Act charged the Secretary of Commerce with fostering air commerce, issuing and enforcing air traffic rules, licensing pilots, certifying aircraft, establishing airways, and operating and maintaining aids to air navigation. Radio, of course, would play an important role in aviation navigation and safety.

A new Aeronautics Branch of the Department of Commerce assumed primary responsibility for aviation oversight. The first head of the Branch was William P. MacCracken, Jr., who played a key part in convincing Congress of the need for this new governmental role.

On April 6, 1927, MacCracken received Pilot License No. 1, becoming the first person to obtain a pilot license from a civilian agency of the U.S. Government.

The FCC

While the FRC was a good beginning, it became clear that the scope of radio regulation needed to be expanded to encompass other forms of electronic communication, including telegraph and telephone services. Passage of the Communications Act of 1934, which expanded on the 1927 Act, replaced the FRC with the Federal Communications Commission (FCC). The President of the United States appoints the five FCC commissioners. No more than three of them may be from the same political party. The five-year terms are staggered so that they all cannot be replaced at one time. Congress continually monitors FCC operations and, because the FCC is no longer a permanent agency, it must be reauthorized by Congress every two years.

From its inception, the FCC has been a very powerful agency. Standards are explicitly laid out and updated continuously to reflect changes in technology. Stations that fail to follow them are subject to criminal prosecution. While the Communications Act of 1934 itself is only about 150 pages long, the regulations set up under the law today occupy many thousands of pages.

Key duties of the FCC include allocating radio frequency bands along international guidelines, assigning frequencies for various radio services and individual stations, and determining the operational and technical qualifications of radio operators. No radio or television station in the United States can be sold, moved, shut down, or change its operating hours or power level without express permission from the FCC. Because of licensing, the FCC's word is law on whether or not a station can legally exist.

As technology changes, the FCC works steadily to keep up. Cable television, digital data transmission over the telephone lines, satellite broadcasting, high-definition television (HDTV), and other advances in radio communication technique all fall under its jurisdiction. Whenever a new format is invented and brought to the FCC's attention, the FCC examines it and makes rules appropriate for its regulation. And, importantly, the FCC also works to make sure that U.S. radio regulations are in harmony with those of the International Telecommunciation Union.

Early Aeronautics Branch Responsibilities

In fulfilling its civil aviation responsibilities, the Department of Commerce initially concentrated on functions such as safety rulemaking and the certification of pilots and aircraft. It took over the building and operation of the nation's system of lighted airports, a task begun by the Post Office Department. The Department of Commerce improved aeronautical radio communications, and introduced radio beacons as an effective aid to air navigation.

As commercial flying increased, the Bureau encouraged a group of airlines to establish the first three centers for providing air traffic control (ATC) along the airways. In 1936, the Bureau itself took over the centers and began to expand the ATC system. The pioneer air traffic controllers used maps, blackboards, and mental calculations to ensure the safe separation of aircraft traveling along designated routes between cities.

The Civil Aeronautics Act

In 1938, the Civil Aeronautics Act transferred the Federal civil aviation responsibilities from the Commerce Department to a new independent agency, the Civil Aeronautics Authority. The legislation also expanded the government's role by giving the Authority the power to regulate airline fares and to determine the routes that air carriers would serve.

In 1940, President Franklin Roosevelt split the Authority into two agencies, the Civil Aeronautics Administration (CAA) and the Civil Aeronautics Board (CAB). CAA was responsible for ATC, airman and aircraft certification, safety enforcement, and airway development. CAB was entrusted with safety rulemaking, accident investigation, and economic regulation of the airlines. Both organizations were part of the Department of Commerce. Unlike CAA, however, CAB functioned independently of the Secretary.

The DC-3 first flew on December 17, 1935, exactly 32 years after the first flight of the Wright brothers. More than 10,000 were built, and many are still flying.

On the eve of America's entry into World War II, CAA began to extend its ATC responsibilities to takeoff and landing operations at airports. This expanded role eventually became permanent after the war. The application of RADAR to ATC helped controllers in their drive to keep abreast of the postwar boom in commercial air transportation. In 1946, Congress gave CAA the added task of administering the Federal-aid airport program, the first peacetime program of financial assistance aimed exclusively at promoting development of the nation's civil airports.

The Birth of FAA

Today, air travel is carefully managed by air traffic controllers to ensure passenger safety.

The approaching introduction of jet airliners and a series of midair collisions spurred passage of the Federal Aviation Act of 1958. This legislation transferred CAA's functions to a new independent body, the Federal Aviation Agency (FAA) that had broader authority to combat aviation hazards. The act took safety rulemaking from CAB and entrusted it to the new FAA. It also gave FAA sole responsibility for developing and maintaining a common civil-military system of air navigation and air traffic control, a responsibility CAA previously shared with others.

The scope of the Federal Aviation Act owed much to the leadership of Elwood "Pete" Quesada, an Air Force general who served as President Eisenhower's principle advisor on civil aeronautics. After becoming the first Administrator of the agency he helped to create, Quesada mounted a vigorous campaign for improved airline safety.

The FCC's Aviation Radio Services

The aviation radio service is an internationally-allocated radio service providing for safety of life and property in air navigation. There are two types of aviation radio services:

- Aircraft Radio Stations are stations in the aeronautical mobile service that use radio equipment, such as two-way radiotelephones, RADAR, radionavigation equipment, and emergency locator transmitters (ELTs), on board aircraft for the primary purpose of ensuring safety of aircraft in flight.
- Ground Radio Stations are usually of two types. The Aeronautical and Fixed Service includes stations used for ground-to-air communications with aircraft about aviation safety, navigation, or preparation for flight. The Aeronautical Radionavigation Service is made up of stations used for navigation, obstruction warning, instrument landing, and measurement of altitude and range.

Like all other radio services, the FCC coordinates the U.S. aviation radio services in compliance with the ITU regulations. Part of this work involves issuing licenses for the operation and maintenance of aircraft radio and RADAR equipment.

The ITU

Almost every nation in the world belongs to the International Telecommunication Union (ITU). This United Nations agency is headquartered in Geneva, Switzerland. Since radio waves do not respect international boundaries, global telecommunications standards are agreed upon through the ITU's Radiocommunication Sector (ITU-R).

The ITU-R plays a vital role in the global management of the radio-frequency spectrum and satellite orbits. The appropriate use of these limited natural resources ensures safety of life on land, at sea, and in the skies. Use of portions of the radio spectrum is increasingly in demand from a large and growing number of services, such as fixed, mobile, broadcasting, amateur, space research, emergency telecommunications, meteorology, global positioning systems, environmental monitoring and communication services.

The ITU-R's mission is to ensure the rational, equitable, efficient and economical use of the radio-frequency spectrum by all radiocommunication services, including those using satellite orbits, and to carry out studies and approve recommendations on radio communication matters.

In implementing this mission, ITU-R aims at creating the conditions for harmonized development and efficient operation of existing and new radio systems.

The ITU-R's primary objective is to ensure interference free operations of radio systems. This is accomplished through implementation of the Radio Regulations and Regional Agreements, and the efficient and timely update of these instruments through the processes of World and Regional Radiocommununication Conferences. Furthermore, radio standardization establishes "Recommendations" intended to assure the necessary performance and quality in operating radio systems. It also seeks ways and means to conserve spectrum and ensure flexibility for future expansion and new technological developments.

The ITU divides the world into three general regions. North and South America are located in ITU Region 2, as shown below. As an aid to enforcement of radio laws, radio stations identify themselves using call signs that begin with certain ITU assigned prefixes. All U.S. stations begin with K, N, W and certain A prefixes.

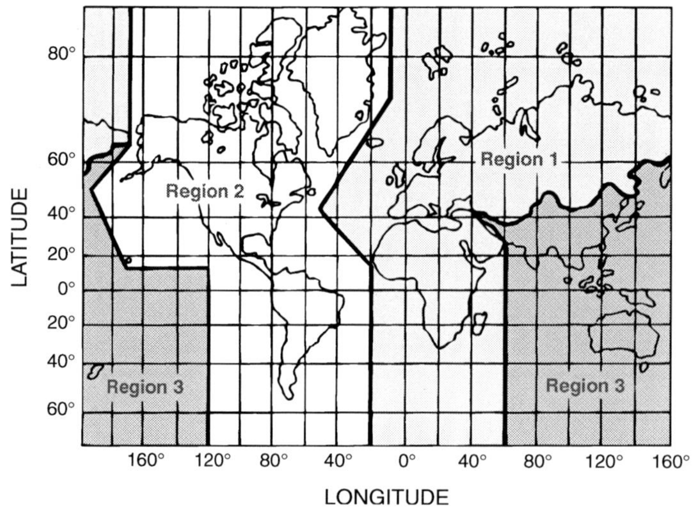

ITU Regions

ITU agreements are fine-tuned by individual governments. In the United States, the FCC regulates all non-government radio frequencies. Federal government spectrum is under the jurisdiction of the President and managed by the National Telecommunications and Information Administration (NTIA).

Summary

The growth of radio technology since its invention more than 100 years ago has required a parallel growth of regulation and spectrum management, both in the United States and internationally. Through the work of the ITU and FCC, we enjoy the benefits of radio in all its forms, from the safety communications systems on sea, land, and in the air, to the entertainment media that bring news, music, sports programming, and so much more into our daily lives.

The purpose of this book is to help you earn your Commercial Radio Operator or Maintainer License from the Federal Communications Commission. Understanding the history of radio licensing provides you with a framework for your study, and an understanding of the necessity of FCC licensing.

CHAPTER B

Commercial Radio Operator/ Maintainer Licenses & Examination Requirements

The United States of America

Federal Communications Commission

GENERAL RADIOTELEPHONE OPERATOR LICENSE

(General Radiotelephone Certificate)

This certifies that the individual named below is a licensed radio operator and is authorized to operate licensed radio stations for which this class of license is valid. The authority granted is subject to any endorsement placed on this license. The authority granted is also subject to the orders, rules, and regulations of the Federal Communications Commission, the statutes of the United States, and the provision of any treaties to which the United States is a party, which are binding upon radio operators.

This license may not be assigned or transferred to any other person. This license is valid for the lifetime of the holder unless suspended by the FCC.

Endorsement: SHIP RADAR ENDORSEMENT

Licensee: GORDON V. WEST

Place of Issuance: LONG BEACH, CA.

Date of Birth	Issuance Date	License Number
JUNE 22, 1942	JANUARY 2, 1985	PG-11-30673

Signature of Licensee

FCC 756-E
January 1985

The Federal Communications Commission grants two basic classes of radio operator licenses –commercial and amateur. Both licensing programs exist for the same reasons: to prevent interference to other radio stations; to assure technician and operator qualifications, and to bring order to the use of the radio spectrum.

The FCC issues a commercial radio operator or maintainer license when an applicant passes an examination demonstrating adequate knowledge of radio rules, operating procedures, and electronics. The holder of a commercial radio operator license or permit is not authorized to operate amateur radio stations. Only a person holding an amateur radio operator license may operate an amateur radio station. Only a person holding a commercial radio operator license may operate a commercial radio station.

Besides being able to pass the necessary radio law and technical examinations, applicants for commercial radio licenses must be citizens of the United States, or be eligible for employment in the United States, and must be able to transmit and receive messages in English.

WHO NEEDS A COMMERCIAL RADIO OPERATOR LICENSE?

Here is a summary of when you need – and do not need – an FCC commercial radio operator license to operate and/or repair and maintain radio stations.

Radio Operating

You ***need*** a commercial radio license to operate the following:

1. Ship radio stations if:
    a. the vessel carries more than six passengers for hire; or
    b. the radio operates on medium or high frequencies (300 kHz to 30 MHz); or

c. the ship sails to foreign ports; or
d. the ship is larger than 300 gross tons and is required to carry a radio station for safety purposes.

2. Coast (land) stations that operate on medium or high frequencies, or operate with more than 1500 watts of peak envelope power.
3. Aircraft radio stations (except those that use only VHF frequencies [higher than 30 MHz] on domestic flights).
4. Civil Air Patrol stations that operate on medium or high frequencies.
5. International fixed public radiotelephone and radiotelegraph stations.
6. Coast and ship stations transmitting radiotelegraphy.

You ***do not need*** a commercial radio operator license to operate the following:

1. Coast stations operating on VHF frequencies with 250 watts or less of carrier power.
2. Ship stations operating only on VHF frequencies while sailing on domestic voyages.
3. Aircraft stations which operate only on VHF frequencies and do not make foreign flights.

Radio Maintenance and Repair

You ***need*** a commercial radio license to maintain and repair the following:

1. All ship radio and RADAR stations.
2. All coast radio stations.
3. All hand-carried radio units used to communicate with ships and coast stations on marine frequencies.
4. All aircraft stations and aeronautical ground stations (including hand-carried portable units) used to communicate with aircraft.
5. All aircraft RADAR stations.
6. International fixed public radiotelephone and radiotelegraph stations.

You ***do not need*** a commercial radio license to operate, repair or maintain any of the following types of stations:

1. Two-way land mobile radio equipment, such as that used by police and fire departments; taxicabs and truckers; businesses and industries; ambulances and rescue squads; local, state and federal government agencies.
2. Personal radio equipment used in the Citizens Band, Family Radio Service, Radio Control, and General Mobile Radio Services.
3. Auxiliary broadcast stations, such as remote pickup stations.
4. Domestic public fixed and mobile radio systems, such as mobile telephone systems, cellular systems, rural radio systems, point-to-point microwave systems, multipoint distribution systems, etc.
5. Stations operated in the Cable Television Relay Service.
6. Satellite stations, both uplink and downlink, of all types.
7. AM, FM, and TV Broadcast stations. (Note: While not required, many licensees of broadcast stations require their engineers and technicians to hold a General Radiotelephone Operator License.)

Important Cautions!

These listings only describe when radio operator licenses are necessary. Before you operate any radio station, ***make certain that the station is licensed*** as required.

TYPES OF LICENSES, PERMITS & ENDORSEMENTS

Prior to 1984, broadcast service radio and TV engineers needed the FCC First Class commercial license. Land mobile radio techs were required to have at least the Second Class commercial license. The Third Class license granted "on air" privileges, but did not allow equipment adjustments.

In 1984, the FCC discontinued the First Class and Second Class Radiotelephone Operator Licenses and replaced them with the General Radiotelephone Operator License (GROL). The Third Class Radiotelephone

Operator Permit, Aircraft Radiotelegraph Endorsement, and the Broadcast Endorsement were eliminated and not replaced.

Effective May 20, 2013, the FCC created a new Radiotelegraph Operator License (T), and eliminated 3 previous certificates, the Third Class (T3), Second Class (T2), and First Class (T1) Radiotelegraph Operator Certificates, which were valid for 5 year terms, and which must be renewed every five years. Upon renewal, those holding T1 and T2 licenses will be issued a new Radiotelegraph Operator License (T), which is valid for the lifetime of the holder. Upon renewal, those holding a Third Class (T3) Radiotelegraph Operator Certificate will be issued a Marine Radio Operator Permit (MP), which also is a lifetime license.

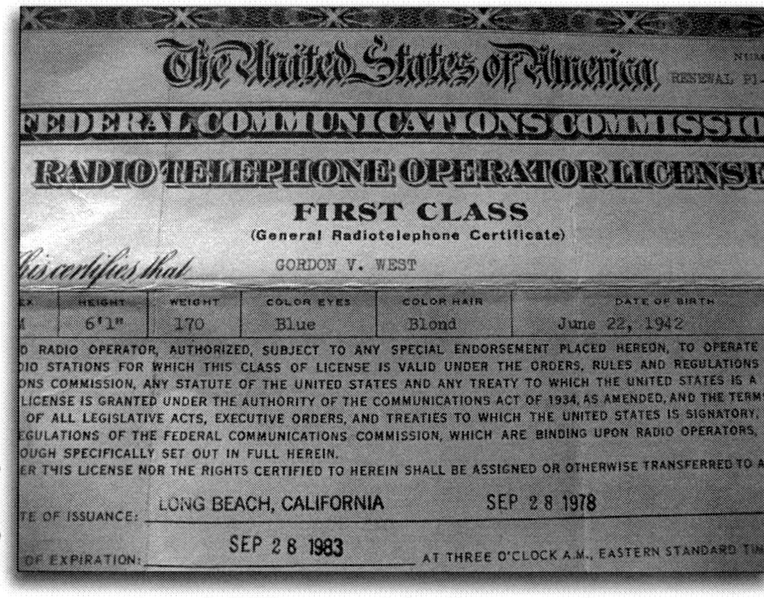

An early FCC commercial radio operator license.

Today, the Federal Communications Commission issues six types of commercial radio operator / maintainer licenses; three types of permits, and two types of endorsements. These are listed below in Tables 2-1 through 2-9. Effective May 20, 2013, all new commercial radio licenses issued by the FCC are valid for the lifetime of the holder.

Table 2-1. Restricted Radiotelephone Operator Permit (RR) / (RL)
Restricted Radiotelephone Operator Permit (RR) holders are authorized to operate most aircraft and aeronautical ground stations. They also can operate marine radiotelephone stations aboard pleasure craft (other than those carrying six or more passengers for hire on the Great Lakes or bays or tidewaters or in the open sea) when operator licensing is required. RRs are issued to Residents of the U. S. for the lifetime of the holder. RLs are issued to Non-Resident aliens for the lifetime of the holder.

Residents (RR)

An RR is issued without any examination. To qualify for an RR, you must meet all four of the following requirements:

1. Be either a legal resident of (or otherwise eligible for employment in) the United States, or hold an aircraft pilot certificate valid in the United States, or hold an FCC radio station license in your own name (see limitation on validity below).
2. Be able to speak and hear.
3. Be able to keep at least a rough, written log.
4. Be familiar with provisions of applicable treaties, laws and rules that govern the radio station you will operate.

Non-Residents (RL)

If you are a non-resident alien, you must hold at least one of the following three documents to be eligible for an RL:

1. A valid United States pilot certificate issued by the Federal Aviation Administration.
2. A foreign aircraft pilot certificate that is valid in the United States on the basis of reciprocal agreements with foreign governments.
3. A valid radio station license issued by the FCC in your own name. (An RL issued on this basis will authorize you to operate only your own station.)

You ***do not need*** an RR to operate the following:

1. A voluntarily-equipped ship or aircraft station, including a Civil Air Patrol (CAP) station, which operates only on VHF frequencies and ***does not*** make foreign voyages or flights.
2. An aeronautical ground or coast station that operates only on VHF frequencies.
3. On-board stations.
4. A marine utility station, unless it is taken aboard a vessel that makes a foreign voyage.
5. A survival craft station, when using telephony or an emergency position indicating radio beacon (EPIRB).
6. A ship RADAR station, if the operating frequency is determined by a fixed-tuned device and if the RADAR is capable of being operated by only external controls.
7. Shore RADAR, shore radiolocation, maritime support or shore radio-navigation stations.

Table 2-2. Marine Radio Operator Permit (MP)

A Marine Radio Operator Permit (MP) is required to operate radiotelephone stations aboard certain vessels that sail the Great Lakes. It also is required to operate radiotelephone stations aboard vessels of more than 300 gross tons, and vessels that carry six or more passengers for hire in the open sea or any tidewater area of the United States. It also is required to operate certain aviation radiotelephone stations and certain coast radiotelephone stations. MPs are issued for the lifetime of the holder.

You must meet all three of the following requirements to be eligible for a MP:

1. Be a legal resident of, or eligible for employment in, the United States.
2. Be able to receive and transmit spoken messages in English.
3. Pass a written examination covering basic radio law and operating procedures (Element 1).

Table 2-3. General Radiotelephone Operator License (PG)

A General Radiotelephone Operator License (PG) is required to adjust, maintain or internally repair FCC-licensed radiotelephone transmitters in the aviation, maritime and international fixed public radio services. A GROL also conveys all of the operating authority of the Marine Radio Operator Permit (MP). A GROL is issued for the lifetime of the holder. A GROL is also required to operate the following:

1. Any maritime land radio station or compulsorily equipped ship radiotelephone station operating with more than 1500 watts of peak envelope power (PEP).
2. Voluntarily-equipped (pleasure) ship and aeronautical (including aircraft) stations with more than 1000 watts of peak envelope power (PEP).

You must meet all three of the following requirements to be eligible for a GROL:

1. Be a legal resident of, or otherwise eligible for employment in, the United States.
2. Be able to receive and transmit spoken messages in English.
3. Pass a written examination covering basic radio laws and operating procedures (Element 1), and electronics fundamentals and techniques required to repair and maintain radio transmitters and receivers (Element 3).

Table 2-4. GMDSS Radio Operator's License (DO)

The GMDSS Radio Operator's License (DO) qualifies personnel as Global Maritime Distress and Safety System (GMDSS) radio operators for the purposes of operating GMDSS radio installations, including some basic equipment adjustments. It does not authorize repair and maintenance of GMDSS equipment. The GMDSS Radio Operator's License also confers all of the operating authority of the Restricted Radiotelephone Operator's Permit (RR), and the Marine Radio Operator's Permit (MP). The DO is valid for the lifetime of the holder.

You must meet all three of the following requirements to be eligible for a DO:

1. Be a legal resident of, or otherwise eligible for employment in, the United States.
2. Be able to receive and transmit spoken messages in English.
3. Pass written examinations covering basic radio law and maritime operating procedures (Element 1), and GMDSS radio operating practices (Element 7).

Table 2-5. Restricted GMDSS Radio Operator's License (RG)

The Restricted GMDSS Radio Operator's License qualifies personnel as GMDSS radio operators serving aboard compulsory ships that operate exclusively within Sea Area A1. Sea Area A1 is the area within VHF radiotelephone coverage of at least one coast station at which continuous DSC (digital selective calling) is available. The RG is valid for the lifetime of the holder.

You must meet all three of the following requirements to be eligible for this license:

1. Be a legal resident of, or otherwise eligible for employment in, the United States.
2. Be able to receive and transmit spoken messages in English.
3. Pass written examinations covering basic radio law and maritime operating procedures (Element 1) and GMDSS radio equipment required to be fitted on ships sailing solely within Sea Area A1 (Element 7R).

IMPORTANT NOTE:

U.S. Coast Guard Requirements for GMDSS and GMDSS-R Radio Operators

In order to be certified for employment as a GMDSS radio operator on board a mandatory ship, individuals must take and pass a U.S. Coast Guard-approved course in addition to obtaining the FCC GMDSS License (DO) or GMDSS Restricted License (RG). This requirement is part of the USGC's Standards of Training, Certification and Watchkeeping for Seafarers (STCW) program. Generally speaking, the exam taken at the end of the Coast Guard-approved course satisfies the FCC exam requirement.

You can learn more about the STCW requirements by visiting the USCG National Maritime Center (NMC) website at: **http://www.uscg.mil/nmc/stcw_unlicguide.asp#1**. You can also find a listing of schools offering USCG approved courses at the same website. The specific USCG requirements are spelled out in USCG Rules, Part 46, section 10.603, "Requirements for radio officers' licenses, and STCW certificates or endorsements for GMDSS radio operators." To view the full rules section, go to: **http://www.access.gpo.gov/nara/cfr/waisidx_02/46cfr10_02.html**

Table 2-6. GMDSS Radio Maintainer's License (DM)

The DM qualifies personnel as GMDSS radio maintainers to perform at-sea repair and maintenance of GMDSS equipment. It also confers all of the operating authority of the Restricted Radiotelephone Operator's Permit (RR), the Marine Radio Operator's Permit (MP), and the General Radiotelephone Operator License (PG). The DM is valid for the lifetime of the holder.

You must meet all three of the following requirements to be eligible for a DM:

1. Be a legal resident of, or otherwise eligible for employment in, the United States.
2. Be able to receive and transmit spoken messages in English.
3. Pass written examinations covering basic radio law and maritime operating procedures (Element 1), technical examination on electronic fundamentals and techniques (Element 3), and GMDSS radio maintenance practices and procedures (Element 9).

Note: A GMDSS Radio Maintainers License (DM) does not authorize operation of GMDSS equipment. In instances where an applicant qualifies for both a GMDSS Radio Operator's License (DO) and a GMDSS Radio Maintainers License (DM) at one sitting, a GMDSS Radio Operator/Maintainer License (DB) will be issued.

Table 2-7. Radiotelegraph Operator's License (T)

The Radiotelegraph Operator's License (T) authorizes the holder to operate, repair and maintain ship and coast radiotelegraph stations in the maritime services. It also confers all of the operating authority of the Restricted Radiotelephone Operator's Permit (RP), and the Marine Radio Operator's Permit (MP). The T is valid for the lifetime of the holder.

You must meet all four of the following requirements to be eligible for a T:

1. Be a legal resident of, or otherwise eligible for employment in, the United States.
2. Be able to receive and transmit spoken messages in English.
3. Pass Morse code examinations at 16 code groups per minute (16 CG, Telegraphy Element 1) and 20 words-per-minute plain language (20 PL, Telegraphy Element 2), both receiving and transmitting by hand.

4. Pass written examinations covering basic radio law and operating procedures for radiotelephony (Element 1), and electronics technology applicable to radiotelegraph stations (Element 6).

Notes on the Morse code requirements for the new T License

A Morse code hand-sending test will not be required for the new Radiotelegraph Operator's License (T). The FCC has taken the position that applicants who can receive telegraphy by ear can also send code manually.

Current or previous Amateur Extra Class licensees (those holding a current or expired license) receive examination credit for Telegraphy Element 1 (16 Code Groups per minute) and Telegraphy Element 2 (20 Plain Language words per minute) towards the Radiotelegraph Operator's License (T) providing the applicant passed the 20 words-per-minute Amateur Extra Class Morse code telegraphy examination (Element 1C) prior to April 15, 2000. In order to receive this credit, a photocopy of the Amateur Extra Class license dated prior to April 15, 2000 must be attached to the FCC Form 605 application.

Notes on the Renewal of T1, T2, and T3 Certificates

The FCC Report & Order effective May 20, 2013, created the new Radiotelegraph Operator License (T), and eliminated 3 previous certificates, the Third Class (T3), Second Class (T2), and First Class (T1) Radiotelegraph Operator Certificates. These certificates are valid for a 5 year term, and must be renewed. Upon renewal, those holding T1 and T2 licenses will be issued a new Radiotelegraph Operator License (T), which is a lifetime license. Upon renewal, those holding a Third Class (T3) Radiotelegraph Operator Certificate will be issued a Marine Radio Operator Permit (MP). These individuals will retain credit for having passed Telegraphy Elements 1 and 2, so they are able to upgrade to a Radiotelegraph Operator License by passing written Element 6. All 3 Certificates may be renewed from anytime in the last year of their term up to five years following expiration.

Table 2-8. Ship RADAR Endorsement
The Ship RADAR Endorsement may be placed on the General Radiotelephone Operator License (PG), GMDSS Radio Maintainer's License (DM), GMDSS Radio Operator / Maintainer license (DB), on the First or Second Class Radiotelegraph Operator's Certificate (T1 or T2), or the new Radio Telegraph Operator's License (T). Only persons whose commercial radio operator license bears this endorsement may repair, maintain or internally adjust ship and aircraft RADAR equipment. The Ship RADAR Endorsement is valid for the term of the license.

You must meet both of the following requirements to be eligible for a Ship RADAR Endorsement:

1. Hold (or qualify for) a General Radiotelephone Operator License (PG), a First or Second Class Radiotelegraph Operator's Certificate (T1 or T2), a GMDSS Radio Maintainer's license (DM), or a GMDSS Radio Operator / Maintainer license (DB), or a new Radio Telegraph Operator's License (T).
2. Pass a written examination (Element 8) covering special rules applicable to ship RADAR stations and the technical fundamentals of RADAR and RADAR maintenance procedures.

Table 2-9. Six-Months Service Endorsement
A Six Months Service Endorsement must be obtained by any holder of a T1, T2, or T who wishes to serve as the sole radio operator on board a large U.S. cargo ship sailing on the high seas.

You must meet all four of the following requirements to be eligible for a Six Months Service Endorsement:

1. You have been employed as a radio operator on board ships of the United States for a period totaling at least six months; and
2. The ships were in service during the applicable six-month period; and
3. You held a T1, T2, or T issued by the FCC during this entire six-month qualifying period; and
4. You hold a radio officer's license issued by the U.S. Coast Guard at the time the six months service endorsement is requested.

How to Obtain a Six Month Service Endorsement

For those applicants who are qualified by having at least 180 days of creditable service, the following is to be submitted to the Federal Communications Commission, 1270 Fairfield Road, Gettysburg, PA 17325-7245:

1. A certification letter signed by the vessel's master or owner/agent specifying the vessel name, vessel call sign, dates of shipment and discharge, total number of days served (minus any portion of any single in-port period exceeding seven days), and the names(s) and certificate number of the chief radio officer holding a six months service endorsement on the vessel during shipment; and
2. A completed FCC Form 605 Commercial Radio Operator License Application and Schedule E; and
3. An original or copy of the T License, or T1 or T2 Certificate; and
4. A valid copy of a U.S. Coast Guard license; and
5. Certificate(s) of Discharge to Merchant Seaman.

EXAMINATIONS & TESTS REQUIRED FOR LICENSES

There are a total of 8 written examinations and 2 telegraphy tests required for the various FCC commercial radio operator licenses, permits, certificates, and endorsements. *Table 2-10* summarizes the exams and tests required for each.

Table 2-10. FCC Commercial Radio License Examinations & Tests by License Type

Name of License Permit or Endorsement	FCC Acronym	Examination Elements									Term
		Written							Telegraphy		
		1	3	6	7	7R	8	9	1	2	
Restricted Radiotelephone Operator's Permit	RR / RL	No Exam Required									Lifetime
Marine Radio Operator's Permit	MP	•									Lifetime
GROL General Radiotelephone Operator's License	PG	•	•								Lifetime
GMDSS Radio Operator's License	DO	•			•						Lifetime
Restricted GMDSS Radio Operator's License	RG	•				•					Lifetime
GMDSS Radio Maintainer's License	DM	•	•					•			Lifetime
Radiotelegraph Operator's License	T	•		•					○	○	Lifetime
Ship RADAR Endorsement							•				
Six Months Service Endorsement		No Written Exam – Documentation Is Required									

Written Examination Content

All FCC written examinations are multiple-choice format. The question pools for each examination element are public information, and can be obtained from the FCC website. Here is a brief description of the content covered in each FCC written examination:

- **Element 1:** Basic radio law and operating practice with which every maritime radio operator should be familiar. To pass, an examinee must correctly answer at least 18 out of 24 questions. The Element 1 question pool contains a total of 144 questions.
- **Element 3:** General Radiotelephone. Electronic fundamentals and techniques required to adjust, repair, and maintain radio transmitters and receivers. To pass, an examinee must correctly answer 75 out of 100 questions, from the following topics: operating procedures; radio wave propagation; radio practice; electrical principles; circuit components; practical circuits; signals and emissions; and antennas and feed lines. The Element 3 question pool contains a total of 600 questions.
- **Element 6:** Advanced Radiotelegraph. Technical, legal, and other matters applicable to the operation of all classes of radiotelegraph stations. To pass, an examinee must correctly answer at least 75 out of 100 questions. The Element 6 question pool contains a total of 616 questions.

- **Element 7:** GMDSS Radio Operating Practices. GMDSS radio operating procedures and practices sufficient to show detailed practical knowledge of the operation of all GMDSS sub-systems and equipment. The exam consists of questions from the following categories: general information, narrow-band direct-printing, INMARSAT, NAVTEX, digital selective calling, and survival craft. To pass, an examinee must correctly answer at least at least 75 out of 100 questions. The Element 7 question pool contains a total of 600 questions.
- **Element 7R:** Restricted GMDSS Radio Operating Practices. Fifty questions concerning those GMDSS radio operating procedures and practices that are applicable to ship stations on vessels that sail exclusively in sea area A1, as defined in sections 80.1069 and 80.1081 of the Commission's Rules. To pass, an examinee must correctly answer at least 38 out of 50 questions. The Element 7R question pool contains a total of 300 questions.
- **Element 8:** Ship RADAR Techniques. Specialized theory and practice applicable to the proper installation, servicing, and maintenance of aircraft and ship RADAR equipment in general use for aviation and marine navigation purposes. To pass, an examinee must correctly answer at least 38 out of 50 questions. The Element 8 question pool contains a total of 300 questions.
- **Element 9:** GMDSS Radio Maintenance Practices and Procedures. Requirements set forth in IMO Assembly on Training for Radio Personnel (GMDSS), Annex 5 and IMO Assembly on Radio Maintenance Guidelines for the Global Maritime Distress and Safety System related to Sea Areas A3 and A4. The exam consists of questions from the following categories: radio system theory, amplifiers, power sources, troubleshooting, digital theory, and GMDSS equipment and regulations. To pass, an examinee must correctly answer at least 38 out of 50 questions. The Element 9 question pool contains a total of 250 questions.

Telegraphy Tests Content

There are 2 Morse code telegraphy tests that apply to the Radiotelegraph Operator's License (T):

- Element 1 - 16 code groups per minute.
- Element 2 - 20 words per minute.

Telegraphy exams may consist of both transmitting and receiving tests, or just a receiving test. A Morse code hand-sending examination probably will not be required. The FCC has taken the position that applicants who can receive telegraphy by ear can also send code manually. However, the testing company (COLEM) can require a sending segment in a telegraphy examination.

Examinees must copy by ear and, if subject to a sending test, send by hand, plain text and code groups in the international Morse code using all the letters of the alphabet, numerals 0-9, period, comma, question mark, slant mark, and prosigns $\overline{\text{AR}}$, $\overline{\text{BT}}$, and $\overline{\text{SK}}$. Examinees must copy and send at the required speeds for one continuous minute without making any errors. Each test lasts approximately five minutes. The failing of any code test automatically terminates the examination. Code speeds are computed using five letters per word or code group. Punctuation symbols and numbers count as two letters each.

EXAMINATION ADMINISTRATION

Responsibility for the Federal Communication Commission's commercial radio operator license testing program rests with the Wireless Telecommunications Bureau – WTB. Today, all FCC commercial license examinations are administered by private organizations known as COLEMs – Commercial Operator License Examination Managers. A complete list of COLEMs appears in the Appendix on page 286. COLEMs are responsible for:

- Coordination of examination sessions.
- Verifying the identity of each examinee.
- Preparing, administering and grading examinations.
- Notifying examinees of examination results (pass/fail).
- Certifying that an applicant has passed the test elements required to qualify for a commercial operator license.

- Electronically filing all new commercial radio operator licenses using the FCC's Universal Licensing System.
- Issuing a Proof-of-Passing Certificate (PPC) within 3 days of the examination and providing a copy of the PPC to the examinee.
- Ensuring that no activity takes place that would compromise the integrity of the examination and that no unauthorized material is permitted in the examination room.
- Handling post-examination questions and problems.
- Treating all applicants equally regarding fees and services rendered.

Examiners are prohibited from administering an examination to an employee of their COLEM organization, a relative of their own, or a relative of an employee of their COLEM.

The COLEM program, implemented by the FCC in 1993, has made access to commercial licensing sessions much easier. Prior to the COLEM system, applicants wishing to take an FCC commercial license examination had to travel to one of 25 FCC Field Offices located around the country. Today, FCC commercial examinations are offered by COLEMs throughout the United States and in some foreign countries, as well.

PAPERWORK TO OBTAIN COMMERCIAL RADIO OPERATOR LICENSES

REQUIRED IDENTIFICATION

All applicants for an FCC Commercial radio operator or maintainer license who appear at a examination session are required by the FCC to show ***two forms of identification,*** one showing identity and the other showing employment eligibility.

The following documents establish both identity and employment authorization:

(a) a United States passport (even if expired);
(b) an alien registration receipt card or green card;
(c) an unexpired foreign passport with appropriate INS endorsements; or
(d) an unexpired work permit issued by the INS.

The following documents are acceptable to establish identity only:

(a) a driver's license;
(b) a state-issued identification card with the individual's photograph;
(c) a school identification card with photograph;
(d) a voter's registration card;
(e) a U.S. military card or draft record;
(f) an identification card issued by a federal, state, or local government agency;
(g) a military dependent's identification card;
(h) Native American tribal documents; or
(i) a U.S. Coast Guard Merchant Mariner Card.

The following documents are acceptable to establish employment authorization only:

(a) a social security card;
(b) a Certification of Birth Abroad issued by the State Department;
(c) an original or certified copy of a birth certificate issued by a State, county, municipal authority or outlying possession of the United States bearing an official seal;
(d) a Native American tribal document; or
(e) an INS-issued United States Citizen Identification Card (as well as certain other INS-issued documents).

UNIVERSAL LICENSING SYSTEM

The FCC uses a streamlined application processing system called the Universal Licensing System. ULS is an automated process that uses a single, integrated licensee database for all wireless radio services. ULS uses just five forms and emphasizes electronic filing of all licensee information. The five forms designed specifically for ULS use are FCC Forms 601 through 605. FCC Form 605, along with its Schedules E and F, is used when applying for or amending Restricted and Commercial Radio Operator permits and licenses.

FCC Registration Number

The collection of the applicant's taxpayer identification number (Social Security Number) is mandated by Congress under the requirements of the Debt Collection Act of 1996. The TIN information is kept confidential by a registration procedure and issuance of an FCC Registration Number (FRN). Applicants for new licenses have the option of obtaining an FRN by registering their TIN with the FCC beforehand, or simply supplying their Social Security Number to the examiners on application Form 605.

The FRN can be obtained electronically through the FCC web page at **www.fcc.gov**, then click on Commission Registration System, CORES. If the SSN is given, the FCC automatically converts this number to an FRN which appears on the license when it is issued by the Commission. You also will receive a CORES password by letter which ensures that only you and your authorized representatives are able to update your CORES registration information and make changes to your licenses. If you forget your password or have other password-related questions, contact FCC technical support on-line at **cores@fcc.gov**, or by calling the FRN Helpline at 877-480-3201.

New Licenses

When you take an examination for a new FCC commercial license, the examination will be administered by a test team certified by a COLEM. Under normal circumstances, the test team will provide all of the required forms to be completed for your new license.

Before the examination begins, you will pay an exam fee to the examiner team. There is no license application fee. You will complete FCC Form 605 and Schedule E, where you will identify which license(s) you are applying for. When you complete Form 605, it asks for your Social Security Number. Once the paperwork from the exam session is filed with the FCC, the FCC will assign you an FCC Registration Number (FRN). If you already have an FRN, you should use it on the Form 605.

If you pass your written exam or Morse code test, you will receive a copy of your Proof of Passing Certificate (PPC) issued on behalf of the COLEM by the test team. The test team will forward the original PPC, your payment, and your exam results to the COLEM. In turn, the COLEM will certify your examination results and electronically file your license application with the FCC. You probably will not have to file any paperwork directly with the FCC. You should be sure to obtain a copy of your PPC and keep it for your records.

NATIONAL RADIO EXAMINERS • POB 200065 Arlington, Texas 76006-0065 • 800-669-9594

Proof-of-Passing Certificate

This Certifies that: Examinee's Name Telephone

DATE OF EXAM:
CITY/STATE (Test Center Location)

SAMPLE

WRITTEN EXAMINATION ELEMENTS ☐1 ☐3 ☐6 ☐7 ☐7R ☐8

☐2

☐ Ship Radar Endorsement

☐ Restricted GMDSS ☐ Radiotelegraph Operator

COMMERCIAL RADIO EXAMINERS

Signatures Examiner No.

(1) Accountable Test Center Manager

(2) Commercial Radio Examiner

(3) Commercial Radio Examiner

Signature of Examinee

Registration No: W13

Larry P. Pollock - Commercial Operator License Examination Manager

When you pass your exam, you'll receive a Proof of Passing certificate like this one!

Application Procedures for Name Changes, Renewals, Lost, Stolen or Duplicate Licenses

Paper filings to the FCC can take weeks or months for manual processing. Electronic filing by a COLEM usually results in an immediate, real-time application for a name change, renewal, or duplicate license. If you need to renew a license, obtain a revised license because of a name change, replace a lost or stolen license, or obtain a duplicate of your commercial radio license, you will find it easiest to contact the COLEM that processed your original license for instructions and assistance with these needs. You also may contact National Radio Examiners (NRE) at 800-669-9594 to obtain the proper forms and then have NRE process your application electronically with the FCC. NRE charges a nominal fee for these services.

You MUST apply for a replacement license if your legal name changes.

You may renew your commercial radio license beginning 90 days prior to its expiration date. If you file an application to renew your license before it expires, you may continue to operate under the authority of your license while the FCC processes your renewal application. However, if you fail to renew your license before it expires, you cannot operate or maintain equipment that requires that license until it is renewed. You may file to renew your expired license any time during the five-year grace period after your license expires. If you fail to renew your license within the grace period, you must apply for a new license and re-take the required examination(s).

Obtaining a Reference Copy or Original Copy of Your FCC License

In order to streamline the license process and reduce government costs the FCC discontinued printing and mailing official paper license grants to licensees on a routine basis as of February 17, 2015. According to the FCC, the fact that a license grant appears on the FCC Universal Licensing System (ULS) database is proof enough that a person holds that license.

However, there are circumstances when it is either necessary or desirable to have a paper copy of your license. There are several ways that a license holder can obtain a paper copy of their FCC license, and there are two different versions of a paper license copy available to a licensee.

Reference Copy

You may print an unofficial "reference copy" of your license from the FCC's ULS database at http://wireless.fcc.gov/uls/. It displays the FCC logo and the watermark "Reference Copy" across each page.

Note that the FCC does not consider the "reference copy" valid proof of holding the license. In addition, many other entities – most importantly the U.S. Coast Guard – as well as government contractors, U.S. Military, and others, only recognize an "Official Copy" of the license.

Official Copy

Prior to February 17, 2015, the FCC printed official license copies on its distinctive paper stock. The FCC explains that the special water mark paper it had used was much more expensive than the plain paper it now uses, and the entire reason for the new "paperless license" policy is to reduce government costs.

The "Official Copy" of the license issued by the FCC displays the FCC logo and the watermark "Official Copy" is now printed across each page of an official authorization from FCC. These Official Copies are printed on plan, standard copy paper.

Accessing Your "Paperless" FCC License Record CORES, FRN, ULS and You

The FCC uses computer databases to issue and maintain all Commercial and Amateur Radio licenses. If you already hold an FCC radio license, you are familiar with this system. If you are a new licensee, you need to have a basic understanding of how this system works and how to access your FCC license record.

CORES is the FCC's Commission Registration System. This database maintains your personal record. When you apply as a new licensee, you fill out Form 605 and provide your Social Security Number (SSN) to the FCC. When your information is entered into CORES, it issues you a unique Federal Registration Number (FRN). Your FRN is the number you use to access CORES.

At the same time the CORES system creates your FRN, your new license is automatically generated on the Commission's Universal License System (ULS). The ULS database is where your license information resides. Once these initial steps are completed, the FCC automatically sends you two letters in the mail. The first letter tells you that an FRN has been established by the FCC. The second letter provides you with a temporary password so you may logon to CORES. You should use the temporary password to logon to establish a personalized password. When resetting your password you will be prompted to establish Personal Security Questions (PSQ) and answers for your CORES FRN access. The FCC requires that you update your FRN information in CORES should there be a change to your name, address, e-mail or other information.

Use the same FRN and password to access the ULS system to view your license information and to print out a Reference or Official Copy of your license. Note that any changes you make in your CORES record are NOT automatically changed on your license record in ULS. The CORES and ULS databases are separate, and you must make the changes two times, once in CORES and again in your ULS record.

Access the CORES website at https://apps.fcc.gov/coresWeb/publicHome.do.

Access the ULS website at http://wireless.fcc.gov/uls/.

Instructions

1. Download and Print a License Reference Copy: Visit the FCC's ULS License Manager System at http://wireless.fcc.gov/uls/. Search for the license either by license number or name. Print the unofficial "reference copy" of a license for your records.

2. Download and Print a License Official Copy: Login to the FCC's ULS License Manager System at http://wireless.fcc.gov/uls/ using your FRN and password. Select "Download the Electronic Authorization" of the official license. On the "Download Authorizations" page the license holder will add their call sign to the "Authorizations to Download" and then click download. The PDF of the license can be saved to a computer and printed later or the file can be opened and printed immediately.

3. Login and Set Paper Preferences: Login to the FCC's ULS License Manager System at http://wireless.fcc.gov/uls/ using your FRN and password. Select "Set Paper Authorization Preferences." This option determines whether or not a user will receive paper authorizations (printed license and FRN information) from the FCC for future mailings. On the "Set Paper Authorization Preferences" page, select either YES or NO and then click SAVE. By selecting YES, you will continue to receive paper authorizations printed and mailed to you by the FCC. By selecting NO, you will not receive authorizations printed and mailed from the FCC. The selected preference will affect *all* granted authorizations for this FRN.

4. Receive License via E-mail: When changing your address, renewing or requesting a duplicate copy of the license, a licensee who already has an FCC FRN and provides a valid e-mail address under "Applicant Information" while logged in to the ULS system will receive an official ULS-generated electronic authorization via e-mail.

5. Contact FCC: Licensees may contact FCC Support via the web, by telephone (1-877-480-3201) or by mail to request paper licenses.

CHAPTER C

Global Maritime Distress & Safety System

The most significant advancement in life and safety at sea was the SOLAS Convention, which was adopted in 1914 in response to the Titanic disaster. The main objective of the Safety Of Life At Sea (SOLAS) Convention was to specify minimum standards for the construction, equipping, and operation of ships to help assure their safety. These standards also specified radio equipment and personnel requirements for safety communications purposes.

For the remainder of the 20th Century, ship safety communications relied on Radio Officers trained in telegraphy (Morse code) and the advancements in traditional, terrestrial radio technology. Mostly, this was a system of ship-to-ship communications in which a vessel in distress would make a broadcast call for help, hoping it would be heard and that help would be on its way.

The second most significant advancement in life and safety at sea was made possible by the advent of satellite communications, the development of digital radio techniques, the creation of the global position system, and advances in other communications technology. The Global Maritime Distress and Safety System has revolutionized ship safety communications. GMDSS relies on ship-to-shore communications in which vessels in distress can immediately contact designated rescue centers that coordinate rescue efforts around the world.

Today, specific radio equipment requirements depend upon a ship's area of operation rather than its tonnage. The system also provides redundant means of distress alerting, and requires emergency sources of power to assure communications capability in a dire emergecny. The GMDSS system is able to reliably perform the following functions:

- alerting, including position determination of the unit in distress;
- search and rescue coordination;
- locating (homing);
- maritime safety information broadcasts;
- general communications, and
- bridge-to-bridge communications.

This system has changed international distress communications from being primarily ship-to-ship based to a ship-to-shore system. GMDSS provides automatic distress alerting and locating in cases where a radio operator doesn't have time to send a detailed SOS or MAYDAY call, and also requires ships to receive maritime safety information broadcasts, which are intended to help prevent a distress from happening in the first place.

Sea Areas

Under GMDSS, there are absolutely no geographic gaps or "black holes" that a ship could sail into without any way to signal distress. The world's oceans are divided into four communications areas. GMDSS sea areas serve two purposes: to describe areas where GMDSS services are available, and to define what GMDSS equipment ships must carry. Prior to GMDSS, the number and type of radio safety equipment ships had to carry depended upon their tonnage. With GMDSS, the number and type of radio safety equipment ships must carry depends upon the areas in which they travel.

The sea areas described in *Table 3-1* are established by individual countries, which equip their shore stations with appropriate VHF, MF, HF or satellite facilities to cover their particular segments of ocean.

Under GMDSS, the safety communications equipment that must be carried by large ships is defined by where a vessel sails the world's oceans.

Table 3-1. Ocean Communications Areas

Sea Area	Description
Sea Area A1	Sea area A1 is the area within VHF radiotelephone coverage of at least one coast station at which continuous DSC (digital selective calling) is available (approximately 20-30 nautical miles). A1 ships must carry VHF equipment and a satellite EPIRB.
Sea Area A2	Sea area A2 is the area within MF (medium frequency) radiotelephone coverage of at least one coast station at which continuous DSC is available (approximately150 to 200 nautical miles), excluding Sea Area A1. A2 ships must carry VHF and MF equipment and a satellite EPIRB. The FCC may grant a waiver to non-SOLAS Sea Area 2 vessels to use satellite phone systems in lieu of MF-DSC where satellite coverage assures 24/7 connectivity.
Sea Area A3	Sea area A3 is the area within the coverage of an INMARSAT geostationary satellite in which continuous alerting is available (approximately 70° North to 70° South), excluding Sea Areas A1 and A2. A3 ships must carry VHF and MF/HF equipment, a satellite EPIRB, and either HF or satellite communications. (Most of the world's ocean area is in sea area A3.)
Sea Area A4	Sea area A4 is the remainder of the seas of the world (essentially the polar regions) and relies primarily on HF communications. A4 ships must carry VHF, MF and HF equipment and a satellite EPIRB.

THE DEVELOPMENT OF GMDSS

In 1959, the United Nations established the International Maritime Organization (IMO) to oversee the safety of shipping worldwide, and to prevent ships from polluting the oceans. As a U.N. agency, the IMO is ***the*** international governing body for the maritime service. Its member nations account for more than 97 percent of the world's ocean shipping. Among the IMO's duties is the specification of equipment to be carried aboard certain classes of ships.

In the early 1970s, the IMO began looking at ways to improve maritime distress and safety communications. In 1979, a group of its experts drafted the International Convention on Maritime Search and Rescue, which called for development of a global search and rescue plan. This group also adopted a resolution calling for development by IMO of a Global Maritime Distress and Safety System to provide the communication support needed to implement the search and rescue plan. GMDSS represents more than a decade of cooperative work by the IMO and the International Telecommunication Union (ITU).

Nearly 200 nations strong, the ITU meets regularly to agree on radio operating procedures and the allocation of radio frequencies throughout the world. At conferences held in 1983, 1987, and 1992, the ITU adopted GMDSS-related amendments to the International Radio Regulations to establish the frequencies, operational procedures, and radio personnel requirements for GMDSS.

GMDSS was officially adopted in November, 1988, when world shipping leaders met in London for an IMO conference. A year later, world shipping leaders approved the introduction of new automatic communications equipment that would mean the end of Morse code for ships at sea. In 1974, the IMO first amended the SOLAS Convention to implement the Global Maritime Distress and Safety System internationally. The amendments were initially effective on February 1, 1992, and GMDSS was fully implemented by February 1, 1999. In fact, the satellite-based GMDSS has almost completely replaced shipboard radiotelegraphers.

All ships subject to SOLAS Chapter IV have to carry GMDSS equipment. In general, this means all cargo ships over 300 gross registered tons and all passenger vessels on international voyages. An overview of how the GMDSS system works is illustrated in *Figure 3-1*.

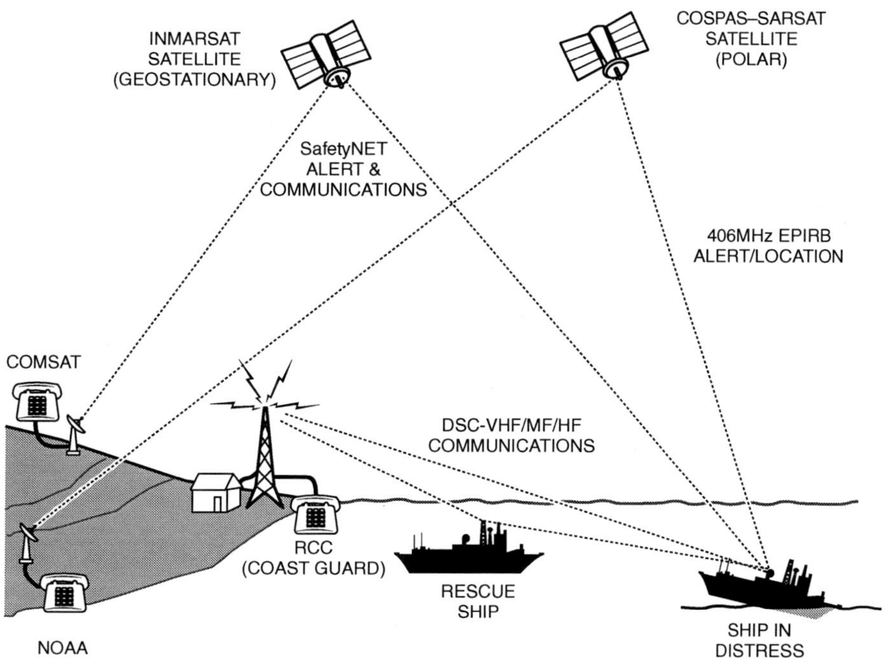

Figure 3-1. The GMDSS System
(Courtesy of FCC Aviation & Marine Branch)

FCC Accepts RTCM's GMDSS Task Force Recommendations

GMDSS is a continually-evolving system. The Radio Technical Commission for Maritime Services (RTCM) GMDSS Task Force monitors the development of new technologies and recommends improvements to the system. In early 2017, the FCC accepted and implemented rules changes on a wide range of new and improved technologies, including:

- EPIRBs must broadcast position information
- New standards for Personal Locator Beacons (PLBs)
- New standards for small Satellite Emergency Notification Devices (SENDs)
- Permit the use of Maritime Survivor Locating devices
- Allow use of AIS-SART devices
- Permit the use of VHF hand-held radios from shore to ship communications.

See the full list of RTCM recommendations at www.RTCM.org and click on "publications."

What is an EPIRB?

An Emergency Position Indicating Radio Beacon is carried aboard ships to alert shore and ship stations of a distress situation. They are required on large passenger and freight ships, and soon may be required on offshore recreational vessels, too. EPIRBs must comply with FCC rules, usually based on RTCM standards, and now must be sold with built-in global positioning system (GPS) capabilities.

EPIRBs transmit two signals: a 5-watt data signal on 406.0 - 406.1 MHz to alert shore and ship stations of a distress; and a ¼-watt local analog homing signal on 121.5 MHz. The modern EPIRB's 406 MHz data signal also carries Global Navigation Satellite System coordinates. The EPIRB may soon transmit VHF Automatic Identification System (AIS) signals, as well.

Once activated, the EPIRB signals are received by multiple satellite systems that are all part of our own GPS constellations:

- COSPAS-SARSAT geostationary satellites (GEOSAR)
- COSPAS-SARSAT low earth orbit satellites for doppler position determinations (LEOSAR)
- INMARSAT, and
- a new system of Mid Earth Orbits satellites MEOSAR

as well as Galileo (European) and GLONASS (Russian) GNSS constellations.

A network of ground stations – including Local User Terminals (LUTs), Mission Control Centers (MCCs), Search and Rescue Points of Contact (SPOCs), and Rescue Control Centers (RCCs) – receive these signals and work to mobilize help, usually in less than 60 seconds. Even Personal Locator Beacon (PLB) signals on sea and on land are received loud and clear.

Source: ACR

406 MHz EPRIB with built in GPS for distress signaling.

EPIRB REGISTRATION REQUIRED!

An Emergency Position Indicating Radio Beacon (EPIRB) or any Personal Locator Beacon (PLB) must be registered with the National Oceanic and Atmospheric Administration (NOAA). If you are inspecting a marine radio station, closely examine the registration information.

All beacon owners should compare their 15 character identification code, printed on the beacon, with their current registration sticker received from NOAA. They should match exactly. If there is a discrepancy, the EPIRB and/or PLB owner should contact NOAA immediately to correct the information.

Despite the requirement to register all EPIRBs and PLBs, there are some reports showing that up to 40% of EPIRB activations are from unregistered beacons, a deadly mistake when minutes can mean the difference between rescue or lost at sea.

In an emergency, the EPIRB and PLB will transmit on 406 MHz with the sender's unique, registered, digitally-coded distress signal. The code allows emergency officials monitoring the system to tell who is sending the signal. The rescue center officials can obtain critical data about a boat's owner, home port, emergency contacts, description, and other information to begin a search, even before a satellite gets a fix on the beacon's location.

The 406 MHz EPIRB, with the now-required built-in GPS, will dramatically speed up the rescue process. These dual capability units are sometimes nicknamed "GPIRBs," and are the ultimate safety signaling device an a catastrophic emergency.

Remember – all EPIRBs must be currently registered! To register an EPIRB call

1-888-212-SAVE
or visit
www.beaconregistration.noaa.gov

Advantages of GMDSS

Among biggest advantages of GMDSS is that it eliminates reliance on a single person for communications and the need for manual watchkeeping. GMDSS requires a minimum of two licensed operators and typically two maintenance methods on "mandatory ships" to ensure distress communications capability at all times. The Channel 16 VHF and 2182 kHz MF networks have been upgraded to include GMDSS Digital Selective Calling (DSC) on Channel 70 VHF and on 2187.5 kHz MF.

The US Coast guard continues to maintain the shore watch on Channel 16 VHF, part of its "Rescue 21" upgrade project (see page 28). Under Rescue 21, the VHF Channel 16 watch also includes sophisticated radio direction finding equipment to help speed up rescue efforts. **The US Coast Guard no longer monitors 2182 kHz voice or 2187.5 kHz DSC. However, most other countries do monitor these two international distress and safety channels, which provide communications coverage along their coastlines close to shore.**

HOW DOES GMDSS WORK?

GMDSS is a sophisticated ship-to-shore alerting system with ship-to-ship capability. Its purpose is to automate and improve distress communications and search and rescue (SAR) operations for the ocean shipping industry on a global basis. GMDSS is made up of several communications systems, some of which have been in operation for many years, and others that rely on newer technologies.

- First, there is COSPAS-SARSAT, a joint international satellite-based search and rescue system established by Canada, France, the former USSR, and the United States to locate emergency radio beacons transmitting on 406 MHz. The COSPAS-SARSAT satellite system, which has been operational since 1982, provides distress alerting when signals from Emergency Position Indicating Radio Beacons (EPIRBs) are detected. The COSPAS-SARSAT system consists of three types of satellites working in conjunction with one another. First, there are GEOSAR satellites in geosynchronous orbits, which provide nearly instant alerting of rescue coordination centers when an EPIRB is activated. If the EPIRB contains a GPS unit, the signal from the EPIRB also provides the location of the unit in distress. Second, there are LEOSAR satellites in low Earth orbit. These satellites provide Doppler position measurements to identify the location of activated EPIRBs that do not contain built-in GPS units. The third system now in operation is MEOSAR, to further complement COSPAS-SARSAT position finding.
- Second, there is the International Maritime Satellite Organization's maritime mobile satellite system (INMARSAT), also in operation since 1982, that forms a major component for distress alerting and communications. The four INMARSAT satellites operate in the 1.5 and 1.6 GHz "L-bands." They are in geostationary orbits 22,300 miles above the equator and cover most of the world's ocean regions, or "sea areas." Expect IRIDIUM Communications, Inc., satellites to soon become an official part of GMDSS, with its January 2017 launch of 10 new satellites in low earth orbit.
- Third, under GMDSS, automated terrestrial data systems and existing communications systems have been combined into one overall communications system which, together with the satellites, make up the Global Maritime Distress and Safety System.

Source: Telenor Networks Maritime Radio, Norway

A typical GMDSS radio station on board a passenger ship.

Digital Selective Calling

Digital Selective Calling (DSC) allows mariners to instantly send an automatically formatted distress alert to the US Coast Guard, or other rescue authority, anywhere in the world. The SOLAS Convention of 1999 requires all passenger ships, and most other ships over 300 gross tons and larger, on international voyages, including all cargo ships, to carry DSC equipped radios.

All GMDSS systems provide a connection for the NMEA 0183 and NMEA 2000 systems so the GPS receiver can provide position data.

The Coast Guard requires the interconnection between GPS and DSC radio equipment, as well as the programming of all DSC equipment with a Maritime Mobile Service Identity number. This is a nine digit number used by marine digital selective calling systems as well as AIS. MMSI numbers are regulated and managed internationally by the International Telecommunications Union (ITU), just as radio call signs are regulated.

All vessels required to carry a shipboard radio, or which travel outside the United States, are required to have a ship station license, and the Federal Communications Commission assigns a ship identity number along with the license grant. Recreational vessels which cruise domestically may obtain an MMSI number through approved organizations, such as Boat US, SeaTow, and the U.S. Power Squadron.

As a marine electronics technician, make sure the radio equipment on board the vessel you are inspecting, or servicing, has the MMSI number entered into the radio equipment. Also, double check that the radio systems and GPS equipment are exchanging data correctly. This way, in an emergency activation, the Coast Guard will know who they are looking for and where they are located!

Automatic Identification System

The Automatic Identification System (AIS) transponder operates in the VHF maritime band, and acts like a shipboard transponder capable of sending and receiving over 4,500 reports per minute and can transmit updates as often as every 2 seconds. AIS transmissions use 9.6 kb GMSK FM modulation using HDLC Packet protocols. The system provides for automatic contention resolution between itself and other stations; and communications integrity is maintained even in ultra-busy harbor locations. The Class A AIS unit broadcasts ship information every 2 to 10 seconds while underway, and every 3 minutes while at anchor. It has been found that AIS is more effective than 9 GHz search and rescue transponders. AIS Class A equipment meets ITU-R recommendation M.1371-1, "all ships of 300 gross tonnage and upwards, engaged on international voyages, and cargo ships of 500 gross tonnage and upwards, not engaged on international voyages, and passenger ships irrespective of size shall be fitted with AIS."

Anybody who can receive and detect an AIS signal will also detect an AIS-SART. The transmission signal from an AIS-SART consists of an MMSI-like ID code, where the first three digits will be "970." The ID code consists of 9 digits and the AIS-SART uses the remaining 6 digits to indicate a manufacturer code (2 digits) in addition to the unit's unique serial number (4 digits). AIS may soon be included in SART devices, as well as new types of man-overboard equipment, since so many marine VHF radios now have AIS receivers built into the equipment. Learn all you can about AIS!

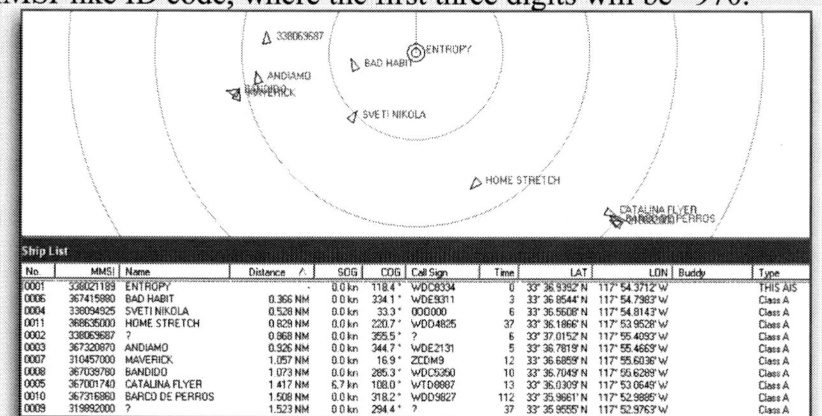

No.	MMSI	Name	Distance	SOG	COG	Call Sign	Time	LAT	LON	Buddy	Type
0001	338021189	ENTROPY	-	0.0 kn	118.4 °	WDC8334	0	33° 36.9352' N	117° 54.3712' W		THIS AIS
0006	367415880	BAD HABIT	0.366 NM	0.0 kn	334.1 °	WDE9311	3	33° 36.8544' N	117° 54.7983' W		Class A
0004	338094925	SVETI NIKOLA	0.528 NM	0.0 kn	33.3 °	000000	6	33° 36.5608' N	117° 54.8143' W		Class A
0011	368635000	HOME STRETCH	0.829 NM	0.0 kn	220.7 °	WDD4825	37	33° 36.1866' N	117° 53.9528' W		Class A
0002	338069687	?	0.868 NM	0.0 kn	355.5 °	?	6	33° 37.0152' N	117° 55.4093' W		Class A
0003	367320870	ANDIAMO	0.926 NM	0.0 kn	344.7 °	WDE2131	5	33° 36.7819' N	117° 55.4669' W		Class A
0007	310457000	MAVERICK	1.057 NM	0.0 kn	16.9 °	ZCDM9	12	33° 36.6859' N	117° 55.6036' W		Class A
0008	367039780	BANDIDO	1.073 NM	0.0 kn	285.3 °	WDC5350	10	33° 36.7049' N	117° 55.6289' W		Class A
0005	367001740	CATALINA FLYER	1.417 NM	6.7 kn	108.0 °	WTD8887	13	33° 36.0309' N	117° 53.0649' W		Class A
0010	367316860	BARCO DE PERROS	1.508 NM	0.0 kn	318.2 °	WDD9827	112	33° 35.9661' N	117° 52.9885' W		Class A
0009	319892000	?	1.523 NM	0.0 kn	294.4 °	?	37	33° 35.9555' N	117° 52.9763' W		Class A

An AIS RADAR display showing the ENTROPY surrounded other by boats.

- GMDSS provides for digital selective calling (DSC) services on the high-frequency (HF), medium-frequency (MF), or very-high-frequency (VHF) bands, depending upon the location of a ship in distress. These DSC services are used for ship-to-ship, ship-to-shore, and shore-to-ship automatic alerting, while existing terrestrial HF, MF and VHF radiotelephony equipment provides distress, urgent, and safety-related communications.
- GMDSS enhances search and rescue operations at sea through the use of the 9-GHz search and rescue RADAR transponder (SART), as well as certified VHF AIS-SART, an approved alternative to RADAR SART.
- GMDSS creates a global network for the dissemination of Maritime Safety Information (MSI) using three systems: NAVTEX, INMARSAT's enhanced group calling (EGC), and HF narrow-band direct-printing (NBDP) radiotelegraphy. NAVTEX is an English language direct printing meteorological, warning and urgent safety information dissemination service that operates on 518 kHz. A selective message-rejection feature allows only certain communications to be received. The enhanced group call (EGC) system was developed by INMARSAT to enable pre-determined groups of ships (or all vessels) to receive warning and distress messages. All ships must carry equipment for receiving Maritime Safety Information broadcasts.

Table 4-1. GMDSS Shipboard Radio Equipment

Equipment:	Function:
406-MHz EPIRB	Ship-to-shore alerts via COSPAS-SARSAT satellite
VHF radio (DSC and voice)	SAR (Search and Rescue) communications
MF Radio or Satellite Phone required for sea area A2	Ship-to-shore alerts and communications
HF radio (required for sea areas A3 and A4)	Ship-to-shore alerts and communications
INMARSAT ship Earth station plus EGC capability	Ship-to-shore alerts, communications and MSI (SafetyNET)
NAVTEX receiver	MSI (Safety NET) 518 kHz
9-GHz SART or certified AIS-SART	SAR locating beacon
2-way VHF portable radios	SAR communications

EGC=Enhanced group calling MSI=Maritime Safety Information
SAR=Search and rescue SART=Search and rescue transponder

GMDSS LICENSES & EXAMINATION REQUIREMENTS

See chapter B for detailed information on GMDSS operator and maintainer licenses, examinations, and U.S. Coast Guard approved course requirements.

U.S.C.G. discontinues 2182 kHz and 2187.5 kHz distress and safety watch!

Effective August 1, 2013, the U.S. Coast Guard has dropped its radio watch on the 2182 kHz voice and 2187.5 kHz digital selective calling (DSC) International Distress and Safety frequencies. The Coast Guard also discontinued its 2670 kHz marine information and weather broadcasts. These frequencies provide relatively short-range communications to about 30 miles of shore.

While the USCG has discontinued radio watches on these frequencies, any Element 1 exam question and answer regarding 2182 kHz and/or 2187.5 kHz remains the same indicated answer because FCC rules have not changed.

The USCG bulletin announcing this change explained: "This termination decision was made after a review of Coast Guard medium frequency (MF) communications sites revealed significant antenna and infrastructure support degradation that put the Coast Guard at risk of not being able to receive and respond to calls for assistance on the 2 MHz distress frequencies."

The USCG watch on VHF Channel 16, 156.800 MHz, (voice) and VHF Channel 70 (DSC) will continue to offer the Coast Guard's Rescue 21 coverage throughout close-to-shore sea areas of the United States from Coast Guard units on land, at sea and in the air. See page 70 for a full discussion of this change in Coast Guard watch policy.

Rescue 21

The Coast Guard continuously monitors channel 16.

The National Distress and Response System (NDRS) was installed and deployed by the Coast Guard during the 1970s to protect the nation's coasts and rescue mariners at sea. To address the limitations of the NRDS, the USCG has implemented a new communications system entitled Rescue 21.

This major systems acquisition program is harnessing global positioning and cutting-edge communications technology that enables the Coast Guard to perform all missions with greater agility and efficiency. Rescue 21 has closed 88 known coverage gaps in coastal areas of the United States, enhancing the safety of life at sea. The system's expanded system frequency capacity enables greater coordination with the Department of Homeland Security, as well as other federal, state and local agencies and first responders.

Rescue 21 provides an updated, leading-edge Very High Frequency FM (VHF-FM) communications system, replacing the NDRS. Rescue 21 covers coastline, navigable rivers and waterways in the continental U.S., Alaska, Hawaii, Guam and Puerto Rico. The system is on the air now!

By replacing outdated legacy technology with a fully integrated system, Rescue 21 provides the Coast Guard with upgraded tools and technology to protect the nation's coasts and rescue mariners at sea. Rescue 21 revolutionizes how the Coast Guard uses command, control, and communications for all missions within the coastal zone. The system:

- where feasible, incorporates automatic direction-finding equipment to improve locating mariners in distress
- improves interoperability amongst federal, state, and local agencies
- enhances clarity of distress calls
- allows simultaneous channel monitoring
- upgrades the recording and playback feature of distress calls
- reduces coverage gaps for coastal communications and along navigable rivers and waterways
- supports Digital Selective Calling for registered users
- in the contiguous 48 states, provides portable towers for restoration of communications during emergencies or natural disasters

Rescue 21 and Recreational Boats

Rescue 21 is designed to serve both commercial and recreational vessels. Recreational boaters who have DSC-capable VHF radios can take advantage of Rescue 21 by **registering their radios for MMSI numbers, and interconnecting the radio with the GPS unit** on their boats. Once registered and interconnected to the GPS, an emergency call will allow the Coast Guard to quickly identify the boat in distress and know its precise location, speeding rescue efforts.

To learn more about Rescue 21, visit: **www.uscg.mil/acquisition/rescue21**.

Always monitor channel 16 after leaving the slip.

CHAPTER D

Marine & Aviation Radio Law

Element 1 Question Pool

The FCC Commercial Element 1 written examination covers basic marine and aviation radio law and operating practice with which every maritime and aviation radio operator should be familiar. The Element 1 examination is used to prove that the examinee possesses the qualifications to operate marine and aircraft licensed radio stations that can only be operated by a person holding a Marine Radio Operator Permit (MP).

The MP is the basic FCC commercial radio operator license. Passing an Element 1 exam is required for virtually all other FCC commercial licenses, including GROL, GMDSS, GMDSS-R, GMDSS Maintainer, and the Radiotelegraph license.

The Element 1 question pool contains a total of 144 questions covering four Subelement Topic areas. Each Key Topic contains 6 questions. One question is taken at random from each Key Topic to create an examination with 24 questions. To pass, you must correctly answer at least 18 out of the 24 questions on the written exam. *Table 4-1* summarizes the Subelement topics and number of questions in each topic area.

Table 4-1. Summary of Element 1 Question Pool

Subelement Topic	Key Topics	No. of Questions	Examination Questions
Rules and Regulations	6	36	6
Communications Procedures	6	36	6
Equipment Operations	6	36	6
Other Equipment	6	36	6
Total	24	144	24

Each Element 1 examination is administered by a Commercial Operator License Examination Manger (COLEM). The COLEM must construct the examination by selecting 24 questions from the Element 1 question pool contained in this book, as explained above. Each question in the pool contains the question, and 4 multiple-choice answers – one correct answer and three wrong answers (distracters). The COLEM may change the order of the four answer choices, but cannot change any of the question or answer content.

In this book, official Element 1 questions and 4 answer choices are followed by an explanation of why the correct answer is right, and the identification of the correct answer.

At the end of some of the question explanations in this chapter, you will see a rules reference (such as §80.177). This identifies the FCC rule on which the particular question is based, taken from the Code of Federal Regulations Title 47, Part 80. If you wish to explore the full Part 80 rules, you can find them at: http://tinyurl.com/q82kzgn.

Subelement A – Rules & Regulations

6 Key Topics, 6 Exam Questions

Key Topic 1: Equipment Requirements

1-1A1 What is a requirement of all marine transmitting apparatus used aboard United States vessels?

A. Only equipment that has been certified by the FCC for Part 80 operations is authorized.
B. Equipment must be type-accepted by the U.S. Coast Guard for maritime mobile use.
C. Certification is required by the International Maritime Organization (IMO).
D. Programming of all maritime channels must be performed by a licensed Marine Radio Operator.

For marine transmitting equipment to be authorized for use aboard United States vessels, it must be certified by the Federal Communications Commission under Part 80 rules. All such transmitting equipment must bear a label which indicates the FCC equipment certification ID number. This may be a series of letters, numbers, or letters and numbers, and the label may not necessarily say "Part 80!" If you really want to check whether or not the gear on board is certified, the FCC offers public information on equipment certification numbers and Rules Parts for which a particular radio is certified. Technical standards for radio equipment are covered in Subpart E of the Rules. §80.43 and § 80.203 (a). **ANSWER A**

1-1A2 What transmitting equipment is authorized for use by a station in the maritime services?

A. Transmitters that have been certified by the manufacturer for maritime use.
B. Unless specifically excepted, only transmitters certified by the Federal Communications Commission for Part 80 operations.
C. Equipment that has been inspected and approved by the U.S. Coast Guard.
D. Transceivers and transmitters that meet all ITU specifications for use in maritime mobile service.

Amateur radio equipment certified for Part 97 use is not permitted as a substitute for a marine SSB or VHF use. You will jeopardize your own license by transmitting on marine channels over any piece of equipment that is not Part 80 certified. Only Part 80 equipment for marine radio use! **ANSWER B**

This is the FCC ID tag on a marine transceiver.

Want to check FCC Part 80 Certification?
Go to http://www.fcc.gov/oet/ea/fccid/ on the Internet. The FCC ID Search Form appears. In the first box enter the "Grantee (manufacturer) Code" – the first 3 characters of the FCC ID found on the label of the piece of equipment you're checking. Next, enter the "Product Code" in the second box. This is the remaining characters and numbers on the FCC ID. Click the "search" button and you should find the information you're looking for.

1-1A3 Small passenger vessels that sail 20 to 100 nautical miles from the nearest land must have what additional equipment?

A. Inmarsat-B terminal.
B. Inmarsat-C terminal.
C. Aircraft Transceiver with 121.5 MHz.
D. MF SSB Transceiver.

Picture a small passenger vessel sailing in Sea Area A2, which extends from 20 nautical miles to 100 nautical miles from the nearest land. Sea Area A2 requires VHF equipment, PLUS longer-range medium- and high-frequency equipment, too. This added MF and HF SSB transceiver is capable of sending out a voice distress call on 2182 kHz, as well as a digital distress call on 2187.5 kHz. All new marine SSB radios deliver both MF and HF frequencies, and that is why I say MF-HF. So in addition to a VHF radio, the SSB MF-HF transceiver is also required for passenger ships sailing in Sea Area A2. The US Coast Guard no longer monitors MF 2 MHz channels 2182 kHz voice or 2187.5 kHz DSC. **ANSWER D**

Sea Area	Distance from Shore	Required Radio Equipment
Sea Area A1	Coast to 20nm out	VHF + EPIRB
Sea Area A2	20nm to 100nm	VHF, MF SSB or Satellite phone + EPIRB
Sea Area A3	High Seas	VHF + MF and HF radio equipment + EPIRB
Sea Area A4	Polar Regions	VHF + MF and HF radio equipment + EPIRB

1-1A4 What equipment is programmed to initiate transmission of distress alerts and calls to individual stations?

A. NAVTEX.
B. GPS.
C. DSC controller.
D. Scanning Watch Receiver.

Digital Selective Calling (DSC) is an automated way of signaling specific stations on marine VHF Channel 70. All 25 watt marine VHF equipment manufactured after 1999 includes this digital selective calling capability. You can spot a VHF and marine SSB with DSC because it will have a red plastic cover protecting the distress push button. VHF radios with DSC also contain a Channel 70 silent-watch receiving capability, either by sampling or by continuous running a Channel 70 watch. When someone pushes the distress button, valuable information may be received by everyone receiving the station calling the digital Mayday. This same DSC controller may also be employed in non-distress situations as an alternate means of dialing up a menu of fellow boaters and their DSC information to call them digitally without needing to pick up the microphone. **ANSWER C**

1-1A5 What is the minimum transmitter power level required by the FCC for a medium-frequency transmitter aboard a compulsorily fitted vessel?

A. At least 100 watts, single-sideband, suppressed-carrier power.
B. At least 60 watts PEP.
C. The power predictably needed to communicate with the nearest public coast station operating on 2182 kHz.
D. At least 25 watts delivered into 50 ohms effective resistance when operated with a primary voltage of 13.6 volts DC.

The international radiotelephone distress frequency, 2182 kHz, is classified in the medium-frequency range. Since all compulsorily fitted vessels must carry radio equipment capable of transmitting on this frequency, 60 watts peak envelope power (PEP) is considered the minimum power level necessary to ensure contact with another station. §80.807(c)(2) and §80.855(d)(2). **ANSWER B**

1-1A6 Shipboard transmitters using F3E emission (FM voice) may not exceed what carrier power?

A. 500 watts.
B. 250 watts.
C. 100 watts.
D. 25 watts.

FM radio equipment aboard a vessel cannot exceed 25 watts carrier power. It also must be able to reduce output power to one watt. Such a wide range ensures that ship-to-ship communications will not interfere with other stations, and that a distress message has a good chance of being received elsewhere. VHF shore stations may run 50 watts of power. Coast Guard stations can run even more! §80.215(g). **ANSWER D**

Key Topic 2: License Requirements

1-2A1 Which commercial radio operator license is required to operate a fixed-tuned ship RADAR station with external controls?

A. A radio operator certificate containing a Ship RADAR Endorsement.
B. A Marine Radio Operator Permit or higher.
C. Either a First or Second Class Radiotelegraph certificate or a General Radiotelephone Operator License.
D No radio operator authorization is required.

Small boat RADAR has been automated to the point that no technical adjustments are necessary from the outside controls. If the RADAR installation, along with marine VHF, is installed aboard a small recreational vessel that cruises only within local, domestic waters, no radio operator license is required, and no station license is required as long as the recreational vessel is for local pleasure use only. §80.177 **ANSWER D**

1-2A2 When is a Marine Radio Operator Permit or higher license required for aircraft communications?

A. When operating on frequencies below 30 MHz allocated exclusively to aeronautical mobile services.
B. When operating on frequencies above 30 MHz allocated exclusively to aeronautical mobile services.
C. When operating on frequencies below 30 MHz not allocated exclusively to aeronautical mobile services.
D. When operating on frequencies above 30 MHz not assigned for international use.

The majority of aeronautical communications take place on regular aeronautical channels, and no marine radio operator permit is required for domestic private aircraft use. However, aircraft and helicopter pilots using radios for fish spotting, using frequencies NOT allocated EXCLUSIVELY to aeronautical mobile services, must obtain the marine radio operator permit (MROP) when operating below 30 MHz. **ANSWER C**

1-2A3 Which of the following persons are ineligible to be issued a commercial radio operator license?

A. Individuals who are unable to send and receive correctly by telephone spoken messages in English.
B. Handicapped persons with uncorrected disabilities which affect their ability to perform all duties required of commercial radio operators.
C. Foreign maritime radio operators unless they are certified by the International Maritime Organization (IMO).
D. U.S. Military radio operators who are still on active duty.

Persons who cannot correctly transmit or receive voice messages in English are not eligible for a commercial operator license (§13.9). This includes the mute and deaf. Persons afflicted with other physical handicaps may be issued a commercial radio license, if found qualified, with certain restrictive endorsements. **ANSWER A**

1-2A4 What are the radio operator requirements of a passenger ship equipped with a GMDSS installation?

A. The operator must hold a General Radiotelephone Operator License or higher-class license.
B. The operator must hold a Restricted Radiotelephone Operator Permit or higher-class license.
C. The operator must hold a Marine Radio Operator Permit or higher-class license.
D. Two operators on board must hold a GMDSS Radio Operator License or a Restricted GMDSS Radio Operator License, depending on the ship's operating areas.

Operating a radiotelephone station aboard a GMDSS-equipped passenger ship requires that at least 2 persons hold an appropriate GMDSS Radio Operator License. If a passenger ship operates exclusively within 20 nautical miles of shore, one of the two GMDSS radio operator license holders may hold the Restricted GMDSS. Beyond 20 nautical miles both radio operators must hold the FULL GMDSS Radio Operator License. §80.159(d). **ANSWER D**

1-2A5 What is the minimum radio operator requirement for ships subject to the Great Lakes Radio Agreement?

A. Third Class Radiotelegraph Operator's Certificate.
B. General Radiotelephone Operator License.
C. Marine Radio Operator Permit.
D. Restricted Radiotelephone Operator Permit.

Interestingly, the FCC gives this information in a rule (§80.161) by itself: "Each ship subject to the Great Lakes Radio Agreement must have on board an officer or member of the crew who holds a Marine Radio Operator Permit or higher class license." The Great Lakes carry heavy marine traffic, and effective radio communications help prevent navigation mishaps. **ANSWER C**

1-2A6 What is a requirement of every commercial operator on duty and in charge of a transmitting system?

A. A copy of the Proof-of-Passing Certificate (PPC) must be in the station's records.
B. The original license or a photocopy must be posted or in the operator's personal possession and available for inspection.
C. The FCC Form 605 certifying the operator's qualifications must be readily available at the transmitting system site.
D. A copy of the operator's license must be supplied to the radio station's supervisor as evidence of technical qualification.

The person in charge of maintaining correct operation of a commercial radio station must be able to provide proof of technical expertise and competence by prominently displaying his or her FCC commercial radio license while on duty. §13.19(c). **ANSWER B**

Commercial ships, including ferries like this one, must have FCC-licensed operators to run their radio equipment.

Key Topic 3: Watchkeeping

1-3A1 Radio watches for compulsory radiotelephone stations will include the following:

A. VHF channel 22a continuous watch at sea.
B. 121.5 MHz continuous watch at sea.
C. VHF channel 16 continuous watch.
D. 500 kHz.

Compulsory vessels are required to maintain a watch on VHF Channel 16 (and on MF 2182 kHz) when underway. However, the Channel 16 watch may not be required on a vessel operating within the Vessel Traffic Service (VTS), or when the ship is operating with hand-held bridge-to-bridge VHF equipment under 80.143(c). And since their marine VHF equipment most likely contains a DSC controller, they are also guarding Channel 70, 156.525 MHz, for any digital calls. §80.147 and §80.148. **ANSWER C**

1-3A2 All compulsory equipped cargo ships (except those operating under GMDSS regulations or in a VTS) while being navigated outside of a harbor or port, shall keep a continuous radiotelephone watch on:

A. 2182 kHz and Ch-16.
B. 2182 kHz.
C. Ch-16.
D. Cargo ships are exempt from radio watch regulations.

When you are far out at sea, always maintain an active radiotelephone watch. Channel 16 and 2182 kHz should always be monitored with the squelch just slightly closed. This would allow a weak station – perhaps a sailor overboard with a submersible marine VHF set – to call you with his emergency as you pass by. **ANSWER A**

1-3A3 What channel must all compulsory, non-GMDSS vessels monitor at all times in the open sea?

A. Channel 8.
B. Channel 70.
C. Channel 6.
D. Channel 16.

Channel 16, 156.800 MHz, is an international radiotelephone distress frequency in the VHF range. All vessels using radio equipment are required by law to maintain a watch on this frequency. VHF Channel 16 is also a calling channel. If your marine VHF suddenly emits a warbling tone from the speaker, check your VHF Channel 16 watch, because it is likely the DSC controller has received an incoming digital call. If it was a distress call, your VHF Channel 16 display may contain valuable information about the calling ship's location. Before transmitting, be sure to write down the other ship's latitude and longitude! §80.148. **ANSWER D**

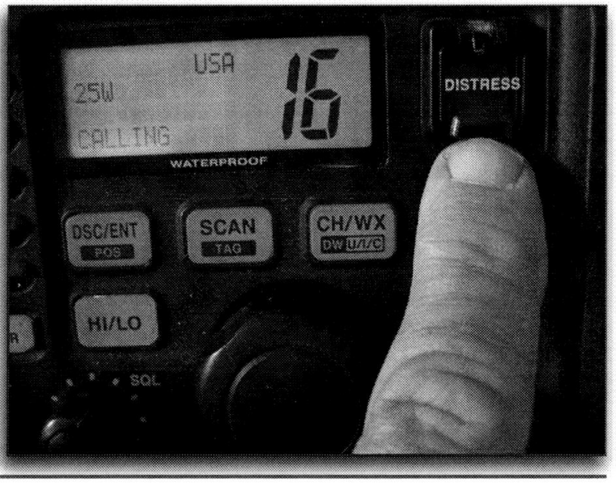

Always monitor channel 16 for distress calls!

1-3A4 When a watch is required on 2182 kHz, at how many minutes past the hour must a 3 minute silent period be observed?

A. 00, 30.
B. 15, 45.
C. 10, 40.
D. 05, 35.

The medium frequency 2182 kHz is still operational, and a silent period is required on the hour and the half hour where you carefully listen to 2182 kHz for any voice distress call. While the U.S. Coast Guard no longer monitors 2182 kHz, all FCC rules regarding 2182 kHz are still in effect. **ANSWER A**

1-3A5 Which is true concerning a required watch on VHF Ch-16?

A. It is compulsory at all times while at sea until further notice, unless the vessel is in a VTS system.
B. When a vessel is in an A1 sea area and subject to the Bridge-to-Bridge act and in a VTS system, a watch is not required on Ch-16, provided the vessel monitors both Ch-13 and VTS channel.
C. It is always compulsory in sea areas A2, A3 and A4.
D. All of the above.

Read over all of the correct answers – the VHF Channel 16 radio watch is not required in a few cases, but is always required when cruising beyond sight of land. **ANSWER D**

1-3A6 What are the mandatory DSC watchkeeping bands/channels?

A. VHF Ch-70, 2 MHz MF DSC, 6 MHz DSC and 1 other HF DSC.
B. 8 MHz HF DSC, 1 other HF DSC, 2 MHz MF DSC, and VHF Ch-70.
C. 2 MHz MF DSC, 8 MHz DSC, VHF Ch-16 and 1 other HF DSC.
D. None of the above.

Mandatory DSC watch-keeping for most compulsory equipped vessels is DSC VHF Channel 70 plus DSC medium frequency 2187.5 kHz, high frequency 8414.5 kHz, and one additional high frequency channel, such as 12 MHz, 12577 kHz. The medium and high frequency DSC watch-keeping are usually accomplished by silent channel scanning of the DSC controller. However, if the marine SSB transceiver has a separate DSC scanning receive antenna receptacle, there MUST be an antenna connected to this outlet in order to receive a distress call! On VHF Channel 70, the scanning or dual receiver normally uses the same antenna as the principle marine VHF antenna. Although the U.S. Coast Guard no longer monitors 2187.5 kHz, all FCC rules regarding this DSC frequency are still in effect. **ANSWER B**

Radio Watchkeeping Regulations

In general, any vessel equipped with a VHF marine radiotelephone (whether required or voluntarily) must maintain a watch on channel 16 (156.800 MHz) whenever the radiotelephone is not being used to communicate. (Source: FCC 47 CFR § 80.148, § 80.310, NTIA Manual 8.2.29.6.c(2)(e), ITU RR 31.17, 33.18, AP13 §25.2)

In addition, every power-driven vessel of 20 meters or over in length or of 100 tons and upwards carrying one or more passengers for hire, or a towing vessel of 26 feet or over in length, as well as every dredge and floating plant operating near a channel or fairway, must also maintain a watch on channel 13 (156.650 MHz) – or channel 67 (156.375 MHz) if operating on the lower Mississippi River – while navigating on U.S. waters (which include the territorial sea, internal waters that are subject to tidal influence, and, those not subject to tidal influence but that are used or are determined to be capable of being used for substantial interstate or foreign commerce). Sequential monitoring techniques (scanners) alone cannot be used to meet this requirement; two radios (including portable radios, i.e. handhelds) or one radio with two receivers, are required. These vessels must also maintain a watch on the designated Vessel Traffic Service (VTS) frequency, in lieu of maintaining watch on channel 16, while transiting within a VTS area. (See 33 CFR §§ 2.36, 26, and 161; 47 CFR §§ 80.148, 80.308-309; NTIA: NTIA Manual Chapter 8.2.29.7.)

Digital Selective Calling

Ships, where so equipped, shall, while at sea, maintain an automatic digital selective calling watch on the appropriate distress and safety calling frequencies [e.g. channel 70] in the frequency bands in which they are operating. If operating in a GMDSS Sea Area A1 may discontinue their watch on channel 16. However, ships, where so equipped, shall also maintain watch on the appropriate frequencies for the automatic reception of transmissions of meteorological and navigational warnings and other urgent information for ships.

Ship stations complying with these provisions should, where practicable, maintain a watch on channel 13 (156.650 MHz) for communications related to the safety of navigation. (Source: ITU RR 31.17, 33.18, AP13 §25.2)

Who regulates whom?

Three U.S. government agencies, the Federal Communications Commission, the National Telecommunications and Information Administration, and the U.S. Coast Guard; and two international organizations, the International Telecommunications Union and the International Maritime Organization; have each established marine radio watch keeping regulations. Regulations on radio watch keeping exist for all boats and ships – commercial, recreational, government and military, domestic and foreign – carrying marine radios.

International Telecommunications Union (ITU). ITU regulates all use of radio spectrum by any person or vessel outside U.S. waters. ITU rules affecting radio, which have treaty status in the U.S. and most other nations, are published in the ITU Radio Regulations. The ITU has established three VHF marine radio channels recognized worldwide for safety purposes:

1. Channel 16 (156.800 MHz) - Distress, safety and calling
2. Channel 13 (156.650 MHz) - Intership (bridge-to-bridge) navigation
3. Channel 70 (156.525 MHz) - Digital Selective Calling

International Maritime Organization (IMO). IMO regulates the outfitting and operation of most vessels engaged on international voyages, except warships. Most IMO radio regulations affect all passenger ships and other ships of 300 gross tonnage and upward. IMO rules affecting radio are promulgated in the Safety of Life at Sea (SOLAS) Convention which has been ratified in the U.S.

Federal Communications Commission (FCC). The FCC regulates all sales, marketing, and use of radios in the U.S., including those onboard any recreational, commercial, state and local government, and foreign vessel in U.S. territorial waters. These regulations are contained in Title 47, Code of Federal Regulations.

National Telecommunications and Information Administration (NTIA). The NTIA regulates all use of radio onboard any federal government vessel, including military vessels. NTIA rules do not apply outside the federal government.

U.S. Coast Guard (USCG). The USCG regulates carriage of radio on most commercial vessels, foreign vessels in U.S. waters, survival craft, and vessels subject to the Bridge-to-Bridge Act (generally all vessels over 20m length) and operating in a Vessel Traffic Service (VTS) area.

Key Topic 4: Logkeeping

1-4A1 Who is required to make entries in a required service or maintenance log?

A. The licensed operator or a person whom he or she designates.
B. The operator responsible for the station operation or maintenance.
C. Any commercial radio operator holding at least a Restricted Radiotelephone Operator Permit.
D. The technician who actually makes the adjustments to the equipment.

The FCC rules need to be studied for this correct answer. Rule §80.409 (a)(4) states: "The station licensee and the radio operator in charge of the station are responsible for the maintenance of station logs." And remember "the operator RESPONSIBLE for the station operation or maintenance" is the person REQUIRED to make entries in the service log. **ANSWER B**

1-4A2 Who is responsible for the proper maintenance of station logs?

A. The station licensee.
B. The commercially-licensed radio operator in charge of the station.
C. The ship's master and the station licensee.
D. The station licensee and the radio operator in charge of the station.

FCC Rule §80.409: The station licensee and the radio operator in charge of the station are responsible for the maintenance of station logs. These persons must keep the log in an orderly manner. Key letters or abbreviations may be used if their proper meaning or explanation is contained elsewhere in the same log." No erasures. **ANSWER D**

1-4A3 Where must ship station logs be kept during a voyage?

A. At the principal radiotelephone operating position.
B. They must be secured in the vessel's strongbox for safekeeping.
C. In the personal custody of the licensed commercial radio operator.
D. All logs are turned over to the ship's master when the radio operator goes off duty.

Station logs (both radiotelephone and radiotelegraph) must be kept at the principal radio operating position during the entire voyage. §80.409. **ANSWER A**

1-4A4 What is the proper procedure for making a correction in the station log?

A. The ship's master must be notified, approve and initial all changes to the station log.
B. The mistake may be erased and the correction made and initialized only by the radio operator making the original error.
C. The original person making the entry must strike out the error, initial the correction and indicate the date of the correction.
D. Rewrite the new entry in its entirety directly below the incorrect notation and initial the change.

FCC Rule §80.409: "Erasures, obliterations or willful destruction within the retention period are prohibited. Corrections may be made only by the person originating the entry by striking out the error, initialing the correction and indicating the date of correction." While not required, it is a good idea to keep station logs in ink. **ANSWER C**

1-4A5 How long should station logs be retained when there are entries relating to distress or disaster situations?

A. Until authorized by the Commission in writing to destroy them.
B. For a period of three years from the last date of entry, unless notified by the FCC.
C. Indefinitely, or until destruction is specifically authorized by the U.S. Coast Guard.
D. For a period of one year from the last date of entry.

"Routine" logs are required to be retained by the station licensee for at least two years. But a log that contains distress traffic must be retained for at least three years if a distress message was received or transmitted. §80.409(b). **ANSWER B**

1-4A6 How long should station logs be retained when there are no entries relating to distress or disaster situations?

A. For a period of three years from the last date of entry, unless notified by the FCC.
B. Until authorized by the Commission in writing to destroy them.
C. For a period of two years from the last date of entry.
D. Indefinitely, or until destruction is specifically authorized by the U.S. Coast Guard.

Your radio station log is an official record of when your radio was turned on, general types of traffic passed, and notations of any distress traffic received. If there were no entries relating to distress traffic, your radio station log is retained for 2 years from the date of the last entry. The log should also include any changes of your radio station equipment. If you switched the old antenna for a new one, note this in the log. If your vessel is inspected for its radiotelephone equipment, have the radio inspector note this in your radio log. Keep the log at the principle operating location. No longer are you required to list each and every radio transmission. §80.409(b). **ANSWER C**

Key Topic 5: Log Entries

1-5A1 Radiotelephone stations required to keep logs of their transmissions must include:
A. Station, date and time.
B. Name of operator on duty.
C. Station call signs with which communication took place.
D. All of these.

And more! In addition to all of these, logs must contain the following and the time of their occurrence: a summary of all distress, urgency and safety traffic; the position of the ship at least once a day; the time the watch is discontinued, the reason for it, and the time the watch resumes; the times when storage batteries are placed on charge and taken off charge; results of required equipment tests; and a daily statement of the condition of the required radiotelephone equipment. §80.409. **ANSWER D**

1-5A2 Which of the following is true?
A. Battery test must be logged daily.
B. EPIRB tests are normally logged monthly.
C. Radiotelephone tests are normally logged weekly.
D. None of the above.

Newer-model, 406-MHz EPIRBs incorporate self-test circuitry to validate the equipment is operational. An EPIRB test should be performed before any voyage, and tests are normally logged monthly in your ship's radiotelephone station log. **ANSWER B**

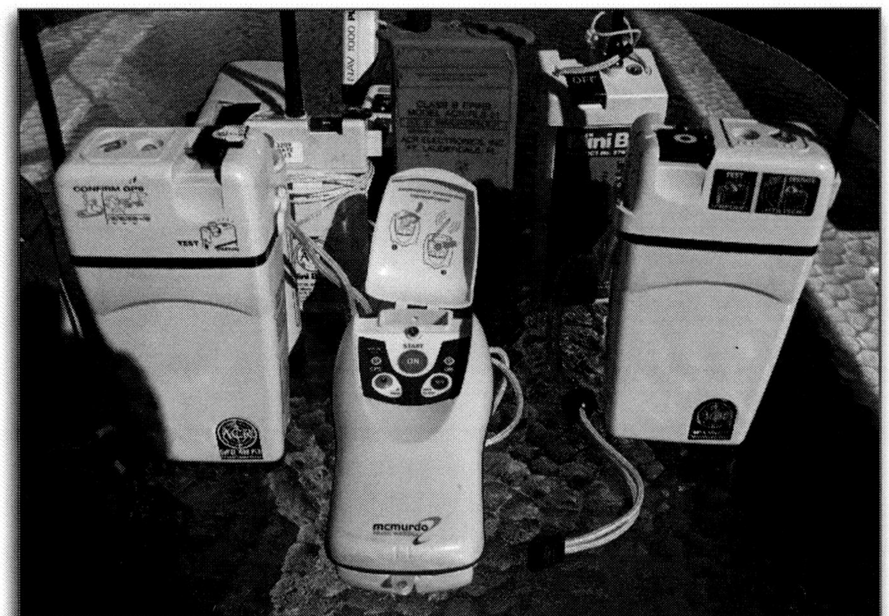

EPIRBs may soon be required on seagoing recreational vessels. All new EPIRBs must now have GNS positioning capabilities built in.

Source: ACR

Keep a log of when you test your on-board EPIRBs. Be sure and check battery expiration dates, too.

1-5A3 Where should the GMDSS radio log be kept on board ship?
A. Captain's office.
B. Sea cabin.
C. At the GMDSS operating position.
D. Anywhere on board the vessel.

The big equipment GMDSS radio log is normally part of the GMDSS operating position, and may be a computer log if there are safety measures to insure it cannot be accidentally erased. Many radio operators print a hard copy of their computer radio logs at the end of the day. **ANSWER C**

1-5A4 Which of the following statements is true?
A. Key letters or abbreviations may be used in GMDSS Radio Logbooks if their meaning is noted in the log.
B. Key letters or abbreviations may not be used in GMDSS Radio Logbooks under any circumstances.
C. All Urgency communications must be entered in the logbook.
D. None of the above.

It is acceptable for abbreviations to be used in the radio log book as long as the meaning is noted somewhere in that same log book. **ANSWER A**

1-5A5 Which of the following logkeeping statements is true?

A. Entries relating to pre-voyage, pre-departure and daily tests are required.
B. Both a) and c)
C. A summary of all required Distress communications heard and Urgency communications affecting the station's own ship. Also, all Safety communications (other than VHF) affecting the station's own ship must be logged.
D. Routine daily MF-HF and Inmarsat-C transmissions do not have to be logged.

The radio log shall contain entries relating to pre-voyage and pre-departure operating sessions, as well as daily tests that may have been performed. Of course, the log shall contain a summary of all distress and urgency communications affecting the station's own ship, as well as high-frequency, long-range safety communications that would affect the station's own ship. **ANSWER B**

1-5A6 Which of the following statements concerning log entries is false?

A. All Safety communications received on VHF must be logged.
B. All required equipment tests must be logged.
C. The radio operator must log on and off watch.
D. The vessels daily position must be entered in the log.

Here is a question where you find the one answer that is false. An easy way to discover the correct answer is to see what distracters are TRUE, like logging daily position, on and off watch, and logging equipment tests. But what they are looking for is the INCORRECT requirement to log VHF safety calls – this not required ! **ANSWER A**

Key Topic 6: Miscellaneous Rules & Regulations

1-6A1 What regulations govern the use and operation of FCC-licensed ship stations in international waters?

A. The regulations of the International Maritime Organization (IMO) and Radio Officers Union.
B. Part 80 of the FCC Rules plus the international Radio Regulations and agreements to which the United States is a party.
C. The Maritime Mobile Directives of the International Telecommunication Union.
D. Those of the FCC's Wireless Telecommunications Bureau, Maritime Mobile Service, Washington, DC 20554.

Part 80 specifies rules for operating procedures, technical standards, safety watch requirements, emission classes, transmitter power, transmitter licensing, frequencies and tolerances, and more. These rules must be followed by all FCC-licensed ship stations wherever they travel in the world. **ANSWER B**

1-6A2 When may the operator of a ship radio station allow an unlicensed person to speak over the transmitter?

A. At no time. Only commercially-licensed radio operators may modulate the transmitting apparatus.
B. When the station power does not exceed 200 watts peak envelope power.
C. When under the supervision of the licensed operator.
D. During the hours that the radio officer is normally off duty.

In some cases a telephone call must be made from ship to shore; e.g., a passenger may want to call home. The passenger does not own the station license, therefore, the station operator supervises the communication to keep it legal. A marine operator is contacted over a VHF frequency and the call is placed just like any other phone call. The holder of the ship station license is charged for the call. §80.156. **ANSWER C**

1-6A3 Where do you make an application for inspection of a ship GMDSS radio station?

A. To a Commercial Operator Licensing Examination Manager (COLE Manager).
B. To the Federal Communications Commission, Washington, DC 20554.
C. To the Engineer-in-Charge of the FCC District Office nearest the proposed place of inspection.
D. To an FCC-licensed technician holding a GMDSS Radio Maintainer's License.

An application for the inspection of a ship GMDSS radio station begins with the FCC Engineer-in-Charge at the FCC District Office that is nearest to the proposed place of the ship inspection. If entering a port without "local knowledge" of where to obtain an inspection, start with the FCC District Office. This is the correct answer for this exam.

Recently, the FCC has made ship owners responsible for this annual inspection by requiring them to arrange their own inspections using local, FCC-licensed technicians who hold the required GMDSS Maintainer license. Still, the FCC is in the loop, so go with answer C, the FCC Engineer-in-charge, but know that the Commission requests going direct to a licensed technician, which helps reduce the workload at the FCC District Office. **ANSWER C**

1-6A4 Who has ultimate control of service at a ship's radio station?

A. The master of the ship.
B. A holder of a First Class Radiotelegraph Certificate with a six months' service endorsement.
C. The Radio Officer-in-Charge authorized by the captain of the vessel.
D. An appointed licensed radio operator who agrees to comply with all Radio Regulations in force.

The master of the ship (usually the owner or "captain" of the vessel) owns the station license and enjoys ultimate authority on the air. He is permitted to designate a qualified operator to handle radio messages. § 80.114. **ANSWER A**

1-6A5 Where must the principal radiotelephone operating position be installed in a ship station?

A. At the principal radio operating position of the vessel.
B. In the chart room, master's quarters or wheel house.
C. In the room or an adjoining room from which the ship is normally steered while at sea.
D. At the level of the main wheel house or at least one deck above the ship's main deck.

In order to assure that the person in command receives correct and timely information, it makes sense to keep the radio close to the steering room at all times, particularly in a busy shipping channel or at sea during a rough storm. §80.853(d). **ANSWER C**

1-6A6 By international agreement, which ships must carry radio equipment for the safety of life at sea?

A. All ships traveling more than 100 miles out to sea.
B. Cargo ships of more than 100 gross tons and passenger vessels on international deep-sea voyages.
C. All cargo ships of more than 100 gross tons.
D. Cargo ships of more than 300 gross tons and vessels carrying more than 12 passengers.

Cargo ships travel all over the world, most of the time far away from shore, and may require immediate assistance. Passenger ships can be in danger at a moment's notice. For these reasons, these ships must carry radio equipment. §80.851 and §80.901. **ANSWER D**

International Convention for the Safety of Life at Sea (SOLAS)

The SOLAS Convention is generally regarded as the most important of all international treaties concerning the safety of merchant ships. The first version was adopted in 1914, in response to the Titanic disaster; the second in 1929; the third in 1948, and the fourth in 1960.

Chapter IV of the SOLAS Convention covers Radiocommunications. It incorporates the Global Maritime Distress and Safety System (GMDSS). All passenger ships and all cargo ships of 300 gross tonnage and upwards on international voyages are required to carry equipment designed to improve the chances of rescue following an accident, including satellite emergency position indicating radio beacons (EPIRBs) and search and rescue transponders (SARTs) for the location of the ship or survival craft.

Regulations in Chapter IV cover undertakings by contracting governments to provide radiocommunciation services as well as ship requirements for carriage of specific types of radiocommunications equipment. The SOLAS Chapter is closely linked to the Radio Regulations of the International Telecommunication Union.

Hand-Held VHF Radios Permitted On Shore

The FCC now permits the use of VHF hand-held radios on shore to communicate back to their associated ship station. This use is only permitted from docks, piers, sea shore, or areas adjacent to local harbors and lakeside. The hand-held radios may not be used on shore to talk to one another - only back to the associated ship station.

Become a small passenger ship radio inspector with your GROL license!
See page 69 at the end of this chapter for details.

Subelement B – Communications Procedures

6 Key Topics, 6 Exam Questions

Key Topic 7: Bridge-to-Bridge Operations

1-7B1 What traffic management service is operated by the U.S. Coast Guard in certain designated water areas to prevent ship collisions, groundings and environmental harm?

A. Water Safety Management Bureau (WSMB).
B. Vessel Traffic Service (VTS).
C. Ship Movement and Safety Agency (SMSA).
D. Interdepartmental Harbor and Port Patrol (IHPP).

The Vessel Traffic Service improves vessel transit safety by providing big ship operators with advance information of other reported marine traffic in the area. Large commercial shipping vessels are tracked by shore side RADAR as well as the automatic identification system, and actively participate in VHF radio calls to the Vessel Traffic Service. VTS provides advice and recommendations over the VHF airwaves to help maintain vessel safety within the VTS area. VHF Channel 14 is a common channel for Vessel Traffic Service agencies throughout the country, but VTS comms may be found on other port operations channels as well. Passive vessels, such as smaller passenger ferries, or dredges, will also "squawk" their positions via AIS, and do more listening than transmitting on VTS channels. Non-participating vessels are encouraged to monitor Vessel Traffic System announcements, allowing smaller, recreational-type vessels a "heads up" on large ships coming into and out of a port or waterway. Think of the VTS as Air Traffic Control for ships, but slightly different in that the information provided by VTS centers is based on reports of participating vessels and can be no more accurate than the information they receive. Vessel Traffic Centers may not know all of the hazardous circumstances within the VTS area; and unlike Air Traffic Control, VTS does not usually issue orders to incoming or departing ships. The VTS operates in the 156-MHz range and protects, among other regions, the Seattle (Puget Sound), New York, New Orleans and Houston shipping channels. Other regions use VTS when there is no interference to the regions just described. §80.5. **ANSWER B**

VTS helps manage traffic in crowded waters, such as the lower Mississippi near New Orleans.

1-7B2 What is a bridge-to-bridge station?

A. An internal communications system linking the wheel house with the ship's primary radio operating position and other integral ship control points.
B. An inland waterways and coastal radio station serving ship stations operating within the United States.
C. A portable ship station necessary to eliminate frequent application to operate a ship station on board different vessels.
D. A VHF radio station located on a ship's navigational bridge or main control station that is used only for navigational communications.

A bridge-to-bridge station has nothing to do with draw bridges! Rather, the word "bridge" refers to the pilothouse where the ship's captain or the port pilot navigates the vessel in congested harbor or port waterways. VHF Channel 13 is reserved for bridge-to-bridge communications only, allowing large ships to always have a common channel to radio their navigational intentions. §80.5. **ANSWER D**

1-7B3 When may a bridge-to-bridge transmission be more than 1 watt?

A. When broadcasting a distress message and rounding a bend in a river or traveling in a blind spot.
B. When broadcasting a distress message.
C. When rounding a bend in a river or traveling in a blind spot.
D. When calling the Coast Guard.

When a 25-watt marine VHF radio is switched to Channel 13, internal software drops the power output to 1 watt. This can be seen as "low power" on the display. This keeps the communications confined to a small area on purpose. But in an emergency, or when rounding a bend in a river, 25 watts may be selected to insure your Channel 13 communications will be heard farther away. **ANSWER A**

1-7B4 When is it legal to transmit high power on Channel 13?

A. Failure of vessel being called to respond.
B. In a blind situation such as rounding a bend in a river.
C. During an emergency.
D. All of these.

Channel 13 messages must be about navigation; for example, passing or meeting other vessels. Your power must not be more than one watt unless you are declaring an emergency, the other vessel fails to respond, or if an obstruction prevents low-power communication. In such cases, higher output power is allowed. §80.331(c). **ANSWER D**

1-7B5 A ship station using VHF bridge-to-bridge Channel 13:

A. May be identified by the name of the ship in lieu of call sign.
B. May be identified by call sign and country of origin.
C. Must be identified by call sign and name of vessel.
D. Does not need to identify itself within 100 miles from shore.

Calling another ship station by that ship's name is perfectly legal because it is often impossible to know the other ship's call sign in advance. When you make contact on Channel 13 (156.65 MHz) the other ship station will identify itself with its own call sign. §80.102(c). **ANSWER A**

Navigating large vessels like this cruise ship requires expert bridge-to-bridge communications.

1-7B6 The primary purpose of bridge-to-bridge communications is:

A. Search and rescue emergency calls only.
B. All short-range transmission aboard ship.
C. Navigational communications.
D. Transmission of Captain's orders from the bridge.

Bridge-to-bridge communications are only for navigation. VHF Channel 13 is known as the bridge-to-bridge channel. It is available to all ships. **ANSWER C**

Key Topic 8: Operating Procedures-1

1-8B1 What is the best way for a radio operator to minimize or prevent interference to other stations?

A. By using an omni-directional antenna pointed away from other stations.
B. Reducing power to a level that will not affect other on-frequency communications.
C. Determine that a frequency is not in use by monitoring the frequency before transmitting.
D. By changing frequency when notified that a radiocommunication causes interference.

Even a few watts of power used on board a ship can be heard clearly thousands of miles away. This can disrupt other radio services, perhaps even masking a distress call from another ship. You may only be able to hear one of two stations, so listen for a few minutes before transmitting – it's not only polite, but also the law. Answer A is not correct because an omni-directional antenna transmits equally well in all directions; you can't point it away from other stations. Answers B and D try to solve a problem after causing it. §80.92 and §80.87. **ANSWER C**

1-8B2 Under what circumstances may a coast station using telephony transmit a general call to a group of vessels?

A. Under no circumstances.
B. When announcing or preceding the transmission of Distress, Urgency, Safety or other important messages.
C. When the vessels are located in international waters beyond 12 miles.
D. When identical traffic is destined for multiple mobile stations within range.

Maritime stations on shore are permitted to transmit general calls to all ships at sea in the case of distress, urgency, and safety. The distress announcement is "Mayday" spoken three times. The urgency call is "Pan-Pan" spoken three times followed by "all stations" spoken three times, followed by the designator of the shore station. A safety call, to announce important navigation information is preceded with "Securite" spoken three times, followed by "all stations" spoken three times, followed by the shore station identification. §80.111(a). **ANSWER B**

1-8B3 Who determines when a ship station may transmit routine traffic destined for a coast or government station in the maritime mobile service?

A. Shipboard radio officers may transmit traffic when it will not interfere with ongoing radiocommunications.
B. The order and time of transmission and permissible type of message traffic is decided by the licensed on-duty operator.
C. Ship stations must comply with instructions given by the coast or government station.
D. The precedence of conventional radiocommunications is determined by FCC and international regulation.

Ensure the smooth flow of information by following a coast or government station's requests to the letter. Making sudden demands on the coast station could upset their methods of handling paperwork, with the possible result that your message gets damaged or even lost. §80.116(g). **ANSWER C**

1-8B4 What is required of a ship station which has established initial contact with another station on 2182 kHz or Ch-16?

A. The stations must change to an authorized working frequency for the transmission of messages.
B. The stations must check the radio channel for Distress, Urgency and Safety calls at least once every ten minutes.
C. Radiated power must be minimized so as not to interfere with other stations needing to use the channel.
D. To expedite safety communications, the vessels must observe radio silence for two out of every fifteen minutes.

After making contact on the calling frequency, move to another frequency to keep the distress channel clear. § 80.116(c). **ANSWER A**

1-8B5 How does a coast station notify a ship that it has a message for the ship?

A. By making a directed transmission on 2182 kHz or 156.800 MHz.
B. The coast station changes to the vessel's known working frequency.
C. By establishing communications using the eight-digit maritime mobile service identification.
D. The coast station may transmit, at intervals, lists of call signs in alphabetical order for which they have traffic.

Coast stations use a "bulletin-board" method of traffic notification, which works well. It allows ships to find out if they have messages without querying the coast station. You should listen several times during the day, as the message queue constantly changes. §80.108. **ANSWER D**

1-8B6 What is the priority of communications?

A. Safety, Distress, Urgency and radio direction-finding.
B. Distress, Urgency and Safety.
C. Distress, Safety, radio direction-finding, search and rescue.
D. Radio direction-finding, Distress and Safety.

How to place a radio call. See page 68 at the end of this chapter.

DUSt off your logbook, and always log in distress calls. I also note urgency calls, and sometimes will even log safety calls regarding missing aids to navigation. **ANSWER B**

Key Topic 9: Operating Procedures-2

1-9B1 Under what circumstances may a ship or aircraft station interfere with a public coast station?

A. In cases of distress.
B. Under no circumstances during on-going radiocommunications.
C. During periods of government priority traffic handling.
D. When it is necessary to transmit a message concerning the safety of navigation or important meteorological warnings.

Since distress messages have the ultimate priority, operators may break in on any station at any time. If the operator of the distress station feels that the best way to attract attention is by disrupting another station, then (and only then!) it is legal to do so. All other messages must be sent in such a manner as to not disrupt other stations. §80.312. **ANSWER A**

1-9B2 Ordinarily, how often would a station using a telephony emission identify?

A. At least every 10 minutes.
B. At the beginning and end of each transmission and at 15-minute intervals.
C. At 15-minute intervals, unless public correspondence is in progress.
D. At 20-minute intervals.

Stations engaged in conversation must identify their stations at the beginning and end of each transmission and at 15-minute intervals. This not only lets listeners know who is communicating, but it also satisfies the FCC's rule that only licensed operators are allowed to use the airwaves. Operators must give their station call sign in English. §80.102. **ANSWER B**

1-9B3 When using a SSB station on 2182 kHz or VHF-FM on channel 16:

A. Preliminary call must not exceed 30 seconds.
B. If contact is not made, you must wait at least 2 minutes before repeating the call.
C. Once contact is established, you must switch to a working frequency.
D. All of these.

The idea is to take up as little time on the distress or calling frequency as possible. Keep in mind that some vessel in distress may need the frequency at any time. §80.116. **ANSWER D**

1-9B4 What should a station operator do before making a transmission?

A. Except for the transmission of distress calls, determine that the frequency is not in use by monitoring the frequency before transmitting.
B. Transmit a general notification that the operator wishes to utilize the channel.
C. Check transmitting equipment to be certain it is properly calibrated.
D. Ask if the frequency is in use.

Always listen before transmitting; however, it is legal to transmit a distress call immediately. **ANSWER A**

1-9B5 On what frequency should a ship station normally call a coast station when using a radiotelephony emission?

A. On a vacant radio channel determined by the licensed radio officer.
B. Calls should be initiated on the appropriate ship-to-shore working frequency of the coast station.
C. On any calling frequency internationally approved for use within ITU Region 2.
D. On 2182 kHz or Ch-16 at any time.

The FCC grants Private Coast Station Licenses to marine businesses to operate land-based stations, and assigns specific channels to such stations. To minimize unnecessary calling of these stations on the distress channels (2128 kHz and Channel 16), it is best to call the shore station on its assigned frequency. §80.116(a). **ANSWER B**

1-9B6 In the International Phonetic Alphabet, the letters E, M, and S are represented by the words:

A. Echo, Michigan, Sonar.
B. Equator, Mike, Sonar.
C. Echo, Mike, Sierra
D. Element, Mister, Scooter

A phonetic alphabet question! If you have forgotten the International Phonetic Alphabet, use this table to refresh your memory! **ANSWER C**

ITU International Phonetic Alphabet

A – Alfa – **AL** fah	H – Hotel – **HOH** tel	O – Oscar – **OSS** car	V – Victor – **VICK** tah
B – Bravo – **BRAH** voh	I – India – **IN** dee ah	P – Papa – **PAH** pah	W – Whiskey – **WISS** key
C – Charlie – **CHAR** lee	J – Juliet – **JEW** lee ett	Q – Quebec – **KEH** beck	X – X-ray – **ECKS** ray
D – Delta – **DELL** tah	K – Kilo – **KEY** loh	R – Romeo – **ROW** me o	Y – Yankee – **YANG** kee
E – Echo – **ECK** ohh	L – Lima – **LEE** mah	S – Sierra – **SEE** air rah	Z – Zulu – **ZOO** loo
F – Foxtrot – **FOKS** trot	M – Mike – **MIKE**	T – Tango – **TANG** go	
G – Golf – **GOLF**	N – November – **NO** vem ber	U – Uniform – **YOU** nee form	

Key Topic 10: Distress Communications

1-10B1 What information must be included in a Distress message?

A. Name of vessel.
B. Location.
C. Type of distress and specifics of help requested.
D. All of the above.

People can become so scared during an emergency that they may panic and forget to include even the most basic information needed to receive assistance. On land, a 911 operator may be able to learn the location from which a phone call is received, but no such luxury may exist at sea when using voice. The more information you provide over the air, the better your chances of receiving help. §80.316. **ANSWER D**

1-10B2 What are the highest priority communications from ships at sea?

A. All critical message traffic authorized by the ship's master.
B. Navigation and meteorological warnings.
C. Distress calls are highest and then communications preceded by Urgency and then Safety signals.
D. Authorized government communications for which priority right has been claimed.

The ultimate priority in radio traffic goes to those messages whose reception could mean the difference between life and death. In such cases, rules and regulations pertaining to the legality of frequency use and duration are temporarily suspended. §80.312 and §80.91. **ANSWER C**

1-10B3 What is a Distress communication?

A. Communications indicating that the calling station has a very urgent message concerning safety.
B. An internationally recognized communication indicating that the sender is threatened by grave and imminent danger and requests immediate assistance.
C. Radio communications which, if delayed, will adversely affect the safety of life or property.
D. An official radio communication notification of approaching navigational or meteorological hazards.

Grave and imminent danger defines a distress communication – it has top priority over messages of urgency and safety. § 80.5 and §80.314. **ANSWER B**

1-10B4 What is the order of priority of radiotelephone communications in the maritime services?

A. Alarm and health and welfare communications.
B. Navigation hazards, meteorological warnings, priority traffic.
C. Distress calls and signals, followed by communications preceded by Urgency and Safety signals and all other communications.
D. Government precedence, messages concerning safety of life and protection of property, and traffic concerning grave and imminent danger.

Distress calls always take precedence over all other types of communication. Distress means lives are in danger and immediate assistance is required. Urgency calls concern the safety of a ship or person. Safety traffic concerns navigation problems and weather warnings. §80.91. **ANSWER C**

1-10B5 The radiotelephone Distress call and message consists of:

A. MAYDAY spoken three times, followed by the name of the vessel and the call sign in phonetics spoken three times.
B. Particulars of its position, latitude and longitude, and other information which might facilitate rescue, such as length, color and type of vessel, and number of persons on board.
C. Nature of distress and kind of assistance required.
D. All of the above.

Many people become so frightened during an emergency that they forget how to call for help. It is important to stay calm and help the authorities help you. Provide them with as much information as possible to make their job easier. "MAYDAY MAYDAY MAYDAY" captures their attention; your call sign and vessel name tell them who you are; your latitude and longitude tell them where you are. State the nature of your distress so the authorities can bring along the necessary equipment. Give the number of persons aboard and conditions of any injured. Briefly describe your boat's length and color of the hull. If a ship is swamped, count heads immediately. If someone is missing, tell the authorities. §80.315(b). **ANSWER D**

1-10B6 What is Distress traffic?

A. All messages relative to the immediate assistance required by a ship, aircraft or other vehicle threatened by grave or imminent danger, such as life and safety of persons on board, or man overboard.
B. In radiotelephony, the speaking of the word, "Mayday."
C. Health and welfare messages concerning property and the safety of a vessel.
D. Internationally recognized communications relating to important situations.

While choice D describes the entire spectrum of possible situations in which distress traffic may occur, choice A is the correct answer to this question. §80.5 and §80.314. **ANSWER A**

Key Topic 11: Urgency and Safety Communications

1-11B1 What is a typical Urgency transmission?

A. A request for medical assistance that does not rise to the level of a Distress or a critical weather transmission higher than Safety.
B. A radio Distress transmission affecting the security of humans or property.
C. Health and welfare traffic which impacts the protection of on-board personnel.
D. A communications alert that important personal messages must be transmitted.

Let's first start with the highest priority, a Distress call, where safety of life or protection of the vessel from sinking or fire is at hand. Next is the Urgency transmission, preceded by the words, "Pan-Pan, Pan-Pan, Pan-Pan." It could apply to a passenger or crewman on board in need of medical attention or air evacuation, or a man overboard situation where the vessel is returning to pick up the overboard mariner. Third is the Safety call, preceded by the word "Securite" spoken three times. It usually deals with weather or navigational information. §80.5 and §80.327. **ANSWER A**

1-11B2 What is the internationally recognized Urgency signal?

A. The letters "TTT" transmitted three times by radiotelegraphy.
B. The words "PAN PAN" spoken three times before the Urgency call.
C. Three oral repetitions of the word "Safety" sent before the call.
D. The pronouncement of the word "Mayday."

Remember that "Mayday" is reserved for Distress traffic only; "PAN PAN" – spoken three times – is for Urgency. §80.327. **ANSWER B**

1-11B3 What is a Safety transmission?

A. A communications transmission which indicates that a station is preparing to transmit an important navigation or weather warning.
B. A radiotelephony warning preceded by the words "PAN PAN."
C. Health and welfare traffic concerning the protection of human life.
D. A voice call proceeded by the words "Safety Alert."

Choices B and C are incorrect because they pertain to Urgency traffic. Safety traffic concerns only navigation and weather information. §80.5, also §80.329. **ANSWER A**

1-11B4 The Urgency signal concerning the safety of a ship, aircraft or person shall be sent only on the authority of:

A. Master of ship.
B. Person responsible for mobile station.
C. Either Master of ship or person responsible for mobile station.
D. An FCC-licensed operator.

The FCC rules state this clearly: "The urgency signal must be sent only on the authority of the master or person responsible for the mobile station." The person sending urgency traffic has a "very urgent message to transmit concerning the safety of a ship, aircraft, or other vehicle, or the safety of a person." It is important that such a message be made by a dependable person. If the master of the ship is busy, as often happens in urgency situations, then the responsibility falls upon the person so designated. §80.327 (a). **ANSWER C**

1-11B5 The Urgency signal has lower priority than:

A. Ship-to-ship routine calls.
B. Distress.
C. Safety.
D. Security.

DUSt off your thinking and remember Distress has the highest priority, followed by the Urgency signal, then followed by Safety signal. **ANSWER B**

1-11B6 What safety signal call word is spoken three times, followed by the station call letters spoken three times, to announce a storm warning, danger to navigation, or special aid to navigation?

A. PAN PAN.
B. MAYDAY.
C. SAFETY.
D. SECURITY.

Distress messages, Urgency messages and Safety messages each have their own special call word. The word for Safety messages is SECURITY. Be careful not to choose answer C. When you hear "SECURITY SECURITY SECURITY," you know that a message concerning a storm warning or danger to navigation is about to be sent. §80.5 and §80.329. **ANSWER D**

DISTRESS CALL AND MESSAGE

Sending Distress Call and Message
Transmit (in this order):

1. Distress signal MAYDAY (spoken three times)
2. The words THIS IS (spoken once)
3. Name of vessel in distress (spoken three times) and call sign (spoken once)
4. Position of vessel in distress by latitude and longitude or by bearing (true or magnetic, state which) and distance to a well-known landmark such as a navigational aid or small island, or in any terms which will assist a responding station in locating the vessel in distress. Include any information on vessel movement such as course, speed and destination,
5. Nature of distress (sinking, fire, etc.)
6. Kind of assistance desired
7. Any other information which might facilitate rescue, such as: length or tonnage of vessel number of persons on board and number needing medical attention, color of hull, cabin, masts, etc.
8. The word OVER

Example: Distress Call and Message
MAYDAY-MAYDAY-MAYDAY
THIS IS BLUE DUCK - BLUE DUCK - BLUE DUCK - WA 1234 – MAYDAY - BLUE DUCK
DUNGENESS LIGHT BEARS 185 DEGREES MAGNETIC - DISTANCE 2 MILES
STRUCK SUBMERGED OBJECT NEED PUMPS - MEDICAL ASSISTANCE - AND TOW
THREE ADULTS-TWO CHILDREN ABOARD, ONE PERSON COMPOUND FRACTURE OF ARM
ESTIMATE CAN REMAIN AFLOAT TWO HOURS
BLUE DUCK IS THIRTY-TWO FOOT CABIN CRUISER - BLUE HULL - WHITE DECK HOUSE - OVER
NOTE: Repeat at intervals until answer is received. If no answer is received on the Distress frequency, repeat using any other available channel on which attention might be attracted.

ACKNOWLEDGEMENT OF DISTRESS MESSAGE
If you hear a Distress Message from a vessel and it is not answered, then YOU must answer. If you are reasonably sure that the distressed vessel is not in your vicinity, you should wait a short time for others to acknowledge.

Sending Acknowledgement of Receipt of Distress Message
Acknowledgement of receipt of a Distress Message usually includes the following:

1. The distress signal MAYDAY
2. Name of vessel sending the Distress Message (spoken three times)
3. The words THIS IS (spoken once)
4. Name of your vessel (spoken three times) and your call sign (spoken once)
5. The words RECEIVED MAYDAY (spoken once)
6. The word OVER (spoken once)

Example: Acknowledge Message
MAYDAY BLUE DUCK - BLUE DUCK - BLUE DUCK - WA 1234
THIS IS-WHITE WHALE - WHITE WHALE - WHITE WHALE - WZ 4321
RECEIVED MAYDAY - OVER

Distress, Urgency & Safety Operating Procedures

If you are in distress, you may use any means at your disposal to attract attention and obtain assistance. You are not limited to the use of your marine radiotelephone. Often, visual signals, including flags, flares, lights, smoke, etc., or audible signals such as your boat's horn or siren, or a whistle or megaphone, will get the attention and help you need.

For boats equipped with marine radiotelephone, help is just a radio signal away. Channel 16 (156.800 MHz) the VHF-FM Distress, Urgency and Safety calling frequency, is the primary emergency channel in the VHF marine band. Vessels equipped with marine MF single sideband equipment should use safety frequencies 4125 kHz or 8291 kHz for calling a U.S. Coast Guard long-range communications facility. The 2182 kHz (voice) and 2187.5 kHz (DSC) International Distress, Safety and Calling frequencies are no longer monitored by the U.S. Coast Guard due to this frequency's limited range capabilities. In an emergency you may use any frequency to call for help!

There are other types of marine stations located ashore that are listening to Channel 16 and 2182 kHz along with the marine radio equipped vessels operating in the area. Because of this coverage, almost any kind of a call for assistance on Channel 16, or 2182 kHz, will probably get a response. There are times, however, when the situation demands immediate attention; when you just can't tolerate delay. These are the times when you need to know how to use (or respond to) the Distress and Urgency signals and how to respond to the Safety signal.

Spoken Emergency Signals
There are three spoken emergency signals:

1. Distress Signal: MAYDAY The distress signal MAYDAY is used to indicate that a mobile station is threatened by grave and imminent danger and requests immediate assistance. MAYDAY has priority over all other communications.

2. Urgency Signal: PAN-PAN This signal is used when the safety of the vessel or person is in jeopardy. "Man overboard" messages are sent with the Urgency signal. PAN-PAN (properly pronounced PAHN-PAHN) has priority over all other communications with the exception of distress traffic.

3. Safety Signal: SECURITY This call is used for messages concerning the safety of navigation or giving important meteorological warnings (properly pronounced SAY-CURITAY).

Any message headed by one of the emergency signals (MAYDAY, PAN-PAN, or SECURITY) must be given precedence over routine communications. This means listen. Don't transmit. Be prepared to help if you can. The decision of which of these emergency signals to use is the responsibility of the person in charge of the vessel.

Key Topic 12: GMDSS

1-12B1 What is the fundamental concept of the GMDSS?

A. It is intended to automate and improve existing digital selective calling procedures and techniques.
B. It is intended to provide more effective but lower cost commercial communications.
C. It is intended to provide compulsory vessels with a collision avoidance system when they are operating in waters that are also occupied by non-compulsory vessels.
D. It is intended to automate and improve emergency communications in the maritime industry.

When the Titanic sent out its radio distress call, CQD, a nearby vessel had already closed down its radio station, and people on deck could see distant fireworks, thinking some sort of celebration was underway on the doomed ship. This disaster in 1912 prompted basic rules for guarding the radio distress channel. Nearly 100 years later, major upgrades in maritime wireless communications were mandated by the International Maritime Organization to insure no radio call for help would ever go unanswered. The Global Marine Distress Safety System (GMDSS) requires commercial vessel compliance, with small recreational boats encouraged to install simple marine VHF radios to take part in this worldwide system of automated distress messaging. **ANSWER D**

1-12B2 The primary purpose of the GMDSS is to:

A. Allow more effective control of SAR situations by vessels.
B. Provide additional shipboard systems for more effective company communications.
C. Automate and improve emergency communications for the world's shipping industry.
D. Provide effective and inexpensive communications.

A key concept of the GMDSS system is automated emergency (distress) communications on universal frequencies anywhere in the world. **ANSWER C**

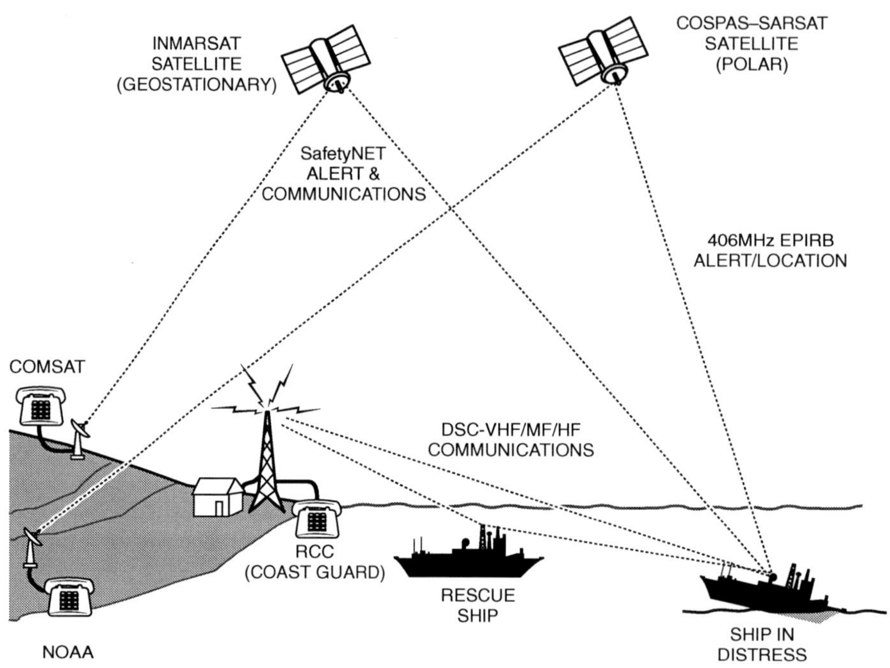

The GMDSS System
(Courtesy of FCC Aviation & Marine Branch)

1-12B3 What is the basic concept of GMDSS?

A. Shoreside authorities and vessels can assist in a coordinated SAR operation with minimum delay.
B. Search and rescue authorities ashore can be alerted to a Distress situation.
C. Shipping in the immediate vicinity of a ship in Distress will be rapidly alerted.
D. All of these.

You are on a small sailboat, participating in an ocean race, 500 miles from shore. In heavy wind and seas, your keel sheers off and your sailboat instantly turns turtle. Everyone on board scrambles in the air pocket inside the hull. The Captain turns on the 406 MHz EPIRB, and the GMDSS COSPAS/SARSAT system goes into action. In less than 4 hours, an offshore US Coast Guard vessel comes along side, sends down a rescue swimmer, and all lives are saved. Other nearby vessels arrive on scene as well, all part of the GMDSS system. **ANSWER D**

1-12B4 GMDSS is primarily a system based on?

A. Ship-to-ship Distress communications using MF or HF radiotelephony.
B. VHF digital selective calling from ship to shore.
C. Distress, Urgency and Safety communications carried out by the use of narrow-band direct printing telegraphy.
D. The linking of search and rescue authorities ashore with shipping in the immediate vicinity of a ship in Distress or in need of assistance.

A large container ship, sailing from Alaska to Japan, may be so much on automatic pilot that they miss the distress flare coming from 5 survivors huddled in a life raft, after their long-line fishing vessel went under. A DSC handheld sends out a distress alert, and they now get noticed on radio and all lives are saved. The key words here are "immediate vicinity of a ship in Distress." **ANSWER D**

1-12B5 What is the responsibility of vessels under GMDSS?

A. Vessels over 300 gross tons may be required to render assistance if such assistance does not adversely affect their port schedule.
B. Only that vessel, regardless of size, closest to a vessel in Distress, is required to render assistance.
C. Every ship is able to perform those communications functions that are essential for the Safety of the ship itself and of other ships.
D. Vessels operating under GMDSS, outside of areas effectively serviced by shoreside authorities, operating in sea areas A2, and A4 may be required to render assistance in Distress situations.

One of the laws at sea is to always help a fellow mariner in distress. The Global Marine Distress Safety System is not only for the safety of the ship itself, but also for the safety of other ships in the immediate vicinity. **ANSWER C**

1-12B6 GMDSS is required for which of the following?

A. All vessels capable of international voyages.
B. SOLAS Convention ships of 300 gross tonnage or more.
C. Vessels operating outside of the range of VHF coastal radio stations.
D. Coastal vessels of less than 300 gross tons.

SOLAS stands for Safety Of Life At Sea, and ships of 300 gross tonnage or more are required to meet GMDSS capability, with more elaborate equipment required as they venture farther from shore:

Sea area A-1 Within range of a coastal VHF-DSC equipped shore station, approximately 20-30 nm.
Sea area A-2 Beyond A-1 and within range of a coastal MF-DSC equipped shore station, approximately 150-200 nm.
Sea area A-3 Beyond A-1 and A-2 between 70 degrees N and S latitude, within range of Inmarsat.
Sea area A-4 Beyond A-1, A-2, and A-3. Essentially beyond 70 degrees N and S latitude, the Polar regions.

ANSWER B

GMDSS TASK FORCE

The U. S. Coast Guard, with the support of other government maritime organizations, chartered the GMDSS Task Force in 1993 to assist the private sector in implementing the GMDSS. In 2005, the Coast Guard passed direct sponsorship of the Task Force to the Radio Technical Commission for Maritime Services (RTCM), a non-profit public interest group which provides secretarial support and hosts Task Force meetings. Learn more at www.rtcm.org. The National Marine Electronics Association (NMEA) a non-profit trade organization representing marine electronic manufacturers and service agents, also hosts a Task Force meeting annually. Learn more at www.nmea.org.

The GMDSS Task Force is responsible for the following activities:

1. Provide a forum for interested parties to consider all aspects of GMDSS implementation and recommend action to appropriate authorities. This includes considering modernization of GMDSS systems and procedures and participation in the U.S. SOLAS Communications, Search and Rescue (COMSAR) Working Group, which formulates U.S. positions on IMO/COMSAR issues.
2. Maintain a Public Relations program to inform all sectors of the maritime community on the benefits of GMDSS participation and to solicit feedback on GMDSS problems.
3. Maintain an overview of GMDSS training to encourage well-trained GMDSS operators on compulsory vessels, indoctrinate operators of non-compulsory vessels with an appropriate level of voluntary GMDSS training, and advocate programs to minimize false alerts in order to improve the effectiveness of Search and Rescue operations.
4. Maintain liaison with equipment manufacturers and service agents to facilitate proper equipping of compulsory vessels and providing advice to non-compulsory vessels that use GMDSS systems on a voluntary basis.
5. Maintain liaison with the recreational vessel community, which is by far the largest group in the U.S. maritime sector, to which very little regulation applies and for which there is significant demand for manuals and voluntary training aids.
6. Monitor progress of government and private sector shore radio system modernization to support the GMDSS and offer new services.
7. Monitor developments and promulgate information concerning non-GMDSS radio systems of interest to the maritime community such at the Automatic Identification System (AIS), the Ship Security and Alerting System (SSAS); Voyage Data Recorders (VDR), Long Range Identification and Tracking systems (LRIT), Vessel Monitoring Systems (VMS) used by fishing vessels, and developments in E-Navigation.

Subelement C – Equipment Operations

6 Key Topics, 6 Exam Questions

Key Topic 13: VHF Equipment Controls

1-13C1 What is the purpose of the INT-USA control settings on a VHF?

A. To change all VTS frequencies to Duplex so all vessels can receive maneuvering orders.
B. To change all VHF channels from Duplex to Simplex while in U.S. waters.
C. To change certain International Duplex channel assignments to simplex in the U.S. for VTS and other purposes.
D. To change to NOAA weather channels and receive weather broadcasts while in the U.S.

Some VHF channels have dual roles throughout the world. For instance, VHF Channel 7A (156.350 MHz) is for simplex communications here in North America. However, in other parts of the world, Channel 7 (no A after it) splits the receiver 4.6 MHz up for duplex operation, receiving at 160.950 MHz. For operation here in North America always make sure the VHF radio is in the USA domestic mode. **ANSWER C**

1-13C2 VHF ship station transmitters must have the capability of reducing carrier power to:

A. 1 watt.
B. 10 watts.
C. 25 watts.
D. 50 watts.

FCC rules require radio operators to use as little output power as possible to maintain reliable communications. Restricting VHF transmissions to only 1 watt is common when calling for another ship, because it is usually within eyeshot and 1 watt is more than powerful enough. If the station does not respond, then try calling with higher power. §80.873(c). **ANSWER A**

1-13C3 The Dual Watch (DW) function is used to:

A. Listen to Ch-70 at the same time while monitoring Ch-16.
B. Sequentially monitor 4 different channels.
C. Sequentially monitoring all VHF channels.
D. Listen on any selected channel while periodically monitoring Ch-16.

A marine VHF may also incorporate a dual watch mode, allowing the radio to periodically sample Channel 16 for activity. Out on the high seas, this would allow a nearly continuous watch of the distress channel when the dial is tuned to an alternate channel, such as calling Channel 9. **ANSWER D**

1-13C4 Which of the following statements best describes the correct setting for manual adjustment of the squelch control?

A. Adjust squelch control to the minimum level necessary to barely suppress any background noise.
B. Always adjust squelch control to its maximum level.
C. Always adjust squelch control to its minimum level.
D. Adjust squelch control to approximately twice the minimum level necessary to barely suppress any background noise.

Easy logical question here. Turn the squelch control to just above the point it silences background white noise. If you turn it any further, you may squelch out a weak, distant signal. **ANSWER A**

1-13C5 The "Scan" function is used to:

A. Monitor Ch-16 continuously and switching to either Ch-70 or Ch-13 every 5 seconds.
B. Scan Ch-16 for Distress calls.
C. Scan Ch-70 for Distress alerts.
D. Sequentially scan all or selected channels.

The VHF radio scan function allows you to tag specific VHF channels and let the receiver do the searching for channel activity. **ANSWER D**

1-13C6 Why must all VHF Distress, Urgency and Safety communications (as well as VTS traffic calls) be performed in Simplex operating mode?

A. To minimize interference from vessels engaged in routine communications.
B. To ensure that vessels not directly participating in the communications can hear both sides of the radio exchange.
C. To enable an RCC or Coast station to only hear communications from the vessel actually in distress.
D. To allow an RCC or Coast station to determine which transmissions are from other vessels and which transmissions are from the vessel actually in distress.

In the United States, the marine VHF channels listed below are operated in the Simplex mode. If the VHF transceiver inadvertently gets switched to the "INT" International mode, the receiver offsets 4.6 MHz higher on "A" channels, precluding simplex ship-to-ship and ship-to-shore communications. This is a big problem with the US Coast Guard on simplex liaison Channel 22A – if the mariner has his radio in the international mode, the Coast Guard shore station will hear him loud and clear, but the mariner will never hear the Coast Guard. **ANSWER B**

The modern marine VHF transceiver automatically selects simplex when the radio indicates "Domestic" or "USA" mode, on the following channels:

USA Channel	ITU Channel	Frequency (MHz)	USA Channel	ITU Channel	Frequency (MHz)
7A	1007	156.350	78A	1078	156.925
18A	1018	156.900	79A	1079	156.975
19A	1019	156.950	80A	1080	157.025
21A	1021	157.050	81A	1081	157.075
22A	1022	157.100	82A	1082	157.125
23A	1023	157.150	83A	1083	157.175
65A	1065	156.275	88A	1088	157.425
66A	1066	156.325			

Key Topic 14: VHF Channel Selection

1-14C1 What channel must VHF-FM-equipped vessels monitor at all times when the vessel is at sea?

A. Channel 8.
B. Channel 16.
C. Channel 5A.
D. Channel 1A.

FCC rules require each VHF ship station during its hours of operation to maintain a watch on 156.800 MHz (Channel 16) whenever such station is not being used for exchanging communications. This same VHF radio, with an aural watch on Channel 16, is also silently guarding VHF Channel 70 (156.525 MHz), the DSC calling channel. Recreational VHF radios sample Channel 70, but commercial vessels are usually equipped with more expensive marine radios that offer a dedicated receiver for continuous guarding of VHF Channel 70. If the radio is very old, and you don't see a red plastic cover DSC button, this is probably a non-DSC-capable radio and should be retired immediately. §80.148. **ANSWER B**

Always monitor channel 16, at sea, and even when tied-up at the dock.

1-14C2 What is the aircraft frequency and emission used for distress communications?

A. 243.000 MHz - F3E.
B. 121.500 MHz - F3E.
C. 156.525 MHz - F1B.
D. 121.500 MHz - A3E.

Aircraft emergencies, as well as all activated EPIRBs, transmit on 121.500 MHz, A3E. Even though you have heard that there is no longer a 121.500 MHz EPIRB satellite guard, nonetheless, all EPIRBs continue to transmit a low power homing signal on 121.500 MHz, in addition to their data transmission on 406 MHz. **ANSWER D**

1-14C3 Which VHF channel is used only for digital selective calling?

A. Channel 70.
B. Channel 16.
C. Channel 22A.
D. Channel 6.

The international VHF digital selective calling frequency is 156.525 MHz, which is Channel 70. With this method of communication, digital codes trigger circuits on specially-designed receivers. Ordinarily, once you make contact with another station set up for DSC, both of you can switch to another frequency. DSC keeps the call-up private, since no one else can hear it. A distress message on DSC is different because it triggers other DSC radios. Routine messages trigger only those radios set up to receive messages that include a specific DSC code. **ANSWER A**

In a distress, lift the cover and press the button for five seconds to sound a distress signal on channel 70.

1-14C4 Which channel is utilized for the required bridge-to-bridge watch?

A. DSC on Ch-70.
B. VHF-FM on Ch-16.
C. VHF-FM on Ch-13 in most areas of the continental United States.
D. The vessel's VHF working frequency.

On bridge-to-bridge Channel 13, you will often hear port pilots talking to the large commercial ships they are bringing into the harbor. **ANSWER C**

1-14C5 Which channel would most likely be used for routine ship-to-ship voice traffic?

A. Ch-16.
B. Ch-08.
C. Ch-70.
D. Ch-22A.

For this question, let's eliminate the incorrect answers. Channel 16 is for distress and calling, not routine communications. Channel 70 is non-voice for digital selective calling. Channel 22A is a liaison channel with US Coast Guard, leaving VHF Channel 08 as a channel for commercial ships to communicate routine traffic. Channel 8 is not permitted for recreational vessel communications.

PLEASURE CRAFT

Channel 68 – Routine pleasure craft comms
Channel 69 – Routine pleasure craft comms
Channel 70 – Digital Selective Calling
Channel 71 – Routine pleasure craft comms
Channel 72 – Routine pleasure craft comms
Channel 78A – Routine pleasure craft comms (ITU 1078)

As you can see, pleasure craft operators only have a few ship-to-ship and ship-to-shore (yacht club) simplex channels.

FOR-HIRE CRAFT

Channel 7A – Commercial ship comms (ITU 1007)
Channel 8 – Commercial ship comms
Channel 10 – Commercial ship comms
Channel 11 – Commercial ship comms
Channel 18A – Commercial ship comms (ITU 1018)
Channel 19A – Commercial ship comms (ITU 1019)
Channel 67 – Commercial ship comms
Channel 70 – Digital Selective Calling
Channel 77 – Commercial ship comms
Channel 79A – Commercial ship comms (ITU 1079)
Channel 80A – Commercial ship comms (ITU 1080)
Channel 88A – Commercial ship and Aircraft comms (ITU 1088)

Channel 9 may be used for both commercial and non-commercial calling, as well as for brief communications. It is generally illegal to operate on "any available channel" by announcing "go up one." Every marine VHF channel has a specific purpose. **ANSWER B**

1-14C6 What channel would you use to place a call to a shore telephone?

A. Ch-16.
B. Ch-70.
C. Ch-28.
D. Ch-06.

The local VHF marine operator channels are Channel 24 to 28, and Channels 84 to 87. This makes Channel 28 a public correspondence channel to place a call to a shore telephone. The majority of public correspondence marine operators have vanished from the airwaves, so if you raise anyone on Channel 28, be sure to let me know! Public correspondence telephone channels are duplex, with the VHF radio receiver automatically shifting up 4.6 MHz from the transmit frequency. With the right type of expensive shipboard equipment, you could operate full duplex when making a telephone call from ship to shore. The duplex telephone channels also may offer the capability of an automated radio check! Try channels 26, 27, or 28. Depress the microphone P-T-T transmit button, hold the mic up to your lips, and say TEST ONE TWO THREE, give your vessel name, and then release the mic button. On one of these three channels you may hear a recording of your voice transmitter test played back to you, along with an ad for the maritime organization providing this free service on marine VHF telephone channels. **ANSWER C**

U.S./ITU VHF Marine Radio Channels and Frequencies

(Source: www.navcen.USCG.gov/marcomms/vhf.htm)

U.S./ITU Channel Number	Ship Trans (MHz)	Ship Recv (MHz)	Use
01A (1001)	156.050	156.050	Port Operations and Commercial, VTS. (Available only in New Orleans/ Lower Mississippi area.)
05A (1005)	156.250	156.250	Port Operations or VTS in the Houston, New Orleans and Seattle areas.
6	156.300	156.300	Intership Safety
07A (1007)	156.350	156.350	Commercial
8	156.400	156.400	Commercial (Intership only)
9	156.450	156.450	Boater Calling. Commercial and Non-Commercial.
10	156.500	156.500	Commercial
11	156.550	156.550	Commercial. VTS in selected areas.
12	156.600	156.600	Port Operations. VTS in selected areas.
13	156.650	156.650	Intership Navigation Safety (Bridge-to-bridge). Ships >20m length maintain a listening watch on this channel in US waters.
14	156.700	156.700	Port Operations. VTS in selected areas.
15	--	156.75	Environmental (Receive only). Used by Class C EPIRBs.
16	156.800	156.800	International Distress, Safety and Calling. Ships required to carry radio. USCG, and most coast stations maintain a listening watch on this channel.
17	156.850	156.850	State Control
18A (1018)	156.900	156.900	Commercial
19A (1019)	156.950	156.950	Commercial
20	157.000	161.600	Port Operations (duplex)
20A (1020)	157.000	157.000	Port Operations
21A (1021)	157.050	157.050	U.S. Coast Guard only
22A (1022)	157.100	157.100	Coast Guard Liaison and Maritime Safety Information Broadcast
23A (1023)	157.150	157.150	U.S. Coast Guard only
24	157.200	161.800	Public Correspondence (Marine Operator)
25	157.250	161.850	Public Correspondence (Marine Operator)
26	157.300	161.900	Public Correspondence (Marine Operator)
27	157.350	161.950	Public Correspondence (Marine Operator)
28	157.400	162.000	Public Correspondence (Marine Operator)
63A (1063)	156.175	156.175	Port Operations and Commercial, VTS. (Available only in New Orleans/ Lower Mississippi area.)
65A (1065)	156.275	156.275	Port Operations
66A (1066)	156.325	156.325	Port Operations
67	156.375	156.375	Commercial. Used for Bridge-to-bridge communications in I
68	156.425	156.425	Non-Commercial
69	156.475	156.475	Non-Commercial
70	156.525	156.525	Digital Selective Calling (voice communications not allowed)
71	156.575	156.575	Non-Commercial
72	156.625	156.625	Non-Commercial (Intership only)
73	156.675	156.675	Port Operations
74	156.725	156.725	Port Operations
77	156.875	156.875	Port Operations (Intership only)
78A (1078)	156.925	156.925	Non-Commercial
79A (1079)	156.975	156.975	Commercial. Non-Commercial in Great Lakes only
80A (1080)	157.025	157.025	Commercial. Non-Commercial in Great Lakes only
81A (1081)	157.075	157.075	U.S. Government only - Environmental protection operations
82A (1082)	157.125	157.125	U.S. Government only
83A (1083)	157.175	157.175	U.S. Coast Guard only
84	157.225	161.825	Public Correspondence (Marine Operator)
85	157.275	161.875	Public Correspondence (Marine Operator)
86	157.325	161.925	Public Correspondence (Marine Operator)
87A (1087)	157.375	157.375	Public Correspondence (Marine Operator)
88A (1088)	157.425	157.425	Commercial, Intership only.
AIS 1	161.975	161.975	Automatic Identification System (AIS)
AIS 2	162.025	162.025	Automatic Identification System (AIS)

NOAA Weather Radio Frequencies

Channel	Frequency (MHz)
WX1	162.550
WX2	162.400
WX3	162.475
WX4	162.425
WX5	162.450
WX6	162.500
WX7	162.525

Note that the letter "A" indicates simplex use of the ship station transmit side of an international duplex channel, and that operations are different than international operations on that channel. Some VHF transceivers are equipped with an "International - U.S." switch for that purpose. "A" channels are generally only used in the United States, and use is normally not recognized or allowed outside the U.S. The letter "B" indicates simplex use of the coast station transmit side of an international duplex channel. The U.S. does not currently use "B" channels for simplex communications in this band.

Pleasure boaters should normally use channels listed as Non-Commercial. Channel 16 is used for calling other stations or for distress alerting. Channel 13 should be used to contact a ship when there is danger of collision. All ships of length 20m or greater are required to guard VHF channel 13, in addition to VHF channel 16, when operating within U.S. territorial waters. Users may be fined by the FCC for improper use of these channels.

Key Topic 15: MF-HF Equipment Controls

1-15C1 Which modes could be selected to receive vessel traffic lists from high seas shore stations?

A. AM and VHF-FM.
B. ARQ and FEC.
C. VHF-FM and SSB.
D. SSB and FEC.

High frequency high seas shore stations transmit voice and data traffic lists using SSB for voice, and forward error correction (FEC) for data. **ANSWER D**

1-15C2 Why must all MF-HF Distress, Urgency and Safety communications take place solely on the 6 assigned frequencies and in the simplex operating mode?

A. For non-GMDSS ships, to maximize the chances for other vessels to receive those communications.
B. Answers a) and c) are both correct.
C. For GMDSS or DSC-equipped ships, to maximize the chances for other vessels to receive those communications following the transmission of a DSC call of the correct priority.
D. To enable an RCC or Coast station to only hear communications from the vessel actually in distress.

Distress, Urgency, and Safety communications take place only on the 6 assigned frequencies, simplex, to insure that both non-GMDSS ships as well as commercial GMDSS ships may simultaneously receive these emergency DSC calls. All shore stations monitor all six frequencies **ANSWER B**

HF DIGITAL SELECTIVE CALLING
Portsmouth/NMN, Boston/NMF, Miami/NMA, New Orleans/NMG, Pt. Reyes/NMC, Honolulu HI/NMO, Kodiak AK/NOJ

Upper Sideband	
4207.5 kHz	Coast Guard will normally respond to DSC test calls if acknowledgment is requested. Reports of uncancelled or unacknowledged inadvertently transmitted distress calls will be forwarded to the FCC.
6312 kHz	
8414.5 kHz	
12577 kHz	
16804.5 kHz	

Note: For radiotelex and digital selective calling, frequencies listed are assigned. Carrier frequency is located 1700 Hz below the assigned frequency.

1-15C3 To set-up the MF/HF transceiver for a voice call to a coast station, the operator must:

A. Select J3E mode for proper SITOR operations.
B. Select F1B mode or J2B mode, depending on the equipment manufacturer.
C. Select J3E mode for proper voice operations.
D. Select F1B/J2B modes or J3E mode, depending on whether FEC or ARQ is preferred.

Mode J3E is single sideband, suppressed carrier, using upper sideband for voice operations. **ANSWER C**

1-15C4 MF/HF transceiver power levels should be set:

A. To the lowest level necessary for effective communications.
B. To the level necessary to maximize the propagation radius.
C. To the highest level possible so as to ensure other stations cannot "break-in" on the channel during use.
D. Both a) and c) are correct.

Medium frequency and high frequency SSB transceiver may all have power output selections. Use the lowest power possible to minimize interference to other stations on the same frequency.
ANSWER A

All SSB radios have a low/mid/high power output selection. Run low power, whenever possible.

1-15C5 To set-up the MF/HF transceiver for a TELEX call to a coast station, the operator must:

A. Select J3E mode for proper SITOR operations.
B. Select F1B mode or J2B mode, depending on the equipment manufacturer.
C. Select F1B/J2B modes or J3E mode, depending on whether ARQ or FEC is preferred.
D. None of the above.

On medium frequency/high frequency transceivers operating TELEX, the equipment is normally set to the F1B or J2B mode, depending on the type of equipment being operated. **ANSWER B**

1-15C6 What is the purpose of the Receiver Incremental Tuning (RIT) or "Clarifier" control?
A. It acts as a "fine-tune" control on the receive frequency.
B. It acts as a "fine-tune" control on the transmitted frequency.
C. It acts as a "fine-tune" control on both the receive and transmitted frequencies.
D. None of the above.
The RIT allows the SSB operator to fine tune the receive frequency without affecting the transmit frequency. **ANSWER A**

Key Topic 16: MF-HF Frequency & Emission Selection

1-16C1 On what frequency would a vessel normally call another ship station when using a radiotelephony emission?
A. Only on 2182 kHz in ITU Region 2.
B. On 2182 kHz or Ch-16, unless the station knows that the called vessel maintains a simultaneous watch on another intership working frequency.
C. On the appropriate calling channel of the ship station at 15 minutes past the hour.
D. On the vessel's unique working radio channel assigned by the Federal Communications Commission.
Not only are 2182 kHz and 156.800 MHz (Ch-16) distress frequencies, but they are also calling frequencies. If you do not know the intership working frequency in advance, make your call on the calling frequency. When you get a response, ask about the intership frequency and move to it. A radio "watch" is the act of listening on a designated frequency for any possible incoming messages. §80.5 and §80.116(b). **ANSWER B**

1-16C2 What is the MF radiotelephony calling and Distress frequency?
A. 2670 kHz.
B. Ch-06 VHF.
C. 2182 kHz.
D. Ch-22 VHF.
Since medium and high frequencies are usually given in kHz or MHz, answers B and D are obviously wrong. Since this is telephony, answer C is correct. Also note that kilohertz is abbreviated kHz, and megahertz is abbreviated MHz. See where each one is capitalized, as this will make a big difference in your log keeping. The U.S. Coast Guard no longer monitors 2182 kHz. §80.313. **ANSWER C**

1-16C3 For general communications purposes, paired frequencies are:
A. Normally used with private coast stations.
B. Normally used between ship stations.
C. Normally used between private coast and ship stations.
D. Normally used with public coast stations.
Paired frequencies are called duplex channels, and we only operate duplex with public coast stations. **ANSWER D**

1-16C4 What emission must be used when operating on the MF distress and calling voice frequency?
A. J3E – Single sideband telephony.
B. A1A – On-off keying without modulation by an audio frequency.
C. F3E – Frequency modulation telephony.
D. A3E – Amplitude modulation telephony, double sideband.
Only emission J3E (single sideband voice) may be transmitted on 2182 kHz. Emission J3E is technically defined as "Single sideband, amplitude modulated, suppressed carrier, single channel containing analog information." §80.313 **ANSWER A**

1-16C5 Which of the following defines high frequency "ITU Channel 1212"?
A. Ch-12 in the 16 MHz band.
B. Ch-1216 in the MF band.
C. The 12th channel in the 12 MHz band.
D. This would indicate the 1st channel in the 12 MHz band.
ITU (International Telecommunications Union) high frequency channels, usually duplex, are abbreviated with a 3 digit or 4 digit number. The first or the second digit(s) indicate the MHz band. ITU Channel 1212 would indicate the 12th channel in the 12 MHz band. Simplex high frequency channels are another story – the high frequency distress channels may illustrate the band MHz, simply followed by the letter "S" for simplex. Keep your ITU channel list handy, next to the high frequency radiotelephone operating point, and this will help you identify both channel number as well as the exact frequency. All marine MF and HF channels are operated UPPER sideband. **ANSWER C**

1-16C6 For general communications purposes, simplex frequencies are:
A. Normally used between ship stations and private coast stations.
B. Normally used with public coast stations.
C. Normally used between ship stations.
D. Both a) and c) are correct.
On both SSB as well as marine VHF, simplex frequencies are normally used between ship stations, as well as private coast stations. Most public coast stations operate duplex. **ANSWER D**

HF DISTRESS & SAFETY WATCHKEEPING SCHEDULE

HF Radiotelephone (Single Sideband) – Distress and Initial Contact

Authorized for the handling of Distress message traffic and initial contact with United States Coast Guard Long Range Communication facilities.

kHz SHIP STATION	kHz COAST STATION	Boston NMF	Chesapeake NMN	Miami NMA	New Orleans NMG
4125	4125	2300-1100Z	2300-1100Z	2300-1100Z	2300-1100Z
6215	6215	24 HRS	24 HRS	24 HRS	24 HRS
8291	8291	24 HRS	24 HRS	24 HRS	24 HRS
12290	12290	1100-2300Z	1100-2300Z	1100-2300Z	1100-2300Z

		Station and Schedule (UTC)		
kHz SHIP STATION	kHz COAST STATION	Pt. Reyes, CA NMC	Honolulu NMO	Kodiak, AK NOJ
4125	4125	24 HRS	0600-1800Z	24 HRS
6215	6215	24 HRS	24 HRS	24 HRS
8291	8291	24 HRS	24 HRS	
12290	12290	24 HRS	1800-0600Z	

kHz SHIP STATION	kHz COAST STATION	Station and Schedule(UTC) NRV Guam
6215	6215	0900-2100Z
12290	12290	2100-0900Z

HF Radiotelphone (single sideband) - Working Channels

These channels are available at all Coast Guard Long Range Communication Facilities for traffic handling purposes after initial contact is established on the HF Radiotelephone (Single Sideband) - Distress and Initial Contact frequencies.

ITU CHANNEL	kHz SHIP STATION	kHz COAST STATION
424	4134	4426
601	6200	6501
816	8240	8764
1205	12242	13089
1625	16432	17314

U.S. COAST GUARD COMMUNICATION STATIONS

The USCG has automated its HF shore stations so that all calls received are routed voice over internet to a centralized communications center. This automation improves response time to calls made to any High Seas Radio SSB Coast Guard station. The transmitter and receiver sites remain in their permanent locations, but the live radio operators are now housed in a central command center.

USCG Station	SELCAL	MARITIME MOBILE SERVICE IDENTITY
Communications Area Master Station Atlantic, Chesapeake VA/NMN	---	003669995
Communications Area Master Station Atlantic, remotely keying transmitters at Boston/NMF	---	003669991
Communications Area Master Station Atlantic, remotely keying transmitters at Miami/NMA	---	003669997
Communications Area Master Station Atlantic, remotely keying transmitters at New Orleans/NMG	---	003669998
Communications Area Master Station Pacific, Pt. Reyes CA/NMC	---	003669990
Communications Area Master Station Pacific, remotely keying transmitters at Guam/NRV	1096	
Communications Area Master Station Pacific, remotely keying transmitters at Honolulu HI/NMO	---	003669993
Communications Station Kodiak AK/NOJ	---	003669899
Marianas Section Guam	---	003669994

Note that except for the digital selective calling channels listed at the bottom of this page, the frequency channels described here are generally not Global Maritime Distress & Safety System (GMDSS) distress and safety channels. The Coast Guard does NOT monitor GMDSS radiotelephone or radiotelex channels.

Key Topic 17: Equipment Tests

1-17C1 What is the proper procedure for testing a radiotelephone installation?

A. A dummy antenna must be used to insure the test will not interfere with ongoing communications.
B. Transmit the station's call sign, followed by the word "test" on the frequency being used for the test.
C. Permission for the voice test must be requested and received from the nearest public coast station.
D. Short tests must be confined to a single frequency and must never be conducted in port.

While answers A and C sound plausible, it is perfectly legal to test a radiotelephone installation on the air, provided that it is done legally. Using the word "test" informs all stations listening that you are simply testing the radio equipment and do not require any emergency assistance. Using the ship's antenna during this test (instead of a dummy load) also verifies the antenna system is working properly. §80.101(a)(2). **ANSWER B**

1-17C2 When testing is conducted on 2182 kHz or Ch-16, testing should not continue for more than ______ in any 5-minute period.

A. 2 minutes.
B. 1 minute.
C. 30 seconds.
D. 10 seconds.

FCC Rule §80.101 states: "Test signals must not exceed ten seconds, and must not be repeated until at least one minute has elapsed. On these distress frequencies, the time between tests must be a minimum of five minutes." It is imperative to monitor the frequency before commencing any test. If the frequency or channel is in use, you must not conduct your test. Only test when the frequency is absolutely clear. **ANSWER D**

1-17C3 Under GMDSS, a compulsory VHF-DSC radiotelephone installation must be tested at what minimum intervals at sea?

A. Daily.
B. Annually, by a representative of the FCC.
C. Weekly.
D. Monthly.

Seagoing vessels required to carry VHF-DSC equipment must test their VHF radios daily. Even though that vessel may be halfway between New York City and London, far out at sea, there could be a mariner in distress in their vicinity hollering Mayday on Channel 16 or pressing the DSC distress button on VHF Channel 70. That radio must be tested daily and monitored constantly, even far out at sea. **ANSWER A**

1-17C4 The best way to test the MF-HF NBDP system is?

A. Make a radiotelephone call to a coast station.
B. Initiate an ARQ call to demonstrate that the transmitter and antenna are working.
C. Initiate an FEC call to demonstrate that the transmitter and antenna are working.
D. Initiate an ARQ call to a Coast Station and wait for the automatic exchange of answerbacks.

A good way to test your medium frequency or high frequency long range digital system would be to initiate an automatic repeat request (ARQ) call to your associated high frequency coast station, and wait for the automatic exchange of answerbacks. If you don't link up with your associated coast station, try another band. **ANSWER D**

1-17C5 The best way to test the INMARSAT-C terminal is?

A. Compose and send a brief message to your own INMARSAT-C terminal.
B. Send a message to a shore terminal and wait for confirmation.
C. Send a message to another ship terminal.
D. If the "Send" light flashes, proper operation has been confirmed.

Large ocean going vessels may carry INMARSAT-C non-voice computer communications on board, and a good way to test your system is to compose, and then send a brief message to your own INMARSAT-C terminal. Many recreational vessels also carry INMARSAT-C equipment, and this equipment is tested the same way. **ANSWER A**

1-17C6 When may you test a radiotelephone transmitter on the air?

A. Between midnight and 6:00 AM local time.
B. Only when authorized by the Commission.
C. At any time (except during silent periods) as necessary to assure proper operation.
D. After reducing transmitter power to 1 watt.

Rather than risk operating with broken equipment, it is better to test it on the air to find out whether or not it is working properly. Don't take any more time than is necessary. To effectively test your transmitting equipment, attempt to make contact with another vessel on a working frequency. This allows you to determine that your modulation is proper. On a marine VHF transceiver, the only way to test modulation would be to listen to your own transmit signal with a companion marine VHF handheld. Licensed technicians may use a deviation meter to check modulation. Just make sure not to conduct your checks on the air during silent periods on the distress channels. §80.101. **ANSWER C**

Key Topic 18: Equipment Faults

1-18C1 Under normal circumstances, what do you do if the transmitter aboard your ship is operating off-frequency, overmodulating or distorting?

A. Reduce to low power.
B. Reduce audio volume level.
C. Stop transmitting.
D. Make a notation in station operating log.

Operating a transmitter out of its specifications is not only illegal, it is also dangerous. It could interfere with other radio services. The operator must constantly be aware of the radio's behavior and repair any problems as soon as possible. A high frequency SSB station may sometimes go into distortion if the battery supply voltage to the equipment is slightly low. Stop transmitting until you solve the problem! If you notice the radio dial light dimming on modulation peaks, it is likely that the radio equipment is going into distortion. §80.90. **ANSWER C**

1-18C2 Which would be an indication of proper operation of a SSB transmitter rated at 60 watt PEP output?

A. In SSB (J3E) voice mode, with the transmitter keyed but without speaking into the microphone, power output is indicated.
B. In SITOR communications, the power meter can be seen fluctuating regularly from zero to the 60 watt relative output reading.
C. In SSB (J3E) mode, speaking into the microphone causes power meter to fluctuate slightly around the 60 watt reading.
D. A steady indication of transmitted energy on an RF Power meter with no fluctuations when speaking into the microphone.

The correct answer to this question requires eliminating the incorrect answers! For answer A, the properly operating single sideband transmitter would show no power output when there is no modulation into the microphone. In answer C, power output during modulation that does not return to 0 watts out during pauses in the modulation is likely RF feedback. Same thing with answer D, saying "hello" into the SSB microphone, without anything else going on, seeing the meter pegged at 60 watts, equals RF feedback. Answer B is correct because when transmitting simple TELEX over radio, power will fluctuate from 0 watts to 60 watts. **ANSWER B**

1-18C3 If a ship radio transmitter signal becomes distorted:

A. Reduce transmitter power.
B. Use minimum modulation.
C. Cease operations.
D. Reduce audio amplitude.

If your radio system is reported as distorted and "splattering," immediately stop transmitting, switch to a dummy load, and investigate the problem. Most important, cease operating on the air. §80.90. **ANSWER C**

1-18C4 What would be an indication of a malfunction on a GMDSS station with a 24 VDC battery system?

A. A constant 30 volt reading on the GMDSS console voltmeter.
B. After testing the station on battery power, the ammeter reading indicates a high rate of charge that then declines.
C. After testing the station on battery power, a voltmeter reading of 30 volts for brief period followed by a steady 26 volt reading.
D. None of the above.

A GMDSS station, operating from a 24 volt DC battery system, should never show a charging voltage much greater than 26 volts DC. If you show a constant 30 volt reading on the GMDSS console volt meter, shut off the equipment immediately. It is likely the charging system has lost regulation, and your batteries have gone into an overcharge situation. Shut down the generator or charging system to prevent a potential 24 volt DC battery system wipeout. **ANSWER A**

1-18C5 Your antenna tuner becomes totally inoperative. What would you do to obtain operation on both the 8 MHz and 22 MHz frequency bands?

A. Without an operating antenna tuner, transmission is impossible.
B. It is impossible to obtain operation on 2 different HF bands, without an operating antenna tuner.
C. Bypass the antenna tuner and shorten the whip to 15 ft.
D. Bypass the antenna tuner. Use a straight whip or wire antenna approximately 30 ft long.

The maritime 8 MHz and 22 MHz bands may be resonated simultaneously using a dipole with each leg approximately 30 feet long, or a 30 ft. wire counterpoised off the metal of your vessel. This antenna has double resonant points, and most single sideband marine transceivers will output a fair amount of power, even with an elevated SWR. Just because your automatic antenna tuner is not functioning, don't think that your radio station is off the air – develop your own resonant antenna system, and get back on the air! **ANSWER D**

1-18C6 Which of the following conditions would be a symptom of malfunction in a 2182 kHz radiotelephone system that must be reported to the Master, then logged appropriately.

A. Much higher noise level observed during daytime operation.
B. No indication of power output when speaking into the microphone.
C. When testing a radiotelephone alarm on 2182 kHz into an artificial antenna, the Distress frequency watch receiver becomes unmuted, an improper testing procedure.
D. Failure to contact a shore station 600 nautical miles distant during daytime operation.

On a marine single sideband transceiver, depressing the mic button and speaking into the microphone should show modulated power output. Even with a bad antenna system, your transmitter will continue to put out a few watts. However, when speaking in the microphone and detecting absolutely no power output, you must report to the master of the ship, make an appropriate log entry, and then begin to check whether or not your microphone needs replacement. **ANSWER B**

New VHF Digital Small Message Service

The FCC has approved the use of some non-safety and non-distress VHF channels for short data bursts containing small messages, complying with RTCM 12301. These SMS data bursts take place when the channel is idle with no voice traffic breaking squelch.

However, the following VHF channels may NOT be used for these SMS data bursts:
1001, 1063, 1005, 1065, 06, 1066, 67, 70 DSC, 12, 13, 73, 14, 74, 15, 75, 16 distress, 17, 77, 1020, 1022, nor AIS 1/2 channels.

Ship Station Licenses May Now Be Transferred

New FCC rules now permit the assignment and transfer of control of ship station licenses. Previously, when a boat was sold the station license including the MMSI number needed to be cancelled by the old owner. The new owner then was required to obtain a new ship station license and bring in a technician to reprogram MMSI numbers in the ship's radio equipment.

There is still considerable back-and-forth paperwork required to accomplish the transfer between the old and new boat owners. The FCC call center is most helpful in sorting out how to accomplish this. Call 888-CALL-FCC (888-225-5322).

Subelement D – Other Equipment

6 Key Topics, 6 Exam Questions

Key Topic 19: Antennas

1-19D1 What are the antenna requirements of a VHF telephony coast, maritime utility or ship station?

A. The shore or on-board antenna must be vertically polarized.
B. The antenna array must be type-accepted for 30-200 MHz operation by the FCC.
C. The horizontally-polarized antenna must be positioned so as not to cause excessive interference to other stations.
D. The antenna must be capable of being energized by an output in excess of 100 watts.

A vertically-polarized antenna takes up less space on board a ship and avoids the potential problem of cross-polarization. By requiring all antennas to be polarized the same way, the chance of receiving ground wave radio signals properly is greatly increased. Also, a non-directional antenna reduces the chance that a distress call could be missed simply because the ship's antenna is pointing the wrong way. §80.72 and §80.81. **ANSWER A**

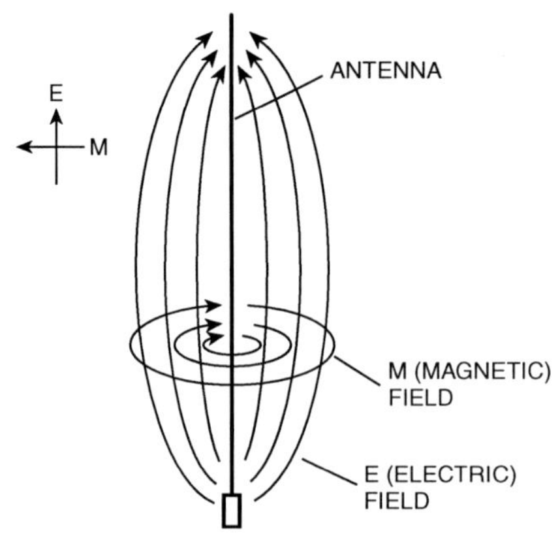

Vertically-Polarized Antenna

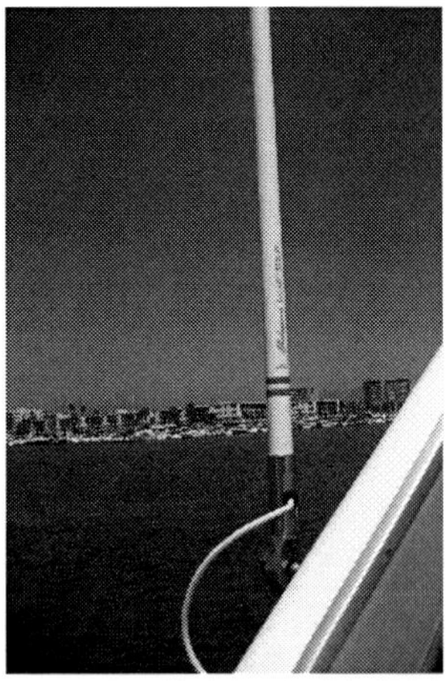

Vertically-polarized antennas are used on marine radio installations.

1-19D2 What is the antenna requirement of a radiotelephone installation aboard a passenger vessel?

A. The antenna must be located a minimum of 15 meters from the radiotelegraph antenna.
B. The antenna must be vertically polarized and as non-directional and efficient as is practicable for the transmission and reception of ground waves over seawater.
C. An emergency reserve antenna system must be provided for communications on 156.800 MHz.
D. All antennas must be tested and the operational results logged at least once during each voyage.

A marine VHF antenna must be vertically polarized for the proper transmission and reception of distant radio signals. If the antenna is operated in the horizontal plane, distant ground wave signal strength will decrease dramatically due to incorrect cross-polarization. **ANSWER B**

1-19D3 What is the most common type of antenna for GMDSS VHF?

A. Horizontally polarized circular antenna.
B. Long wire antenna.
C. Both A and B.
D. Neither A and B.

The most common GMDSS VHF antenna is a white vertical collinear array – a type not listed here. **ANSWER D**

1-19D4 What is the purpose of the antenna tuner?

A. It alters the electrical characteristics of the antenna to match the frequency in use.
B. It physically alters the length of the antenna to match the frequency in use.
C. It makes the antenna look like a half-wave antenna at the frequency in use.
D. None of the above.

A marine SSB automatic antenna tuner alters the electric characteristics of the antenna system to optimize power output to the frequency band in use. **ANSWER A**

1-19D5 What advantage does a vertical whip have over a long wire?

A. It radiates more signal fore and aft.
B. It radiates equally well in all directions.
C. It radiates a strong signal vertically
D. None of the above.

For long range SSB, the vertical whip with an antenna tuner offers the best omni-directional performance. Although standing insulated rigging and long wire antenna systems for marine SSB may indeed work well, they may exhibit lobes of peaks and nulls in the transmitted and received signals. **ANSWER B**

1-19D6 A vertical whip antenna has a radiation pattern best described by?

A. A figure eight.
B. A cardioid.
C. A circle.
D. An ellipse.

The vertical whip antenna transmits and receives omni-directional. This means the radiation pattern is much like a circle. Always inspect the "GTO-15" single conductor high voltage wire connection leading to the high frequency vertical whip. This is a common failure, due to salt water contamination. The lead-in wire is called GTO-15, and should be completely replaced if the whip connection is turning green with corrosion caused by salt water. **ANSWER C**

Key Topic 20: Power Sources

1-20D1 For a small passenger vessel inspection, reserve power batteries must be tested:

A. At intervals not exceeding every 3 months.
B. At intervals not exceeding every 6 months
C. Before any new voyage
D. At intervals not exceeding 12 months, or during the inspection.

The reserve power source batteries must be physically inspected every 12 months and the inspection noted in the radio log. This inspection should show specific gravity tests if flooded batteries are in use, or voltage and current tests for sealed batteries. And the batteries should be free of terminal deposits, too, as the GMDSS radio inspector will be looking over the battery compartment very carefully! §80.808(a)(9). **ANSWER D**

1-20D2 What are the characteristics of the Reserve Source of Energy under GMDSS?

A. Supplies independent HF and MF installations at the same time.
B. Cannot be independent of the propelling power of the ship.
C. Must be independent of the ship's electrical system when the RSE is needed to supply power to the GMDSS equipment.
D. Must be incorporated into the ship's electrical system.

It is of critical importance that a Reserve Source of Energy (RSE) is available to your emergency radio system. This energy source must be absolutely independent of other electrical systems, and will rely on batteries or an electrical generating system that is well maintained, and high enough out of the bilge to prevent damage in case of flooding. **ANSWER C**

1-20D3 Which of the following terms is defined as a back-up power source that provides power to radio installations for the purpose of conducting Distress and Safety communications when the vessel's main and emergency generators cannot?

A. Emergency Diesel Generator.
B. Reserve Source of Energy.
C. Reserve Source of Diesel Power.
D. Emergency Back-up Generator.

The reserve source of energy will be carefully inspected by anyone conducting the required inspection of your GMDSS station. **ANSWER B**

1-20D4 In the event of failure of the main and emergency sources of electrical power, what is the term for the source required to supply the GMDSS console with power for conducting distress and other radio communications?

A. Emergency power.
B. Ship's emergency diesel generator.
C. Reserve source of energy.
D. Ship's standby generator

The term describing an alternate power source for your GMDSS console is Reserve Source of Energy (RSE). **ANSWER C**

1-20D5 What is the requirement for emergency and reserve power in GMDSS radio installations?

A. An emergency power source for radio communications is not required if a vessel has proper reserve power (batteries).
B. A reserve power source is not required for radio communications.
C. Only one of the above is required if a vessel is equipped with a second 406 EPIRB as a backup means of sending a distress alert.
D. All newly constructed ships under GMDSS must have both emergency and reserve power sources for radio communications.

The reserve power supply must simultaneously energize the reserve transmitter at its required antenna power, and the reserve receiver for at least 1 hour continuously. **ANSWER D**

1-20D6 What is the meaning of "Reserve Source of Energy"?

A. The supply of electrical energy sufficient to operate the radio installations for the purpose of conducting Distress and Safety communications in the event of failure of the ship's main and emergency sources of electrical power.
B. High caloric value items for lifeboat, per SOLAS regulations.
C. Diesel fuel stored for the purpose of operating the powered survival craft for a period equal to or exceeding the U.S.C.G. and SOLAS requirements.
D. None of these.

The reserve source of energy must be regularly inspected and tested to insure its reliability in case a catastrophic emergency occurs on board. **ANSWER A**

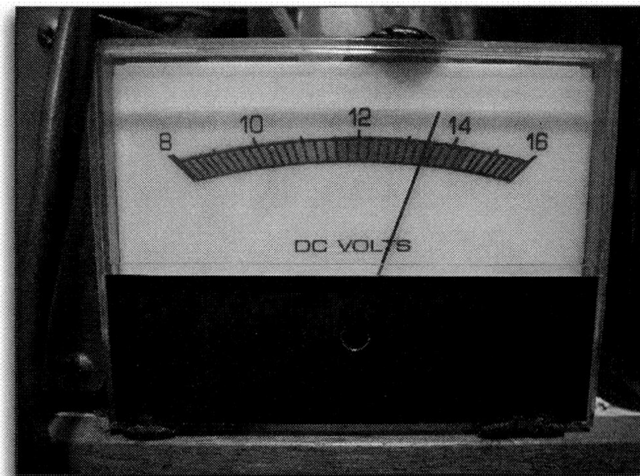

The RSE looks good on this volt meter reading.

Key Topic 21: EPIRBs

1-21D1 What is an EPIRB?

A. A battery-operated emergency position-indicating radio beacon that floats free of a sinking ship.
B. An alerting device notifying mariners of imminent danger.
C. A satellite-based maritime distress and safety alerting system.
D. A high-efficiency audio amplifier.

"EPIRB" is the acronym for Emergency Position Indicating Radio Beacon. It is designed to attract attention by broadcasting a distress signal. The Category 1 EPIRB must be able to float free and ballast itself into upright position; its antenna must deploy automatically; it must contain a visual indicator that shows when it is operating, and it must be waterproof. §80.1053 **ANSWER A**

FCC rules now require all new EPIRBs to have a built-in GPS transponder so the data-burst will include position information. New EPIRBs may also include VHFAIS position sending, in place of the 121.5 MHz homing signal.

SOURCE: ACR

Types of EPIRBs

EPIRBs cost from $200 to about $1500. These devices are designed to save your life if you get into trouble by alerting rescue authorities and indicating your location. There are six EPIRB types that are described below. Category I and II EPIRBS are the current models for use with the GMDSS. We list the old categories of obsolete EPIRBS in case you are doing a vessel inspection and should find any of them on board a vessel. Of the Category I and II EPIRBS, it is recommended that they be purchased with GPS capabilities built in.

Category I – 406/121.5 MHz. Float-free, automatically activated EPIRB. Detectable by satellite anywhere in the world. Recognized by GMDSS.

Category II – 406/121.5 MHz. Similar to Category I, except is manually activated. Some models are also water activated.

Obsolete EPIRBs – There are 4 classes of older EPIRBs that are now obsolete: Classes A, B, C, and S. They all operated on frequencies that are no longer monitored for emergency calls. If you come across one of these older EPIRBs, advise the vessel's owner that it is obsolete and should be replaced immediately with a Category I or II EPIRB, and recommend one that has GPS capability built-in.

1-21D2 When are EPIRB batteries changed?

A. After emergency use; after battery life expires.
B. After emergency use or within the month and year replacement date printed on the EPIRB.
C. After emergency use; every 12 months when not used.
D. Whenever voltage drops to less than 20% of full charge.

All companies that make EPIRB batteries must stamp on the outside of the battery case the month and year of the battery's manufacture and the month and year when 50% of its useful life is due to expire. When the expiration date is reached, the battery must be replaced; but if the EPIRB is actually used before that date, the battery must be replaced immediately afterward. §80.1053(e). **ANSWER B**

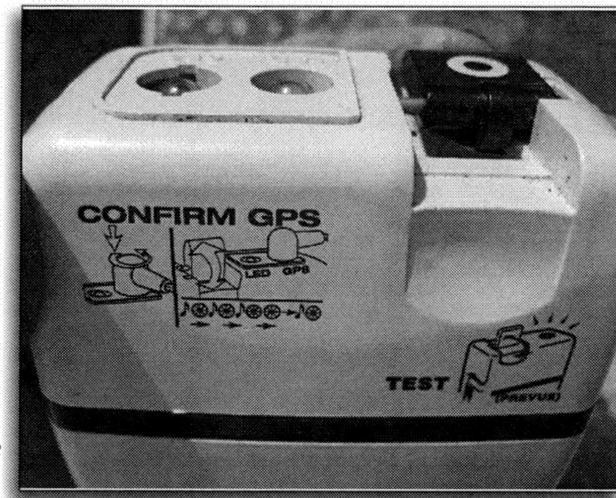

Today's EPIRBs have built-in self-test features, and the battery expiration date is printed on the battery compartment cover.

SOURCE: ACR

1-21D3 If a ship sinks, what device is designed to float free of the mother ship, is turned on automatically and transmits a distress signal?

A. An emergency position indicating radio beacon.
B. EPIRB on 2182 kHz and 405.025 kHz.
C. Bridge-to-bridge transmitter on 2182 kHz.
D. Auto alarm keyer on any frequency.

An EPIRB automatically transmits a distress call on a specially-assigned frequency when a vessel sinks. A water-activated battery is the central power source. The frequencies used by EPIRBs are in the VHF and UHF ranges (making answer B incorrect). An EPIRB is meant to be tracked by orbiting satellites, part of the COSPAS-SARSAT system. Direction-finding techniques help home in on the signal. **ANSWER A**

1-21D4 How do you cancel a false EPIRB distress alert?

A. Transmit a DSC distress alert cancellation.
B. Transmit a broadcast message to "all stations" canceling the distress message.
C. Notify the Coast Guard or rescue coordination center at once.
D. Make a radiotelephony "distress cancellation" transmission on 2182 kHz.

If for any reason an EPIRB is inadvertently activated, immediately contact the nearest U.S. Coast Guard unit or appropriate rescue coordination center by telephone, radio or ship earth station and cancel the distress alert. §80.335(e). **ANSWER C**

1-21D5 What is the COSPAS-SARSAT system?

A. A global satellite communications system for users in the maritime, land and aeronautical mobile services.
B. An international satellite-based search and rescue system.
C. A broadband military satellite communications network.
D. A Wide Area Geostationary Satellite program (WAGS).

The COSPAS-SARSAT system is a joint effort of the U.S., Russia, Canada, France, and other nations. COSPAS-SARSAT consists of two types of satellites. LEOSAR low-Earth-orbit satellites provide Doppler position measurements to identify the location of activated EPRIBs. GEOSAR satellites in geosynchronous orbits provide nearly instant alerting of an activated EPIRB. **ANSWER B**

1-21D6 What is an advantage of a 406 MHz satellite EPIRB?

A. It is compatible with the COSPAS-SARSAT system, and Global Maritime Distress and Safety System (GMDSS) regulations.
B. Provides the fastest and most accurate way the Coast Guard has of locating and rescuing persons in distress.
C. Includes a digitally encoded message containing the ship's identity and nationality.
D. All of the above.

A 406 MHz satellite EPIRB (Emergency Position Indicating Radio Beacon) is a small, battery-powered emergency radio transmitting device carried on vessels for use in cases of distress at sea. Its identification code must be registered with NOAA before use, and installed in a manner so that it will automatically float free and activate in the event of a sinking. They are detectable by satellite anywhere in the world by the COSPAS-SARSAT system. The EPIRB signal provides a unique I.D. number with the registered owner's identifying information, and newer "G-PRIBs" fitted with GPS units also send the location with the alerting signal. This GPS feature eliminates the need for many hours of Doppler position finding. **ANSWER D**

Key Topic 22: SARTs

1-22D1 In which frequency band does a search and rescue transponder operate?

A. 3 GHz.
B. S-band.
C. 406 MHz.
D. 9 GHz.

Search and rescue transponders sweep through the entire 9 GHz ship RADAR band. The 9 GHz signal CANNOT be detected with an S band radar. **ANSWER D**

1-22D2 How should the signal from a Search And Rescue Radar Transponder appear on a RADAR display?

A. A series of dashes.
B. A series of spirals all originating from the range and bearing of the SART.
C. A series of 12 equally spaced dots.
D. A series of twenty dashes.

If you see a series of 12 equally spaced dots on your RADAR display, they will align with the source location of the activated SART. Follow the dotted line to the SART transponder. **ANSWER C**

1-22D3 What is the purpose of the SART's audible tone alarm?

A. It informs survivors that assistance may be nearby.
B. It informs survivors when the battery's charge condition has weakened.
C. It informs survivors when the SART switches to the "standby" mode.
D. It informs survivors that a nearby vessel is signaling on DSC.

Some SART equipment also contains an audible alarm that sounds when assistance may be nearby, and also indicates that the equipment is turned on and transmitting. **ANSWER A**

1-22D4 Which statement is true regarding the SART?

A. This is a performance monitor attached to at least one S-band navigational RADAR system.
B. This is a 9 GHz transponder capable of being received by another vessel's S-band navigational RADAR system.
C. This is a performance monitor attached to at least one X-band navigational RADAR system.
D. This is a 9 GHz transponder capable of being received by vessel's X-band navigational RADAR system.

It is important to remember that an X-band SART system cannot be received by an S-band RADAR – only an X-band RADAR can receive the SART signal. **ANSWER D**

1-22D5 At what point does a SART begin transmitting?

A. It immediately begins radiating when placed in the "on" position.
B. It must be manually activated.
C. If it has been placed in the "on" position, it will respond when it has been interrogated by a 9-GHz RADAR signal.
D. If it has been placed in the "on" position, it will begin transmitting immediately upon detecting that it is in water.

When the SART equipment is switched on, it will not begin transmitting until it receives a signal from a nearby 9 GHz X-band RADAR. This prolongs battery life. **ANSWER C**

1-22D6 How can a SART's effective range be maximized?

A. The SART should be placed in water immediately upon activation.
B. The SART should be held as high as possible.
C. Switch the SART into the "high" power position.
D. If possible, the SART should be mounted horizontally so that its signal matches that of the searching RADAR signal.

If you are in a life raft with SART equipment, make sure it is held up in the vertical plane and as high as possible when activated. **ANSWER B**

AIS SART Alternative to RADAR SART

The International Maritime Organization (IMO) has amended the GMDSS regulations to permit the use of AIS Search and Rescue Locating Device Transponders as an alternative to 9 GHz SARTs.

Key Topic 23: Survival Craft VHF

1-23D1 Which statement is NOT true regarding the requirements of survival craft portable two-way VHF radiotelephone equipment?

A. Watertight to a depth of 1 meter for 5 minutes.
B. Effective radiated power should be a minimum of 0.25 watts.
C. Operates simplex on Ch-70 and at least one other channel.
D. The antenna is fixed and non-removable.

Portable, survival-craft VHF radio equipment must be submersible, operate with an ERP of at least 0.25 watts, and have a fixed, non-removable antenna. Make sure your batteries are fresh. Looking for the "not true" statement, Ch-70 is a DSC channel, not a voice emergency channel on a hand-held survival radio. **ANSWER C**

Regularly inspect your portable radio gear. If you spot moisture in the display window, replace the unit with a substitute and log your actions.

1-23D2 Which statement is NOT true regarding the requirements of survival craft portable two-way VHF radiotelephone equipment?

A. Operation on Ch-13.
B. Effective radiated power should be a minimum of 0.25 Watts.
C. Simplex voice communications only.
D. Operation on Ch-16.

Portable, survival-craft VHF equipment does not need to operate on VHF Channel 13, the bridge-to-bridge channel. **ANSWER A**

1-23D3 With what other stations may portable survival craft transceivers communicate?

A. Communication is permitted between survival craft.
B. Communication is permitted between survival craft and ship.
C. Communication is permitted between survival craft and rescue unit.
D. All of the above.

The survival craft transceiver may communicate with other survival craft radios, the mother ship, and of course, with the rescue unit. **ANSWER D**

1-23D4 Equipment for radiotelephony use in survival craft stations under GMDSS must have what capability?

A. Operation on Ch-16.
B. Operation on 457.525 MHz.
C. Operation on 121.5 MHz.
D. Any one of these.

Survival equipment MUST have the ability to operate on VHF Channel 16, the international distress and calling frequency, 156.800 MHz, FM. **ANSWER A**

1-23D5 Equipment for radiotelephony use in survival craft stations under GMDSS must have what characteristic(s)?

A. Operation on Ch-16.
B. Watertight.
C. Permanently-affixed antenna.
D. All of these.

Survival craft VHF equipment operates on VHF Channel 16, is submersible (watertight), and the antenna is permanently affixed. **ANSWER D**

1-23D6 What is the minimum power of the SCT

A. Five watts.
B. One watt.
C. ¼ watt.
D. None of the above.

The minimum power for a survival craft transceiver (SCT) is 1 watt. One watt output is reduced to approximately 0.25 watts effective radiated power because the antenna system comprises ERP to the negative because it is relatively shorter than a natural ¼ wavelength. **ANSWER B**

Key Topic 24: NAVTEX

1-24D1 NAVTEX broadcasts are sent:

A. Immediately following traffic lists.
B. In categories of messages indicated by a single letter or identifier.
C. On request of maritime mobile stations.
D. Regularly, after the radiotelephone silent periods.

NAVTEX is a digital Notice to Mariners transmitted on 518 kHz. It takes a dedicated NAVTEX receiver, or a 518 kHz receiver and computer software, to decode NAVTEX messages. The effective range of NAVTEX is approximately 400 nautical miles during the day, and twice that, from skywave propagation, at night. NAVTEX messages are sent by categories indicated by a single letter or identifier. **ANSWER B**

NAVTEX Stations in the United States

There are 12 NAVTEX stations in the United States. The transmitter identification character B(1) is a single unique letter which is allocated to each transmitter. It is used to identify the broadcasts which are to be accepted by the receiver and those which are to be rejected. In order to avoid erroneous reception of transmissions from two stations having the same B(1) character, it is necessary to ensure that such stations have a large geographical separation. NAVTEX transmissions have a designed range of about 400 nautical miles. NAVTEX stations in the U.S. use the following B(1) characters:

B(1) Character	Station	Starting Time	Call Sign	B(1) Character	Station	Starting Time	Call Sign
F	Cape Cod MA	0045Z	NMF	Q	Cambria CA	0045	NMQ
N	Chesapeake VA	0130	NMN	W	Astoria OR	0130	NMW
E	Savannah GA	0040	keyed by NMN	J	Kodiak AK**	0300	NOJ
A	Miami FL	0000*	NMA	X	Kodiak AK**	0340	NOJ
R	San Juan PR	0200	NMR	O	Honolulu HI	0040	NMO
G	New Orleans LA	0300	NMG	V	Guam	0100	NRV
C	Pt. Reyes CA	0000	NMC				

* - Until a planned new automatic broadcast scheduler is installed, Miami's starting time of 0000 will be delayed approximately 5 minutes.

** - Kodiak also broadcasts safety information during time slots previously allocated to Adak.

1-24D2 MSI can be obtained by one (or more) of the following:

A. NAVTEX.
B. SafetyNET.
C. HF NBDP.
D. All of the above.

Maritime safety information (MSI) may be received by any of the following: NAVTEX; the Inmarsat-C safety net, and high frequency digital mode. All of these can receive meteorological as well as emergency messages in a digital format – that is, non-voice. To learn more about MSI, visit the Coast Guard website www.navcen.uscg.gov/marcomms/gmdss/msi.htm. **ANSWER D**

1-24D3 Which of the following is the primary frequency that is used exclusively for NAVTEX broadcasts internationally?

A. 518 kHz.
B. 2187.5 kHz.
C. 4209.5 kHz.
D. VHF channel 16 when the vessel is sailing in Sea Area A1, and 2187.5 kHz when in Sea Area A2.

Back to NAVTEX – messages are broadcast only on 518 kHz. **ANSWER A**

1-24D4 What means are used to prevent the reception of unwanted broadcasts by vessels utilizing the NAVTEX system?

A. Operating the receiver only during daytime hours.
B. Coordinating reception with published broadcast schedules.
C. Programming the receiver to reject unwanted broadcasts.
D. Automatic receiver de-sensitization during night hours.

The modern NAVTEX receiver may be pre-programmed to accept only certain routine NAVTEX broadcasts, while rejecting unwanted routine NAVTEX broadcasts. **ANSWER C**

1-24D5 When do NAVTEX broadcasts typically achieve maximum transmitting range?

A. Local noontime.
B. Middle of the night.
C. Sunset.
D. Post sunrise.

The ultimate range in receiving a local harbor's NAVTEX broadcast would be in the middle of the night via skywave propagation. Of course, when closer than 100 miles to a major seaport, the NAVTEX broadcast should come in loud and clear any time of day or night. **ANSWER B**

1-24D6 What is the transmitting range of most NAVTEX stations?

A. Typically 50-100 nautical miles (90-180 km) from shore.
B. Typically upwards of 1000 nautical miles (1800 km) during the daytime.
C. Typically 200-400 nautical miles (360-720 km).
D. It is limited to line-of-sight or about 30 nautical miles (54 km).

Two hundred to four hundred nautical miles is the approximate range of a NAVTEX broadcast. However, the greatest range reducer is onboard noise. If your NAVTEX receiver does not offer good range, listen with a good communications receiver and try to track down the onboard noise source. **ANSWER C**

NAVTEX MARITIME SAFETY BROADCASTS

NAVTEX in the United States

The International Maritime Organization has designated NAVTEX as the primary means for transmitting coastal urgent marine safety information to ships worldwide. The Coast Guard began operating NAVTEX from Boston in 1983. Today, in the United States, NAVTEX is broadcast from Coast Guard facilities in Cape Cod, Chesapeake VA, Savannah GA, Miami FL, New Orleans LA, San Juan PR, Cambria CA, Pt. Reyes CA, Astoria OR, Kodiak AK, Honolulu HI, and Guam.

NAVTEX coverage is reasonably continuous in the east, west and Gulf coasts of the United States, as well the area around Kodiak Alaska, Guam and Puerto Rico. The U.S. has no coverage in the Great Lakes, though coverage of much of the Lakes is provided by the Canadian Coast Guard. Since the U.S. Coast Guard originally only installed NAVTEX at sites where Morse code telegraphy transmissions were made previously, propagation analyses show some coverage gaps, particularly in the southeast United States, Alaska, and Guam. NAVTEX broadcasts from Adak were permanently terminated in December 1996 due to closure of the Naval facility there.

NAVTEX Message Selection

Every NAVTEX message is preceded by a four character header B(1)B(2)B(3)B(4). B(1) is an alpha character identifying the station, and B(2) is an alpha character used to identify the subject of the message. Receivers use these characters to reject messages from stations or concerning subjects of no interest to the user. B(3)B(4) are two-digit numerics identifying individual messages, used by receivers to keep already received messages from being repeated. For example, a message preceded by the characters FE01 from a U.S. NAVTEX Station indicate that this is a weather forecast message from Boston MA.

Subject indicator character B(2)

The subject indicator character is used by the receiver to identify different classes of messages below. The indicator is also used to reject messages concerning certain optional subjects which are not required by the ship (e.g. LORAN C messages might be rejected in a ship which is not fitted with a LORAN C receiver). Receivers also use the B(2) character to identify messages which, because of their importance, may not be rejected (designated by an asterisk).

NAVTEX broadcasts use following subject indicator characters:

A = Navigational warnings *
B = Meteorological warnings *
C = Ice reports
D = Search & rescue information, and pirate warnings *
E = Meteorological forecasts
F = Pilot service messages
G = DECCA messages
H = LORAN messages
I = OMEGA messages (note OMEGA has been discontinued)
J = SATNAV messages (i.e. GPS or GLONASS)
L = Navigational warnings - additional to letter A (Should not be rejected by the receiver)
V = Notice to Fishermen (U.S. only - currently not used)
W = Environmental (U.S. only - currently not used)
X } Special services - allocation by IMO NAVTEX Panel
Y }
Z = No message on hand

Note: The subject indicator characters B, F and G are normally not used in the U.S. Since the National Weather Service normally includes meteorological warnings in forecast messages, meteorological warnings are broadcast using the subject indicator character E. U.S. Coast Guard District Broadcast Notices to Mariners affecting ships outside the line of demarcation, and inside the line of demarcation in areas where deep draft vessels operate, use the subject indicator character A.

Technical Information

All NAVTEX broadcasts are made on 518 kHz, using narrow-band direct printing 7-unit forward error correcting (FEC or Mode B) transmission. This type of transmission is also used by Amateur Radio service (AMTOR). Broadcasts use 100 baud FSK modulation, with a frequency shift of 170 Hz. The center frequency of the audio spectrum applied to a single sideband transmitter is 1700 Hz. The receiver 6 dB bandwidth should be between 270-340 Hz.

Each character is transmitted twice. The first transmission (DX) of a specific character is followed by the transmission of four other characters, after which the retransmission (RX) of the first character takes place, allowing for time-diversity reception of 280 ms.

For more information, see ITU Recommendations M.540-2 and M.476-5, available from the ITU Radiocommunications Sector at www. itu.int/ITU-R/index.
Practical Instruction for the Use of a NAVTEX Receiver
Chart of NAVTEX Stations Worldwide
Worldwide NAVTEX Broadcast Schedule
NOAA/National Weather Service NAVTEX Page
Other Maritime Safety Broadcast Information

Coast Guard 406 MHz Direction Finding Program

The U.S. Coast Guard is equipping all of its search and rescue aircraft with new direction finding (DF) equipment designed to locate 406 MHz EPIRBs that are sending emergency messages. Depending on the aircraft's altitude, this new equipment is capable of locating DF beacons at distances of more than 100 miles. In addition, Rescue 21 coastal VHF towers are being evaluate for their 406 MHz DF capability, as a cruising cutters. Portable DFs with this capability are also being evaluated, due in part to the rapidly-expanding deployment of Personal Locator Beacons (PLBs) for both maritime and land use.

Source: ACR

This Personal Locator Beacon (PLB) is approved for emergency use on both land and sea.

406 MHz EPIRBs

The 406 MHz EPIRB was designed to operate with satellites. The signal frequency (406 MHz) has been designated internationally for use only for distress. Other communications and interference, such as on 121.500 MHz, is not allowed on this frequency. Its signal allows a satellite local user terminal to accurately locate the EPIRB (much more accurately -- 2 to 5 km vice 25 km -- than 121.5/243 MHz devices), and identify the vessel (the signal is encoded with the vessel's identity) anywhere in the world (there is no range limitation). These devices are detectable not only by COSPAS-SARSAT satellites which are polar orbiting, but also by geostationary GOES weather satellites. EPIRBs detected by the GEOSTAR system, consisting of GOES, MEOSARs, and other geostationary satellites, send rescue authorities an instant alert, but without location information unless the EPIRB is equipped with an integral GPS receiver. EPIRBs detected by COSPAS-SARSAT (e.g. TIROS N) satellites provide rescue authorities location of distress, but location and sometimes alerting may be delayed as much as an hour or two. Although these EPIRBs also include a low power 121.5 MHz homing signal, homing on the more powerful 406 MHz frequency has proven to be a significant aid to search and rescue aircraft. These are the only EPIRB types which can be sold in the United States.

The new FCC rules will phase out the sale and distribution of EPRIBs that do not contain a GPS transponder. While existing non-GPS EPIRBs remain legal to carry on board, all new units sold should now have built-in GPS.

406 MHz emergency locating transmitters (ELTs) for aircraft are also available. 406 MHz personnel locating beacons (PLBs) are available.

The Coast Guard recommends mariners purchase a 406 MHz EPIRB making sure it has an integral GPS transponder. Mariners should phase out any on-board EPRIBs that do not have GPS capabilities.

406 MHz GEOSAR System

The major advantage of the 406 MHz low earth orbit system is the provision of global Earth coverage using a limited number of polar-orbiting satellite. Coverage is not continuous, however, and it may take up to a couple of hours for an EPIRB alert to be received. To overcome this limitation, COSPAS-SARSAT has 406 MHz EPIRB repeaters aboard several geostationary satellites.

Note that GEOSAR cannot detect 121.5 MHz alerts, nor can it route unregistered 406 MHz alerts to a rescue authority. GEOSAR cannot calculate the location of any alert it receives, unless the beacon has an integral GPS receiver.

The COSPAS-SARSAT System

COSPAS-SARSAT is an international satellite-based search and rescue system established by the U.S., Russia, Canada and France to locate emergency radio beacons transmitting on the frequencies 121.5, 243 and 406 MHZ.
COSPAS stands for Space System for Search of Distress Vessels (a Russian acronym). SARSAT stands for Search and Rescue Satellite-Aided Tracking.

Testing EPIRBs

406 MHz EPIRBs can be tested through its self-test function, which is an integral part of the device. 406 MHz EPIRBs can also be tested inside a container designed to prevent its reception by the satellite. Testing a 406 MHz EPIRB by allowing it to radiate outside such a container is illegal.

Battery Replacement
406 MHz EPIRBs use a special type of lithium battery designed for long-term low-power consumption operation. Batteries must be replaced by the date indicated on the EPIRB label using the model specified by the manufacturer. It should be replaced by a dealer approved by the manufacturer. If the replacement battery is not the proper type, the EPIRB will not operate for the duration specified in a distress.

Registration of 406 MHz EPIRBs
Proper registration of your 406 MHz satellite emergency position-indicating radio beacon (EPIRB) is intended to save your life, and is mandated by Federal Communications Commission regulations. The Coast Guard is enforcing this FCC registration rule.

Your life may be saved as a result of registered emergency information. This information can be very helpful in confirming that a distress situation exists, and in arranging appropriate rescue efforts. Also, GOES, a geostationary National Oceanic & Atmospheric Administration weather satellite system can pick up and relay an EPIRB distress alert to the Coast Guard well before the international COSPAS-SARSAT satellite can provide location information. If the EPIRB is properly registered, the Coast Guard will be able to use the registration information to immediately begin action on the case. If the EPIRB is unregistered, a distress alert may take as much as two hours longer to reach the Coast Guard over the international satellite system. If an unregistered EPIRB transmission is abbreviated for any reason, the satellite will be unable to determine the EPIRB's location, and the Coast Guard will be unable to respond to the distress alert. Unregistered EPIRBs have needlessly cost the lives of several mariners since the satellite system became operational.

What happens to your registration form?
The registration sheet you fill out and send in is entered into the U.S. 406 Beacon Registration Database maintained by NOAA/NESDIS. If your EPIRB is activated, your registration information will be sent automatically to the appropriate USCG SAR Rescue Coordination Center (RCC) for response. One of the first things the RCC watchstanders do is attempt to contact the owner/operator at the phone number listed in the database to determine if the vessel is underway (thus ruling out the possibility of a false alarm due to accidental activation or EPIRB malfunction), the intended route of the vessel if underway, the number of people on board, etc., from a family member. If there is no answer at this number, or no information, the other numbers listed in the database will be called to attempt to get the information described above needed to assist the RCC in responding appropriately to the EPIRB alert.

When RCC personnel contact the emergency phone numbers you provide, they will have all the information you have provided on the registration form. You should let these contacts know as much about your intended voyage as possible (i.e., intended route, stops, area you normally sail/fish/recreate, duration of trip, number of people going, etc.). The more information these contacts have, the better prepared our SAR personnel will be to react. The contacts can ask the RCC personnel contacting them to be kept informed of any developments, if they so desire.

Registration regulations
You may be fined for false activation of an unregistered EPIRB. The U.S. Coast Guard routinely refers cases involving the non-distress activation of an EPIRB (e.g., as a hoax, through gross negligence, carelessness or improper storage and handling) to the Federal Communications Commission. The FCC will prosecute cases based upon evidence provided by the Coast Guard, and will issue warning letters or notices of apparent liability for fines up to $10,000.

However, the Coast Guard has suspended forwarding non-distress activations of properly registered 406 MHz EPIRBs to the FCC, unless activation was due to hoax or gross negligence, since these search and rescue cases are less costly to prosecute.

If you purchase a new or a used 406 MHz EPIRB, you MUST register it with NOAA. If you change your boat, your address, or your primary phone number, you MUST re-register your EPIRB with NOAA. If you sell your EPIRB, make sure the purchaser re-registers the EPIRB, or you may be called by the Coast Guard if it later becomes activated.
An FCC ship station license is no longer required to purchase or carry an EPIRB.

How to register
You may register by visiting the SARSAT Beacon Registration page.
There is no charge for this service. IT MAY SAVE YOUR LIFE.
For more information see the NOAA SARSAT Homepage.

To register an EPIRB call:
1-888-212-SAVE
or visit
www.beaconregistration.noaa.gov

MAKING THE CALL (General Operating Procedures)

Calling Intership

Turn your radiotelephone on and listen on the appropriate distress and calling frequency, Channel 16 or 2182 kHz, to make sure it is not being used. If it is clear, put your transmitter on the air. This is usually done by depressing the "push to talk" button on the microphone. (To hear a reply, you must release this button.)

Speak directly into the microphone in a normal tone of voice. Speak clearly and distinctly. Call the vessel with which you wish to communicate by using its name; then identify your vessel with its name and FCC-assigned call sign. Do not add unnecessary words and phrases such as "come in Bob" or "do you read me." Limit the use of phonetics to poor transmission conditions.

This preliminary call must not exceed 30 seconds. If contact is not made, wait at least 2 minutes before repeating the call. After this time interval, make the call in the same manner. This procedure may be repeated no more than three times. If contact is not made during this period, you must wait at least 15 minutes before making your next attempt.

Once contact is established on Channel 16 or 2182 kHz, you must switch to an appropriate working frequency for further communication. You may only use Channel 16 and 2182 kHz for calling and in emergency situations.

Since switching to a working frequency is required to carry out the actual communications, it is often helpful to briefly monitor the working frequency you wish to use before initiating the call on Channel 16 or 2182 kHz. This will help prevent you from interrupting other users of the working frequency channel.

All communications should be kept as brief as possible. At the end of the communications, each vessel is required to give its call sign, after which both vessels switch back to the distress and calling channel in order to reestablish the watch.

Two examples of acceptable forms for establishing communication with another vessel follow:

EXAMPLE 1

Vessel	**Voice Transmission**
BLUE DUCK (on Channel 16)	"MARY JANE - THIS IS - BLUE DUCK - WA 1234" (The name of the vessel being called may be said two or three times if conditions demand.)
MARY JANE (on Channel 16)	"BLUE DUCK - THIS IS - MARY JANE - WA 5678 - REPLY 68" (or some other proper working channel)
BLUE DUCK (on Channel 16)	"68" or "ROGER" (If unable to reply on the channel selected, an appropriate alternate should be selected.)
BLUE DUCK (on working channel)	"BLUE DUCK"
MARY JANE (on working channel)	"MARY JANE"
BLUE DUCK (on working channel)	(Continue with message and terminate communication within 3 minutes. At the end of the communication, each vessel gives its call sign.)

EXAMPLE 2 (A short form most useful when both parties are familiar with it.)

Vessel	**Voice Transmission**
BLUE DUCK (on Channel 16)	"MARY JANE - BLUE DUCK - WA 1234 - REPLY 68"
MARY JANE (on Channel 68)	"MARY JANE - WA 5678"
BLUE DUCK (on Channel 68)	"BLUE DUCK" (Continues message and terminates communications as indicated in Example 1.)
If unable to reply on the channel	
BLUE DUCK (on working channel)	"MARY JANE"
MARY JANE (on working channel)	(Continue with message and terminate communication
BLUE DUCK (on working channel)	within 3 minutes. At the end of the communication, each vessel gives its call sign.)

Calling Ship to Coast (Other than U.S. Coast Guard)

The procedures for calling coast stations are similar to those used in making intership calls with the exception that you normally initiate the call on the assigned working frequency of the coast station.

Calling From Onboard Stations and Associated Ship Units

If there are onboard stations in addition to the main radio installation, the onboard stations will use the name of the vessel followed by "Mobile 1" or "Mobile 2," etc. For example: "BLUE DUCK - THIS IS BLUE DUCK MOBILE 1 - OVER."

If radios are being used from associated ship units (e.g. Tender or Dinghy), the call sign will be composed of the call sign of the ship station followed by "Unit 1" or "Unit 2," etc. For example: "WA 1234 - THIS IS WA 1234 UNIT 1 - OVER."

ROUTINE RADIO CHECK

Radio checks may be initiated on Channel 16 (156.8 MHz) but should be completed by immediately shifting to a working channel.

Listen to make sure that the Distress and Calling frequency is not busy. If it is free, put your transmitter on the air and call a specific station or vessel. For example, "MARY JANE - THIS IS BLUE DUCK - WA 1234 - SHIFT TO CHANNEL (names working channel) OVER." After the reply by Mary Jane, Blue Duck would then say "HOW DO YOU HEAR ME? - OVER." The proper response by Mary Jane, depending on the respective conditions, would be:

"I HEAR YOU LOUD AND CLEAR," or "I HEAR YOU WEAK BUT CLEAR," or "YOU ARE LOUD BUT DISTORTED," etc.

It is not permitted to call a Coast Guard Station on Channel 16 (156.8 MHz or on 2182 kHz) for a radio check. This prohibition does not apply to tests conducted during investigations by FCC representatives or when qualified radio technicians are installing equipment or correcting deficiencies in the station radiotelephone equipment.

How to Conduct a Radio Equipment / GMDSS Inspection of a Small Passenger Vessel

By definition, small passenger vessels are vessels that are less than 100 gross tons and carry more than six passengers for hire. A passenger for hire is defined as a person who pays money or any other kind of material goods or services as compensation for being carried on a vessel. Small passenger vessels are required to carry radio equipment to comply with the requirements of the Communications Act, sections 381-386 also known as Part III of Title III and the requirements are specified in 47 CFR Part 80 Subpart S. Radio carriage requirements for small passenger vessels depend on the area of operation and the distance from the nearest land. A small passenger vessel's area of operation is specified on the Coast Guard's Certificate of Inspection (COI). Generally, a small passenger vessel must carry radio equipment to meet the communication requirements in the area of operation specified by the Coast Guard.

1. Small passenger vessels that sail only on inland lakes and waterways (other than the Great Lakes) are exempt from radio carriage regulations. Likewise, small passenger vessels of less than 50 gross tons that sail in the open ocean or in bays, sounds, and other tidewater areas bordering on the open sea but never more than 300 meters (1000 feet) from shore are also exempt from radio carriage regulations. If vessels of this class carry a radio, no inspection of the radio is required and, if the radio operates only on VHF frequencies and if the vessel does not sail to a foreign port, the radio is exempt from the licensing requirement.
2. Small passenger vessels that sail on the Great Lakes must meet the radio carriage requirements of the Great Lakes Agreement. This is a treaty between the United States and Canada governing radio carriage requirements for ships navigating on the Great Lakes. Those rules are contained in Subpart T of Part 80 of FCC Rules, Sections 80.951 through 80.971. The Coast Guard also requires carriage of an EPIRB if the vessel sails more than 3 miles from shore on the Great Lakes.
3. Small passenger vessels that sail in bays, harbors, rivers and sounds adjacent to the open ocean or in the open ocean not more than 20 nautical miles from the nearest land but always within communication of a VHF coast station that maintains a continuous watch on VHF Channel 16 (156.8 MHz) must carry a VHF radio installation and a Navigation receiver as specified in 80.1085(c). The Coast Guard also requires carriage of an EPIRB if the vessel sails more than 3 nautical miles from shore in the open sea.
4. Small passenger vessels that sail in the open sea more than 20 nautical miles but not more than 100 nautical miles from the nearest land must also carry a medium frequency (MF) radio installation providing communication capability on 2182 kHz, 2638 kHz, 2670 kHz and a public coast station frequency in the 1710-2850 kHz band.
5. Small passenger vessels sailing more than 100 nautical miles but not more than 200 nautical miles from shore must, in addition to the EPIRB, VHF, Navigational Receiver and MF installations mentioned above, carry either: a single sideband radiotelephone installation capable of operating on all of the medium frequency (MF) and high frequency (HF) channels used for distress and safety communications listed in Section 80.905(a)(3)(iii)(A) and capable of DSC operation or an INMARSAT ship earth station through which continuous distress alerting by satellite is available. The vessel must also carry: A NAVTEX receiver for receipt of maritime safety information. A reserve source of power capable of powering all fitted equipment including the navigation receiver. If a ship earth station is elected in lieu of the single sideband combined MF/HF installation described above, the reserve source of power must be capable of powering the associated peripheral equipment necessary for the full functioning of the ship earth station. The vessel must participate in the AMVER System.
6. Small passenger vessels operating more than 200 nautical miles from shore must carry, in addition to all of the equipment specified above: A second VHF.
7. The Small Passenger Vessel inspection will also include a check for the MMSI number programmed in the VHF radio, as well as a test to make sure the MMSI number is programmed to a companion VHF radio.

As per 47 CFR Part 80.59 (a) (1), the following table illustrates the minimum licensing requirements for Inspectors (only one license required in case of multiples):

	General Radiotelephone Operator License (PG)	GMDSS Radio Maintainer License (DM)	Radiotelegraph Operator's Certificate, or Radiotelegraph Operator License (T)
Radiotelephone equipped vessels subject to 47 CFR part 80, subpart R or S	x	x	x
GMDSS equipped vessels subject to 47 CFR part 80, subpart W		x	

A complete checklist for these inspections can be found on the FCC's Enforcement Bureau website at: www.fcc.gov/eb/ShipInsp/spv_checklist.pdf

U.S.C.G. discontinues 2182 kHz and 2187.5 kHz distress and safety watch

Effective August 1, 2013, the U.S. Coast Guard has dropped their radio watch on the 2182 kHz voice and 2187.5 kHz digital selective calling (DSC) International Distress and Safety frequencies. The Coast Guard also discontinued its 2670 kHz marine information and weather broadcasts.

While the USCG has discontinued radio watches on these frequencies, any Element 1 exam question and answer regarding 2182 kHz and/or 2187.5 kHz remains the same indicated answer because FCC rules have not changed.

The USCG bulletin announcing this change explained: "This termination decision was made after a review of Coast Guard medium frequency (MF) communications sites revealed significant antenna and infrastructure support degradation that put the Coast Guard at risk of not being able to receive and respond to calls for assistance on the 2 MHz distress frequencies."

This decision by the Coast Guard is a good call! Atmospheric noise on 2 MHz causes even the best of radio systems to not hear much beyond 30 miles ground wave. And 30 miles to shore is the typical maximum range of the Coast Guard's excellent Rescue 21 VHF channel 16 coverage.

U.S. Coast Guard Communication Stations and Communications Area Master Stations will continue their watch on the following High Frequency safety frequencies:

VOICE	DSC
4125.0 kHz	4207.5 kHz
6215.0 kHz	6312.0 kHz
8291.0 kHz	8414.5 kHz
12290.0 kHz	12577.0 kHz
	16804.5 kHz

VHF Channel 16, 156.800 MHz, will continue to offer the Coast Guard's Rescue 21 coverage throughout boating areas of the United States from Coast Guard units on land, at sea and in the air. VHF Channel 70 will continue to be the DSC call up channel to the Coast Guard.

Internationally, 2182 kHz remains the distress and calling channel. It will still be an on scene distress working channel, and continues to be an authorized calling channel. "2182 kHz remains a frequency designated by the ITU Radio Regulations for distress and safety communications under the GMDSS system. This U.S. Coast Guard announcement means that if a mariner transmits on 2182 kHz, a Coast Guard shore station will not answer," explains the Radio Technical Commission for Maritime Services (www.rtcm.org).

Just know that a distress call to the U.S. Coast Guard is best placed on VHF channel 16, or on any of the High Frequency channels that they continue to monitor!

Only 2 MHz US Coast Guard distress watches are being eliminated.

CHAPTER E

Electronic Fundamentals & Techniques

Element 3 Question Pool

The FCC Commercial Element 3 written examination covers electronic fundamentals and techniques. The Element 3 examination is used to prove that the examinee possesses the qualifications required to adjust, repair and maintain marine and aircraft radio transmitters and receivers that can only be serviced by a person holding a General Radiotelephone Operator's License (PG).

The Element 3 question pool contains a total of 600 questions covering 17 Subelement topic areas. Each Key Topic contains 6 questions. One question is taken at random from each Key Topic to create an examination with 100 questions. To pass, you must correctly answer at least 75 of the 100 questions on the written exam. *Table 5-1* summarizes the Subelement topics and number of questions in each topic area.

Table 5-1. Summary of Element 3 Question Pool

Subelement Topic	Key Topics	No. of Questions	Examination Questions
Principles	8	48	8
Electrical Math	10	60	10
Components	10	60	10
Circuits	4	24	4
Digital Logic	8	48	8
Receivers	10	60	10
Transmitters	6	36	6
Modulation	3	18	3
Power Sources	3	18	3
Antennas	5	30	5
Aircraft	6	36	6
Installation, Maintenance & Repair	8	48	8
Communications Technology	3	18	3
Marine	5	30	5
RADAR	5	30	5
Satellite	4	24	4
Safety	2	12	2
Total	**100**	**600**	**100**

Each Element 3 examination is administered by a Commercial Operator License Examination Manger (COLEM). The COLEM must construct the examination by selecting 100 questions from the Element 3 question pool contained in this book, as explained above. Each question in the pool contains the question, and 4 multiple-choice answers – one correct answer and three wrong answers (distracters). The COLEM may change the order of the four answer choices, but cannot change any of the question or answer content.

Subelement A – Principles

8 Key Topics, 8 Exam Questions

Key Topic 1: Electrical Elements

3-1A1 The product of the readings of an AC voltmeter and AC ammeter is called:

A. Apparent power.
B. True power.
C. Power factor.
D. Current power.

If you multiply ac volts by ac amps, the result is called apparent power. In this question, you must assume there is a phase angle of zero. **ANSWER A**

3-1A2 What is the basic unit of electrical power?

A. Ohm.
B. Watt.
C. Volt.
D. Ampere.

The unit of power is the WATT, with a capitol letter W as its symbol. When we reference the WATT in an equation, we use the capitol letter P, for power. **ANSWER B**

3-1A3 What is the term used to express the amount of electrical energy stored in an electrostatic field?

A. Joules.
B. Coulombs.
C. Watts.
D. Volts.

The joule is a measure of energy, or the capacity to do work. The amount of electrical energy stored in an electrostatic field associated with capacitors is expressed in joules. If a capacitor is charged with one coulomb of electrons at a voltage of one volt, one joule of energy is stored in the capacitor's electrostatic field. **ANSWER A**

3-1A4 What device is used to store electrical energy in an electrostatic field?

A. Battery.
B. Transformer.
C. Capacitor.
D. Inductor.

A capacitor is made up of parallel plates separated by a dielectric (non-conductor). When a voltage is placed across a capacitor, energy is stored in the electrostatic field developed between the capacitor plates. As a reminder, notice the letters "AC" in the word capacitor, and the letters "ATIC" in the word electrostatic. It should lead you to the correct answer. **ANSWER C**

3-1A5 What formula would determine the inductive reactance of a coil if frequency and coil inductance are known?

A. $X_L = \pi f L$
B. $X_L = 2\pi f L$
C. $X_L = 1 / 2\pi f C$
D. $X_L = 1 / R2+X2$

To calculate the inductive reactance of a coil, where the frequency in hertz and coil inductance in henrys are already known, multiply 6.28(2 π) times the frequency in hertz and inductance in henrys. You can save yourself some extra math by remembering "megs" (MHz) and "micros" (μH) will cancel. Same thing with kilohertz and millihenrys! This way you won't need to calculate decimal points. Inductive reactance increases as the frequency of AC is increased. **ANSWER B**

3-1A6 What is the term for the out-of-phase power associated with inductors and capacitors?

A. Effective power.
B. True power.
C. Peak envelope power.
D. Reactive power.

If it is non-productive, it is reactive power in an ac circuit. It is energy stored temporarily in a coil's magnetic field or a capacitor's electric field and then returned to the circuit. It is power not converted to heat and dissipated. **ANSWER D**

Key Topic 2: Magnetism

3-2A1 What determines the strength of the magnetic field around a conductor?

A. The resistance divided by the current.
B. The ratio of the current to the resistance.
C. The diameter of the conductor.
D. The amount of current.

Electromagnetic fields develop around conductors — the more current passing through a conductor, the greater the magnetic field strength around the conductor. **ANSWER D**

3-2A2 What will produce a magnetic field?

A. A DC source not connected to a circuit.
B. The presence of a voltage across a capacitor.
C. A current flowing through a conductor.
D. The force that drives current through a resistor.

In marine and aeronautical installations, you always route your power cable well away from the magnetic or fluxgate compass assembly. As current passes through a conductor, the current produces a magnetic field as a force around the conductor. Stronger current will produce a stronger magnetic field, which could affect the reading of a magnetic or fluxgate compass. **ANSWER C**

3-2A3 When induced currents produce expanding magnetic fields around conductors in a direction that opposes the original magnetic field, this is known as:

A. Lenz's law.
B. Gilbert's law.
C. Maxwell's law.
D. Norton's law.

It is Lenz's law that describes induced currents producing expanding magnetic fields around conductors in a direction that actually opposes the original magnetic field. A classroom demonstration of Lenz's Law involves dropping a rare Earth round magnet down a copper or aluminum tube. It seems like an eternity before the magnet drops out of the bottom end! This is a great demonstration of Lenz's Law. **ANSWER A**

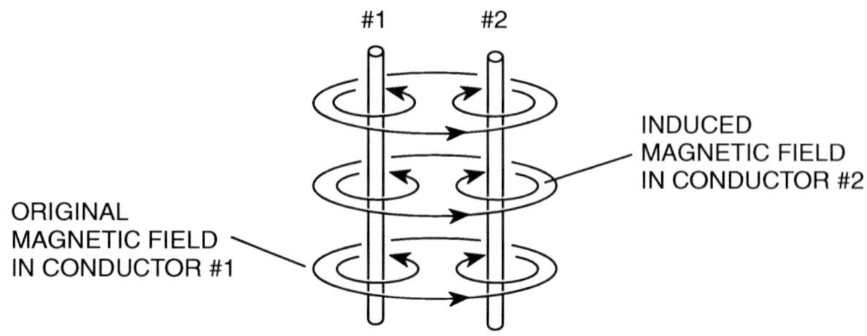

Illustration of Lenz's Law

3-2A4 The opposition to the creation of magnetic lines of force in a magnetic circuit is known as:

A. Eddy currents.
B. Hysteresis.
C. Permeability.
D. Reluctance.

The opposition to the creation of magnetic lines of force in a magnetic circuit is called reluctance. Iron has low reluctance because it has a very high μ. A segment of vacuum or air one centimeter long and one square centimeter of area with μ = 1 has a reluctance of one unit. **ANSWER D**

The formula for reluctance is:

$$R = \frac{L}{\mu A}$$

where: R = Reluctance in **reluctance units**
L = Length in **cm** or **inches**
A = Area in **cm²** or **in²**

3-2A5 What is meant by the term "back EMF"?

A. A current equal to the applied EMF.
B. An opposing EMF equal to R times C (RC) percent of the applied EMF.
C. A voltage that opposes the applied EMF.
D. A current that opposes the applied EMF.

A back EMF (electro-motive force) that opposes the applied EMF is developed in an RL circuit by the inductor. This reduces the current flow in the inductor and the circuit. **ANSWER C**

3-2A6 Permeability is defined as:

A. The magnetic field created by a conductor wound on a laminated core and carrying current.
B. The ratio of magnetic flux density in a substance to the magnetizing force that produces it.
C. Polarized molecular alignment in a ferromagnetic material while under the influence of a magnetizing force.
D. None of these.

Permeability is a certain flux density that develops in a core for a certain value of current. If iron is inserted into the coil, more flux and flux density develops. Permeability is represented by the Greek letter μ (pronounced "mu"). **ANSWER B** The formula for permeability is.

$$\mu = \frac{\beta}{H}$$

where: β = Flux density in guass or **lines per square inch**
H = Oersted of magnetizing force in **ampere turns per inch**

Key Topic 3: Materials

3-3A1 What metal is usually employed as a sacrificial anode for corrosion control purposes?

A. Platinum bushing.
B. Lead bar.
C. Zinc bar.
D. Brass rod.

To protect underwater metals from galvanic corrosion, boatyards install a sacrificial zinc bar to a transom mounting pad with a tie-in to other bonded underwater metals. A small amount of current will be developed in seawater, with the zinc bar slowly depleting itself to protect other underwater metals. **ANSWER C**

This badly-decayed, transom-mounted zinc plate still has a few months of life left in it!

Zinc sacrificial anode "fish" for temporary use at the dock, along with a zinc collar for the shaft.

3-3A2 What is the relative dielectric constant for air?

A. 1
B. 2
C. 4
D. 0

The dielectric constant for air is approximately 1. For atmospheric propagation studies at VHF frequencies, the dielectric constant of air at 1 is normally abbreviated "n", for normal. Glass has a dielectric constant of 8; Bakelite® 5; and Teflon® 2. Teflon is used extensively at VHF and UHF frequencies because it can firmly hold components in place, and has approximately the dielectric constant of air. **ANSWER A**

3-3A3 Which metal object may be least affected by galvanic corrosion when submerged in seawater?

A. Aluminum outdrive.
B. Bronze through-hull.
C. Exposed lead keel.
D. Stainless steel propeller shaft.

Galvanic corrosion – the result of dissimilar metals submerged in seawater – can be devastating without the use of a zinc sacrificial anode. Aluminum outdrives may be the first to go, followed by bronze through-hulls and exposed lead keels. The metal object LEAST affected by galvanic corrosion would be a stainless steel propeller shaft. It usually lasts the longest, well after a bronze propeller has melted away. **ANSWER D**

3-3A4 Skin effect is the phenomenon where:

A. RF current flows in a thin layer of the conductor, closer to the surface, as frequency increases.
B. RF current flows in a thin layer of the conductor, closer to the surface, as frequency decreases.
C. Thermal effects on the surface of the conductor increase the impedance.
D. Thermal effects on the surface of the conductor decrease the impedance.

If you ever had a chance to inspect an old wireless station, you would have seen that the antenna "plumbing" leading out of the transmitter was usually constructed of hollow copper tubing. Radio-frequency current always travels along the thinner outside layer of a conductor, and as frequency increases, there is almost no current in the center of the conductor. The higher the frequency, the greater the skin effect. This is why we use wide copper ground foil to minimize the impedance to AC current that we need to pass to ground. **ANSWER A**

3-3A5 Corrosion resulting from electric current flow between dissimilar metals is called:

A. Electrolysis.
B. Stray current corrosion.
C. Oxygen starvation corrosion.
D. Galvanic corrosion.

Galvanic corrosion occurs between dissimilar metals in a seawater electrolyte. Stray Current corrosion in electrical circuits is called electrolysis, and is different than galvanic corrosion. This question addresses a common maritime problem of dissimilar metals acting as a battery, and the correct answer is galvanic corrosion. **ANSWER D**

3-3A6 Which of these will be most useful for insulation at UHF frequencies?

A. Rubber.
B. Mica.
C. Wax impregnated paper.
D. Lead.

At UHF frequencies, Teflon® and mica are good insulators. Lead would be considered a conductor, and rubber is not appropriate as an insulator at UHF or microwave frequencies. Wax paper was used as insulation in capacitors back in the old days, but not at UHF frequencies! **ANSWER B**

Key Topic 4: Resistance, Capacitance & Inductance

3-4A1 What formula would calculate the total inductance of inductors in series?

A. $L_T = L_1 / L_2$
B. $L_T = L_1 + L_2$
C. $L_T = 1 / L_1 + L_2$
D. $L_T = 1 / L_1 \times L_2$

Calculating total inductance with inductors in series is just like calculating resistors in series: $L_1+L_2+L_3$. This only applies when there is no mutual inductance between the coils. **ANSWER B**

Low pass filter coils in series.

3-4A2 Good conductors with minimum resistance have what type of electrons?

A. Few free electrons.
B. No electrons.
C. Some free electrons.
D. Many free electrons.

A good conductor with many free electrons will allow current to pass easily, with minimum resistance. Whether its a plasma, like the ionosphere, or a metallic conductor, the availability of free electrons contributes to conductivity. **ANSWER D**

3-4A3 Which of the 4 groups of metals listed below are the best low-resistance conductors?

A. Gold, silver, and copper.
B. Stainless steel, bronze, and lead.
C. Iron, lead, and nickel.
D. Bronze, zinc, and manganese.

Silver offers excellent conductance, and is sometimes found with gold in relay contacts. Metals listed in order of their relative conductivity are silver, gold, copper, aluminum, nickel, iron, and lead. Gold is used to plate contacts on a relay because gold is resistant to corrosion, which is a metal oxide that can degrade circuit function. **ANSWER A**

3-4A4 What is the purpose of a bypass capacitor?

A. It increases the resonant frequency of the circuit.
B. It removes direct current from the circuit by shunting DC to ground.
C. It removes alternating current by providing a low impedance path to ground.
D. It forms part of an impedance transforming circuit.

Bypass capacitors shunt AC through a low impedance path to ground. **ANSWER C**

3-4A5 How would you calculate the total capacitance of three capacitors in parallel?

A. $C_T = C_1 + C_2 / C_1 - C_2 + C_3$.
B. $C_T = C_1 + C_2 + C_3$.
C. $C_T = C_1 + C_2 / C_1 \times C_2 + C_3$.
D. $C_T = 1 / C_1 + 1 / C_2 + 1 / C_3$.

Capacitors in parallel simply combine. They add up, just like resistors and coils in series. **ANSWER B**

Receiver band pass capacitors.

3-4A6 How might you reduce the inductance of an antenna coil?

A. Add additional turns.
B. Add more core permeability.
C. Reduce the number of turns.
D. Compress the coil turns.

To decrease the amount of loading inductance in an antenna coil, you could reduce the number of turns, or simply short them out, lowering series inductance. **ANSWER C**

Key Topic 5: Semi-conductors

3-5A1 What are the two most commonly-used specifications for a junction diode?

A. Maximum forward current and capacitance.
B. Maximum reverse current and PIV (peak inverse voltage).
C. Maximum reverse current and capacitance.
D. Maximum forward current and PIV (peak inverse voltage).

When choosing a junction diode, you will need to know how much forward current will be passing through the diode, and the amount of peak inverse voltage (PIV) that the diode must withstand in its non-conducting direction. **ANSWER D**

Diode.

3-5A2 What limits the maximum forward current in a junction diode?

A. The peak inverse voltage (PIV).
B. The junction temperature.
C. The forward voltage.
D. The back EMF.

Guess what will kill most electronic components? High temperature! The limit of the maximum forward current in a junction diode is the junction temperature. This is why you will find these diodes mounted to the chassis of your equipment. The chassis acts as a heat sink to keep the junction temperature from exceeding its maximum limit. **ANSWER B**

3-5A3 MOSFETs are manufactured with THIS protective device built into their gate to protect the device from static charges and excessive voltages:

A. Schottky diode.
B. Metal oxide varistor (MOV).
C. Zener diode.
D. Tunnel diode.

The "front-end" transistors of the modern transceiver must sometimes sustain major amounts of incoming signals from nearby transmitters, or maybe even a nearby lightning strike. Static created by handling the device could destroy a MOSFET (Metal-Oxide-Semiconductor-Field-Effect Transistor) if it weren't for the built-in, gate-protective Zener diode. **ANSWER C**

3-5A4 What are the two basic types of junction field-effect transistors?

A. N-channel and P-channel.
B. High power and low power.
C. MOSFET and GaAsFET.
D. Silicon FET and germanium FET.

Be sure to note carefully the direction of the gate arrow to determine whether it is N-channel or P-channel. **ANSWER A**

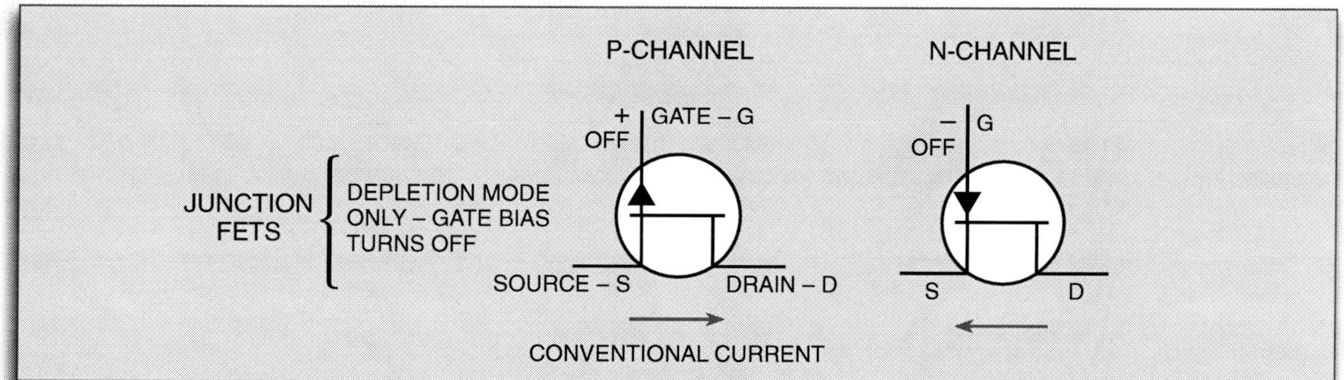

Junction FETs

3-5A5 A common emitter amplifier has:

A. Lower input impedance than a common base.
B. More voltage gain than a common collector.
C. Less current gain than a common base.
D. Less voltage gain than a common collector.

The common emitter amplifier allows small voltage variations in the input to be turned into larger voltage variations at the output, with a multiplication factor called voltage gain. The emitter is at ground, which is a common signal point for input and output. The voltage gain of the amplifier stage is a number obtained by dividing the output voltage by the input voltage. **ANSWER B**

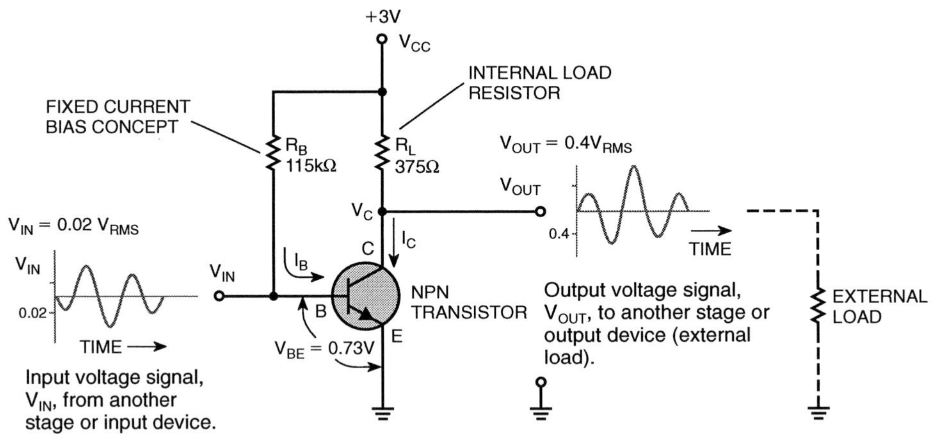

NPN Common-Emitter Voltage Amplifier Stage.

3-5A6 How does the input impedance of a field-effect transistor compare with that of a bipolar transistor?

A. An FET has high input impedance; a bipolar transistor has low input impedance.
B. One cannot compare input impedance without first knowing the supply voltage.
C. An FET has low input impedance; a bipolar transistor has high input impedance.
D. The input impedance of FETs and bipolar transistors is the same.

An FET is much easier to work with in circuits than the simple bipolar transistor. Its higher input impedance is less likely to load down a circuit that it is attached to, and the FET can also handle big swings in signal levels. **ANSWER A**

Key Topic 6: Electrical Measurements

3-6A1 An AC ammeter indicates:

A. Effective (TRM) values of current.
B. Effective (RMS) values of current.
C. Peak values of current.
D. Average values of current.

An ac ammeter shows the effective [root-mean-square (RMS)] value of current in a circuit. **ANSWER B**

AC Ammeter.

3-6A2 By what factor must the voltage of an AC circuit, as indicated on the scale of an AC voltmeter, be multiplied to obtain the peak voltage value?

A. 0.707
B. 0.9
C. 1.414
D. 3.14

To calculate peak voltage from the voltage indicated on an ac voltmeter, which measures V_{RMS}, multiply by 1.414, because $V_{pk} = 1.414\ V_{RMS}$. **ANSWER C**

3-6A3 What is the RMS voltage at a common household electrical power outlet?

A. 331-V AC.
B. 82.7-V AC.
C. 165.5-V AC.
D. 117-V AC.

Household alternating current electrical power is rated at 117 VAC RMS. **ANSWER D**

3-6A4 What is the easiest voltage amplitude to measure by viewing a pure sine wave signal on an oscilloscope?

A. Peak-to-peak.
B. RMS.
C. Average.
D. DC.

Looking at a pure sine wave signal on a scope, it is easy to identify the peak-to-peak voltage measured from the maximum positive excursion of the signal to the maximum negative excursion of the signal. **ANSWER A**

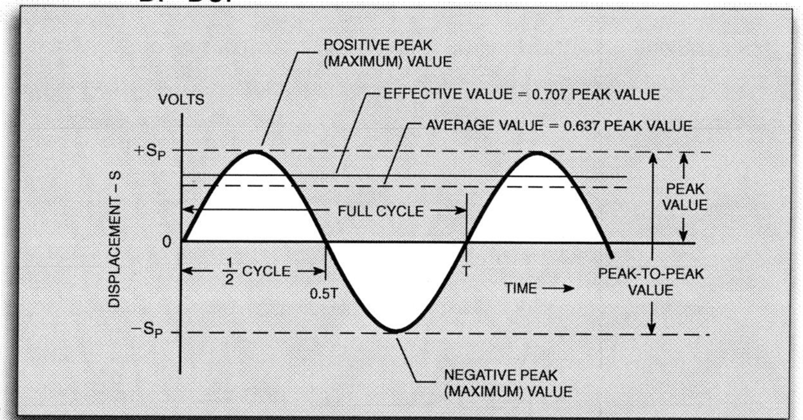

Peak, Peak-to-Peak and RMS Voltages

3-6A5 By what factor must the voltage measured in an AC circuit, as indicated on the scale of an AC voltmeter, be multiplied to obtain the average voltage value?

A. 0.707
B. 1.414
C. 0.9
D. 3.14

To calculate average voltage when monitoring the voltage on the scale of an ac voltmeter, which measures the RMS value, multiply the actual voltage reading by 0.9 because $V_{AVG} = 0.9\ V_{RMS}$. **ANSWER C**

3-6A6 What is the peak voltage at a common household electrical outlet?

A. 234 volts.
B. 117 volts.
C. 331 volts.
D. 165.5 volts.

When you measure house power with a voltmeter, you are measuring RMS. If you were to look at it on an oscilloscope, and look at the electrical peaks, it would be 1.414 times higher than 117 volts, or a total of 165.5 volts peak. **ANSWER D**

Key Topic 7: Waveforms

3-7A1 What is a sine wave?

A. A constant-voltage, varying-current wave.
B. A wave whose amplitude at any given instant can be represented by the projection of a point on a wheel rotating at a uniform speed.
C. A wave following the laws of the trigonometric tangent function.
D. A wave whose polarity changes in a random manner.

A sine wave is as smooth as a turning wheel, giving us a uniform cycle of energy. The point on a wheel is like a vector rotating through 360°. If the sine of the angle is plotted against degrees rotation, we will plot a sinewave. **ANSWER B**

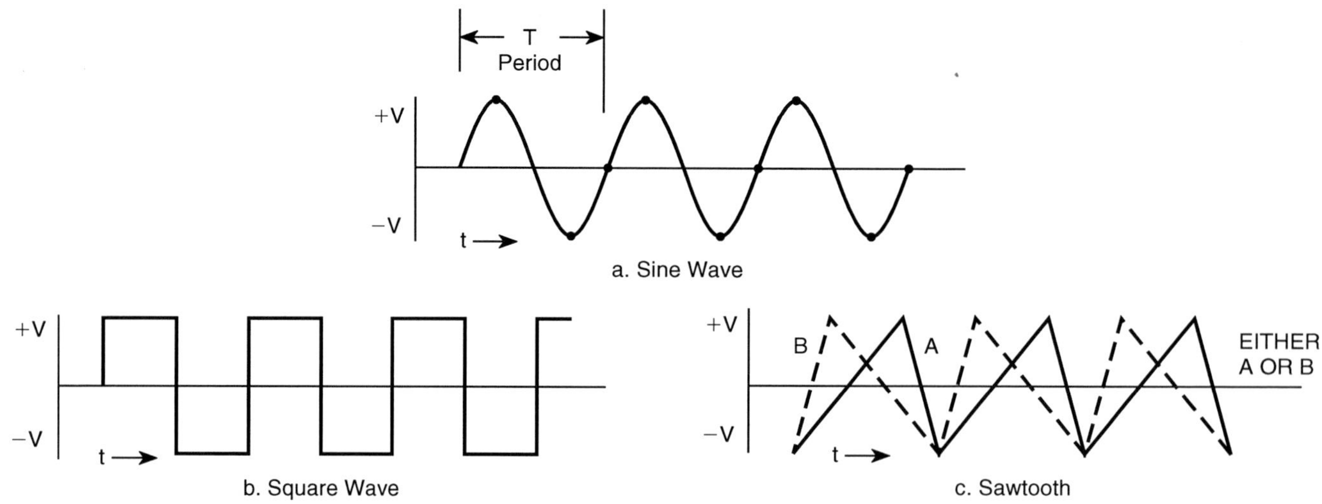

Sine, Square and Sawtooth Waveforms

3-7A2 How many degrees are there in one complete sine wave cycle?

A. 90 degrees.
B. 270 degrees.
C. 180 degrees.
D. 360 degrees.

One complete sine wave cycle is 360 degrees, just like a wheel. **ANSWER D**

3-7A3 What type of wave is made up of sine waves of the fundamental frequency and all the odd harmonics?

A. Square.
B. Sine.
C. Cosine.
D. Tangent.

You get square waves from odd harmonics. **ANSWER A**

3-7A4 What is the description of a square wave?

A. A wave with only 300 degrees in one cycle.
B. A wave whose periodic function is always negative.
C. A wave whose periodic function is always positive.
D. A wave that abruptly changes back and forth between two voltage levels and stays at these levels for equal amounts of time.

Square waves have abrupt changes back and forth between two voltage levels. The abrupt change of a square wave is not desirable as a CW key-attack and key-release wave form. Square waves on CW generate key clicks because of their abrupt changes. **ANSWER D**

3-7A5 What type of wave is made up of sine waves at the fundamental frequency and all the harmonics?

A. Sawtooth wave.
B. Square wave.
C. Sine wave.
D. Cosine wave.

If it has all the harmonics, the answer is a sawtooth wave. **ANSWER A**

3-7A6 What type of wave is characterized by a rise time significantly faster than the fall time (or vice versa)?

A. Cosine wave.
B. Square wave.
C. Sawtooth wave.
D. Sine wave.

The sawtooth wave can have either a very fast rise time and a slow fall time, or a slow rise time and a very fast fall time. A common function generator can generate either a “forward” or a “backward” sawtooth wave. **ANSWER C**

Key Topic 8: Conduction

3-8A1 What is the term used to identify an AC voltage that would cause the same heating in a resistor as a corresponding value of DC voltage?

A. Cosine voltage.
B. Power factor.
C. Root mean square (RMS).
D. Average voltage.

When we compare a dc voltage to an ac voltage that produces the same heating in a resistor, we need to choose the root-mean-square ac voltage. **ANSWER C**

3-8A2 What happens to reactive power in a circuit that has both inductors and capacitors?

A. It is dissipated as heat in the circuit.
B. It alternates between magnetic and electric fields and is not dissipated.
C. It is dissipated as inductive and capacitive fields.
D. It is dissipated as kinetic energy within the circuit.

They sometimes call reactive power “watt-less,” as it is not dissipated when it alternates between the magnetic and the electric fields. **ANSWER B**

3-8A3 Halving the cross-sectional area of a conductor will:

A. Not affect the resistance.
B. Quarter the resistance.
C. Double the resistance.
D. Halve the resistance.

If the cross-sectional area of a piece of wire is reduced by one-half, the resistance will double. **ANSWER C**

3-8A4 Which of the following groups is correct for listing common materials in order of descending conductivity?

A. Silver, copper, aluminum, iron, and lead.
B. Lead, iron, silver, aluminum, and copper.
C. Iron, silver, aluminum, copper, and silver.
D. Silver, aluminum, iron, lead, and copper.

The acronym SCAIL stands for Silver, Copper, Aluminum, Iron, and Lead for conductivity. On a SCAIL of 1 to 10, you are going to do well on your examination, if you will use this simple reminder. **ANSWER A**

3-8A5 How do you compute true power (power dissipated in the circuit) in a circuit where AC voltage and current are out of phase?

A. Multiply RMS voltage times RMS current.
B. Subtract apparent power from the power factor.
C. Divide apparent power by the power factor.
D. Multiply apparent power times the power factor.

When an AC electrical circuit contains both resistance and reactance, the phase angle must be known in order to compute the actual power dissipated (true power). Simply multiplying the voltage by the current, without considering the phase angle, gives the apparent power. Depending on the phase angle, the true power can range anywhere from 0 up to the apparent power. To find the true power, we must multiply the apparent power by the power factor. The power factor is the cosine of the phase angle between the voltage and the current. It is always less than or equal to 1. (In a DC circuit, the power factor is always 1.) **ANSWER D**

Apparent and True Power

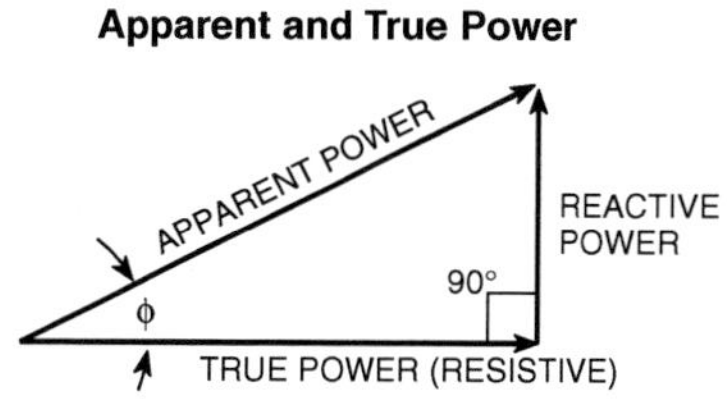

True power is power dissipated as heat, while reactive power of a circuit is stored in inductances or capacitances and then returned to the circuit.

Apparent power is voltage times current without taking into account the phase angle between them. According to right angle trigonometry,

$$\text{cosine } \phi = \frac{\text{side adjacent}}{\text{hypotenuse}}$$

therefore,

$$\text{cosine } \phi = \frac{\text{True Power}}{\text{Apparent Power}}$$

therefore,

True Power = Apparent Power × Cosine ϕ

Cosine ϕ is called the *power factor* (PF) of a circuit.

Since apparent power is E × I,

True Power = E × I × Cos ϕ

or

True Power = E × I × PF

3-8A6 Assuming a power source to have a fixed value of internal resistance, maximum power will be transferred to the load when:

A. The load impedance is greater than the source impedance.
B. The load impedance equals the internal impedance of the source.
C. The load impedance is less than the source impedance.
D. The fixed values of internal impedance are not relative to the power source.

When a load impedance equals the internal impedance of the source, maximum power is transferred. This is why we pay close attention to impedance matching in antenna and transceiver systems. Maximum power transfer theorem: Z load = Z generator. **ANSWER B**

Subelement B – Electrical Math

10 Key Topics, 10 Exam Questions, 3 Drawings

Key Topic 9: Ohm's Law-1

3-9B1 What value of series resistor would be needed to obtain a full scale deflection on a 50 microamp DC meter with an applied voltage of 200 volts DC?

A. 4 megohms.
B. 2 megohms.
C. 400 kilohms.
D. 200 kilohms.

It is standard practice to use a microammeter movement as a voltmeter by the use of a series multiplier resistor. Ohm's Law applies when calculating current through the meter. Current is voltage divided by resistance. (In this application the resistance of the meter can be ignored, since it is insignificant compared to the multiplier resistance.) Your calculator key strokes are 200 ÷ 0.000050 = 4,000,000 which is 4 megohms. **ANSWER A**

3-9B2 Which of the following Ohms Law formulas is incorrect?

A. $I = E / R$
B. $I = R / E$
C. $E = I \times R$
D. $R = E / I$

The Ohm's Law Circle is shown on the facing page. The INCORRECT formula is: **ANSWER B**

3-9B3 If a current of 2 amperes flows through a 50-ohm resistor, what is the voltage across the resistor?

A. 25 volts.
B. 52 volts.
C. 200 volts.
D. 100 volts.

The relationship between voltage (E), current (I), and resistance (R) in an electronic circuit is described by Ohm's Law, which states: the applied electromotive force, E, in volts, is equal to the circuit current, I, in amperes, times the circuit resistance, R, in ohms. It is expressed by the equation $E = I \times R$.

A simple way to remember how to calculate Ohm's Law is to use the circle. The circle shows E, I, and R in position so that it provides the correct equation for your problem. In this question, you are asked to solve for E (voltage), which is equal to I (current) times R (resistance).

To use the circle, cover the letter that you are solving for with your finger. Now, plug in the other two values that they give you in the examination question. Solve the problem by performing the mathematical operation indicated by the position of the remaining letters, as shown here:

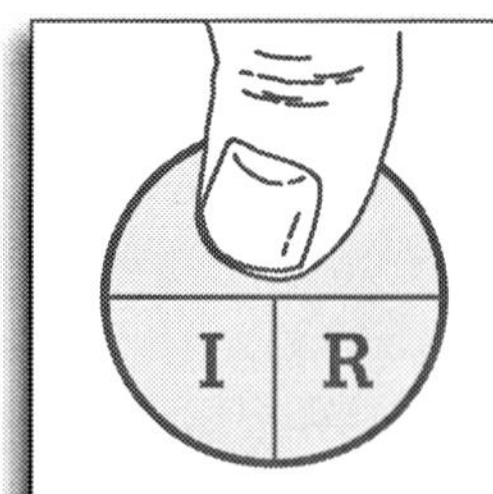

Since we are looking for E, the applied voltage, cover E with your finger and you now have I (2 amperes) times R (50 Ohms). Multiply these two together to obtain your answer of 100 volts. The calculator key strokes are: CLEAR 2 × 50 =. **ANSWER D**

3-9B4 If a 100-ohm resistor is connected across 200 volts, what is the current through the resistor?

A. 2 amperes.
B. 1 ampere.
C. 300 amperes.
D. 20,000 amperes.

In this problem, you are looking for I. Using the Ohm's Law circle, cover I with your finger. You now have E over R, or 200 over 100. Do the division, and you will end up with 2 amps. See how simple this is! Calculator key strokes are: CLEAR 200 ÷ 100 = and your answer is 2. **ANSWER A**

3-9B5 If a current of 3 amperes flows through a resistor connected to 90 volts, what is the resistance?

A. 3 ohms.
B. 30 ohms.
C. 93 ohms.
D. 270 ohms.

Again, use the circle. In this problem, you want to find R. Covering R with your finger leaves E over I. 90 divided by 3 gives 30 ohms. See how simple it is to use Ohm's Law. Calculator key strokes are: CLEAR 90 ÷ 3 =. Your answer is 30. **ANSWER B**

3-9B6 A relay coil has 500 ohms resistance, and operates on 125 mA. What value of resistance should be connected in series with it to operate from 110 V DC?

A. 150 ohms.
B. 220 ohms.
C. 380 ohms.
D. 470 ohms.

Here is another Ohm's Law problem where voltage divided by current equals resistance ($E \div I = R$). But you have to figure the voltage drop before you can apply the formula. The voltage drop across the relay is 500 ohms × 0.125 amps = 62.5 volts. Now, subtract 62.5 from 110 to arrive at the remaining 47.5 volts that must be dropped across the series resistance. Finally, use Ohm's Law to solve for resistance to get the answer (47.5 Volts ÷ 0.125 Amps = 380 Ohms). **ANSWER C**

Ohm's Law & Power Circle

Georg Simon Ohm (1789-1854) experimented with electricity and discovered that the resistance (R) of a conductor deptends on its length in feet, cross-sectional area in circular mils, and the resistivity, which is a parameter that depends on the molecular structure of the conductor and its temperature. Ohm's Law states:
The current in an electrical circuit is directly proportional to the voltage and inversely proportional to the resistance.
The Ohm's Law and Power Circle shown here includes 12 equations that allow us to solve for voltage,(E), current (I), resistance (R), and power (P) if we know the other values. Keep this page bookmarked to help you solve common electrical formulas.

To solve for Voltage (E):

E = I (current) × R (resistance)
E = √P (power) × R (resistance)
E = P (power) ÷ I (current)

To solve for Current (I):

I = P (power) ÷ E (voltage)
I = √P (power) ÷ R (resistance)
I = E (voltage) ÷ R (resistance)

To solve for Resistance (R):

R = E^2 (voltage squared) ÷ P (power)
R = P (power) ÷ I^2 (current squared)
R = E (voltage) ÷ I (current)

To solve for Power (P):

P = E (voltage) × I (current)
P = I^2 (current squared) × R (resistance)
P = E^2 (voltage squared) ÷ R (resistance)

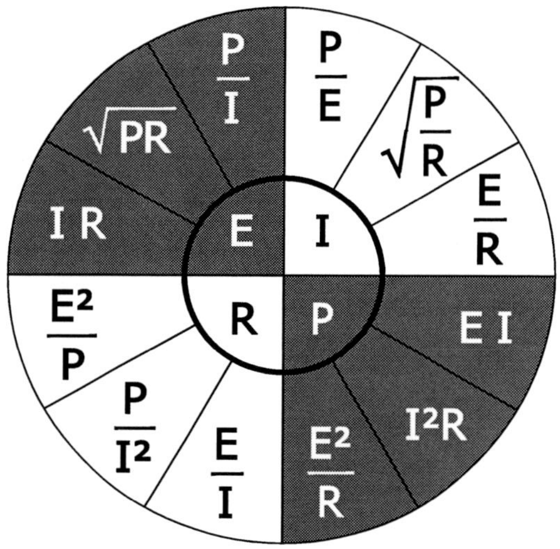

Source: The ARRL Handbook, © 2006, American Radio Relay League

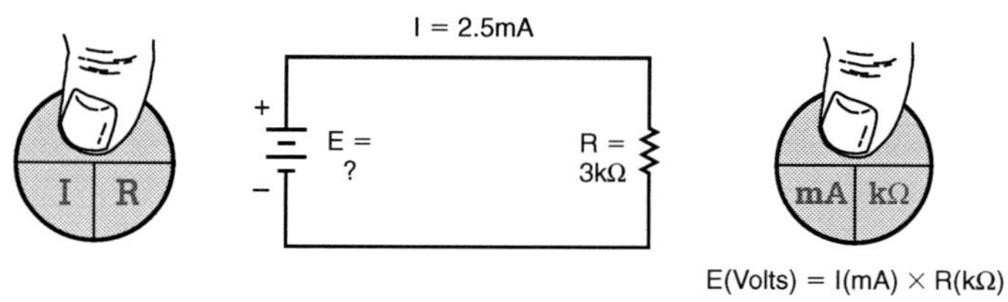

a. Unknown Voltage

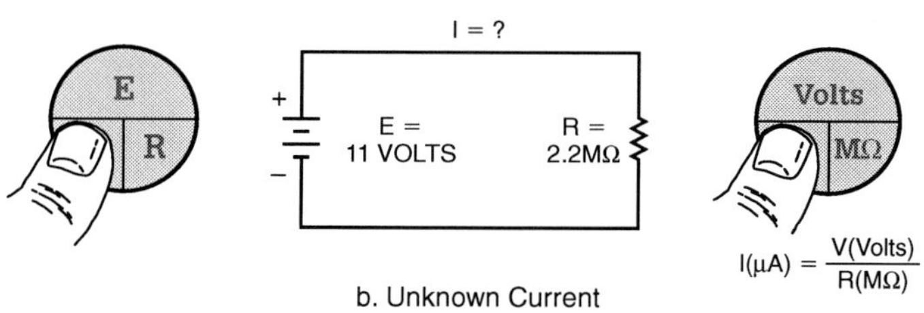

b. Unknown Current

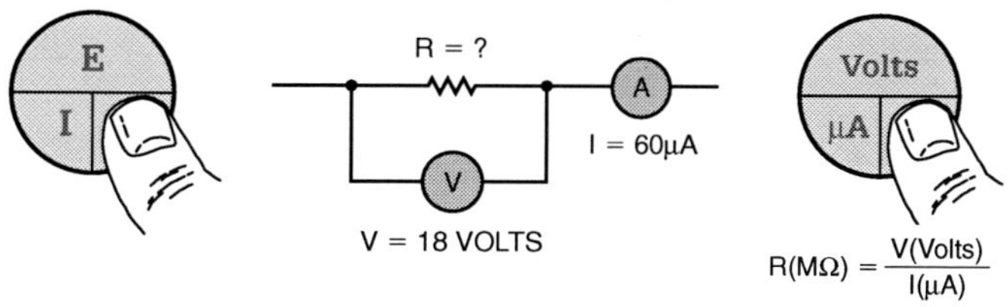

c. Unknown Resistance

The Three Rules of a Series Circuit

1. Current has the same value at any point within a series circuit.
2. The values of the resistances of individual components add up to the total circuit resistance, which is called the equivalent resistance, Req.
3. Voltage drops across the individual component resistances add up to the total applied voltage.

Key Topic 10: Ohm's Law-2

3-10B1 What is the peak-to-peak RF voltage on the 50 ohm output of a 100 watt transmitter?

A. 70 volts.
B. 100 volts.
C. 140 volts.
D. 200 volts.

Ohm's Law gives us the square root of power times resistance equals voltage. But the problem asks peak to peak voltage. Here are the key strokes:
$50 \times 100 = \sqrt{5000}$ x 1.414 = × 2 = 199.96, rounded to 200. **ANSWER D**

3-10B2 What is the maximum DC or RMS voltage that may be connected across a 20 watt, 2000 ohm resistor?

A. 10 volts.
B. 100 volts.
C. 200 volts.
D. 10,000 volts.

Your formula here is $P \times R = E^2$. When you insert the values for this problem, this is what you have: 20 watts × 2000 ohms = 40,000. Then, take the square root of 40,000 to get the answer, 200 volts. **ANSWER C**

3-10B3 A 500-ohm, 2-watt resistor and a 1500-ohm, 1-watt resistor are connected in parallel. What is the maximum voltage that can be applied across the parallel circuit without exceeding wattage ratings?

A. 22.4 volts.
B. 31.6 volts.
C. 38.7 volts.
D. 875 volts.

To solve this question, you must calcuate the maximum voltage that can be applied before its power rating is exceeded. Use the formula $P \times R = E^2$. For the 500-ohm resistor, 2 × 500 = 1000. The square root of 1000 gives the maximum voltage of 31.6 volts. For the 1500-ohm resistor, 1 × 1500 = 1500. The square root of 1500 gives a maximum voltage of 38.72. So the 31.6 V for the 500-ohm resistor limits the voltage value. **ANSWER B**

The Three Rules of a Parallel Circuit

1. The same voltage is applied across each individual branch.
2. The total current is equal to the sum of the individual branch currents.
3. The equivalent (or effective) resistance is equal to the applied voltage divided by the total current, and this value is always less than the smallest resistance contained in any one branch.

3-10B4 In Figure 3B1, what is the voltage drop across R_1?

A. 9 volts.
B. 7 volts.
C. 5 volts.
D. 3 volts.

Look at Figure 3B1 and notice that the diode is reverse biased, and will look like an open circuit to this series resistance circuit problem. Current will be the same throughout all of the circuit, and voltage drops will equal the source voltage. Combine R_3 and R_2 as a total of 300 ohms, $R_3 + R_2$ equals the resistance of R_1, 300 ohms, so each will drop 5 volts. The voltage drop across R_1 equals 5 volts. **ANSWER C**

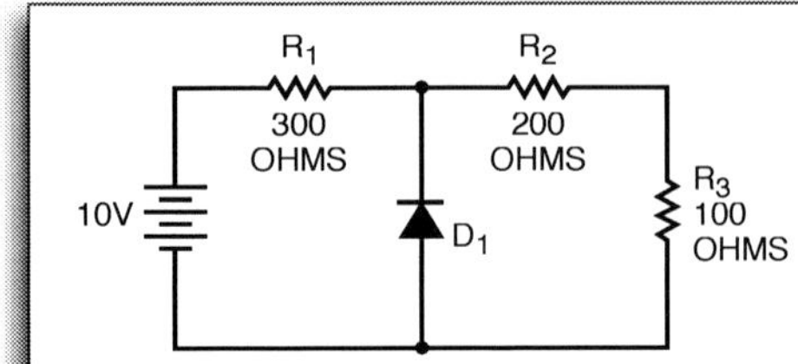

Figure 3B1

3-10B5 In Figure 3B2, what is the voltage drop across R_1?

A. 1.2 volts.
B. 2.4 volts.
C. 3.7 volts.
D. 9 volts.

Look at Figure 3B2 and note that the Zener diode will drop 3 volts from the supply 12 volts, and 12 minus 3 volts equals a 9-volt drop at R_1.
ANSWER D

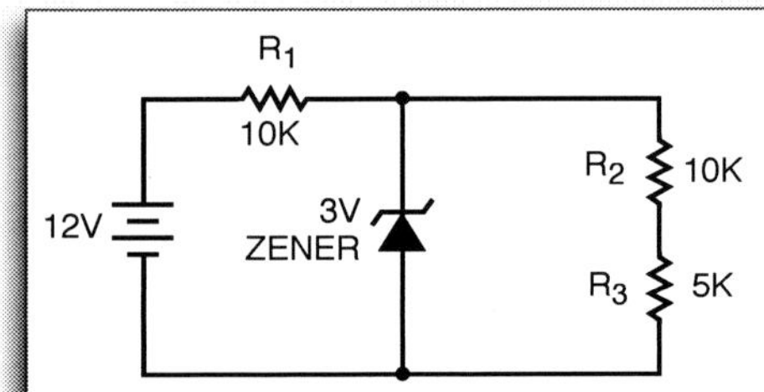

Figure 3B2

3-10B6 What is the maximum rated current-carrying capacity of a resistor marked "2000 ohms, 200 watts"?

A. 0.316 amps.
B. 3.16 amps.
C. 10 amps.
D. 100 amps.

The formula here is $P = I^2 \times R$, or $I^2 = P \div R$. This gives us 200 ÷ 2000 = 0.1. Then, take the square root of 0.1 to get the correct answer: 0.316 amps. **ANSWER A**

Key Topic 11: Frequency

3-11B1 What is the most the actual transmit frequency could differ from a reading of 462,100,000 hertz on a frequency counter with a time base accuracy of ± 0.1 ppm?

A. 46.21 Hz.
B. 0.1 MHz.
C. 462.1 Hz.
D. 0.2 MHz.

To solve this problem, remember to multiply the frequency in MHz times the time base accuracy. Try this on your calculator: Clear 462.1 × 0.1 = 46.21 Hz. **ANSWER A**

3-11B2 The second harmonic of a 380 kHz frequency is:

A. 2 MHz.
B. 760 kHz.
C. 190 kHz.
D. 144.4 GHz.

To calculate the second harmonic of 380 kHz, simply multiply it by 2, and you get 760 kHz. **ANSWER B**

3-11B3 What is the second harmonic of SSB frequency 4146 kHz?

A. 8292 kHz.
B. 4.146 MHz.
C. 2073 kHz.
D. 12438 kHz.

It is important to know where the second harmonic will occur aboard a ship, as this second harmonic could interfere with ship-to-ship Channel 8 Alpha, 8294 kHz, on a companion receiver. **ANSWER A**

3-11B4 What is the most the actual transmitter frequency could differ from a reading of 156,520,000 hertz on a frequency counter with a time base accuracy of ± 1.0 ppm?

A. 165.2 Hz.
B. 15.652 kHz.
C. 156.52 Hz.
D. 1.4652 MHz.

It's easy to solve these problems with a simple calculator. Multiply the frequency in MHz times the time base accuracy. **ANSWER C**

3-11B5 What is the most the actual transmitter frequency could differ from a reading of 156,520,000 hertz on a frequency counter with a time base accuracy of +/− 10 ppm?

A. 146.52 Hz.
B. 1565.20 Hz.
C. 10 Hz.
D. 156.52 kHz.

Keystroke the following on your calculator: Clear 156.52 × 10 = 1565.2 Hz. You are multiplying the frequency in MHz by the time base accuracy of 10 ppm. **ANSWER B**

Readout error = f × a   Where: f = Frequency in **MHz**
a = Counter accuracy in **parts per million**

Frequency Counter
Courtesy of OPTOELECTRONICS

3-11B6 What is the most the actual transmitter frequency could differ from a reading of 462,100,000 hertz on a frequency counter with a time base accuracy of ± 1.0 ppm?

A. 46.21 MHz.
B. 10 Hz.
C. 1.0 MHz.
D. 462.1 Hz.

From Hz to MHz, we end up with 462.1 MHz, which multiplied by 1.0 ppm gives you the answer. The following are the keystrokes: Clear 462.1 × 1 = 462.1 Hz. **ANSWER D**

Key Topic 12: Waveforms

3-12B1 At pi/3 radians, what is the amplitude of a sine-wave having a peak value of 5 volts?

A. −4.3 volts. C. +2.5 volts.
B. −2.5 volts. D. +4.3 volts.

A radian is the angle within a circle whose arc length along the circumference is equal to the radius. There are 2π radians in a circle; therefore π radians = 180°. For this problem the angle is

$$\frac{\pi}{3} = \frac{180°}{3} = 60°$$

therefore, $V = 5 \sin 60°$

$V = 5 \times 0.866 = 4.33V$ **ANSWER D**

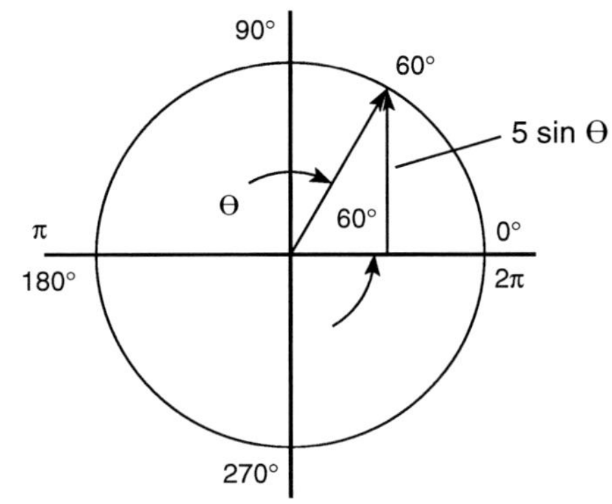

3-12B2 At 150 degrees, what is the amplitude of a sine-wave having a peak value of 5 volts?

A. −4.3 volts. C. +2.5 volts.
B. −2.5 volts. D. +4.3 volts.

Now the angle is 150°. The amplitude of the Vector V = 5 sin Θ is now:

$$V = 5 \sin 30°$$
$$= 5 \times 0.5 = 2.5 \text{ V}$$ **ANSWER C**

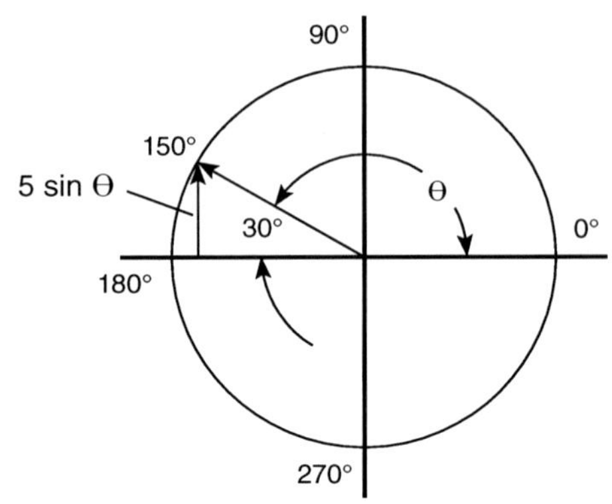

3-12B3 At 240 degrees, what is the amplitude of a sine-wave having a peak value of 5 volts?

A. −4.3 volts. C. +2.5 volts.
B. −2.5 volts. D. +4.3 volts.

To work out these few problems, I suggest you draw a complete sine-wave and note the degrees and recall the following formula V = 5 sin Θ. **ANSWER A**

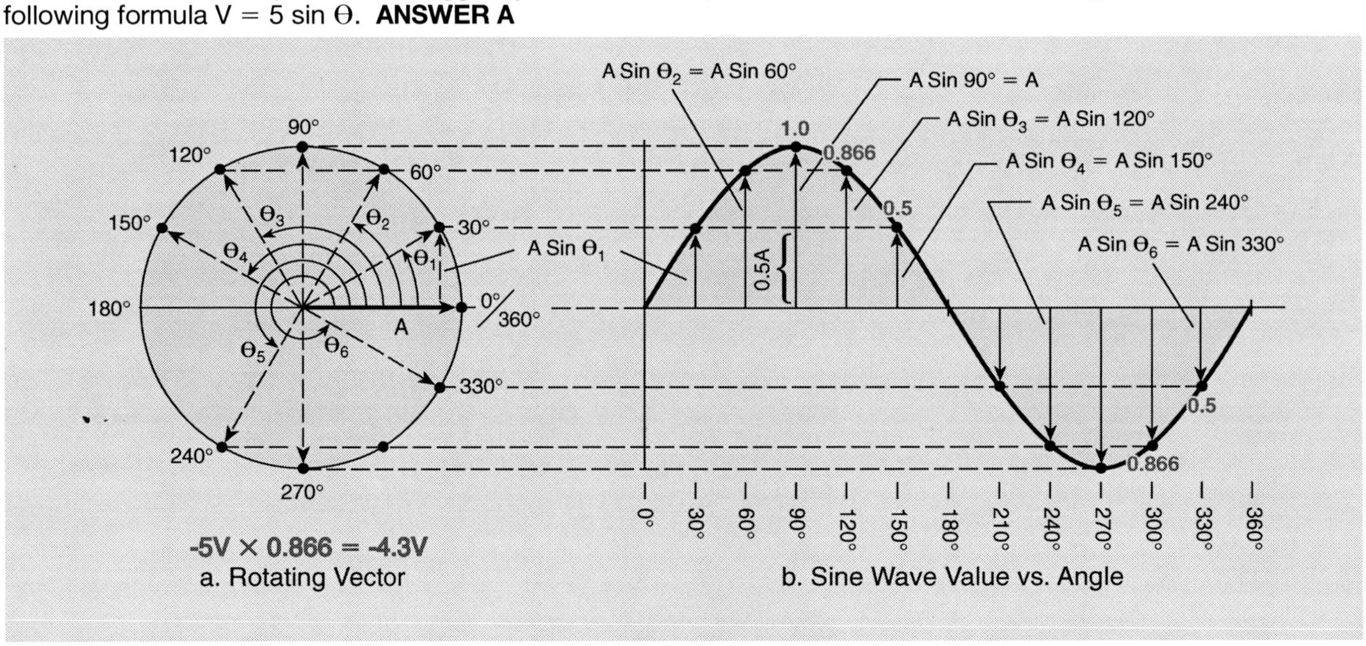

Vector of 5V Generates Sine Wave as It Rotates

3-12B4 What is the equivalent to the root-mean-square value of an AC voltage?
A. AC voltage is the square root of the average AC value.
B. The DC voltage causing the same heating in a given resistor at the peak AC voltage.
C. The AC voltage found by taking the square of the average value of the peak AC voltage.
D. The DC voltage causing the same heating in a given resistor as the RMS AC voltage of the same value.
We can calculate RMS of an AC voltage when it has the same heating of a resistor as if DC voltage of the same value were applied to that resistor. Sometimes AVERAGE is used to define voltage, around 0.65 of a sine wave, a bit less than the 0.707 for RMS. RMS (0.707) is the meaningful figure, since it translates to actual heating power. **ANSWER D**

3-12B5 What is the RMS value of a 340-volt peak-to-peak pure sine wave?
A. 170 volts AC.
B. 240 volts AC.
C. 120 volts AC.
D. 350 volts AC.
First divide 340 in half, and you end up with peak, instead of peak-to-peak. Now, multiply by 0.707 to bring down peak voltage down to RMS. **ANSWER C**

3-12B6 Determine the phase relationship between the two signals shown in Figure 3B3.
A. A is lagging B by 90 degrees.
B. B is lagging A by 90 degrees.
C. A is leading B by 180 degrees.
D. B is leading A by 90 degrees.
As you can see in the figure, the "B" signal is lagging by 90 degrees. **ANSWER B**

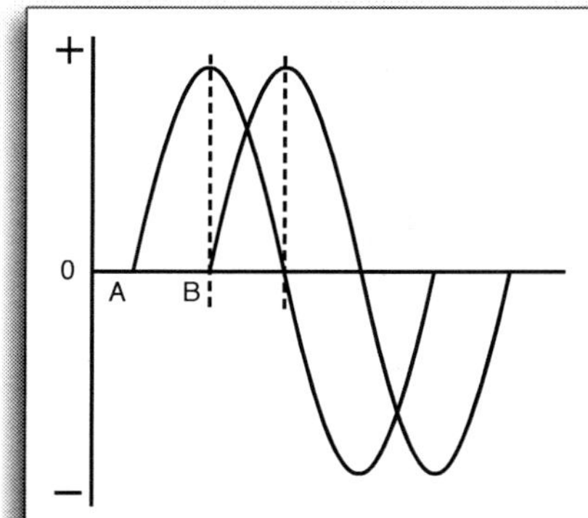

Fig. 3B3

Key Topic 13: Power Relationships

3-13B1 What does the power factor equal in an R-L circuit having a 60 degree phase angle between the voltage and the current?
A. 0.414
B. 0.866
C. 0.5
D. 1.73
If you are using a scientific calculator, you will find the cosine of 60 degrees, given in the problem, comes out 0.5, answer C. The power factor is cos ϕ. **ANSWER C**
If you don't have a scientific calculator, you might memorize this table:

ϕ	cos ϕ
30°	0.866
45°	0.707
60°	0.5

3-13B2 If a resistance to which a constant voltage is applied is halved, what power dissipation will result?
A. Double.
B. Halved.
C. Quadruple.
D. Remain the same.
If the resistance in a piece of wire is reduced to half, the power dissipation doubles. **ANSWER A**

3-13B3 746 watts, corresponding to the lifting of 550 pounds at the rate of one-foot-per-second, is the equivalent of how much horsepower?

A. One-quarter horsepower.
B. One-half horsepower.
C. Three-quarters horsepower.
D. One horsepower.

One horsepower is equal to 746 watts, which corresponds to the lifting of 550 pounds at the rate of one foot per second. **ANSWER D**

3-13B4 In a circuit where the AC voltage and current are out of phase, how can the true power be determined?

A. By multiplying the apparent power times the power factor.
B. By subtracting the apparent power from the power factor.
C. By dividing the apparent power by the power factor.
D. By multiplying the RMS voltage times the RMS current.

If we multiply apparent power times the power factor of an AC voltage and current out of phase, we can determine true power. **ANSWER A**

Apparent and True Power

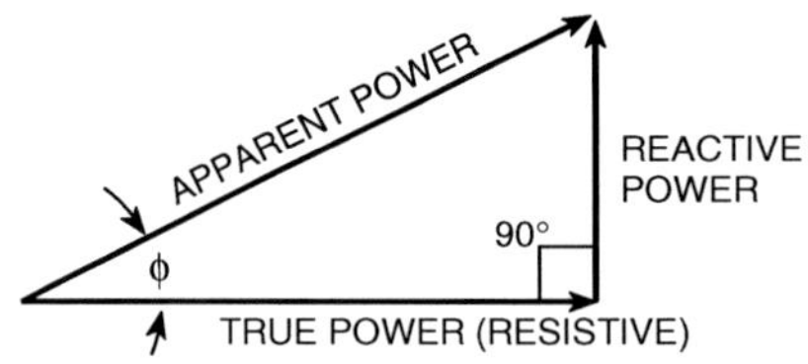

True power is power dissipated as heat, while reactive power of a circuit is stored in inductances or capacitances and then returned to the circuit.

Apparent power is voltage times current without taking into account the phase angle between them. According to right angle trigonometry,

$$\text{cosine } \phi = \frac{\text{side adjacent}}{\text{hypotenuse}}$$

therefore,

$$\text{cosine } \phi = \frac{\text{True Power}}{\text{Apparent Power}}$$

therefore,

True Power = Apparent Power × Cosine ϕ

Cosine ϕ is called the *power factor* (PF) of a circuit.

Since apparent power is E × I,

True Power = E × I × Cos ϕ

or

True Power = E × I × PF

Apparent and True Power

3-13B5 What does the power factor equal in an R-L circuit having a 45 degree phase angle between the voltage and the current?

A. 0.866
B. 1.0
C. 0.5
D. 0.707

Using your scientific calculator, or from memory, remember that the cosine of 45 degrees is 0.707. Think of an airplane, the very popular 707 jet aircraft, taking off at an angle of 45 degrees. See table at 3-13B1. **ANSWER D**

3-13B6 What does the power factor equal in an R-L circuit having a 30 degree phase angle between the voltage and the current?

A. 1.73
B. 0.866
C. 0.5
D. 0.577

At 30 degrees, the power factor is 0.866. If you don't pass your Commercial exam, you could be 86'd out of the room! See table at 3-13B1. **ANSWER B**

Key Topic 14: RC Time Constants-1

3-14B1 What is the term for the time required for the capacitor in an RC circuit to be charged to 63.2% of the supply voltage?

A. An exponential rate of one.
B. One time constant.
C. One exponential period.
D. A time factor of one.

In an RC circuit, assuming there is no initial charge on the capacitor, it takes one time constant to charge a capacitor to 63.2% of its final value. **ANSWER B**

3-14B2 What is the meaning of the term "time constant of an RC circuit"? The time required to charge the capacitor in the circuit to:

A. 23.7% of the supply voltage.
B. 36.8% of the supply voltage.
C. 57.3% of the supply voltage.
D. 63.2% of the supply voltage.

If you have a separate power supply feeding your equipment, when you turn off your power supply with the equipment still turned on, you will notice that the red power supply indicator light dims slowly. The slow decay is because of the voltage still across the large electrolytic filter capacitors on the output of the power supply. This is a visual example that capacitors charge up, and discharge, on a curve. The time required to charge (or discharge) the capacitor in an RC (Resistance-Capacitance) circuit to 63.2% of the supply voltage is called a time constant. **ANSWER D**

$\tau = RC$ where
τ = Greek letter tau, the time constant in **seconds**
R = total resistance in **ohms**
C = capacitance in **farads**

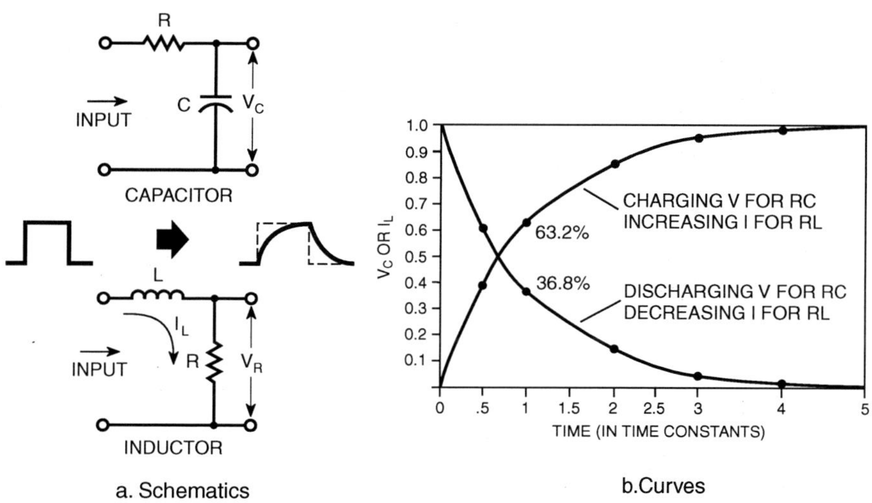

RC (RxC) and RL (L/R) Time Constant

3-14B3 What is the term for the time required for the current in an RL circuit to build up to 63.2% of the maximum value?

A. One time constant.
B. An exponential period of one.
C. A time factor of one.
D. One exponential rate.

In an RL circuit, assuming there is no initial current through the inductance, it takes one time constant for the current in the circuit to build up to 63.2% of its final value. Measuring V_R tracks I_L. **ANSWER A**

3-14B4 What is the meaning of the term "time constant of an RL circuit"? The time required for the:

A. Current in the circuit to build up to 36.8% of the maximum value.
B. Voltage in the circuit to build up to 63.2% of the maximum value.
C. Current in the circuit to build up to 63.2% of the maximum value.
D. Voltage in the circuit to build up to 36.8% of the maximum value.

In an RL circuit, the coil will develop a back EMF that will oppose the flow of current. In the RL circuit, the time required for the current to build up to 63.2% of the maximum value is also called a time constant. **ANSWER C**

$\tau = \frac{L}{R}$ where:
τ = Greek letter tau, the time constant in **seconds**
L = total inductance in **henries**
R = total resistance in **ohms**

3-14B5 After two time constants, the capacitor in an RC circuit is charged to what percentage of the supply voltage?

A. 36.8 %
B. 63.2 %
C. 86.5 %
D. 95 %

To calculate the percentage of charge after two time constants in an RC circuit, write down the percent after a single time constant as 63.2%. The remaining percent of charge is 36.8%. It is found by deducting 63.2% from 100%. In the next time constant the capacitor will charge to 63.2% of the remaining 36.8%. In other words, another amount of charge is added to the capacitor in the second time constant equal to:

$$36.8\% \times 63.2\% = 23.26\% \text{ (rounded to 23.3\%)}$$

This really needs to be calculated in decimal format as:

$$0.368 \times 0.632 = 0.2326$$

Percent values are converted to decimals by moving the decimal point two places to the left. Decimal values are converted to percent values by multiplying by 100 (moving decimal point two places to the right.)

Since the final percent of charge in two time constants is desired, it is only necessary to add the two percent charges together:

	Percent	Decimal
Charge in first time constant =	63.2%	0.632
Charge in second time constant =	23.3%	0.233
Total charge after two time constants	86.5%	0.865

The process can be continued for the third time constant. Since the capacitor has charged to 86.5% in two time constants, if you start from that point, the final value of charge would be another 13.5%. Therefore, in the next time constant, the third, the capacitor would charge another:

$$13.5\% \times 63.2\% = 8.53\% \text{ (rounded to 8.5\%)}$$

After the third time constant, the total charge is:

$$86.5\% + 8.5\% = 95\%$$

If the process is continued through five time constants, the capacitor charge is 99.3%. Therefore, in electronic calculations, the capacitor is considered to be fully charged (or discharged) after five time constants. **ANSWER C**

3-14B6 After two time constants, the capacitor in an RC circuit is discharged to what percentage of the starting voltage?

A. 86.5 %
B. 13.5 %
C. 63.2 %
D. 36.8 %

This question is actually the reverse of the previous question. It is asking to what percent the capacitor has discharged. The capacitor discharges the same percent in a time constant as it charges (for the same RC circuit values, of course.) However, the question is asking for the amount of charge remaining. At one time constant, the capacitor discharges 63.2% (in the previous question, it charged 63.2% in one time constant); therefore, the percent charge remaining on the capacitor is 36.8 % (100%−63.2%). From the previous question, the capacitor charges another 23.3% in the second time constant. Thus, the capacitor will discharge another 23.3% in the second time constant. After two time constants the capacitor has discharged by 86.5%; therefore, the remaining charge is 13.5% (100%-86.5%). In five time constants the capacitor would have discharged 99.3% of its charge, so 0.7% of charge remains – it is considered to be fully discharged. **ANSWER B**

Key Topic 15: RC Time Constants-2

3-15B1 What is the time constant of a circuit having two 220-microfarad capacitors and two 1-megohm resistors all in parallel?

A. 22 seconds.
B. 44 seconds.
C. 440 seconds.
D. 220 seconds.

Here is an RC circuit with all components in parallel. Refer to question 3-15B5. It will be easy if you remember that two equal capacitors in parallel have a total capacitance of twice the value of one of the capacitors, and that two equal value resistors in parallel have a total resistance equal to one half the value of one of the resistors. And more good news – "megs" are being multiplied by "micros" so the powers of ten after the numerical value cancel. Two 220 microfarad capacitors in parallel is a larger 440 microfarad capacitor. Two one-megohm resistors in parallel is a resistance of 0.5 megohms. Solve for the time constant by multiplying 0.5 X 440 ("megs" times "micros") and you end up with 220 seconds as the correct answer. The calculator keystrokes are: Clear, 0.5 × 440 = 220. **ANSWER D**

3-15B2 What is the time constant of a circuit having two 100-microfarad capacitors and two 470-kilohm resistors all in series?

A. 470 seconds.
B. 47 seconds.
C. 4.7 seconds.
D. 0.47 seconds.

For an RC circuit $\tau = RC$, but first we must combine resistor and capacitor values to come up with a single value for each. Since this is a series circuit, the resistor values add ($R_T = R_1 + R_2$) to arrive at a total resistance, but for the capacitance total you have to solve the formula $C_T = (C_1 \text{ X } C_2) \div (C_1 + C_2)$. Therefore,

Powers of 10:

$$C_T = \frac{(1 \times 10^{+2} \times 10^{-6} \times 1 \times 10^{+2} \times 10^{-6})}{(100 \times 10^{-6} + 100 \times 10^{-6})}$$

$$C_T = \frac{(1 \times 10^{-8})}{(2 \times 10^{+2} \times 10^{-6})}$$

$$C_T = \frac{(1 \times 10^{-8})}{2 \times 10^{-4}} = 0.5 \times 10^{-4} = 50 \times 10^{-6}$$

$$C_T = 0.00005 \text{ farads or 50 microfarads}$$

For the resistance:

$$R_T = 470{,}000 + 470{,}000 = 940{,}000$$
$$R_T = 4.7 \times 10^{+5} + 4.7 \times 10^{+5} = 9.4 \times 10^{+5}$$

Now that we have calculated the total series capacitance and the total resistance, we simply multiply 0.00005 X 940,000 = 47 seconds. In powers of ten this is $\tau = 0.5 \text{ X } 10^{-4} \text{ X } 9.4 \text{ X } 10^{+5} = 4.7 \text{ x } 10^{+1} = 47$.

Of course, the old timers that recognize that two equal capacitors in series equals a single capacitor at half the value would immediately jump to τ = 50 microfarads x 940 kilohms, pull out the calculator and press the following: Clear, 0.00005 × 940000 = 47. The time constant is in seconds. Notice if you miss a single decimal point, they have an incorrect answer for you. Watch out! **ANSWER B**

Remember

(For Two Cs)	(For Two Rs)
Capacitors in Series:	Resistors in Series:
$C_T = \frac{C_1 \times C_2}{C_1 + C_2}$	$R_T = R_1 + R_2$
Capacitors in Parallel:	Resistors in Parallel:
$C_T = C_1 + C_2$	$R_T = \frac{R_1 \times R_2}{R_1 + R_2}$
C = Capacitance in **farads**	R = Resistance in **ohms**

Capacitors, Resistors in Series and Parallel

3-15B3 What is the time constant of a circuit having a 100-microfarad capacitor and a 470-kilohm resistor in series?

A. 4700 seconds.
B. 470 seconds.
C. 47 seconds.
D. 0.47 seconds

First of all, this is an RC circuit, so remember the formula:

$\tau = RC$

Time constant (**seconds**) = Resistance (**ohms**) × Capacitance (**farads**)

Remember that "micro" means X 10^{-6} (the multiplication is accomplished by moving the decimal point 6 places to the left), and "kilo" means X 10^{+3} (the multiplication is accomplished by moving the decimal point 3 places to the right). The problem solution is as follows:

$$\tau = 470 \text{ X } 10^{+3} \text{ X } 100 \text{ X } 10^{-6}$$
$$\tau = 4.7 \text{ X } 10^{+2} \text{ X } 10^{+3} \text{ X } 1 \text{ X } 10^{+2} \text{ X } 10^{-6}$$
$$\tau = 4.7 \text{ X } 10^{+1} = 47 \text{ seconds}$$

Although this is fairly easy, you can make it easier by using a calculator. Turn on your calculator, press the clear button (several times for good luck), and then execute the following keystrokes:

470000 × .0001 = and presto, 47 appears on the display.

This is the correct answer. Be sure you understand that 470 kilohms equals 470000 and that 100 microfarads equals 0.0001. **ANSWER C**

3-15B4 What is the time constant of a circuit having a 220-microfarad capacitor and a 1-megohm resistor in parallel?

A. 220 seconds. B. 22 seconds. C. 2.2 seconds. D. 0.22 seconds.

Since it is an RC circuit, use $\tau = RC$.

Using powers of 10:

$$\tau = 1 \times 10^{+6} \times 220 \times 10^{-6} \qquad \tau = 220 \text{ seconds}$$

Using only decimal values:

$$\tau = 1{,}000{,}000 \times .000220 \qquad \tau = 220 \text{ seconds}$$

Remember that when using powers of ten: to multiply, you add the exponents; to divide, you subtract the exponent of the denominator (lower one) from the exponent of the numerator (upper one). Since "meg" = 10^{+6} and "micro" = 10^{-6}, multiplying "megs" times "micros" means you add +6 and -6, and the result is 0 or 10^0, which is 1. In this case, $220 \times 1 = 220$. Actually, this stuff is pretty simple once you get the hang of it. To calculate using decimal values on your calculator, the calculator keystrokes are: Clear, 1000000 x .000220 = 220. **ANSWER A**

3-15B5 What is the time constant of a circuit having two 100-microfarad capacitors and two 470-kilohm resistors all in parallel?

A. 470 seconds. B. 47 seconds. C. 4.7 seconds. D. 0.47 seconds.

Again an RC circuit, only this time everything is in parallel. Since the capacitors are in parallel, the capacitance values add ($C_T = C_1 + C_2$), but for the total resistance you have to solve the formula $R_T = (R_1 \times R_2) \div (R_1 + R_2)$. Therefore,

Powers of 10:

$$R_T = \frac{(4.7 \times 10^{+5} \times 4.7 \times 10^{+5})}{(4.7 \times 10^{+5} + 4.7 \times 10^{+5})}$$

$$R_T = \frac{(22.09 \times 10^{+10})}{9.4 \times 10^{+5}} = 2.35 \times 10^{+5}$$

$$R_T = 235{,}000 \text{ ohms or } 235 \text{ kilohms}$$

for the capacitance:

$$C_T = 0.000100 + 0.000100 = 0.000200$$
$$C_T = 100 \times 10^{-6} + 100 \times 10^{-6} = 200 \times 10^{-6}$$
$$C_T = 200 \text{ microfarads or } 0.0002\ (2 \times 10^{-4}) \text{ farads}$$

Since the total resistance and the total capacitance are known, just multiply $235{,}000 \times .000200$ to end up with 47 seconds. Now that wasn't too hard, was it? This is really easy for old timers. They know that the capacitor values add for parallel capacitors – so they have 200 microfarads. For the resistance, they know that two equal value resistors in parallel result in a single resistor at half the value or 235 kilohms. They would go immediately to $\tau = 200$ microfarads $\times$ 235 kilohms. On their calculator they would press the following: Clear, $.0002 \times 235000 = 47$. The time constant is in seconds, and keep exact track of the decimal points. **ANSWER B**

3-15B6 What is the time constant of a circuit having two 220-microfarad capacitors and two 1-megohm resistors all in series?

A. 220 seconds.
B. 55 seconds.
C. 110 seconds.
D. 440 seconds.

This is another series RC circuit, however, now there are two capacitors in series and two resistors in series. Review the calculations in question 15B2. The procedure is the same. Just remember that two equal capacitors in series have a total capacitance of half of one of the capacitors, and two equal resistors in series have a total resistance that is twice the value of one of the resistors. Now you can do this one in your head. Two 220 microfarad capacitors in series is really one big 110 microfarad capacitor. Write down 0.000110 as you recall that 110 microfarads is 110×10^{-6} — "micro" means move the decimal point six places to the left. Two one-megohm resistors in series would be two megohms. Write down 2,000,000 as you recall that 2 megohms is $2 \times 10^{+6}$ — "meg" means move the decimal point six places to the right. Since $\tau = 2 \times 10^{+6} \times 110 \times 10^{-6}$, you can see by the powers of ten that "megs" times "micros" cancel each other and you end up with $2 \times 110 = 220$ seconds. Anytime you have both "megs" and "micros" multiplied together, all those zeroes to the left or right of the stated numerical value cancel. **ANSWER A**

Intro to Key Topics 16 & 17 Rectangular & Polar Coordinates: The Importance of Impedance

Now that we have reviewed AC waveforms, power factors, phase angles with coils, and time constants with capacitors, let's combine capacitors and coils into AC networks, and study IMPEDANCE.

Impedance is the opposition of circuit elements to the flow of AC current in a circuit with both resistance and reactance, where reactance of either an inductor or a capacitor will create a phase difference between voltage and current. Impedance is a COMPLEX NUMBER, meaning that it is always described by two components, a REAL and an IMAGINARY component. The REAL component is the value of the RESISTANCE and the IMAGINARY component is the value of the REACTANCE.

Understanding how complex impedances work is important to understanding, troubleshooting, and repairing any circuit with alternating current, up to and including microwave frequencies. Antennas of any kind are a primary example of complex impedance at work. Any antenna can be represented by a complex series circuit, consisting of INDUCTANCE, typically in the form of a loading coil, CAPACITANCE, often in the form of a capacitance "hat" or counterpoise, and RADIATION RESISTANCE, the actual "working component" of the antenna. Any of these physical components are vulnerable to damage or destruction in a shipboard installation; knowing how they electrically function will allow you to substitute or improvise electrically equivalent components in emergency conditions. Under more routine conditions, understanding complex impedance is crucial to keeping antennas and other radio equipment tuned to peak performance.

Complex parallel networks are equally prevalent in radio equipment. Parallel tuned stages are used throughout receivers, transmitters, and antenna matching units, and the proper alignment of such circuits has always been a primary task of any radio technician. Again, shipboard installation increases the vulnerability of any of these components by exposing them to much harsher conditions than land-based radios. Knowing how to substitute and improvise such components can be crucial under dire conditions.

Complex impedances can be conveniently calculated by means of the "J Operator." This notation allows rather complex networks to be calculated without the use of trigonometry, or to double check calculations that have been made by trigonometric or other means.

Because Inductive Reactance and Capacitive Reactance are completely COMPLEMENTARY functions, we can always subtract one value from another to arrive at a SINGLE value of reactance, expressed by the letter j. A +j represents a NET inductive reactance, and a -j represents a NET capacitive reactance.

This allows ANY complex impedance to be figured out simply and graphically, by plotting the R (resistance) on a horizontal axis, and the j value (reactance) on a vertical axis. A few simple relationships based on the Right Triangle, primarily the Pythagorean Theorem, will allow you to manipulate complex impedance at will. Phase angle can be directly measured with a protractor, a great relief to anyone allergic to trigonometry.

A special case, called RESONANCE, exists in a complex network when the +j and -j values are equal. In many practical cases, RESONANCE is a highly desirable condition, not only because maximum power is transferred in this state (see POWER FACTOR), but because the problem becomes exceptionally simple to calculate, as the circuit can be replaced with a single resistor. You can simply toss out the reactance as J becomes 0. However, RESONANCE generally only occurs at one frequency, or a few specific frequencies, as in the case of an antenna. Though RESONANCE is indeed a special case, it is a common enough occurrence that it is central to most radio circuits.

Only two questions will be on your exam from these Impedance Networks topics (one from each of the two Key Topics), and working the math may spur your interest in designing radio circuits with coils and capacitors!

Key Topic 16: Impedance Networks-1

3-16B1 What is the impedance of a network composed of a 0.1-microhenry inductor in series with a 20-ohm resistor, at 30 MHz? Specify your answer in rectangular coordinates.

A. 20 −j19 C. 20 +j19
B. 19 +j20 D. 19 -j20

Don't panic, I'm not going to send you back to high school for algebra, trigonometry, and calculus to solve these problems. Here we have a circuit containing both reactance and resistance, giving us impedance, represented by the letter Z. Impedance has both a resistive (R) and a reactive part (X_C or X_L). The parts are represented by vectors that are at right angles (perpendicular) to each other. As a result, $Z = \sqrt{R^2 + X^2}$. Each part forms a leg of a right triangle and contributes to the magnitude of the impedance, which is the hypotenuse of the triangle. You determine the magnitude of the impedance by calculating each part, R and X, and then Z. Let's first calculate the inductive reactance X_L of the circuit in this question with the formula for inductive reactance:

$X_L = 2\pi fL$ Where: L is inductance in **henries**
f is frequency in **hertz**
$\pi = 3.14$

At 30 MHz,

$$X_L = 2 \times 3.14 \times 30 \times 10^{+6} \times 0.1 \times 10^{-6}$$
$$X_L = 18.84 \text{ rounded to 19 ohms}$$

The resistance, R is given as 20 ohms

Therefore,

$$Z = \sqrt{20^2 + 19^2} = \sqrt{400 + 361} = \sqrt{761}$$
$$Z = 27.59 \text{ rounded to 27.6 ohms}$$

But the examination question reads, "Specify your answer in rectangular coordinates." This is good news because it allows us to simplify the problem solving – even to the point of doing them in our heads! Rectangular coordinates consist of the two parts of the impedance discussed above – the first part is the resistance, and the second part is the reactance. The reactance is preceded by a (+j) or a (−j) which indicates whether the circuit's reactance is inductive (+j) or capacitive (−j). The resistance is placed first followed by the reactance; therefore, the impedance of an inductive circuit is $Z = R + jX_L$, and for a capacitive circuit it is $Z = R - jX_C$.

This first problem has an inductor in series with a 20-ohm resistor. Since it's an inductor, its reactance is inductive and a +j precedes it. Therefore, for this problem, the impedance in rectangular coordinates is 20 +j19. That's not so hard, right? **ANSWER C**

3-16B2 In rectangular coordinates, what is the impedance of a network composed of a 0.1-microhenry inductor in series with a 30-ohm resistor, at 5 MHz?

A. 30 −j3 C. 3 −j30
B. 3 +j30 D. 30 +j3

Here is another inductive circuit; therefore, in rectangular coordinates, the reactance is inductive and will have a +j in front of it. Thus, two of the possible answers are already eliminated (A and C because they have a -j reactance). Normally, you would calculate the inductance from $X_L = 2\pi fL$ and place a +j in front of it to determine the correct answer, but in this case, it's much easier. The question gives the resistance as 30 ohms, so you know the answer is 30 +j something. Answer D is the only one that has "30 +j". It is the correct answer. **ANSWER D**

Rectangular Coordinates

A quantity can be represented by points that are measured on an X-Y plane with an X and Y axis perpendicular to each other. The point where the axes cross is the origin, the points on the X and Y axis are called coordinates, and the magnitude is represented by the length of a vector from the origin to the unique point defined by the *rectangular coordinates.*

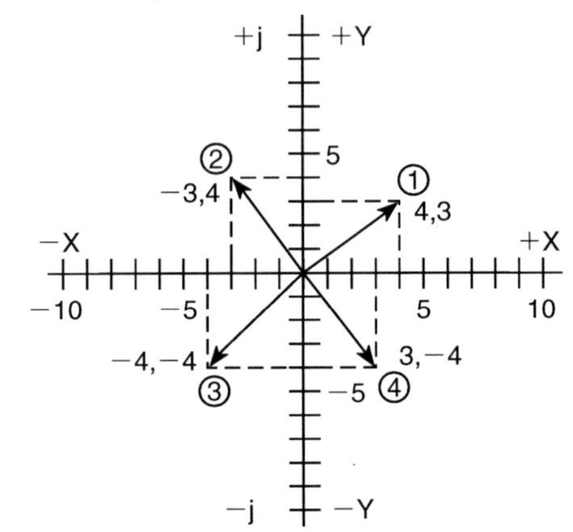

To be able to identify ac quantities and their phase angle, the X axis is called the real axis and the Y axis is called the imaginary axis. There are positive and negative real axis coordinates and positive and negative imaginary axis coordinates. The imaginary axis coordinates have a **j** operator to identify that they are imaginary coordinates.

In rectangular coordinates:
Vector 1 is 4 + j3
Vector 2 is −3 + j4
Vector 3 is −4 − j4
Vector 4 is 3 − j4

3-16B3 In rectangular coordinates, what is the impedance of a network composed of a 10-microhenry inductor in series with a 40-ohm resistor, at 500 MHz?

A. 40 +j31400
B. 40 −j31400
C. 31400 +j40
D. 31400 −j40

This will also be a +j answer because the network is inductive. The resistance is given as 40 ohms, so you already have the resistor value for the rectangular coordinates. The answer will be of the form "40 +j", so look for the answer that begins "40 +j". Only correct answer A has that form. Look, you've picked the correct answer without calculating a thing! You could double-check your work by computing the inductive reactance in ohms, $X_L = 2\pi fL$, to agree with the j operator. Remember, f = 500 MHz. $X_L = 2 \times 3.14 \times 500 \text{ X } 10^{+6} \times 10 \times 10^{-6}$. **ANSWER A**

3-16B4 In rectangular coordinates, what is the impedance of a network composed of a 1.0-millihenry inductor in series with a 200-ohm resistor, at 30 kHz?

A. 200 - j188
B. 200 + j188
C. 188 + j200
D. 188 - j200

Another easy one. This is an inductive circuit with a resistance of 200 ohms; therefore, in rectangular coordinates the answer is 200 +j. Answer B is the only 200 +j answer. Calculate inductive reactance to check your work. $X_L = 2\pi fL$ (where f = 30 kHz) $= 2 \times 3.14 \times 30 \times 10^3 \times 1 \times 10^{-3} = 188.4$. **ANSWER B**

3-16B5 In rectangular coordinates, what is the impedance of a network composed of a 0.01-microfarad capacitor in parallel with a 300-ohm resistor, at 50 kHz?

A. 150 − j159
B. 150 + j159
C. 159 − j150
D. 159 + j150

Here is another one you will need to drag out the calculator on. Come on, isn't this fun? First find capacitive reactance. Refer to question 3-17B1, if the capacitance is in picofarads and the frequency is in megahertz, then

$$X_C = \frac{10^{+6}}{2\pi fC}$$

50 kHz is 0.05 MHz and 0.01 microfarads is 10,000 picofarads, so you will divide 6.28 X .05 X 10,000 into 1 million, and end up with 318.47 ohms.

Now solve for impedance:

$$Z = \frac{Z_1 \times Z_2}{Z_1 + Z_2} \quad \text{or use Special Case: } Z = \frac{RX_C \angle -90^\circ}{\sqrt{R^2 + X_C^2} \angle \arctan \frac{X_C}{R}}$$

$$Z = \frac{300 \angle 0^\circ \times 318.5 \angle -90^\circ}{(300 + j0) + (0 - j318.5)}$$

$$Z = \frac{95{,}550 \angle -90^\circ}{\sqrt{300^2 + 318.5^2} \angle \arctan 318.5/300} = \frac{95{,}550 \angle -90^\circ}{438 \angle -46.5^\circ}$$

$$Z = 218.3 \angle -90^\circ + 46.5^\circ = 218.3 \angle -43.5^\circ \text{ ohms}$$

Your answer comes out 218 ohms. The answer needs to be in rectangular coordinates. To convert to rectangular coordinates, you will be finding the resistance side and the reactance side of a right triangle with a hypotenuse of 218.3 and a contained angle $\Theta = -43.5^\circ$. The parts are:

$$R = 218.3 \cos \Theta = 218.3 \times 0.7253 = 158.3 \text{ (answer calls it 159)}$$

$$X_C = 218.3 \sin \Theta = 218.3 \times 0.6883 = 150.2 \text{ (answer calls it 150)}$$

Therefore, in rectangular coordinates:

$$Z = 158.3 - j150.2$$

Match this to the correct answer C of Z = 159 −j150. **ANSWER C.**

3-16B6 In rectangular coordinates, what is the impedance of a network composed of a 0.001-microfarad capacitor in series with a 400-ohm resistor, at 500 kHz?

A. 318 − j400
B. 400 + j318
C. 318 + j400
D. 400 − j318

Easy one here – do it in your head. It's a 400 ohm resistor in series with a capacitor. You know that gives you a −j, thus, answer D is the correct choice because none of others have "400 −j". If you want to check X_C use the formula in question 3-17B1. **ANSWER D**

Key Topic 17: Impedance Networks-2

3-17B1 What is the impedance of a network composed of a 100-picofarad capacitor in parallel with a 4000-ohm resistor, at 500 KHz? Specify your answer in polar coordinates.

A. 2490 ohms, $\angle 51.5$ degrees
B. 4000 ohms, $\angle 38.5$ degrees
C. 5112 ohms, $\angle -38.5$ degrees
D. 2490 ohms, $\angle -51.5$ degrees

Shift gears now and be alert because this test question requires your answer in polar coordinates. Refer to the figure in the box. Disconnect yourself from the thought process used for rectangular coordinates in the previous questions. Because the resistance is given as 4000, *DO NOT* look at the 4000-ohm resistor as your best choice for the correct answer. Polar coordinate values are calculated differently. In polar coordinates, the ac quantity – let's say impedance Z in this case – is described as a vector at a particular phase angle. You solve for the vector magnitude Z separately and for the phase angle separately. The vector Z (which is the hypotenuse of a right triangle) is defined by a resistance vector (R) and a reactance vector (X) at right angles to each other. If the reactance vector is inductive, it is a $+X_L$; if it is capacitive the reactance is $-X_C$. Positive inductive reactances produce positive angles and negative capacitive reactances produce negative angles. Let's begin the necessary calculations by finding the value of the capacitive reactance with the formula:

$$X_C = \frac{1}{2\pi fC}$$

Where: C is capacitance in **farads**
f is frequency in **hertz**
$\pi = 3.14$

$$X_C = \frac{1}{2 \times 3.14 \times 0.5 \times 10^{+6} \times 100 \times 10^{-12}}$$

Clearing the powers of ten:

$$X_C = \frac{10^{+6}}{6.28 \times 0.5 \times 100}$$

This equation can be used directly when:
f is frequency in **MHz**
C is in **picofarads**

$$X_C = \frac{1{,}000{,}000}{314}$$

$$X_C = 3184.7 \text{ ohms}$$

The resistance is already given at 4000 ohms, so R= 4000.

You now need to find the magnitude of the impedance, Z, in ohms, of the parallel circuit. This must be done with vector (phasor) algebra, as follows:

$$Z = \frac{Z_1 \times Z_2}{Z_1 + Z_2}$$

$$Z = \frac{R\angle 0^\circ \times X_C \angle -90^\circ}{(R + j0) + (0 - jX_C)}$$

Special Case:
When $Z_1 = R\angle 0^\circ$ and $Z_2 = X_C\angle -90^\circ$

$$Z = \frac{RX_C \angle -90^\circ}{\sqrt{R^2 + X_C^2}\ \angle \arctan \frac{-X_C}{R}}$$

$$Z = \frac{4000\angle 0^\circ \times 3185\angle -90^\circ}{(4000 - j3185)} = \frac{4000 \times 3185 \angle -90^\circ}{\sqrt{4000^2 + 3185^2}\ \angle \arctan -3185/4000}$$

$$Z = \frac{12.74 \times 10^{+6} \angle -90^\circ}{\sqrt{16 \times 10^{+6} + 10.144 \times 10^{+6}}\ \angle \arctan -0.7963}$$

$$Z = \frac{12.74 \times 10^{+6} \angle -90^\circ}{5.113 \times 10^3 \angle -38.5^\circ}$$

$$Z = 2.493 \times 10^3 \angle -90^\circ + 38.5^\circ = 2493 \angle -51.5^\circ \text{ ohms}$$

You will need a scientific calculator to look up the tangent of the phase angles. The solution is not easy. You must keep your wits about you because you cannot add and subtract vectors directly. You must deal with the magnitude and the angle in polar coordinates and the real (resistance) and imaginary (reactance) parts in rectangular coordinates. You can further double-check your answer by recognizing that there must be a minus sign in front of the phase angle because the circuit has capacitive reactance. **ANSWER D**

Polar Coordinates

A quantity can be represented by a vector starting at an origin, the length of which is the magnitude of the quantity, and rotated from a zero axis through 360.

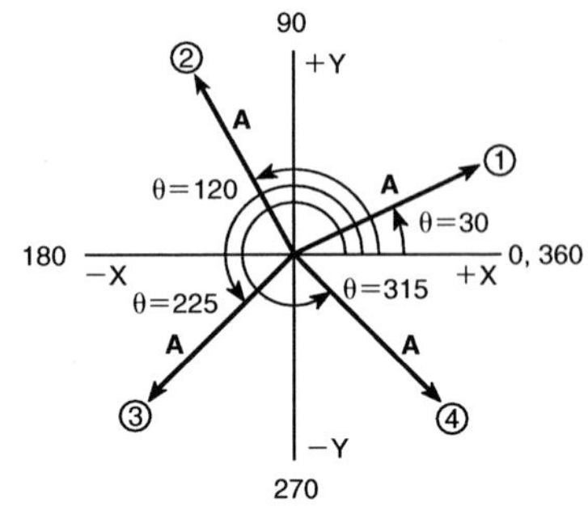

The position of the Vector A can be defined by the angle through which it is rotated. When we do so we are defining the ac quantity in *polar coordinates*.
In polar coordinates:
Vector 1 is $A\angle 30$
Vector 2 is $A\angle 120$
Vector 3 is $A\angle 225$
Vector 4 is $A\angle 315$

3-17B2 In polar coordinates, what is the impedance of a network composed of a 100-ohm-reactance inductor in series with a 100-ohm resistor?

A. 121 ohms, ∠35 degrees
B. 141 ohms, ∠45 degrees
C. 161 ohms, ∠55 degrees
D. 181 ohms, ∠65 degrees

This one is relatively easy – impedance equals the square root of the sum of the resistance squared and the inductive reactance squared. Z in polar coordinates is:

$$Z = \sqrt{100^2 + 100^2}\ \angle \arctan 100/100$$

$$Z = \sqrt{20000}\ \angle \arctan 1$$

$$Z = 141.4\ \angle +45^\circ$$

You might recognize the square root easier if you work in powers of 10. The magnitude of $Z = \sqrt{(1 \text{ X } 10^2)^2 + (1 \text{ X } 10^2)^2} = \sqrt{2 \text{ X } 10^4} = \sqrt{2} \times 10^2 = 1.414 \times 10^2$. You may recognize the $\sqrt{2}$ as 1.414, so the answer is 141.4. From just the magnitude, you could have selected the correct answer, or you might have used the phase angle. A right triangle with equal R and X has a phase angle of 45° because, as you probably recall, the tangent of 45° is 1. **ANSWER B**

3-17B3 In polar coordinates, what is the impedance of a network composed of a 400-ohm-reactance capacitor in series with a 300-ohm resistor?

A. 240 ohms, ∠36.9 degrees
B. 240 ohms, ∠−36.9 degrees
C. 500 ohms, ∠−53.1 degrees
D. 500 ohms, ∠53.1 degrees

Here is a series problem which is easily solved by first calculating the magnitude of the impedance, which equals the square root of resistance squared plus reactance squared. The square root of 300 squared plus 400 squared is the square root of 250,000 or 500 ohms. Since the circuit is capacitive, the phase angle will have a minus sign in front of it. If you remember any trigonometry, you know that a right triangle with a 3-4-5 relationship of the sides is an easy one to determine the side values. With sides of 300 and 400, the hypotenuse will be 500, with a phase angle whose tangent is 3/4 or 4/3 depending on the relationship of the sides. In this case it's 4/3 and the angle is −53.1 degrees. **ANSWER C**

3-17B4 In polar coordinates, what is the impedance of a network composed of a 300-ohm-reactance capacitor, a 600-ohm-reactance inductor, and a 400-ohm resistor, all connected in series?

A. 500 ohms, ∠37 degrees
B. 400 ohms, ∠27 degrees
C. 300 ohms, ∠17 degrees
D. 200 ohms, ∠10 degrees

Here's one that even though the answer is requested in polar coordinates, it is best to start out using rectangular coordinates. $Z = (400 + j600 - j300) = (400 + j300)$. Now you can use the 3-4-5 ratio we talked about in the previous question. With the resistance side of a right triangle at 400 ohms and the reactance side at 300 ohms, you know that the hypotenuse magnitude of the impedance is Z = 500. The angle is a positive one with an arctan of (an angle whose tangent is) 300/400, or 0.75. Grab a trig table or scientific calculator for an angle of 36.8° rounded to 37°. **ANSWER A**

3-17B5 In polar coordinates, what is the impedance of a network comprised of a 400-ohm-reactance inductor in parallel with a 300-ohm resistor?

A. 240 ohms, ∠−36.9 degrees
B. 240 ohms, ∠36.9 degrees
C. 500 ohms, ∠53.1 degrees
D. 500 ohms, ∠−53.1 degrees

We know right off the bat that the answer is going to be either B or C because an inductive reactance will have a positive polar coordinate angle. To come up with 240 ohms at an angle of +36.9° as the correct impedance, let's start out by defining the two impedances that are in parallel:

$$Z_1 = 300\ \angle 0^\circ \text{ since it is a resistance in } \textbf{ohms}$$

$$Z_2 = 400\ \angle +90^\circ \text{ since it is an inductive reactance in } \textbf{ohms}$$

$$Z = \frac{Z_1 \times Z_2}{Z_1 + Z_2} \text{ or Special Case: } Z = \frac{RX_L\ \angle 190^\circ}{\sqrt{R^2 + X_L^2}\ \angle \arctan \frac{X}{R}}$$

Additions and subtractions are easier in rectangular coordinates, and multiplication and division are easier in polar coordinates.

$$Z = \frac{300 \times 400\ \angle 0^\circ + 90^\circ}{(300 + j0) + (0 + j400)} = \frac{120{,}000\ \angle +90^\circ}{500\ \angle \arctan 400/300}$$

$$Z = 240\ \angle +90^\circ - 53.1^\circ$$

$$Z = 240\ \angle 36.9^\circ \text{ ohms}$$

To solve for answer B, we used the 3-4-5 ratio, and both rectangular and polar coordinates. **ANSWER B**

3-17B6 Using the polar coordinate system, what visual representation would you get of a voltage in a sinewave circuit?

A. To show the reactance which is present.
B. To graphically represent the AC and DC component.
C. To display the data on an XY chart.
D. The plot shows the magnitude and phase angle.

The polar coordinate representation of voltage will be a plot showing the magnitude of the voltage and phase angle. **ANSWER D**

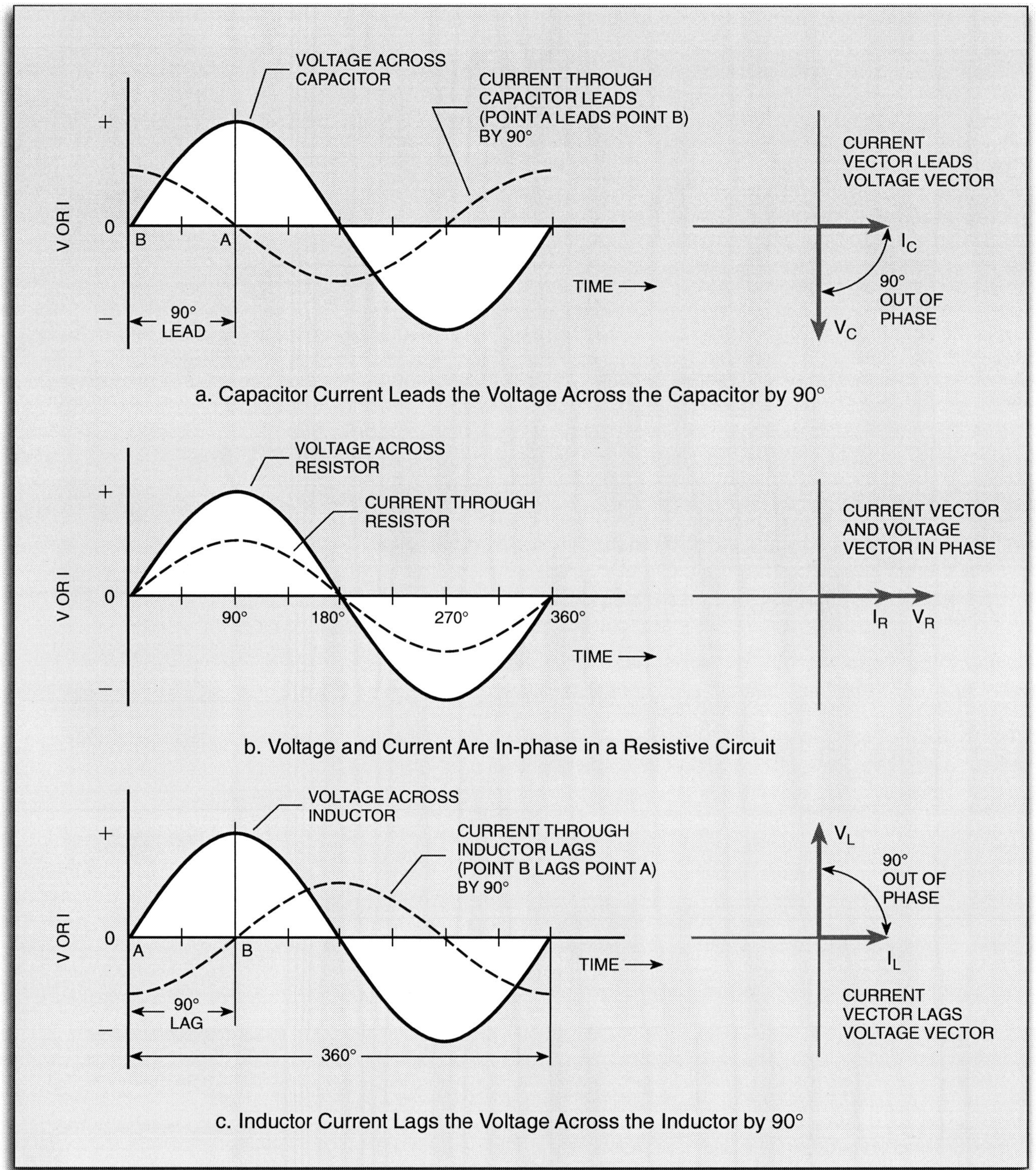

Phase shift – The difference in the angles of the vectors representing voltage and current in a circuit containing capacitance, resistance or inductance.

Key Topic 18: Calculations

3-18B1 What is the magnitude of the impedance of a series AC circuit having a resistance of 6 ohms, an inductive reactance of 17 ohms, and zero capacitive reactance?

A. 6.6 ohms.
B. 11 ohms.
C. 18 ohms.
D. 23 ohms.

The formula for determining impedance is the square root of $R^2 + X^2$. $X = X_L$ when there is zero capacitive reactance. $6^2 = 36$ and $17^2 = 289$. The sum of the two equals 325. The square root of 325 is approximately 18 and the correct answer. **ANSWER C**

3-18B2 A 1-watt, 10-volt Zener diode with the following characteristics: $I_{min.} = 5$ mA; $I_{max.} = 95$ mA; and Z = 8 ohms, is to be used as part of a voltage regulator in a 20-V power supply. Approximately what size current-limiting resistor would be used to set its bias to the midpoint of its operating range?

A. 100 ohms.
B. 200 ohms.
C. 1 kilohms.
D. 2 kilohms.

This is a classic Ohm's Law problem. The supply voltage is 20V. Also assume that the midpoint current is 50 mA. This is calculated by (95 mA - 5 mA) ÷ 2 = 45 + 5 = 50 mA. Now convert mA to amps, and apply Ohm's Law to get the answer as shown: 10 Volts ÷ 0.05 Amps = 200 Ohms. **ANSWER B**

3-18B3 Given a power supply with a no load voltage of 12 volts and a full load voltage of 10 volts, what is the percentage of voltage regulation?

A. 17 %
B. 80 %
C. 20 %
D. 83 %

E_{nl} = no-load voltage; E_{fl} = full-load voltage. The formula to calculate the percentage of voltage regulation is: $[(E_{nl} - E_{fl}) \times 100)] \div E_{fl}$ = % regulation. To work this out, multiply 100 times the difference of the voltages, and divide the answer by the full-load voltage. For our example, this would be $[(12 - 10) \times 100)] \div 10 = 20\%$ regulation. **ANSWER C**

3-18B4 What turns ratio does a transformer need in order to match a source impedance of 500 ohms to a load of 10 ohms?

A. 7.1 to 1.
B. 14.2 to 1.
C. 50 to 1.
D. None of these.

The ratio of turns on a primary to the number of turns on a secondary is equal to the square root of the ratio of the primary impedance to the secondary impedance. Divide 10 ohms into 500 ohms and you get a ratio of 50:1. You can approximate the square root of 50 in your head (7 × 7 = 49, so the square root of 50 is slightly more than 7). Watch out for answer C; be sure to calculate the square root of 50 before you select an answer. **ANSWER A**

Here's proof of the formula used: The power in the primary is equal to the power in the secondary; therefore, $P_P = P_S$

and $$\frac{(V_P)^2}{Z_P} = \frac{(V_S)^2}{Z_S} \text{ and } \left(\frac{V_P}{V_S}\right)^2 = \frac{Z_P}{Z_S}$$

Now $$\frac{V_P}{V_S} = \frac{N_P}{N_S}, \text{ therefore, } \left(\frac{N_P}{N_S}\right)^2 = \frac{Z_P}{Z_S} \text{ and } \frac{N_P}{N_S} = \sqrt{\frac{Z_P}{Z_S}}$$

$$\frac{N_P}{N_S} = \sqrt{\frac{Z_P}{Z_S}}$$

where: N_P = Primary **turns**
N_S = Secondary **turns**
Z_P = Primary Z in **ohms**
Z_S = Secondary Z in **ohms**

3-18B5 Given a power supply with a full load voltage of 200 volts and a regulation of 25%, what is the no load voltage?

A. 150 volts.
B. 160 volts.
C. 240 volts.
D. 250 volts.

Here you must rearrange the formula to compute the no-load voltage. The following 4 steps show how the formula is modified:

$$\frac{(E_{nl} - E_{fl})100}{E_{fl}} = \%\ \text{regulation}$$

$$(E_{nl} - E_{fl})100 = \%(E_{fl})$$

$$(E_{nl} - E_{fl}) = \%\frac{E_{fl}}{100}$$

$$E_{nl} = \%\frac{E_{fl}}{100} + E_{fl}$$

Where:
E_{nl} = No-load voltage in **volts**
E_{fl} = Full-load voltage in **volts**
% = Regulation in **percent**

Now, plug in the numbers to arrive at the answer...250 volts.

$$E_{nl} = \%\frac{25(200)}{100} + E_{fl}$$

$$= \frac{5000}{100} = 50 + 200 = 250 \text{ volts.}$$ **ANSWER D**

3-18B6 What is the conductance (G) of a circuit if 6 amperes of current flows when 12 volts DC is applied?

A. 0.25 Siemens (mhos).
B. 0.50 Siemens (mhos).
C. 1.00 Siemens (mhos).
D. 1.25 Siemens (mhos).

The formula for determining conductance in mhos is $G = 1/R = I \div E$. So, 6 amps ÷ 12 volts = 0.50 Siemens. The Siemen and the mho are units for the same amount of conductance (1 mho = 1 Siemen). **ANSWER B**

Subelement C – Components

10 Key Topics, 10 Exam Questions, 2 Drawings

Key Topic 19: Photoconductive Devices

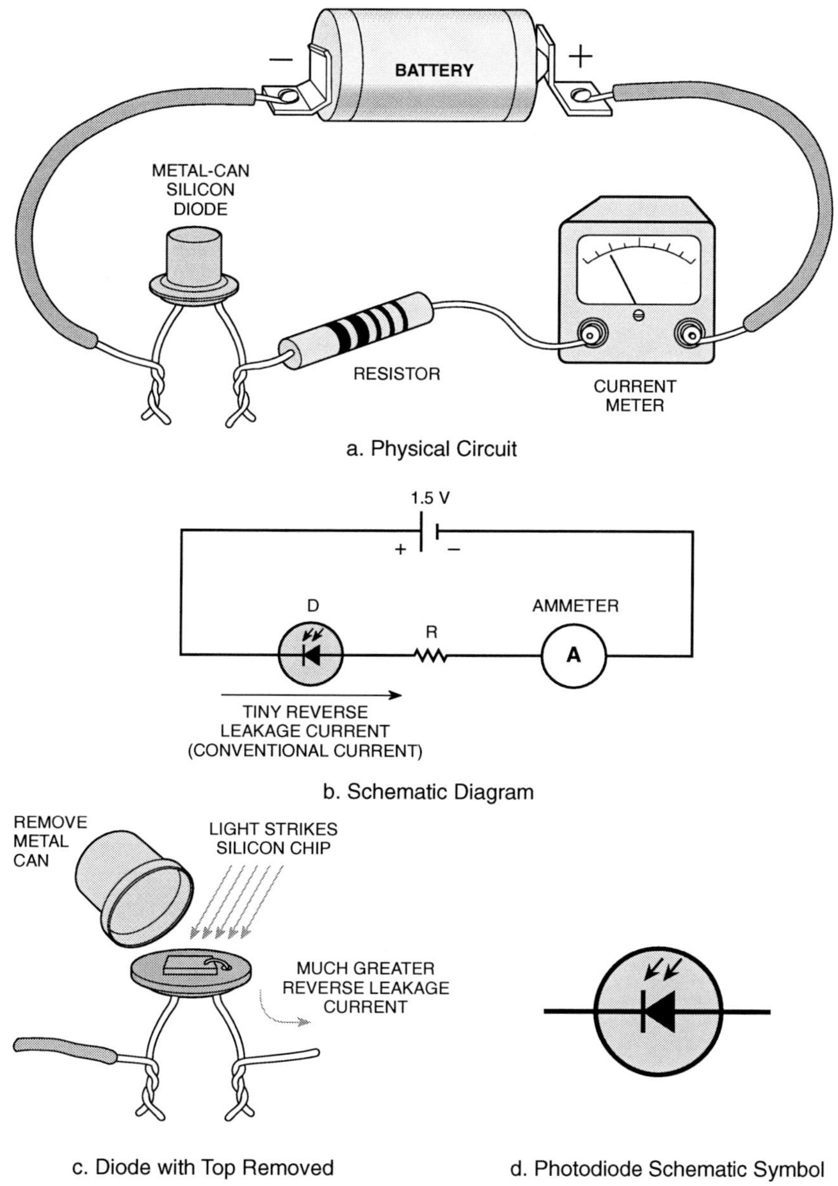

A photodiode is like an ordinary p-n junction diode with a window to admit light. The light greatly increases reverse leakage current.

3-19C1 What happens to the conductivity of photoconductive material when light shines on it?

A. It increases.
B. It decreases.
C. It stays the same.
D. It becomes temperature dependent.

In the dark, a photo cell has high resistance. When you shine light on it, the conductivity increases as the resistance decreases. **ANSWER A**

3-19C2 What is the photoconductive effect?

A. The conversion of photon energy to electromotive energy.
B. The increased conductivity of an illuminated semiconductor junction.
C. The conversion of electromotive energy to photon energy.
D. The decreased conductivity of an illuminated semiconductor junction.

The photo cell has high resistance when no light shines on it, and a varying resistance when light shines on it. It can be used in an ON or OFF state as a simple beam alarm across a doorway, or may be found in almost all modern SLR cameras as a sensitive light meter to judge the amount of light present. **ANSWER B**

3-19C3 What does the photoconductive effect in crystalline solids produce a noticeable change in?

A. The capacitance of the solid.
B. The inductance of the solid.
C. The specific gravity of the solid.
D. The resistance of the solid.

The resistance of the crystalline solid varies when light shines on it because of the photoconductive effect.
ANSWER D

Photo diode within the silver can

3-19C4 What is the description of an optoisolator?

A. An LED and a photosensitive device.
B. A P-N junction that develops an excess positive charge when exposed to light.
C. An LED and a capacitor.
D. An LED and a lithium battery cell.

In an optoisolator, an LED shines through a window at a photosensitive device. This transfers signals from the LED circuit to the photosensitive device without any direct electrical connection between the two. Such an arrangement may also be called an optocoupler. **ANSWER A**

3-19C5 What happens to the conductivity of a photosensitive semiconductor junction when it is illuminated?

A. The junction resistance is unchanged.
B. The junction resistance decreases.
C. The junction resistance becomes temperature dependent.
D. The junction resistance increases

Conductivity increases when light shines on a semiconductor junction because the junction resistance decreases, just like the photo cell. More light; less resistance. **ANSWER B**

3-19C6 What is the description of an optocoupler?

A. A resistor and a capacitor.
B. Two light sources modulated onto a mirrored surface.
C. An LED and a photosensitive device.
D. An amplitude modulated beam encoder.

Optocouplers are found in many modern marine and aviation transceivers. When you tune the dial for a new channel, you are no longer tuning a big variable capacitor. Rather, you are interrupting a light beam on a photosensitive device. **ANSWER C**

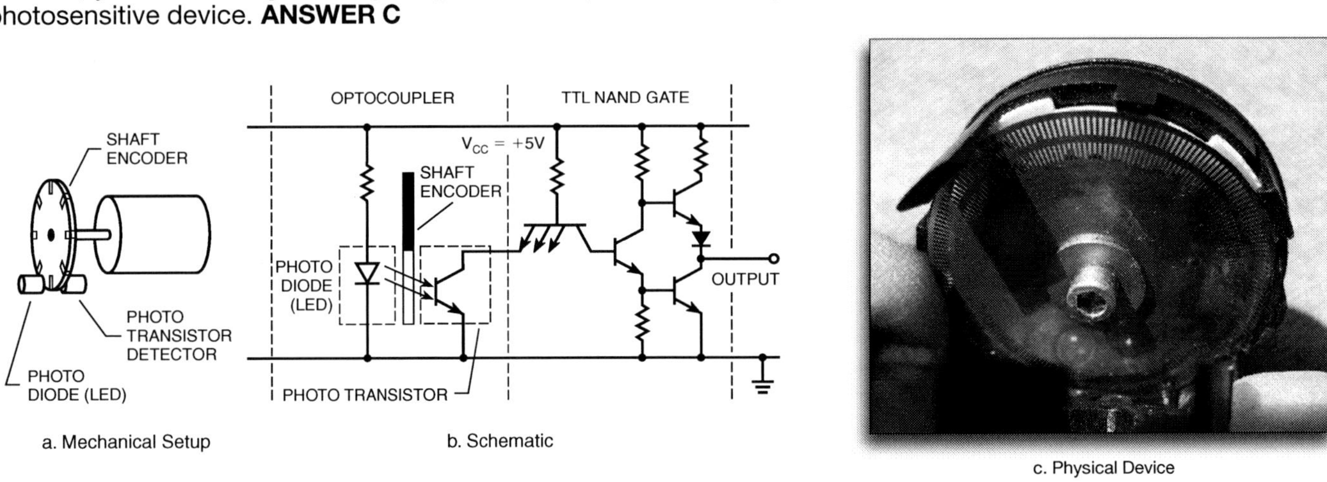

Optocoupler used for shaft encoder.

Key Topic 20: Capacitors

3-20C1 What factors determine the capacitance of a capacitor?
A. Voltage on the plates and distance between the plates.
B. Voltage on the plates and the dielectric constant of the material between the plates.
C. Amount of charge on the plates and the dielectric constant of the material between the plates.
D. Distance between the plates and the dielectric constant of the material between the plates.

Some of the SSB equipment you may work on will still use variable capacitors. The amount of capacitance is determined by the area of the plates as you turn the variable capacitor, the distance between the plates, and the dielectric constant of the material between the plates, usually air in variable capacitors. As you engage more of the plates next to each other, capacitance increases. **ANSWER D**

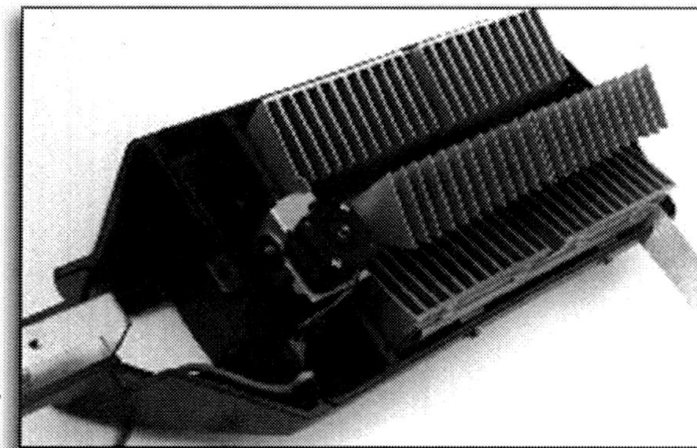

Variable Capacitor
Courtesy of AEA, Inc.

3-20C2 In Figure 3C4, if a small variable capacitor were installed in place of the dashed line, it would?
A. Increase gain.
B. Increase parasitic oscillations.
C. Decrease parasitic oscillations.
D. Decrease crosstalk.

Figure 3C4 is a three-stage intermediate power RF amplifier, sometimes called an IPA. Its purpose is to increase the weak signal from an RF oscillator to a power level sufficient to drive the final power amplifier (PA). The first transistor is in a common base configuration (sometimes called grounded base); the second transistor is a neutralized common emitter configuration; and the third transistor is a common collector configuration, also known as an emitter follower. A small variable capacitor placed where the dotted line is would supply neutralization, which stabilizes the amplifier, decreasing parasitic oscillations. It does this by supplying a small amount of out-of-phase signal from the bottom end of the tank circuit (formed by tank coil (4) and tank capacitor (3), back to the base of the transistor (1). (Note the PLUS signs at the inductor terminals. These show the relative phase of the RF signals.) RF bypass capacitor (2) keeps the midpoint of tank coil (4) at RF signal ground potential, while still allowing collector voltage Vcc to reach the transistor unaffected. The third transistor, in the emitter follower configuration has a low output impedance and no voltage gain. However, the power gain can be considerable. **ANSWER C**

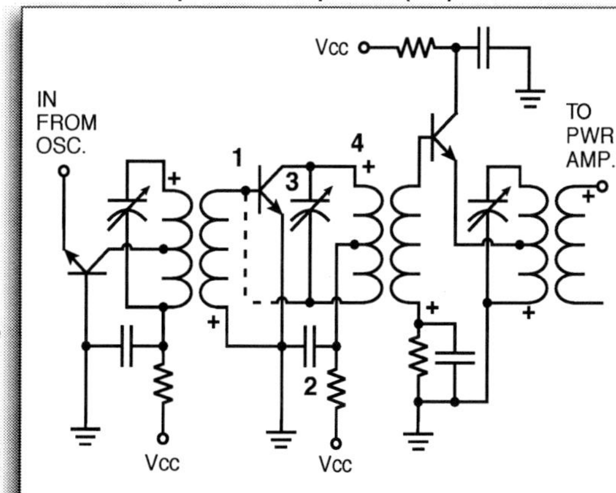

Fig. 3C4

3-20C3 In Figure 3C4, which component (labeled 1 through 4) is used to provide a signal ground?
A. 1
B. 2
C. 3
D. 4

In any amplifier we may speak of signal ground and DC ground, which may be two very different things, as they are in Figure 3C4. The way we supply a DC (non-ground) voltage to a device such as a transistor, yet maintain an RF or signal ground at that point is through a decoupling network usually consisting of a simple series resistor and a bypass capacitor. The reactance of the bypass capacitor at the frequency of interest must be low enough that it appears to be essentially a short circuit. There should be effectively no AC voltage at that point. Sometimes this is easier said than done as real world capacitors may exhibit parasitic inductance, especially at very high frequencies. Sometimes at VHF and UHF frequencies, the lead length of the bypass capacitors must be trimmed to accommodate this parasitic inductance. In low frequency circuits, often there will be several different values of bypass capacitors in parallel to "tame" the power supply lines properly. A great many ills of amplifiers of all frequencies and types can be traced to defective bypass capacitors. An experienced troubleshooter learns to look at all bypass capacitors with suspicion. **ANSWER B**

3-20C4 In Figure 3C5, which capacitor (labeled 1 through 4) is being used as a bypass capacitor?

A. 1
B. 2
C. 3
D. 4

In figure 3C5, we can spot the 0.1 µf capacitor across the 100 Ohm resistor as a capacitor to ground, a bypass capacitor. **ANSWER C**

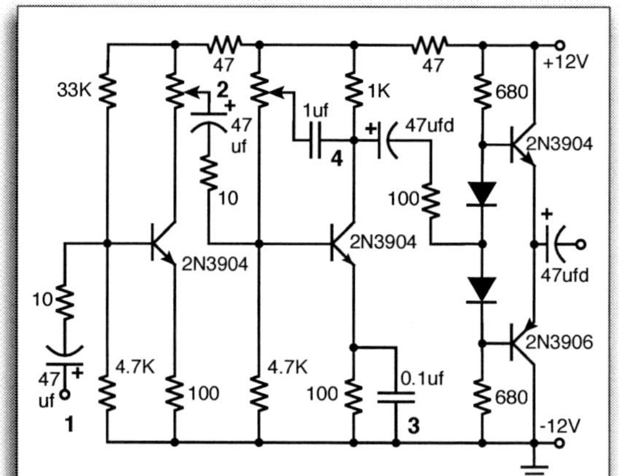

Fig. 3C5

3-20C5 In Figure 3C5, the 1 µF capacitor is connected to a potentiometer that is used to:

A. Increase gain.
B. Neutralize amplifier.
C. Couple.
D. Adjust tone.

You can spot the 1 µf capacitor, connected to the potentiometer, at 4, with the potentiometer adjusting the amplifier tone. **ANSWER D**

3-20C6 What is the purpose of a coupling capacitor?

A. It blocks direct current and passes alternating current.
B. It blocks alternating current and passes direct current.
C. It increases the resonant frequency of the circuit.
D. It decreases the resonant frequency of the circuit.

Coupling capacitors block direct current and pass the AC signal. Put an ohmmeter across a capacitor and you will read an open circuit (after the capacitor charges), yet holding onto one lead of a capacitor and putting a scope on the other lead produces a signal on the scope. **ANSWER A**

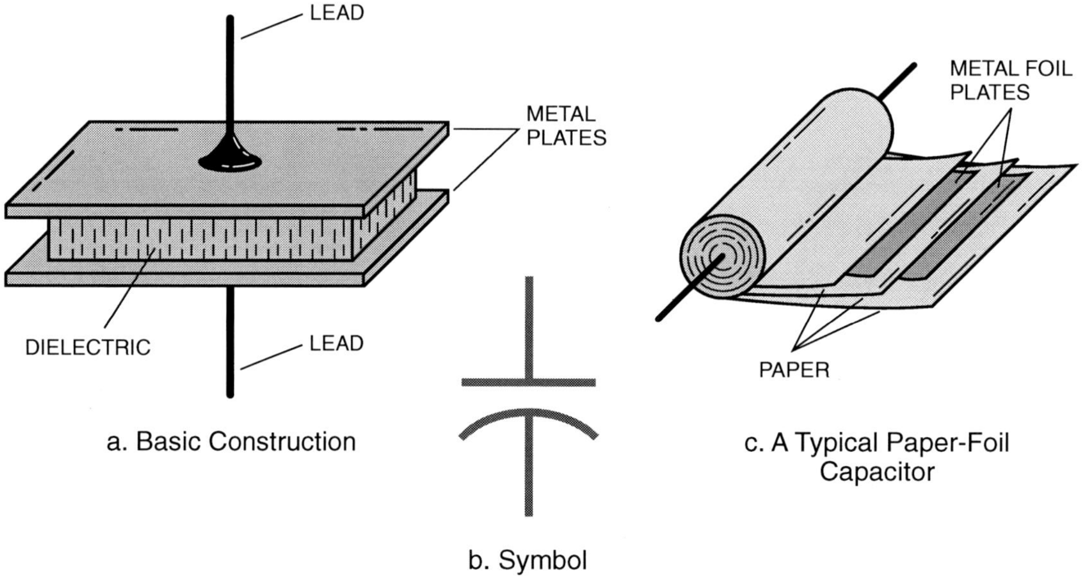

Typical construction and schematic symbol for capacitors.

Key Topic 21: Transformers

3-21C1 A capacitor is sometimes placed in series with the primary of a power transformer to:

A. Improve the power factor.
B. Improve output voltage regulation.
C. Rectify the primary windings.
D. None of these.

The ratio of true-to-apparent power in a circuit is known as the power factor. Power factor is a comparison of the power that is apparently being used in a circuit to the power the circuit is actually using. A capacitor placed in series with the primary of a power transformer improves the power factor. **ANSWER A**

3-21C2 A transformer used to step up its input voltage must have:

A. More turns of wire on its primary than on its secondary.
B. More turns of wire on its secondary than on its primary.
C. Equal number of primary and secondary turns of wire.
D. None of the above statements are correct.

Transformers that step up voltage have more turns on the secondary winding than on the primary. Isolation transformers have the same number of turns on both windings, and step-down transformers have fewer turns on the secondary winding. **ANSWER B**

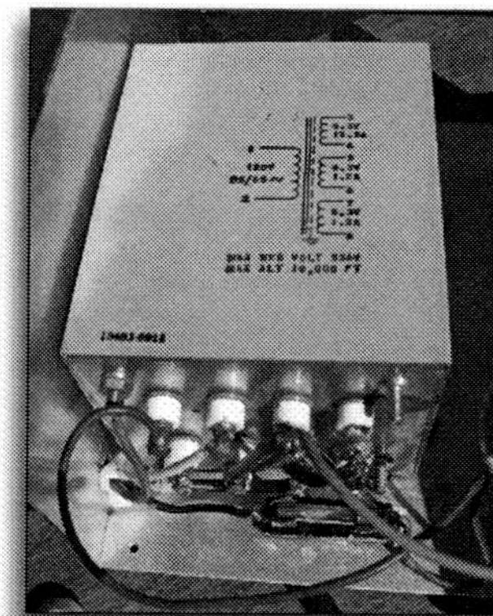

Transformer, from a RADAR.

3-21C3 A transformer primary of 2250 turns connected to 120 VAC will develop what voltage across a 500-turn secondary?

A. 26.7 volts.
B. 2300 volts.
C. 1500 volts.
D. 5.9 volts.

The voltage ratio of any transformer is directly proportional to the turns ratio, whether it is used as a step-up or step-down transformer. If T_1 is the number of primary turns, T_2 is the number of secondary turns, and V_1 is the input voltage, then the output voltage (V_2) can be determined by the simple formula $V_2 = V_1 \times (T_2 \div T_1)$. Applying the formula to the problem at hand, we have $V_2 = 120 \times (500 \div 2250) = 120 \times 0.2222 = 26.7V$. Caution! This formula always assumes no load on the transformer; that is, no power is being transferred. If you forget this, you may conclude that a transformer can give you power gain – an obviously impossible situation! **ANSWER A**

3-21C4 What is the ratio of the output frequency to the input frequency of a single-phase full-wave rectifier?

A. 1:1.
B. 1:2.
C. 2:1.
D. None of these.

A full-wave, single-phase rectifier has twice the output frequency compared to the input frequency. This is much easier to filter at 2:1. **ANSWER C**

3-21C5 A power transformer has a single primary winding and three secondary windings producing 5.0 volts, 12.6 volts, and 150 volts. Assuming similar wire sizes, which of the three secondary windings will have the highest measured DC resistance?

A. The 12.6 volt winding.
B. The 150 volt winding.
C. The 5.0 volt winding.
D. All will have equal resistance values.

You can determine which of three secondary windings may generate the highest voltage by metering each winding for its DC resistance – obviously, out of circuit. The 150-volt winding has the most turns of the secondary windings – the most wire – so it has the highest measured DC resistance. **ANSWER B**

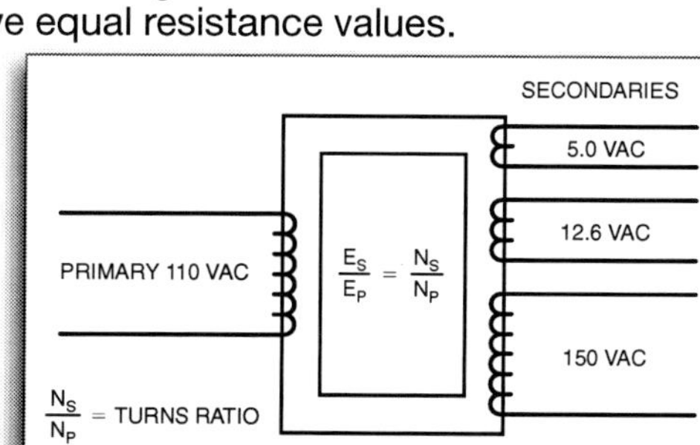

Power Transformer with Multiple Secondaries

3-21C6 A power transformer has a primary winding of 200 turns of #24 wire and a secondary winding consisting of 500 turns of the same size wire. When 20 volts are applied to the primary winding, the expected secondary voltage will be:

A. 500 volts.
B. 25 volts.
C. 10 volts.
D. 50 volts.

This transformer has a turns ratio (N_S/N_P) of $500 \div 200 = 2.5 \div 1$ or 2.5:1, so you simply multiply the primary voltage (20) by the ratio (2.5) to get the answer (50 volts). **ANSWER D**

Key Topic 22: Voltage Regulators, Zener Diodes

3-22C1 In a linear electronic voltage regulator:
- A. The output is a ramp voltage.
- B. The pass transistor switches from the "off" state to the "on" state.
- C. The control device is switched on or off, with the duty cycle proportional to the line or load conditions.
- D. The conduction of a control element is varied in direct proportion to the line voltage or load current.

A linear electronic voltage regulator varies the conduction of a circuit in direct proportion to variations in the line voltage to, or the load current from, the device. In base station power supplies, you will find sophisticated voltage regulation circuits that utilize big, heavy transformers. **ANSWER D**

3-22C2 A switching electronic voltage regulator:
- A. Varies the conduction of a control element in direct proportion to the line voltage or load current.
- B. Provides more than one output voltage.
- C. Switches the control device on or off, with the duty cycle proportional to the line or load conditions.
- D. Gives a ramp voltage at its output.

The switching voltage regulator actually switches the control device completely on or off. **ANSWER C**

3-22C3 What device is usually used as a stable reference voltage in a linear voltage regulator?
- A. Zener diode.
- B. Tunnel diode.
- C. SCR.
- D. Varactor diode.

For voltage regulation and a stable reference voltage, Zener diodes are found in aeronautical handheld transceivers, aircraft radio, marine radio, and business band equipment. The most common cause of voltage regulator failure is a voltage transient or over-voltage. Whenever replacing a defective voltage regulator, monitor other voltage sources to see why the regulator failed. **ANSWER A**

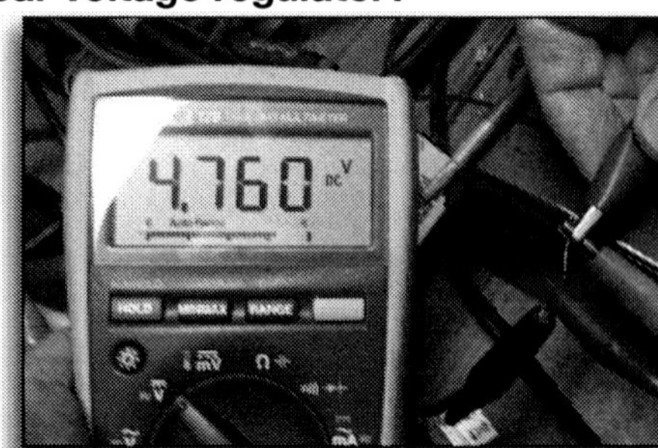

Checking a small Zener diode.

3-22C4 In a regulated power supply, what type of component will most likely be used to establish a reference voltage?
- A. Tunnel Diode.
- B. Battery.
- C. Pass Transistor.
- D. Zener Diode.

The Zener diode is a semiconductor, normally constructed of silicon, which exhibits a constant voltage drop used for reference purpose in a power supply. It may also be used as an AC voltage regulator, with two diodes connected in parallel, reverse polarity. **ANSWER D**

3-22C5 A three-terminal regulator:
- A. Supplies three voltages with variable current.
- B. Supplies three voltages at a constant current.
- C. Contains a voltage reference, error amplifier, sensing resistors and transistors, and a pass element.
- D. Contains three error amplifiers and sensing transistors.

The common voltage regulator found in mobile radio equipment is the 3-terminal regulator. When working on equipment, you can spot these regulators up alongside the metal chassis, using the chassis as a heat sink. Inside that small plastic IC is a voltage regulator containing a voltage reference, an error amplifier, sensing resistors and transistors, and a pass element. These regulators normally run relatively hot, so don't consider them necessarily bad if they seem hot to the touch. Rather, when you find these regulators stone cold you may need to troubleshoot the circuit. **ANSWER C**

3-22C6 What is the range of voltage ratings available in Zener diodes?
- A. 1.2 volts to 7 volts.
- B. 2.4 volts to 200 volts and above.
- C. 3 volts to 2000 volts.
- D. 1.2 volts to 5.6 volts.

The lowest Zener diode voltage available is 2.4 volts. Zener diodes can be found from 2.4 volts up to 200 volts, and higher. **ANSWER B**

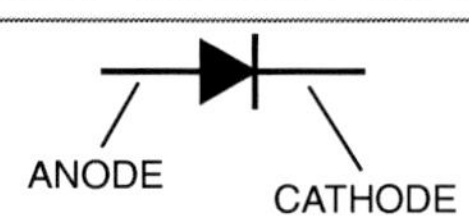

Here is the schematic symbol of a diode. Current will only flow ONE WAY in a diode. You can remember this diode diagram as a one-way arrow (key words).

Semiconductor Diode

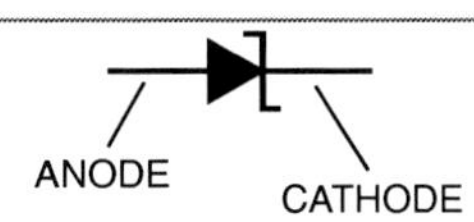

Here is the schematic symbol of a Zener diode. Since a diode only passes energy in one direction, look for that one-way arrow, plus a "Z" indicating it is a Zener diode. Doesn't that vertical line look like a tiny "Z"?

Zener Diode

Key Topic 23: SCRs, Triacs

3-23C1 How might two similar SCRs be connected to safely distribute the power load of a circuit?

A. In series.
B. In parallel, same polarity.
C. In parallel, reverse polarity.
D. In a combination series and parallel configuration.

An SCR can control currents in excess of 100 amps and withstand voltages in excess of 1000 volts. An SCR can be turned on with just a few milliamps of gate current, but can only be switched off by removing or COMMUTATING the anode voltage. Because of this, SCRs are often used in back-to-back pairs, each one controlling one polarity of an AC power sine wave. **ANSWER C**

3-23C2 What are the three terminals of an SCR?

A. Anode, cathode, and gate.
B. Gate, source, and sink.
C. Base, collector, and emitter.
D. Gate, base 1, and base 2.

Like any diode, an SCR has an anode and a cathode, but it also has a special gate element. A control signal on the gate allows current from anode to cathode if the diode is forward biased. **ANSWER A**

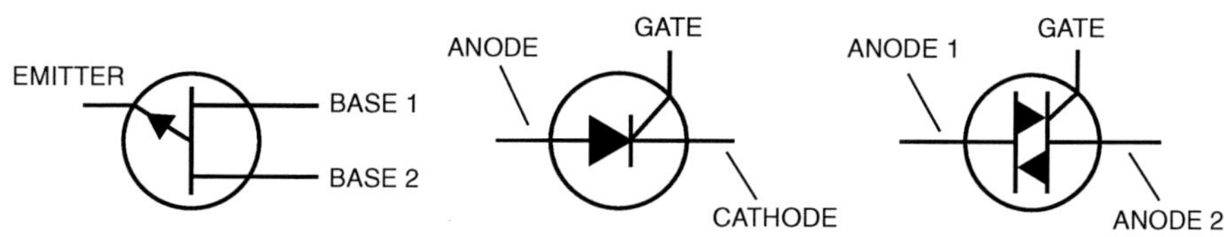

Notice there is an upper and lower line within the unijunction transistor diagram. Think of a uniform—upper and lower—part. There also is only one angled arrowhead line on the opposite side.

The SCR diagram looks a little bit like a switch, doesn't it? The SCR's gate element allows it to control other parts of a circuit. It is a diode with another gate control that turns on or off the SCR.

The triac is best visualized as two diodes connected in parallel, one in one direction, the other in the opposite direction, in a single package. Look at the diagram — this is the way it looks, doesn't it? It is really two SCRs connected head-to-toe.

Unijunction Transistor — *Silicon Controlled Rectifier SCR* — *Bidirectional Triode Thyristor TRIAC*

3-23C3 Which of the following devices acts as two SCRs connected back to back, but facing in opposite directions and sharing a common gate?

A. JFET.
B. Dual-gate MOSFET.
C. DIAC.
D. TRIAC.

The TRIAC acts like two silicon-controlled rectifiers in parallel to control current in either direction. **ANSWER D**

3-23C4 What is the transistor called that is fabricated as two complementary SCRs in parallel with a common gate terminal?

A. TRIAC.
B. Bilateral SCR.
C. Unijunction transistor.
D. Field effect transistor.

The TRIAC can be compared to two complementary SCRs in parallel with a common gate terminal. **ANSWER A**

3-23C5 What are the three terminals of a TRIAC?

A. Emitter, base 1, and base 2.
B. Base, emitter, and collector.
C. Gate, source, and sink.
D. Gate, anode 1, and anode 2.

The TRIAC has one gate and two anodes. The angled line is a gate. Its formal name is bidirectional triode thyristor. **ANSWER D**

3-23C6 What circuit might contain a SCR?

A. Filament circuit of a tube radio receiver.
B. A light-dimming circuit.
C. Shunt across a transformer primary.
D. Bypass capacitor circuit to ground.

One use of an SCR is to allow instrument panel illumination to be adjusted down and dimmed to an appropriate light level. **ANSWER B**

Key Topic 24: Diodes

3-24C1 What is one common use for PIN diodes?
A. Constant current source.
B. RF switch.
C. Constant voltage source.
D. RF rectifier.

PIN diodes may be used in small handheld transceivers for RF switching. This avoids using mechanical relays for switching. **ANSWER B**

3-24C2 What is a common use of a hot-carrier diode?
A. Balanced inputs in SSB generation.
B. Variable capacitance in an automatic frequency control circuit.
C. Constant voltage reference in a power supply.
D. VHF and UHF mixers and detectors.

If your job requires servicing VHF and UHF equipment, you may find a circuit with a hot carrier diode used as a mixer or detector because of its low noise-figure characteristics. **ANSWER D**

3-24C3 Structurally, what are the two main categories of semiconductor diodes?
A. Junction and point contact.
B. Electrolytic and junction.
C. Electrolytic and point contact.
D. Vacuum and point contact.

The semiconductor diode is a common component found in mobile and base station radio equipment. Depending on how the diode is biased, it may look like an open circuit or a closed circuit. Junction and point contact are the two main categories of semiconductor diodes. **ANSWER A**

3-24C4 What special type of diode is capable of both amplification and oscillation?
A. Zener diodes.
B. Point contact diodes.
C. Tunnel diodes.
D. Junction diodes.

The tunnel diode is commonly used in both amplifier and oscillator circuits. The negative resistance region is particularly useful in oscillators.
ANSWER C

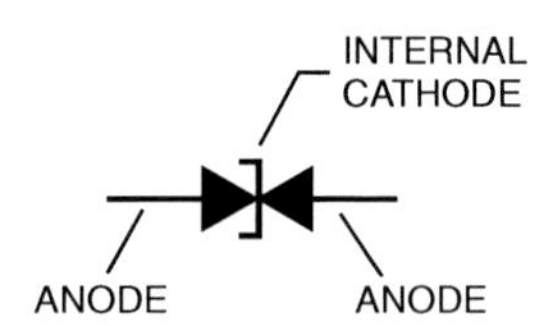

Here is the schematic symbol of a tunnel diode. Memorize the schematic symbol for a tunnel diode. It has arrow-heads tip to tip, as if they are stuck in a tunnel.

Tunnel Diode

3-24C5 What type of semiconductor diode varies its internal capacitance as the voltage applied to its terminals varies?
A. Tunnel diode.
B. Varactor diode.
C. Silicon-controlled rectifier.
D. Zener diode.

We use the varactor diode to tune VHF and UHF circuits by varying the voltage applied to the varactor diode. **ANSWER B**

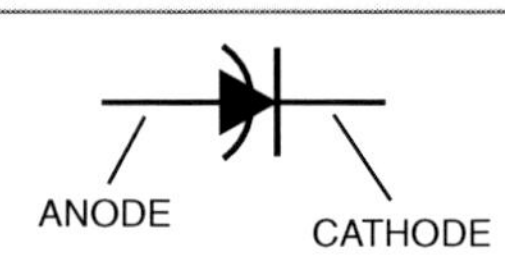

Here is the schematic symbol for a varactor diode. The internal capacitance of a varactor diode varies as the voltage applied to its terminals changes. Your author builds microwave equipment with varactor diodes, so be sure to identify the proper symbol. The symbol is essentially a capacitor and a diode combined.

Varactor Diode

3-24C6 What is the principal characteristic of a tunnel diode?
A. High forward resistance.
B. Very high PIV(peak inverse voltage).
C. Negative resistance region.
D. High forward current rating.

As a tunnel diode is conducting current, there is a spot where current increases as the voltage drop across the diode decreases. This is called a negative resistance region. **ANSWER C**

Key Topic 25: Transistors-1

3-25C1 What is the meaning of the term "alpha" with regard to bipolar transistors? The change of:

A. Collector current with respect to base current.
B. Base current with respect to collector current.
C. Collector current with respect to gate current.
D. Collector current with respect to emitter current.

In bipolar transistors, the term "alpha" is the variation of COLLECTOR current with respect to EMITTER current (CCEC, a key for remembering). This is an important consideration when designing a circuit that will utilize a bipolar transistor. **ANSWER D**

3-25C2 What are the three terminals of a bipolar transistor?

A. Cathode, plate and grid.
B. Base, collector and emitter.
C. Gate, source and sink.
D. Input, output and ground.

Almost like the alphabet, B for base, C for collector, (skip D), and E for emitter. **ANSWER B**

3-25C3 What is the meaning of the term "beta" with regard to bipolar transistors? The change of:

A. Base current with respect to emitter current.
B. Collector current with respect to emitter current.
C. Collector current with respect to base current.
D. Base current with respect to gate current.

When we talk about the term "beta" for a bipolar transistor, think of beta with a "B" for base current. This is the change of COLLECTOR current with respect to BASE current. B for base; B for beta. DC beta = $I_C \div I_B$. **ANSWER C**

3-25C4 What are the elements of a unijunction transistor?

A. Base 1, base 2, and emitter.
B. Gate, cathode, and anode.
C. Gate, base 1, and base 2.
D. Gate, source, and sink.

The two horizontal lines coming out of the unijunction transistor represent base 1 and base 2. Watch out for answer C. The angled arrowhead line is an emitter. **ANSWER A**

3-25C5 The beta cutoff frequency of a bipolar transistor is the frequency at which:

A. Base current gain has increased to 0.707 of maximum.
B. Emitter current gain has decreased to 0.707 of maximum.
C. Collector current gain has decreased to 0.707.
D. Gate current gain has decreased to 0.707.

In a bipolar transistor, the beta cutoff frequency is that frequency at which the grounded emitter current gain has decreased to a certain value – 0.707 of maximum – at a certain frequency. **ANSWER B**

3-25C6 What does it mean for a transistor to be fully saturated?

A. The collector current is at its maximum value.
B. The collector current is at its minimum value.
C. The transistor's Alpha is at its maximum value.
D. The transistor's Beta is at its maximum value.

A transistor is fully saturated when the collector current is at its maximum. **ANSWER A**

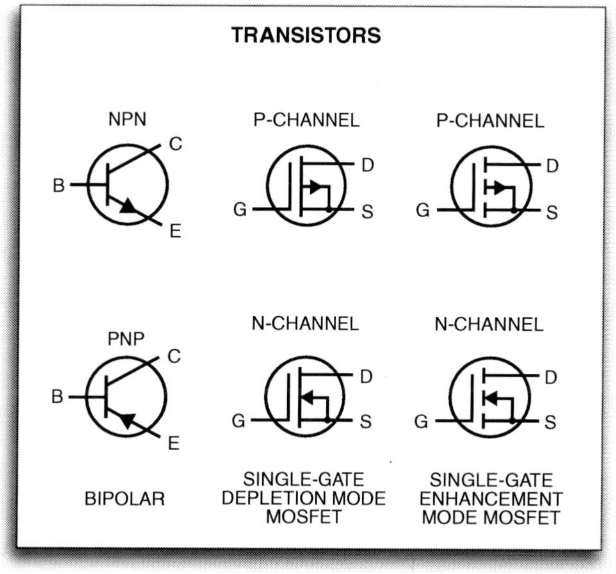

Key Topic 26: Transistors-2

3-26C1 A common base amplifier has:

A. More current gain than common emitter or common collector.
B. More voltage gain than common emitter or common collector.
C. More power gain than common emitter or common collector.
D. Highest input impedance of the three amplifier configurations.

With the common base amplifier, the voltage gain is greater than that of a common emitter or common collector amplifier. **ANSWER B**

3-26C2 What does it mean for a transistor to be cut off?

A. There is no base current.
B. The transistor is at its Class A operating point.
C. There is no current between emitter and collector.
D. There is maximum current between emitter and collector.

At cutoff, in an NPN circuit, there is no current from the emitter to the collector because the base-to-emitter junction is reverse biased. **ANSWER C**

3-26C3 An emitter-follower amplifier has:

A. More voltage gain than common emitter or common base.
B. More power gain than common emitter or common base.
C. Lowest input impedance of the three amplifier configurations.
D. More current gain than common emitter or common base.

An emitter-follower amplifier has more current gain than a common emitter that has power gain or a common base that has voltage gain. Remember, common emitter has greater power gain, common base has greater voltage gain, and emitter-follower has greater current gain. **ANSWER D**

3-26C4 What conditions exists when a transistor is operating in saturation?

A. The base-emitter junction and collector-base junction are both forward biased.
B. The base-emitter junction and collector-base junction are both reverse biased.
C. The base-emitter junction is reverse biased and the collector-base junction is forward biased.
D. The base-emitter junction is forward biased and the collector-base junction is reverse biased.

When a transistor goes into saturation, both the base-emitter junction and collector-base junction are forward biased. Schottky diodes are placed across the collector-base junction to allow the junction to switch out of saturation quickly. **ANSWER A**

3-26C5 For current to flow in an NPN silicon transistor's emitter-collector junction, the base must be:

A. At least 0.4 volts positive with respect to the emitter.
B. At a negative voltage with respect to the emitter.
C. At least 0.7 volts positive with respect to the emitter.
D. At least 0.7 volts negative with respect to the emitter.

For the silicon NPN switching transistor to conduct current between the collector and emitter, the base must be at least 0.7 volts positive with respect to the emitter. Then the NPN switching transistor is considered "on." **ANSWER C**

3-26C6 When an NPN transistor is operating as a Class A amplifier, the base-emitter junction:

A. And collector-base junction are both forward biased.
B. And collector-base junction are both reverse biased.
C. Is reverse biased and the collector-base junction is forward biased.
D. Is forward biased and the collector-base junction is reverse biased.

In a Class A amplifier, the NPN transistor base-emitter junction is forward biased, and the collector-base junction is reverse biased. The biasing within the NPN transistor either adds or subtracts a specific amount of current from the signal at the input. **ANSWER D**

TYPES OF TRANSISTORS

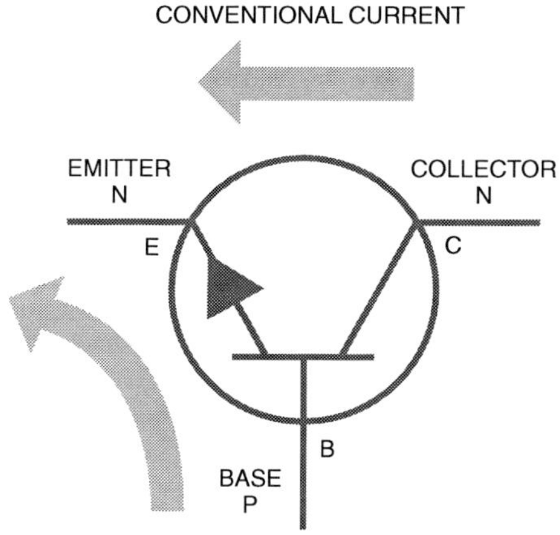

a. Schematic Symbol of NPN Transistor

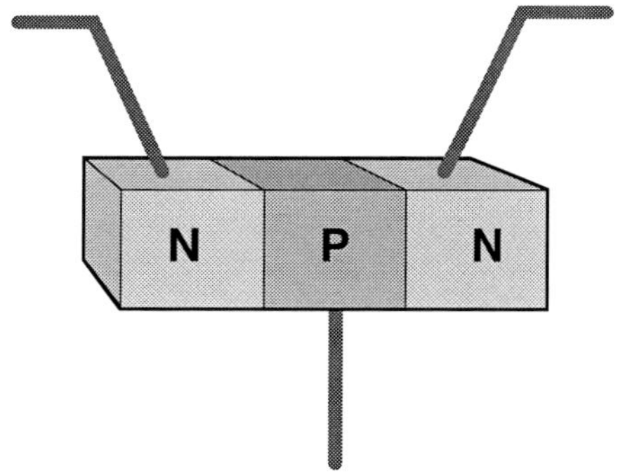

b. Silicon Configuration Suggested by the Symbol

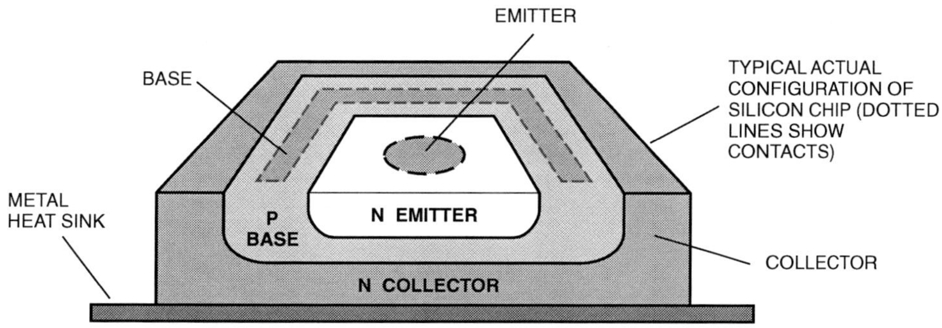

c. Diffused Sandwich Construction

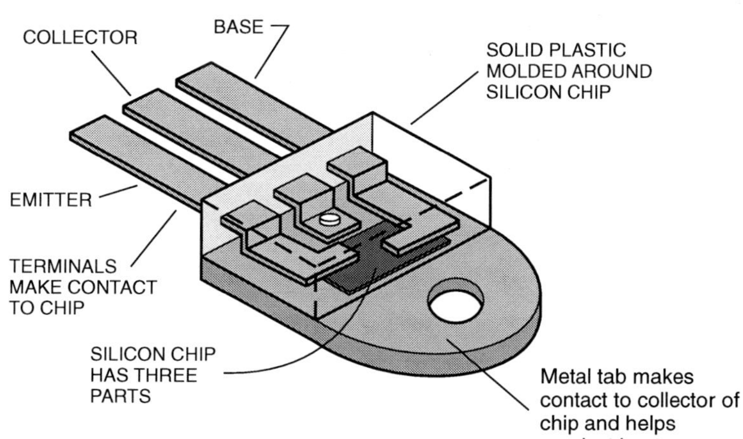

d. Transistor in Package

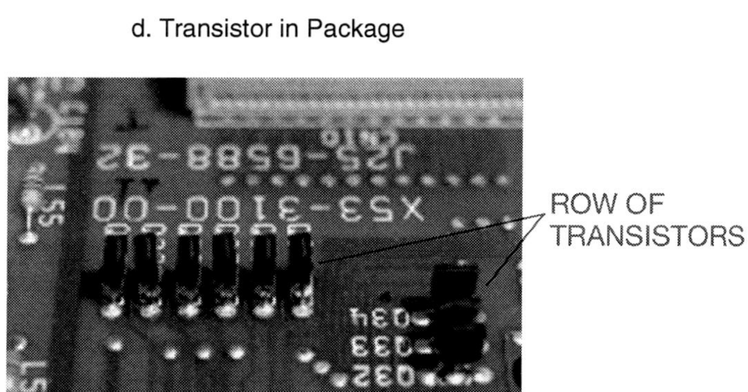

Key Topic 27: Light Emitting Diodes

3-27C1 What type of bias is required for an LED to produce luminescence?

A. Reverse bias. B. Forward bias. C. Logic 0 (Lo) bias. D. Logic 1 (Hi) bias.

The LED requires forward bias in order to illuminate. The LED is a diode, and it must be forward biased to make it give off light. **ANSWER B**

3-27C2 What determines the visible color radiated by an LED junction?

A. The color of a lens in an eyepiece.
B. The amount of voltage across the device.
C. The amount of current through the device.
D. The materials used to construct the device.

The popular colors of light-emitting diodes are red, green, orange, and yellow. Blue is the latest color. Color is determined by the material used to construct the diode. **ANSWER D**

3-27C3 What is the approximate operating current of a light-emitting diode?

A. 20 mA. B. 5 mA. C. 10 mA. D. 400 mA.

The light-emitting diodes (LEDs) found on marine and aviation two-way radio panels are considered LOW POWER. These LEDs draw approximately 20 millliamps of current, and require only 1.7 volts to illuminate. Higher-power LEDs used specifically for illumination can draw as much as 1 ampere or greater. **ANSWER A**

3-27C4 What would be the maximum current to safely illuminate a LED?

A. 1 amp. B. 1 microamp. C. 500 milliamps. D. 20 mA.

This question refers to low-power LEDs used as status indicators. The light-emitting diode is a p-n junction diode that creates light whenever there is a small amount of current in the forward direction. Low-power LEDs used as status indicators draw about 20 mA. We now have red, green, blue, yellow, orange, purple, white, IR, and UV LEDs. Higher power LEDs used for illumination may draw up to 1 ampere of current. The greater their output, the greater the need for increased heat dissipation techniques. **ANSWER D**

			Case 1		Case 2		Case3	
LED	I_F (mA)	V_F(MAX) (V)	V_S (V)	R (kΩ)	V_S (V)	R (kΩ)	V_S (V)	R (kΩ)
A	20	2	12	0.5	9	0.35	5	0.15
B	10	1.6	12	1.04	9	0.74	5	0.34

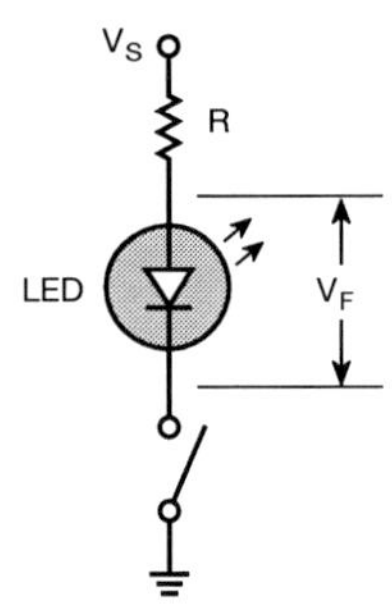

$$R = \frac{(V_S - V_F)}{I_F}$$

where: R is in **kΩ**
V_S and V_F are in **volts**
I_F is in **mA**

How to determine resistor values for LEDs with 3 typical voltage sources.

3-27C5 An LED facing a photodiode in a light-tight enclosure is commonly known as a/an:

A. Optoisolator. B. Seven segment LED. C. Optointerrupter. D. Infra-red (IR) detector.

A light-emitting diode encased in a light-tight enclosure with a photodiode cell forms the optoisolator stage found in many PLL synthesized marine radios. When you adjust the big knob, the optoisolator detects an interrupted light source and counts the number of interruptions for the next digital stage to electronically process the command. It is important never to bang or bump the big knob on any optoisolator system; poor alignment could cause erratic operation. **ANSWER A**

3-27C6 What circuit component must be connected in series to protect an LED?

A. Bypass capacitor to ground.
B. Electrolytic capacitor.
C. Series resistor.
D. Shunt coil in series.

If you closely examine an LED circuit, you will always see it in series with a several-ohm miniature resistor. As resistance is decreased, current will rise, and the LED becomes brighter. At a point where an increase in current causes no greater brilliance, the light emitting diode is said to be in a state of saturation. While an LED has long life, the heat buildup at saturation can quickly destroy an LED. **ANSWER C**

LIGHT EMITTING DIODES

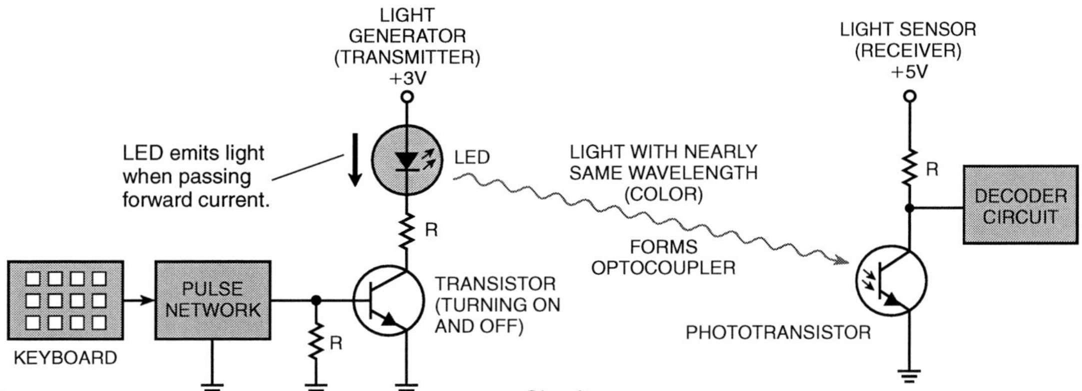

a. Circuit

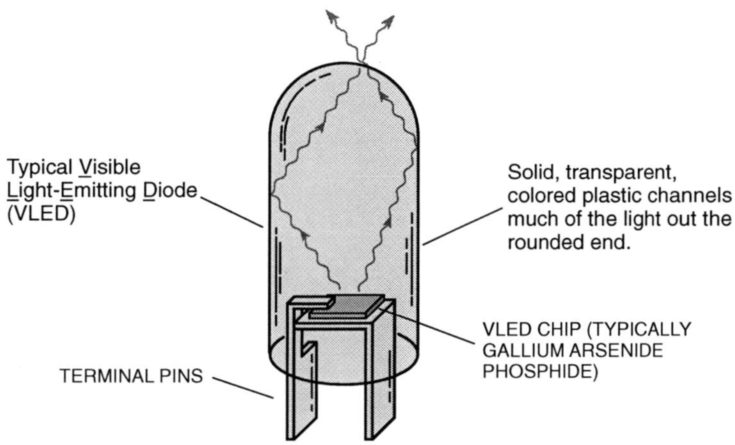

b. Physical Unit for Visible LED

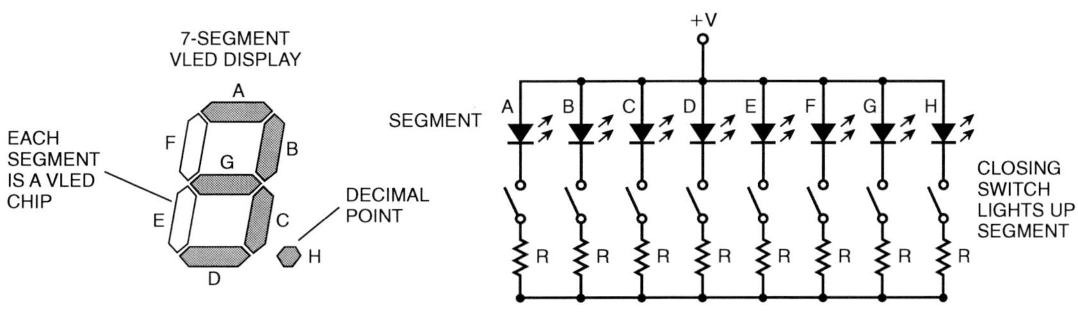

c. VLED Array

An array of LEDs and resistors mounted on a printed circuit board.

Key Topic 28: Devices

3-28C1 What describes a diode junction that is forward biased?

A. It is a high impedance.
B. It conducts very little current.
C. It is a low impedance.
D. It is an open circuit.

A diode junction that is forward biased offers low impedance to the flow of electrons. Think of "forward" as easy movement. **ANSWER C**

3-28C2 Why are special precautions necessary in handling FET and CMOS devices?

A. They have fragile leads that may break off.
B. They are susceptible to damage from static charges.
C. They have micro-welded semiconductor junctions that are susceptible to breakage.
D. They are light sensitive.

Any equipment using FET and CMOS transistors can be easily damaged from a nearby static discharge. **ANSWER B**

3-28C3 What do the initials CMOS stand for?

A. Common mode oscillating system.
B. Complementary mica-oxide silicon.
C. Complementary metal-oxide semiconductor.
D. Complementary metal-oxide substrate.

When we talk about transistorized devices, they are in fact semiconductors. A CMOS is made from layers of complementary metal-oxide semiconductor material. **ANSWER C**

3-28C4 What is the piezoelectric effect?

A. Mechanical vibration of a crystal by the application of a voltage.
B. Mechanical deformation of a crystal by the application of a magnetic field.
C. The generation of electrical energy by the application of light.
D. Reversed conduction states when a P-N junction is exposed to light.

If you apply a voltage to a quartz crystal, it will vibrate at a specific frequency. This is the reason crystal oscillators are so accurate. They oscillate at the frequency, or harmonics of the frequency, of the crystal, and continue to do so with great accuracy unless the temperature changes outside design limits. **ANSWER A**

3-28C5 An electrical relay is a:

A. Current limiting device.
B. Device used for supplying 3 or more voltages to a circuit.
C. Component used mainly with HF audio amplifiers.
D. Remotely controlled switching device.

An electrical relay is usually used as a switch. Its open contacts provide a near infinite resistance open circuit, and its closed contacts provide a near zero resistance closed circuit. **ANSWER D**

3-28C6 In which oscillator circuit would you find a quartz crystal?

A. Hartley.
B. Pierce.
C. Colpitts.
D. All of the above.

Quartz crystals are found in Pierce oscillators. **ANSWER B**

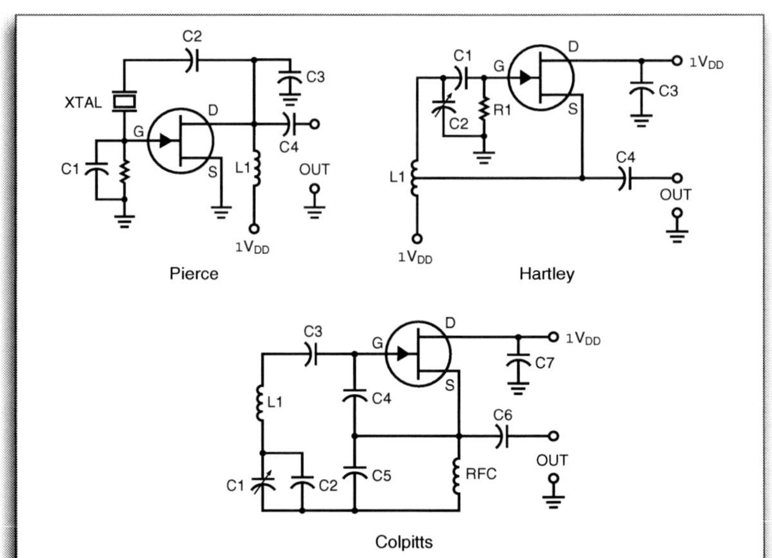

Oscillators

Subelement D – Circuits

4 Key Topics, 4 Exam Questions, 6 Drawings

Key Topic 29: R-L-C Circuits

3-29D1 What is the approximate magnitude of the impedance of a parallel R-L-C circuit at resonance?

A. Approximately equal to the circuit resistance.
B. Approximately equal to X_L.
C. Low, as compared to the circuit resistance.
D. Approximately equal to X_C.

Whether the system is parallel or series R-L-C resonant, the impedance will always be equal to the circuit resistance. **ANSWER A**

3-29D2 What is the approximate magnitude of the impedance of a series R-L-C circuit at resonance?

A. High, as compared to the circuit resistance.
B. Approximately equal to the circuit resistance.
C. Approximately equal to X_L.
D. Approximately equal to X_C.

An antenna is one of the most common examples of a series resonant circuit. At resonance, the antenna appears as a simple resistance. (This resistance is actually a sum of the RADIATION RESISTANCE, and common or "ohmic" resistance found in any conductor.) In this condition, maximum current flows, since the total series reactance is zero. Additionally, the power factor of the equivalent circuit is 1. **ANSWER B**

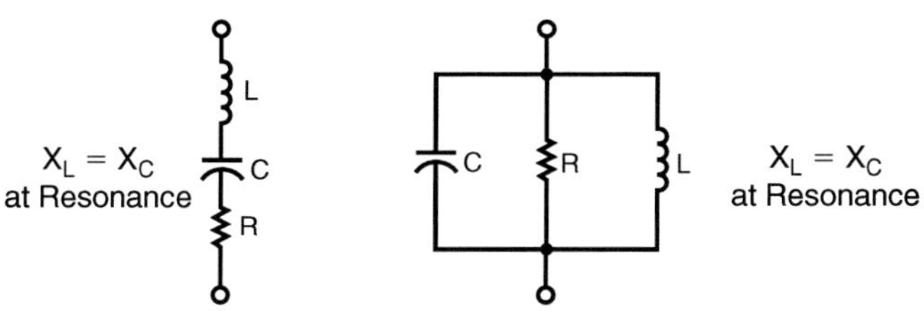

a. Series　　b. Parallel

X_L = Inductive Reactance

$X_L = 2\pi fL$

X_C = Capacitive Reactance

$X_C = \frac{1}{2\pi fC}$

At Resonance $X_L = X_C$

$2\pi f_r L = \frac{1}{2\pi f_r C}$

Solving for f_r

$f_r^2 = \frac{1}{(2\pi)^2 LC}$

The Resonant Frequency, f_r, is:

$f_r = \frac{1}{2\pi \sqrt{LC}}$

Series and Parallel Resonant Circuits

3-29D3 How could voltage be greater across reactances in series than the applied voltage?

A. Resistance.
B. Conductance.
C. Capacitance.
D. Resonance.

In a high frequency series antenna circuit with both inductive and capacitive reactances, voltage greater than the supply voltage exists because the reactances are returning energy previously stored by either the inductive or the capacitive component within the system. At resonance, when inductive reactance and capacitive reactance cancel, this returned energy is added to the supply voltage, creating a greater voltage across the reactances. **ANSWER D**

3-29D4 What is the characteristic of the current flow in a series R-L-C circuit at resonance?

A. Maximum.
B. Minimum.
C. DC.
D. Zero.

A mobile whip antenna will have maximum current in a series R-L-C circuit at resonance. One way you can tell whether or not a mobile whip is working properly is to feel the coil after the transmitter has been shut down. If it's warm, there has been current through the coil and up to the whip tip stinger. **ANSWER A**

3-29D5 What is the characteristic of the current flow within the parallel elements in a parallel R-L-C circuit at resonance?

A. Minimum.
B. Maximum.
C. DC.
D. Zero.

A parallel circuit at resonance is similar to a tuned trap in a multi-band trap antenna. At resonance, the trap keeps power from going any further to the longer antenna elements. But be assured there is maximum current circulating within the parallel R-L-C circuit. Now don't get confused with this question – a parallel circuit has maximum current within it, and minimum current going to the next stage. That's why they call those parallel resonant circuits "traps." **ANSWER B**

3-29D6 What is the relationship between current through a resonant circuit and the voltage across the circuit?

A. The current and voltage are 180 degrees out of phase.
B. The current leads the voltage by 90 degrees.
C. The voltage and current are in phase.
D. The voltage leads the current by 90 degrees.

At resonance, both voltage and current are in phase. **ANSWER C**

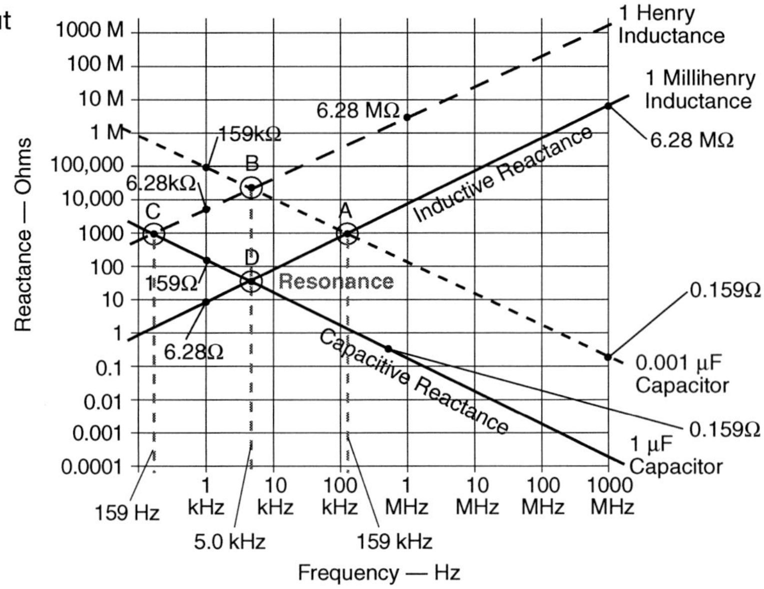

Variation of inductance and capacitive reactance with frequency (illustration not to exact log scale).

Key Topic 30: Op Amps

3-30D1 What is the main advantage of using an op-amp audio filter over a passive LC audio filter?

A. Op-amps are largely immune to vibration and temperature change.
B. Most LC filter manufacturers have retooled to make op-amp filters.
C. Op-amps are readily available in a wide variety of operational voltages and frequency ranges.
D. Op-amps exhibit gain rather than insertion loss.

An op amp exhibits high gain with negligible insertion loss because its input impedance is also high. **ANSWER D**

3-30D2 What are the characteristics of an inverting operational amplifier (op-amp) circuit?

A. It has input and output signals in phase.
B. Input and output signals are 90 degrees out of phase.
C. It has input and output signals 180 degrees out of phase.
D. Input impedance is low while the output impedance is high.

On an inverting op amp, the input and output signals are 180 degrees out of phase. Remember inverting means a 180° phase shift. **ANSWER C**

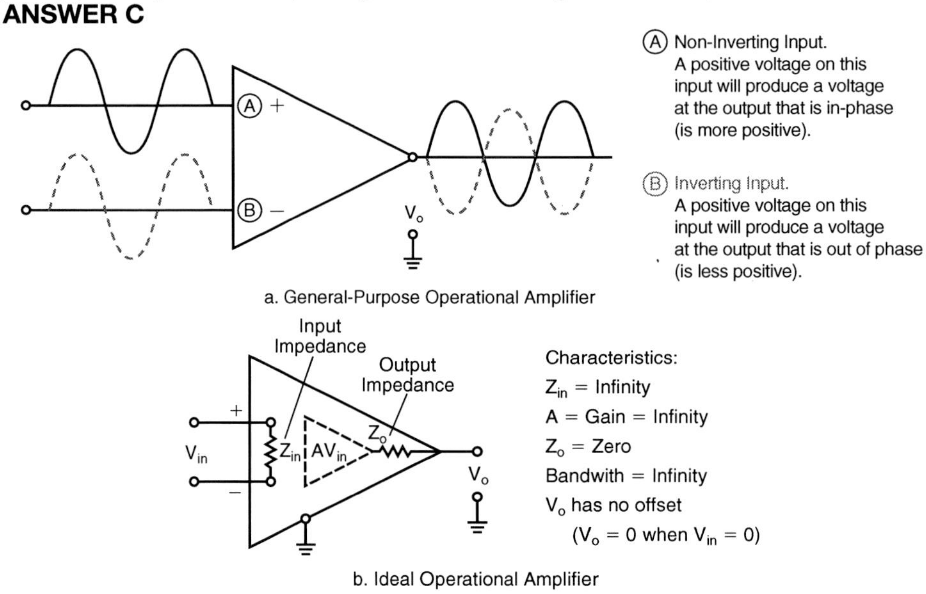

Operational amplifiers (op-amps).

3-30D3 Gain of a closed-loop op-amp circuit is determined by?

A. The maximum operating frequency divided by the square root of the load impedance.
B. The op-amp's external feedback network.
C. Supply voltage and slew rate.
D. The op-amp's internal feedback network.

The op-amp gain without feedback is very high. Feedback, which is applied to set the gain to a particular amount, is supplied by an external feedback network. It may be one or more fixed resistors, or a type of variable circuit. **ANSWER B**

3-30D4 Where is the external feedback network connected to control the gain of a closed-loop op-amp circuit?

A. Between the differential inputs.
B. From output to the non-inverting input.
C. From output to the inverting input.
D. Between the output and the differential inputs.

To control the gain of a closed-loop operational amplifier, external feedback is connected from the output to the inverting input. **ANSWER C**

3-30D5 Which of the following op-amp circuits is operated open-loop?

A. Non-inverting amp.
B. Inverting amp.
C. Active filter.
D. Comparator.

The comparator is a high-gain circuit with an open-loop configuration. The three wrong answers are all closed-loop circuits. **ANSWER D**

3-30D6 In the op-amp oscillator circuit shown in Figure 3D6, what would be the most noticeable effect if the capacitance of C were suddenly doubled?

A. Frequency would be lower.
B. Frequency would be higher.
C. There would be no change. The inputs are reversed, therefore the circuit cannot function.
D. None of the above.

Figure 3D6 is an astable multivibrator. The frequency of oscillation is determined by the reciprocal of the RC time constant formed by C and R. Increasing the value of C will increase the time constant, lowering the frequency of oscillation. **ANSWER A**

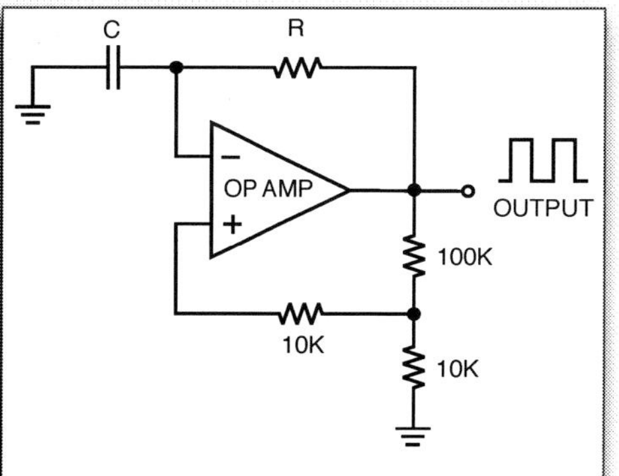

Fig. 3D6

Key Topic 31: Phase Locked Loops (PLLs); Voltage Controlled Oscillators (VCOs); Mixers

3-31D1 What frequency synthesizer circuit uses a phase comparator, look-up table, digital-to-analog converter, and a low-pass antialias filter?

A. A direct digital synthesizer.
B. Phase-locked-loop synthesizer.
C. A diode-switching matrix synthesizer.
D. A hybrid synthesizer.

The latest breed of the modern frequency synthesizer is direct digital synthesis, abbreviated DDS. A single chip may now take up the functions of individual component local oscillators. DDS is found in modern HF/VHF/UHF receivers, and is most always found in software-defined radios. In this question, the low-pass antialias filter is the DDS component which filters out any sampling signals above the desired fundamental, and is a necessary circuit in direct digital synthesis. **ANSWER A**

3-31D2 A circuit that compares the output of a voltage-controlled oscillator (VCO) to a frequency standard and produces an error voltage that is then used to adjust the capacitance of a varactor diode used to control frequency in that same VCO is called what?

A. Doubly balanced mixer.
B. Phase-locked loop.
C. Differential voltage amplifier.
D. Variable frequency oscillator.

In a phase-locked loop (PLL), the "difference between" two frequencies will produce an error voltage to the voltage-controlled oscillator. **ANSWER B**

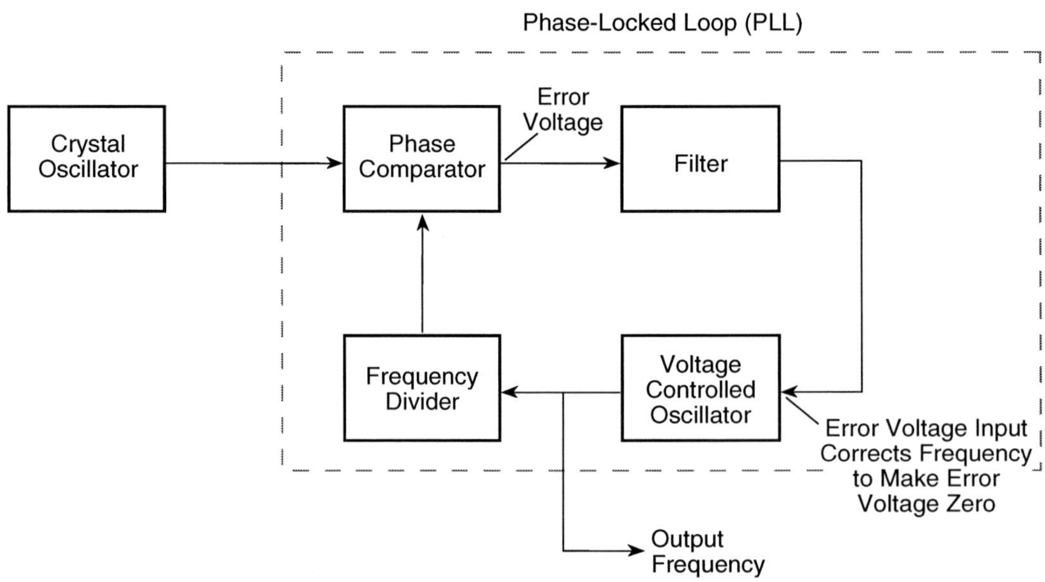

This PLL circuit functions by comparing the crystal oscillator frequency to the VCO frequency and generating an error voltage if they are different. The error voltage is of such a polarity that it forces the VCO to change its frequency to the crystal frequency. However, the VCO does not connect directly to the phase comparator; it is divided down by some integer number before comparison. Changing the amount of frequency division changes the frequency at which the VCO "locks," making the oscillator tunable while also locked to the crystal oscillator. The crystal oscillator is very stable, the divider network is digital so it is stable, making for a very stable circuit for many applications.

Phase-locked loop (PLL).

3-31D3 RF input to a mixer is 200 MHz and the local oscillator frequency is 150 MHz. What output would you expect to see at the IF output prior to any filtering?

A. 50, 150, 200 and 350 MHz.
B. 50 MHz.
C. 350 MHz.
D. 50 and 350 MHz.

The mixer circuit will translate the two input signals to different frequencies, which will be either the sum or difference of the two frequencies. Before filtering out unwanted frequencies, multiple frequencies will result, including the 200 MHz - 150 MHz = 50 MHz; the two local oscillator frequencies of 150 MHz and 200 MHz; and 150 MHz + 200 MHz = 350 MHz. **ANSWER A**

3-31D4 What spectral impurity components might be generated by a phase-locked-loop synthesizer?

A. Spurs at discrete frequencies.
B. Random spurs which gradually drift up in frequency.
C. Broadband noise.
D. Digital conversion noise.

In a phase-locked-loop synthesizer, the biggest challenge is to keep the broadband noise floor as low as possible. It is this broadband noise that DSP manufacturers and designers try to drop to a minimum. **ANSWER C**

3-31D5 In a direct digital synthesizer, what are the unwanted components on its output?
A. Broadband noise.
B. Spurs at discrete frequencies.
C. Digital conversion noise.
D. Nyquist limit noise pulses.
In a direct digital synthesizer, it is NOT broadband noise that appears on the output, but rather spikes from the digital to analog converter resulting in spurs at discrete frequencies, and spurs that re-appear on new frequencies as soon as the receiver tuning dial is changed. Anything passed by the direct digital synthesizer's low pass filter can be a spur problem in the tuning process. **ANSWER B**

3-31D6 What is the definition of a phase-locked loop (PLL) circuit?
A. A servo loop consisting of a ratio detector, reactance modulator, and voltage-controlled oscillator.
B. A circuit also known as a monostable multivibrator.
C. A circuit consisting of a precision push-pull amplifier with a differential input.
D. A servo loop consisting of a phase detector, a low-pass filter and voltage-controlled oscillator.
There are few circuits that have made such a big difference in communications as the phase-locked loop circuit. In fact, frequency selection from a transceiver with PLL synthesis came to a point where the operator actually had too much control of where the set might transmit and receive. Most land mobile equipment now locks out the consumer from getting into and making changes to the PLL. Same thing for marine radios – while mariners may tune in authorized ITU bands, the PLL circuitry has been locked out from the front panel so it can't be changed to go in any other region. **ANSWER D**

Large-scale integrated circuit chips in a PLL section of a communications receiver.

Key Topic 32: Schematics

3-32D1 Given the combined DC input voltages, what would the output voltage be in the circuit shown in Figure 3D7?
A. 150 mV
B. 5.5 V
C. −15 mv
D. −5.5 V

Figure 3D7 is a common summing amplifier, one of the most useful applications of the Operational Amplifier. Its output voltage will be the sum of the input voltages times the gain of the individual channels. Note that this is an inverting amplifier, which means that the output signal is going to be the opposite polarity. The general formula for an inverting summing amplifier is:

$$Vo = -R_F\,(V_1 \div R_1 + V_2 \div R_2 + V_3 \div R_3)$$

where: R_F is the feedback resistance
V_1, V_2, and V_3 are the individual input voltages,
and R_1, R_2, and R_3 are the individual input resistances

Again, notice the minus sign in front of the equation, since this is an INVERTING amplifier. For this particular example, Vo=−100,000 × (0.1/10,000 + 0.2/10,000 + 0.25/10,000) = -100,000 x .000055 = −5.5V. **ANSWER D**

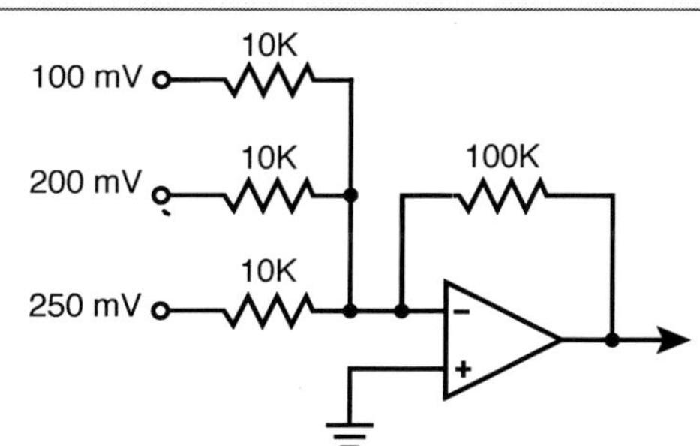

Fig. 3D7

3-32D2 Which lamps would be lit in the circuit shown in Figure 3D8?

A. 2, 3, 4, 5 and 6.
B. 5, 6, 8 and 9.
C. 2, 3, 4, 7 and 8.
D. 1, 3, 5, 7 and 8.

This circuit is a great example of diode logic. And, actually, it's nowhere near as daunting as it looks! (Whew!) Think of this as a simple *maze* puzzle with some gates in the path that only swing one way. Our job is to help Ernie Electron find a path from the bottom of the battery, through our one-way gates, to each of the lamps, and back to the top of the battery. In some cases there is a path, in others there is none. If the path exists, the lamp lights – if it doesn't, it won't. Simple as that.

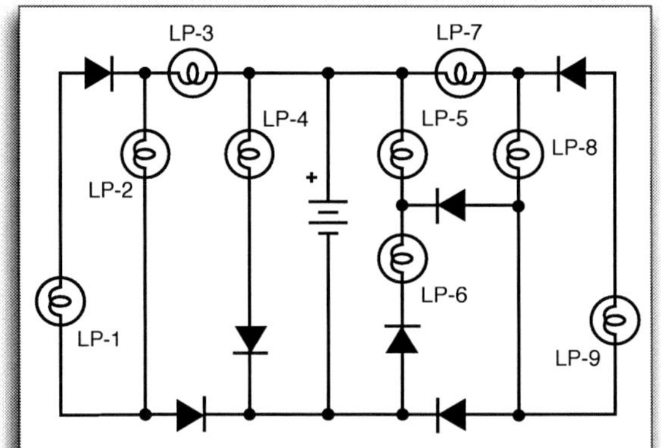

Fig. 3D8

Further simplifying the problem is the fact that the right and left halves of the puzzle (separated by the battery itself) are totally independent. So let's do the left part first. Remember that in a diode, the electrons flow AGAINST the "arrow".

Is there a path for Ernie Electron through LP-1? If he leaves the bottom of the battery, moving left he sees a gate, the horizontal, lower left diode. He can move through that diode because it's pointing toward him. He moves through the lamp, but encounters a BACKWARDS diode at the upper left corner of the diagram. Ernie Electron is in a dead end. LP-1 cannot light.

How about LP-2? Ernie leaves the bottom of the battery, passes through the lower left horizontal diode, makes a right turn, passes through LP-2 AND LP-3, and finds the top of the battery! He kills two birds with one stone, both LP-2 and LP-3 light up.

Ernie once again, leaves the bottom of the battery, and instead of going through the horizontal diode, finds he can also go through the vertical diode, right through LP-4 to the top of the battery. LP-4 lights up. This one was easy. We've now finished the left half of the puzzle.

Now, is there a path for Ernie to LP-5? Well, we might think there's a path to LP-5 via LP-6. But, rats! There's a backwards diode at the bottom end of LP-6. This makes both LP-5 and LP-6 unreachable, they remain dark. How about LP-7? Why, it seems there is a path through LP-7 via LP-8. But can Ernie get to the bottom of LP-8? He has to pass through that lower right horizontal diode. No problem, the "arrow" is pointing towards him; he can pass. LP-7 and LP-8 both light up. Well, we're almost home. What about LP-9? Again, Ernie can pass through the lower right horizontal diode, through LP-9. But, at the TOP of LP-9 he encounters a backwards diode. Alas, his efforts are thwarted. LP-9 cannot light.

Ah yes. There is one diode we haven't addressed yet. It's the one that connects to the junction of LP-6 and LP-7. As it turns out, this diode does nothing for Ernie. It's not a potential "back door" to LP-5, as we might think at first blush, because it's backwards. Now, wasn't that fun? **ANSWER C**

3-32D3 What will occur if an amplifier input signal coupling capacitor fails open?

A. No amplification will occur, with DC within the circuit measuring normal.
B. Improper biasing will occur within the amplifier stage.
C. Oscillation and thermal runaway may occur.
D. An AC hum will appear on the circuit output.

The purpose of a coupling capacitor in any amplifier is to pass an AC signal while leaving the DC operating voltages unaffected. We want a coupling capacitor on the input of an amplifier so the bias voltage will remain constant, while still allowing us to superimpose our input signal on top of it. Likewise, on the output of an amplifier we will usually want a coupling capacitor so as to keep the DC operating voltage isolated from any subsequent stages. A coupling capacitor that fails in the "open" mode will not affect any DC voltages we might have, but will not allow our signal to pass, either. **ANSWER A**

3-32D4 In Figure 3D9, determine if there is a problem with this regulated power supply and identify the problem.

A. R1 value is too low which would cause excessive base current and instantly destroy TR 1.
B. D1 and D2 are reversed. The power supply simply would not function.
C. TR1 is shown as an NPN and must be changed to a PNP.
D. There is no problem with the circuit.

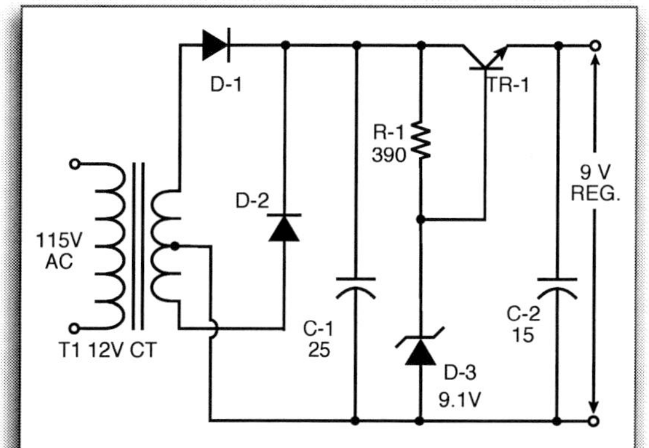

Fig. 3D9

This is an extremely common and useful power supply circuit. The center-tapped secondary transformer and diodes D1 and D2 make this immediately recognizable as a full wave rectifier. C1 is a filter capacitor that levels out the ripple from the rectifier. D3 is a Zener diode, which is used for voltage regulation. Actually the combination of R1 and D3 alone form a perfectly good voltage regulator, all by themselves. The voltage at the top end of D3 will remain at 9.1 Volts regardless of the input voltage or output load, within limits. However, for effective regulation, D3 typically needs to dissipate about 10% of the total supply load. This can require a really fat Zener in high current applications. TR1 effectively makes D3 a much bigger device. However, there will be about a 0.6 volt drop thorough TR1, regardless of current. This can be compensated for by choosing a Zener Diode with a rating that's 0.6 volts higher than you need. Finally, capacitor C-2 is an additional filter capacitor that squashes any residual ripple that happens to make it through the regulator, which should be almost microscopic. This configuration supplies an extremely clean and stable DC voltage. As we commonly say in the field, "If it ain't broke, don't fix it." **ANSWER D**

3-32D5 In Figure 3D10 with a square wave input what would be the output?

A. 1
B. 2
C. 3
D. 4

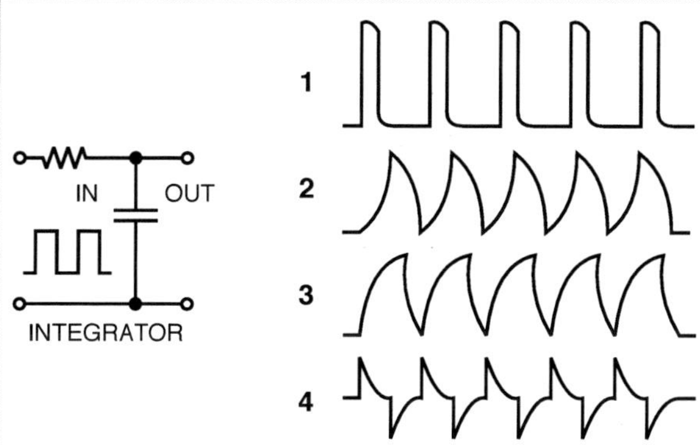

Fig. 3D10

Figure 3D10 is an integrator. Remember our RC time constant from way back? The leading edge of each "sawtooth" is an RC charging curve, and the trailing edge is an RC discharging curve. If the RC time constant is very long compared to the square wave period, we will only use a very small section of those curves, where the voltage change is nearly linear. This will give us a triangle wave. **ANSWER C**

3-32D6 With a pure AC signal input to the circuit shown in Figure 3D11, what output wave form would you expect to see on an oscilloscope display?

A. 1
B. 2
C. 3
D. 4

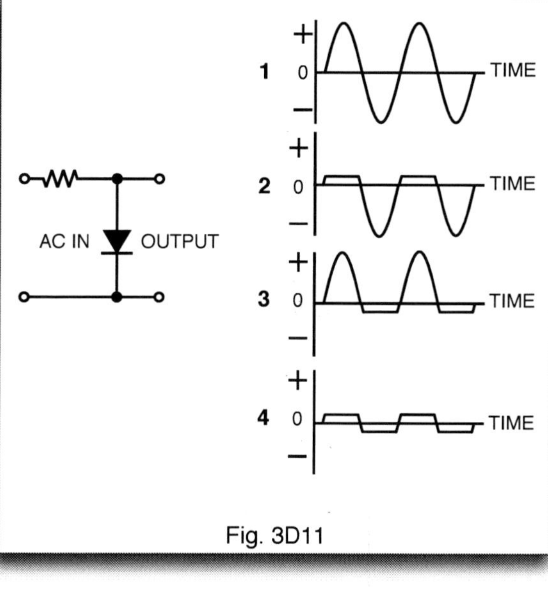

Fig. 3D11

Figure 3D11 is variously known as a "clamper" or a "half-wave clipper" and has a variety of uses in things like video amplifiers, where it's important to establish a DC voltage reference at some place on a complex waveform. In the simplest terms, it's nothing more than an imperfect rectifier. During the negative half of the input signal, the diode is reverse biased, (turned off) and has no effect on the signal. The bottom half of the wave emerges untouched. During the positive half of the cycle, the wave is unaffected until the diode conducts at approximately 0.6 volts (the normal forward junction drop) of a silicon diode. At this point the voltage remains fixed regardless of the current passing through the series resistor. The resulting signal is very nearly a half wave. **ANSWER B**

Subelement E – Digital Logic

8 Key Topics, 8 Exam Questions, 3 Drawings

Key Topic 33: Types of Logic

3-33E1 What is the voltage range considered to be valid logic low input in a TTL device operating at 5 volts?

A. 2.0 to 5.5 volts.
B. -2.0 to -5.5 volts.
C. Zero to 0.8 volts.
D. 5.2 to 34.8 volts.

Transistor-Transistor Logic (TTL) is a binary circuit with two state levels – a 0 state is generally between zero to 0.8 volts, and the 1 state is between 2.4 to 5 volts. There is a small rise time and fall time between each state; this slight delay is called the transition time. **ANSWER C**

3-33E2 What is the voltage range considered to be a valid logic high input in a TTL device operating at 5.0 volts?

A. 2.0 to 5.5 volts.
B. 1.5 to 3.0 volts.
C. 1.0 to 1.5 volts.
D. 5.2 to 34.8 volts.

If the input voltage is at the high level, depending on the number of circuits connected at the point, by specification the logic high level will be between 2.0 to 5.5 volts. Logic low is 0.0 to 0.8 volts. **ANSWER A**

The minimum input voltage, V_{IN}, to keep Q_1 from transistor action when Q_2 is ON is calculated as follows:

When Q_2 is ON:
V_{BE2} of $Q_2 = 0.7V$
V_{BC1} of Q1 $= 0.7V$
$V_{BE2} + V_{BC1} = 1.4V$
Since $V_{BE1} = 0.7V$
$V_{IN} = 1.4V - 0.7V = 0.7V$

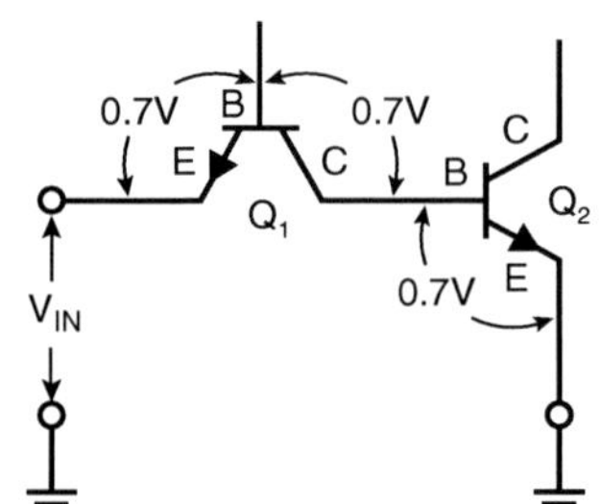

So the absolute minimum V_{IN} is 0.7V. However, to have at least 0.3V noise margin, the minimum $V_{IN} = 1.0V$. Typically, in TTL logic the minimum high-level input voltage is 2V.

3-33E3 What is the common power supply voltage for TTL series integrated circuits?

A. 12 volts.
B. 13.6 volts.
C. 1 volt.
D. 5 volts.

The 5-volt power supply for the TTL series integrated circuits must be well regulated to compensate for the near-instantaneous changes in current within the TTL device. If the power supply momentarily dips, this could falsely trigger other devices down the line! There is usually a 0.01 μF capacitor to help channel any spikes to ground. **ANSWER D**

3-33E4 TTL inputs left open develop what logic state?

A. A high-logic state.
B. A low-logic state.
C. Open inputs on a TTL device are ignored.
D. Random high- and low-logic states.

If a TTL input is left "floating" and is not loaded capacitively through coupled noise from other stages, the OPEN input is normally high. Adding a pull-up resistor will mitigate a high noise floor when a TTL is left open **ANSWER A**

3-33E5 Which of the following instruments would be best for checking a TTL logic circuit?

A. VOM.
B. DMM.
C. Continuity tester.
D. Logic probe.

The common volt/ohm meter, or a continuity tester, or a digital multimeter, on their own cannot display high and low logic states. The well isolated logic probe is the only reliable tester to read ones and zeros. **ANSWER D**

3-33E6 What do the initials TTL stand for?

A. Resistor-transistor logic.
B. Transistor-transistor logic.
C. Diode-transistor logic.
D. Emitter-coupled logic.

If you regularly work with digital modes, transistor-transistor logic is commonplace within this type of transceiver. You will need a logic probe to troubleshoot TTL circuits. **ANSWER B**

Logic Circuit Families

Over the years there have been many types of digital logic circuit families. Their names were based on the interconnection of the components or upon the technologies used. Here are some of them:

DCTL	Direct-Coupled Transistor Logic
RTL	Resistor-Transistor Logic
RCTL	Resistor-Capacitor-Transistor Logic
DTL	Diode-Transistor Logic
ECL	Emitter-Coupled Logic
TTL	Transistor-Transistor Logic
MOS	Metal-Oxide Semiconductor Logic
CMOS	Complementary Metal-Oxide Semiconductor Logic

Because most decision circuits use logic circuits that are made in integrated circuit form so that hundreds of thousands of gates can be made at the same time, TTL, MOS, and CMOS circuits have won out over the other families. TTL uses bipolar junction transistors and MOS uses a field-effect transistor made of a metal-oxide semiconductor sandwich.

Key Topic 34: Logic Gates

3-34E1 What is a characteristic of an AND gate?

A. Produces a logic "0" at its output only if all inputs are logic "1".
B. Produces a logic "1" at its output only if all inputs are logic "1".
C. Produces a logic "1" at its output if only one input is a logic "1".
D. Produces a logic "1" at its output if all inputs are logic "0".

The AND gate has two or more inputs with a single output, and produces a logic "1" at its output only if all inputs are logic "1". Remember AND, all, and the number 1 and the number 1 looking like the letters ll in "all." **ANSWER B**

3-34E2 What is a characteristic of a NAND gate?

A. Produces a logic "0" at its output only when all inputs are logic "0".
B. Produces a logic "1" at its output only when all inputs are logic "1".
C. Produces a logic "0" at its output if some but not all of its inputs are logic "1".
D. Produces a logic "0" at its output only when all inputs are logic "1".

The word NAND reminds me of the word "naw", meaning no or nothing. The NAND gate produces a logic "0" at its output only when all inputs are logic "1". N before AND means negative AND. If an AND gate outputs a "1" when all inputs are "1", a NAND gate outputs a "0", the negative of "1". **ANSWER D**

3-34E3 What is a characteristic of an OR gate?

A. Produces a logic "1" at its output if any input is logic "1".
B. Produces a logic "0" at its output if any input is logic "1".
C. Produces a logic "0" at its output if all inputs are logic "1".
D. Produces a logic "1" at its output if all inputs are logic "0".

The OR gate will produce a logic "1" at its output if any input is, or all inputs are, a logic "1". **ANSWER A**

3-34E4 What is a characteristic of a NOR gate?

A. Produces a logic "0" at its output only if all inputs are logic "0".
B. Produces a logic "1" at its output only if all inputs are logic "1".
C. Produces a logic "0" at its output if any or all inputs are logic "1".
D. Produces a logic "1" at its output if some but not all of its inputs are logic "1".

The NOR gate produces a logic "0" at its output if ANY or all inputs are logic "1". It's a negative OR. With OR gate or NOR gate, spot that word "ANY". **ANSWER C**

3-34E5 What is a characteristic of a NOT gate?

A. Does not allow data transmission when its input is high.
B. Produces a logic "0" at its output when the input is logic "1" and vice versa.
C. Allows data transmission only when its input is high.
D. Produces a logic "1" at its output when the input is logic "1" and vice versa.

The NOT gate will produce a "0" or a "1" at the output – whichever is opposite from the level that is on the input. **ANSWER B**

3-34E6 Which of the following logic gates will provide an active high out when both inputs are active high?

A. NAND. B. NOR. C. AND D. XOR.

As you can see from the truth table, the AND gate has a 1 or high output when all inputs are high. A and B must be "1" for C to be "1". **ANSWER C**

INPUTS		OUTPUT
A	B	C
0	0	0
0	1	0
1	0	0
1	1	1

2-Input AND Gate Truth Table

INPUTS		OUTPUT
A	B	C
0	0	1
0	1	1
1	0	1
1	1	0

2-Input NAND Gate Truth Table

Key Topic 35: Logic Levels

3-35E1 In a negative-logic circuit, what level is used to represent a logic 0?

A. Low level.
B. Positive-transition level.
C. Negative-transition level.
D. High level.

In a negative-logic circuit, the logic "0" is now a high level. H=0, L=1. All 1s and 0s are interchanged from those in a positive logic truth table. **ANSWER D**

3-35E2 For the logic input levels shown in Figure 3E12, what are the logic levels of test points A, B and C in this circuit? (Assume positive logic.)

A. A is high, B is low and C is low.
B. A is low, B is high and C is high.
C. A is high, B is high and C is low.
D. A is low, B is high and C is low.

To keep track of things a little easier, let's give names to the unlabeled parts. The upper left gate is an INVERTER, we'll call it INVERTER 1. To its right is an AND GATE, which we'll call AND 1. Down below we have INVERTER 2, and likewise, AND 2. The last component is an OR GATE; we'll call it OR. Starting at the upper left, we have a HIGH signal (H). INVERTER 1 will give us a LOW signal (L) at its output, which, naturally gives us an L at one input of AND 1. The second input of AND 1 is L, coming right from the lower input signal. The OUTPUT of AND 1 has to be L, because an AND gate has to have to H inputs to give an H output. So, point A is L.

The output of INVERTER 2 is H, the opposite of whatever you put into it. Therefore, one input of AND 2 is also H. The second input of AND 2 is also H, because that comes right from our H input signal. With two H inputs, AND gives us an H output. Point B is H.

Finally if either input of OR is H, we get H out. So point C is also H. There are actually people who do this stuff all in their heads! I'm not one of them. **ANSWER B**

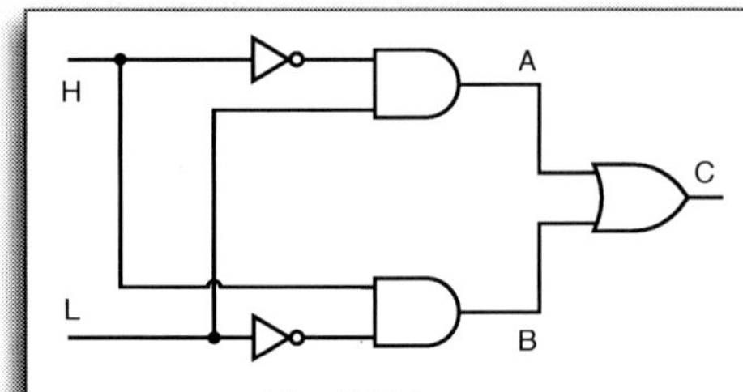

Fig. 3E12

3-35E3 For the logic input levels given in Figure 3E13, what are the logic levels of test points A, B and C in this circuit? (Assume positive logic.)

A. A is low, B is low and C is high.
B. A is low, B is high and C is low.
C. A is high, B is high and C is high.
D. A is high, B is low and C is low.

Now let's take a look at Figure 3E13, and see that A, B, and C are all high. **ANSWER C**

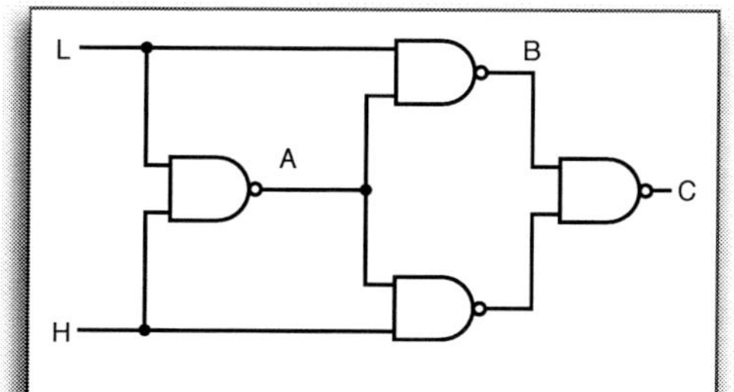

Fig. 3E13

3-35E4 In a positive-logic circuit, what level is used to represent a logic 1?

A. High level
B. Low level
C. Positive-transition level
D. Negative-transition level

In a positive-logic circuit, the logic "1" is a high level, or the most positive level. Think of the four letters in high and the four letters in plus (for positive). **ANSWER A**

3-35E5 Given the input levels shown in Figure 3E14 and assuming positive logic devices, what would the output be?

A. A is low, B is high and C is high.
B. A is high, B is high and C is low.
C. A is low, B is low and C is high.
D. None of the above are correct.

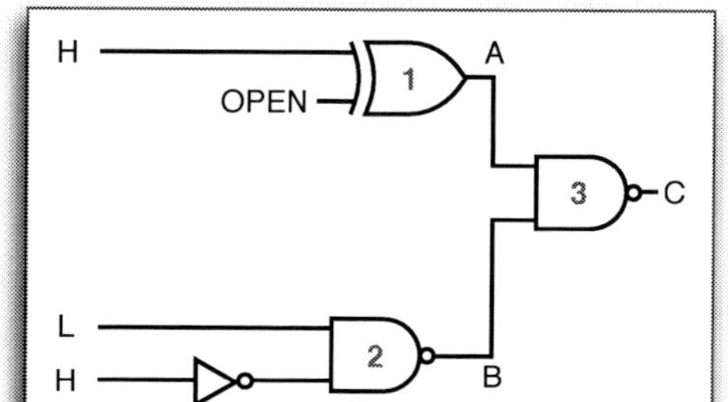

Fig. 3E14

Here is a typical example of *combinational logic*. Working out combinational logic problems is not a difficult task, but it's important to keep track of each individual gate's *output state*. Using the appropriate *truth table* for each gate, jot down the output state of each gate with a 1 or a 0 (or an H or an L, if you prefer) at the output terminal. This will be the *input* state for any following gate. Let's show how this works in the example in this question.

First, look at the output state of gate 1 at point A. This gate is an EXCLUSIVE OR (XOR) gate. The truth table for an XOR gate tells us that if both inputs are HIGH the output is LOW. The upper input is HIGH, designated by H. What about the lower input, labeled OPEN? It is assumed that for all logic gates that an OPEN is HIGH. So, since both inputs are HIGH, the output at point A is LOW. Jot down a 0 or an L at that point.

Gate 2 is a NAND gate. The upper input is L. But the lower input is preceded by an INVERTER, sometimes called a NOT gate (the triangle with a circle on the output). Since the input of the inverter is HIGH, the output of the inverter is LOW. So now our NAND gate has two LOW inputs. The truth table for a NAND gate tells us that with two LOW inputs the output is HIGH (Point B). So you can jot down a 1 or an H at point B.

Finally, we have gate 3, another NAND gate. Since this gate has one HIGH input and one LOW input, the truth table tells us that the output will be HIGH (Point C).

Nothing to it! With a little bit of practice, you'll find that you can work out some of these combinational logic problems in your head. But it's not cheating to jot down 1s and 0s (or Hs and Ls) either. **ANSWER A**

3-35E6 What is a truth table?

A. A list of input combinations and their corresponding outputs that characterizes a digital device's function.
B. A table of logic symbols that indicate the high logic states of an op-amp.
C. A diagram showing logic states when the digital device's output is true.
D. A table of logic symbols that indicates the low logic states of an op-amp.

We use a "truth table" in digital circuitry to characterize a digital device's function. If you buy the owner's technical manual on a piece of radio gear, you will usually find a page describing the truth table of digital devices used in the equipment. **ANSWER A**

AND, OR, and NOT GATES

DIAGRAM SYMBOLS | **TRUTH TABLES**

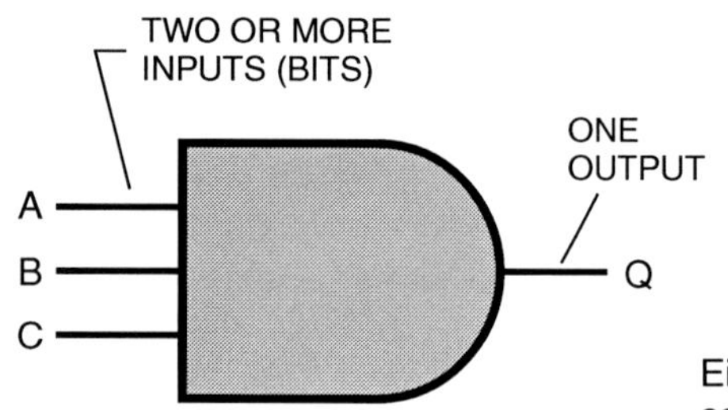

INPUT BITS (A, B, C) — RESULTING OUTPUT BIT (Q)

A	B	C	Q
0	0	0	0
0	0	1	0
0	1	0	0
0	1	1	0
1	0	0	0
1	0	1	0
1	1	0	0
1	1	1	1

Eight possible combinations of three input bits.

Output of an AND gate is 1 only when *all* inputs are 1; that is, Input A AND Input B AND Input C must be 1 for Output Q to be 1.

a. A 3-Input AND Gate

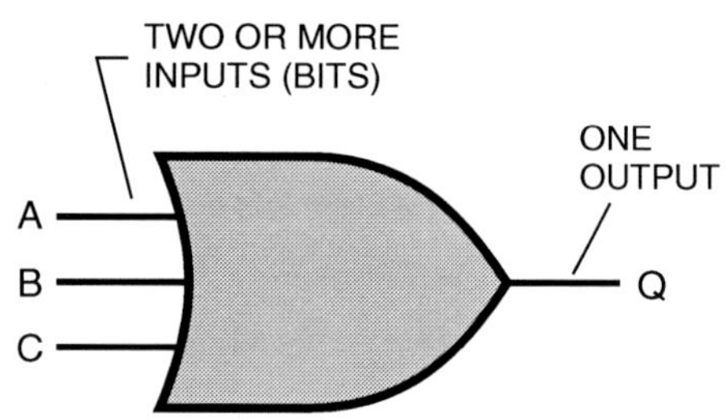

INPUT BITS (A, B, C) — RESULTING OUTPUT BIT (Q)

A	B	C	Q
0	0	0	0
0	0	1	1
0	1	0	1
0	1	1	1
1	0	0	1
1	0	1	1
1	1	0	1
1	1	1	1

Output of an OR gate is 1 when *at least one* input is 1; that is, Input A OR Input B OR Input C, or any combination of A,B,C, must be 1 for Output Q to be 1.

b. A 3-Input OR Gate

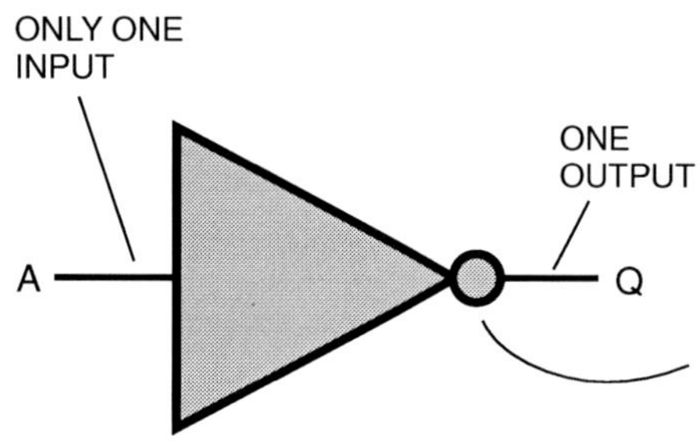

The little circle means *inversion.* This is an inverting digital amplifier.

INPUT BIT (A) — RESULTING OUTPUT BIT (Q)

A	Q
0	1
1	0

An inverter's output is always opposite of the input.

c. A NOT Gate Inverter

In a digital system, all decisions are broken down into the simplest three decisions–AND, OR and NOT–which are handled by circuits called logic gates.

Although the logical designs for digital networks are often worked out with AND and OR gates, the designs usually are modified later to use NAND and NOR gates when they are actually built! This is because NAND and NOR gates normally are simpler and require fewer transistors. When a plain AND or OR gate is needed, designers typically use a NAND gate or a NOR gate, followed by an inverter. The inverter changes the NAND to an AND, and a NOR to an OR.

Key Topic 36: Flip-Flops

3-36E1 A flip-flop circuit is a binary logic element with how many stable states?

A. 1 B. 2 C. 4 D. 8

Flip-flop – two words, two states. A flip-flop circuit is a form of bistable multivibrator – it remains in one of two states until triggered to change to the other state. It is a bistable sequential logic circuit. **ANSWER B**

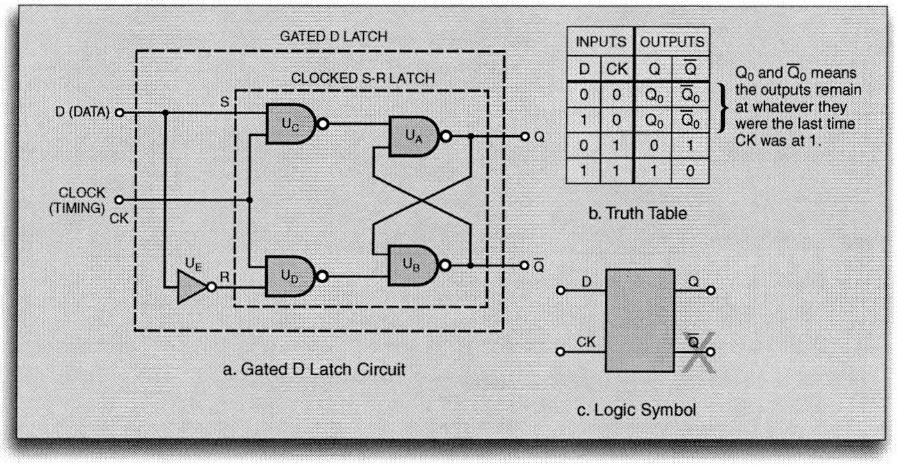

A Clocked R-S Latch (a Type of Flip-Flop) can be Changed to a Gated D Latch by Connecting an Additional Inverter.

3-36E2 What is a flip-flop circuit? A binary sequential logic element with ___stable states.

A. 1 B. 4 C. 2 D. 8

The flip-flop, or bistable multivibrator, is the simplest sequential logic element. By sequential, we mean that it has a memory of a previous logic state. By contrast, combinational logic units like AND and OR gates only care what the present state of the input signals are. This opens up entirely new worlds of digital possibilities.

If a flip flop consists of two transistors, Q1 and Q2, we have two possible states, Q1 ON and Q2 OFF, or Q2 ON and Q1 OFF. A trigger pulse will transfer the device from one state to the other, and it will stay there until it receives another trigger pulse; in other words it is stable in either one or the other possible states, hence "BI-STABLE" In digital shorthand, the first state is often referred to as "Q" while the second state is often referred to as "Q prime" designated by a Q with a bar over it. **ANSWER C**

3-36E3 How many flip-flops are required to divide a signal frequency by 4?

A. 1 B. 4 C. 8 D. 2

The bistable multivibrator, flip-flop, or memory cell – all the same, really – is central to almost everything one does in the digital world. Microprocessors have millions of flip-flops, but you can build one with just two transistors and four resistors. A flip-flop changes states once per trigger pulse. If we extract a trigger pulse from one side of a flip-flop, we will get one trigger pulse out for every two trigger pulses we put in, a two-to-one frequency divider. We can cascade or "daisy chain" as many of these dividers as we want, each dividing the input frequency by two. So, we can divide by 2 or 4 or 8 or 16 or any other power of 2. **ANSWER D**

3-36E4 How many bits of information can be stored in a single flip-flop circuit?

A. 1 B. 2 C. 3 D. 4

One flip-flop circuit can only store a single bit of information. If semiconductor memories were made of flip-flops, one flip-flop would be required for each bit. **ANSWER A**

3-36E5 How many R-S flip-flops would be required to construct an 8 bit storage register?

A. 2 B. 4 C. 8 D. 16

The R-S (Reset-Set) type flip-flop is the simplest logic memory element. It consists of two transistors (Q1 and Q2) and two possible STATES: Q1 on and Q2 off, or Q1 off and Q2 on. It can store 1 bit of information, the bit being either a 1 or a 0. 8 bits of memory would require 8 such flip-flops, containing a total of 16 transistors. **ANSWER C**

3-36E6 An R-S flip-flop is capable of doing all of the following except:

A. Accept data input into R-S inputs with CLK initiated.
B. Accept data input into PRE and CLR inputs without CLK being initiated.
C. Refuse to accept synchronous data if asynchronous data is being input at same time.
D. Operate in toggle mode with R-S inputs held constant and CLK initiated.

An R-S flip-flop doesn't have a toggle mode, so answer D is correct. The R-S flip-flop is capable of all the other functions. **ANSWER D**

FLIP-FLOPS

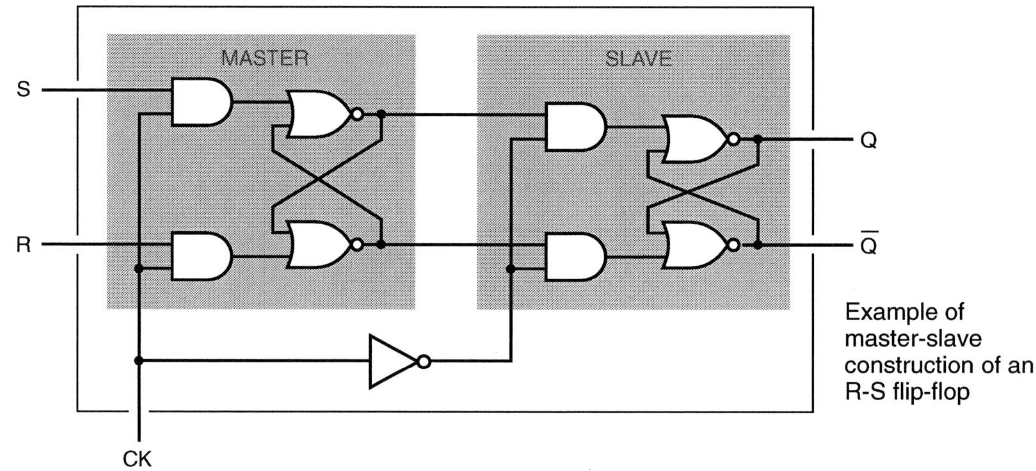

a. Master-Slave Flip-Flop Design

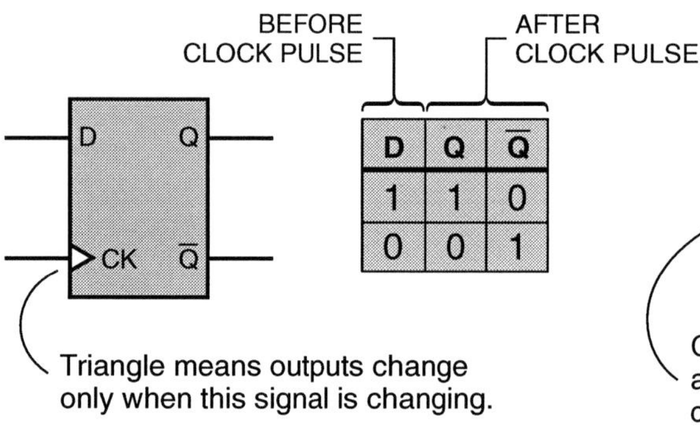

D	Q	$\overline{Q}$
1	1	0
0	0	1

b. D Flip-Flop

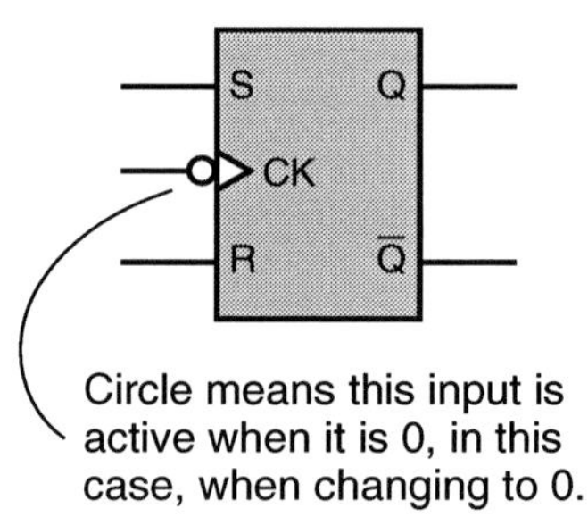

		Q	$\overline{Q}$
0	0	Q_0	$\overline{Q}_0$
0	1	0	1
1	0	1	0
1	1	?	?

Question mark means uncertain.

c. R-S Flip-Flop

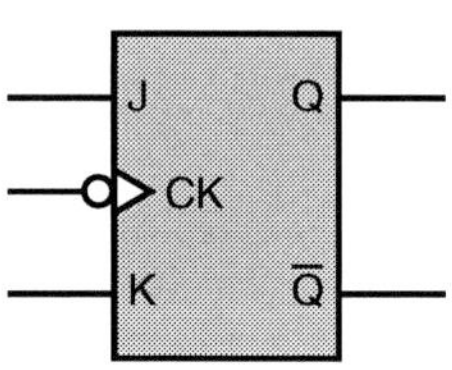

J	K	Q	$\overline{Q}$
0	0	Q_0	$\overline{Q}_0$
0	1	0	1
1	0	1	0
1	1	$\overline{Q}_0$	Q_0

d. J-K Flip-Flop

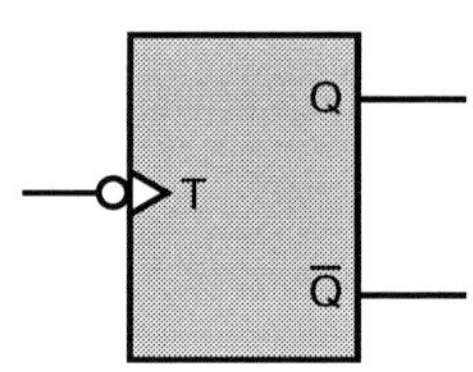

Q	$\overline{Q}$
$\overline{Q}_0$	Q_0

e. T Flip-Flop

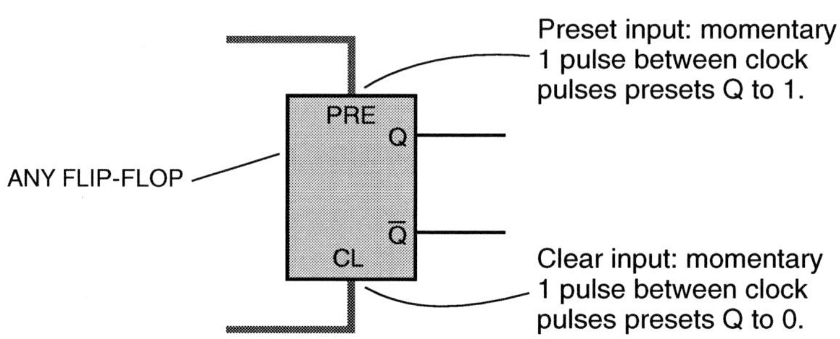

f. Preset and Clear

Several kinds of flip-flop that are widely used. They typically consist of a master stage followed by a slave stage. Any design can be provided with preset and clear inputs.

Key Topic 37: Multivibrators

3-37E1 The frequency of an AC signal can be divided electronically by what type of digital circuit?

A. Free-running multivibrator.
B. Bistable multivibrator.
C. OR gate.
D. Astable multivibrator.

A bistable multivibrator can be used to divide the frequency of an AC signal. **ANSWER B**

3-37E2 What is an astable multivibrator?

A. A circuit that alternates between two stable states.
B. A circuit that alternates between a stable state and an unstable state.
C. A circuit set to block either a 0 pulse or a 1 pulse and pass the other.
D. A circuit that alternates between two unstable states.

An astable multivibrator continuously switches back and forth between two states. It doesn't remain stable for a permanent time in either of its two states. **ANSWER D**

An astable multivibrator constructed using two TTL inverters, two resistors and a capacitor is shown. The output of one inverter is coupled back to the input of the other inverter so the circuit forms a basic oscillator. The pulse will be ON for about 33% of the cycle and its frequency can be approximated as f = 1/(3RC), which is the reciprocal of the period T. Calculate the frequency if R = 10 kilohms and C = 0.001 microfarad.

Solution:

$$RC = 10 \times 10^3 \times 1 \times 10^{-9} = 1 \times 10^{-5} \text{ second}$$

$$3RC = 3 \times 10^{-5} \text{ second}$$

$$f = \frac{1}{3 \times 10^{-5}} = 0.333 \times 10^5 = 33{,}300 \text{ Hz}$$

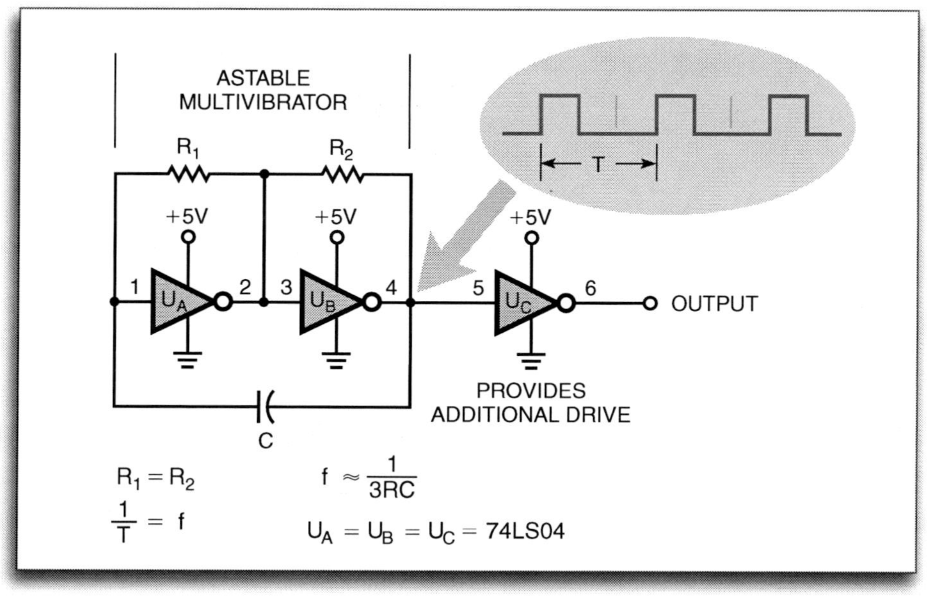

An astable multivibrator

3-37E3 What is a monostable multivibrator?
A. A circuit that can be switched momentarily to the opposite binary state and then returns after a set time to its original state.
B. A "clock" circuit that produces a continuous square wave oscillating between 1 and 0.
C. A circuit designed to store one bit of data in either the 0 or the 1 configuration.
D. A circuit that maintains a constant output voltage, regardless of variations in the input voltage.

This type of multivibrator may momentarily be monostable (MO MO). It stays in its original state until triggered into its other state, where it remains for a time usually determined by external components, after which it returns to the original state. **ANSWER A**

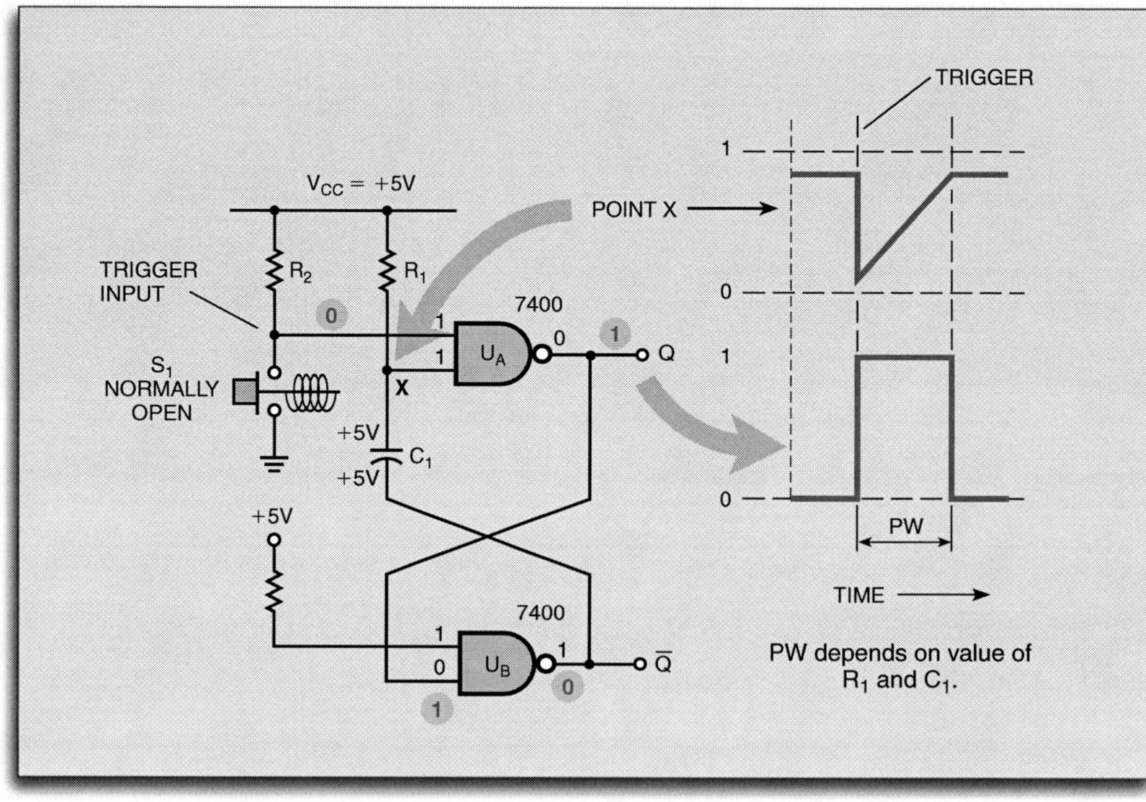

A monostable multivibrator circuit built from two NAND gates.

3-37E4 What is a bistable multivibrator circuit commonly named?
A. AND gate.
B. OR gate.
C. Clock.
D. Flip-flop.

A bistable multivibrator is a flip-flop. Bistable means two stable states. **ANSWER D**

3-37E5 What is a bistable multivibrator circuit?
A. Flip-flop.
B. AND gate.
C. OR gate.
D. Clock.

The bistable multivibrator offers two stable states, and may be controlled to switch from one to the other. In a flip-flop circuit, pulses may drive the bistable multivibrator that will lock into one of two possible positions. **ANSWER A**

3-37E6 What wave form would appear on the voltage outputs at the collectors of an astable, multivibrator, common-emitter stage?
A. Sine wave.
B. Sawtooth wave.
C. Square wave.
D. Half-wave pulses.

Multivibrator pulses are made up of square waves. In the common-emitter stage, the astable multivibrator will cross couple the inputs and outputs. **ANSWER C**

Key Topic 38: Memory

3-38E1 What is the name of the semiconductor memory IC whose digital data can be written or read, and whose memory word address can be accessed randomly?

A. ROM – Read-Only Memory.
B. PROM – Programmable Read-Only Memory.
C. RAM – Random-Access Memory.
D. EPROM – Electrically Programmable Read-Only Memory.

This question answers itself: Random-Access Memory (RAM). **ANSWER C**

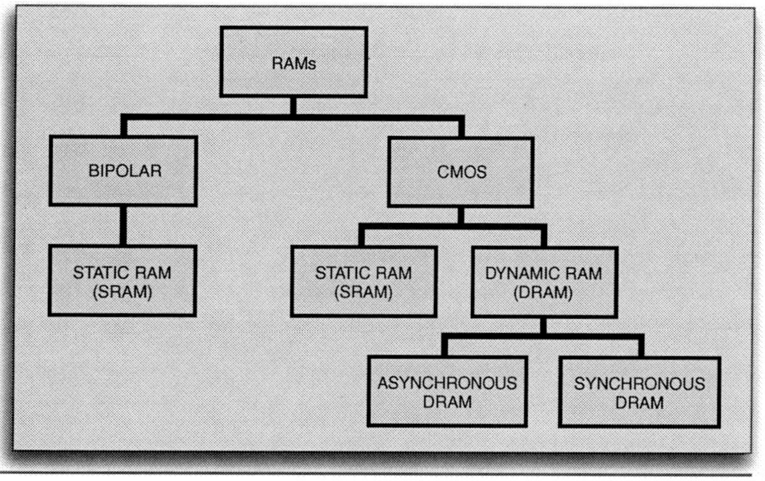

RAM family uses both bipolar and CMOS technology. Bipolar is used only for static RAM.

3-38E2 What is the name of the semiconductor IC that has a fixed pattern of digital data stored in its memory matrix?

A. RAM – Random-Access Memory.
B. ROM – Read-Only Memory.
C. Register.
D. Latch.

ROM – Read-Only Memory – is a semiconductor memory IC that has its digital data placed in the circuitry at the time of manufacture. **ANSWER B**

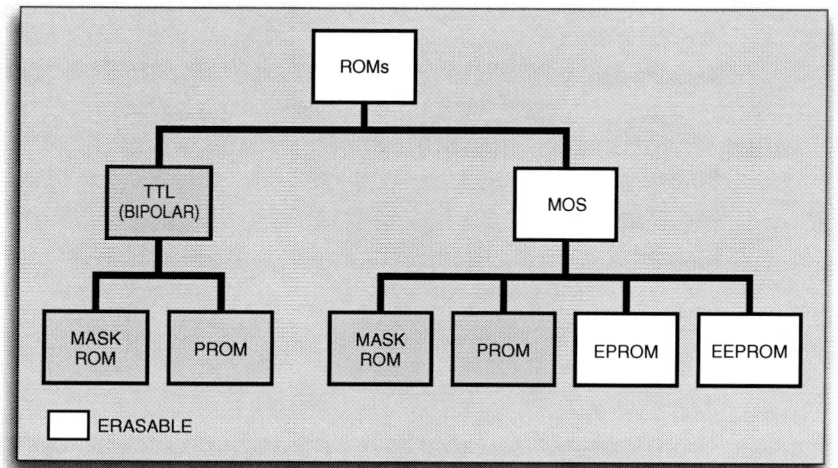

Semiconductor ROM family.

3-38E3 What does the term "IO" mean within a microprocessor system?

A. Integrated oscillator.
B. Integer operation.
C. Input-output.
D Internal operation.

Data in, data out! IO stands for Input-Output in the modern microprocessor circuit. **ANSWER C**

3-38E4 What is the name for a microprocessor's sequence of commands and instructions?

A. Program.
B. Sequence.
C. Data string.
D. Data execution.

In a microprocessor sequence, the program determines the sequence of commands and instructions. **ANSWER A**

3-38E5 How many individual memory cells would be contained in a memory IC that has 4 data bus input/output pins and 4 address pins for connection to the address bus?

A. 8
B. 16
C. 32
D. 64

A 4-bit address means 16 memory addresses. Each memory address contains a 4-bit word. Sixteen words times 4 bits equals 64 individual memory cells. **ANSWER D**

3-38E6 What is the name of the random-accessed semiconductor memory IC that must be refreshed periodically to maintain reliable data storage in its memory matrix?

A. ROM – Read-Only Memory.
B. DRAM – Dynamic Random-Access Memory.
C. PROM – Programmable Read-Only Memory.
D. PRAM – Programmable Random-Access Memory.

A RAM that must be periodically refreshed to maintain reliable data storage is called a dynamic random-access memory – DRAM. **ANSWER B**

Key Topic 39: Microprocessors

3-39E1 In a microprocessor-controlled two-way radio, a "watchdog" timer:

A. Verifies that the microprocessor is executing the program.
B. Assures that the transmission is exactly on frequency.
C. Prevents the transmitter from exceeding allowed power out.
D. Connects to the system RADAR presentation.

Here's a slang word that defied my lookup for its origin. A "watchdog timer" is a circuit that keeps track of the radio's microprocessor in the execution of a program. **ANSWER A**

Addressable microprocessor chip.

3-39E2 What does the term "DAC" refer to in a microprocessor circuit?

A. Dynamic access controller.
B. Digital to analog converter.
C. Digital access counter.
D. Dial analog control.

Since more and more communications products are going with digital circuits, the digital-to-analog converter (DAC) is a necessary stage to take the digital information and convert it to an analog signal for the receiver output. **ANSWER B**

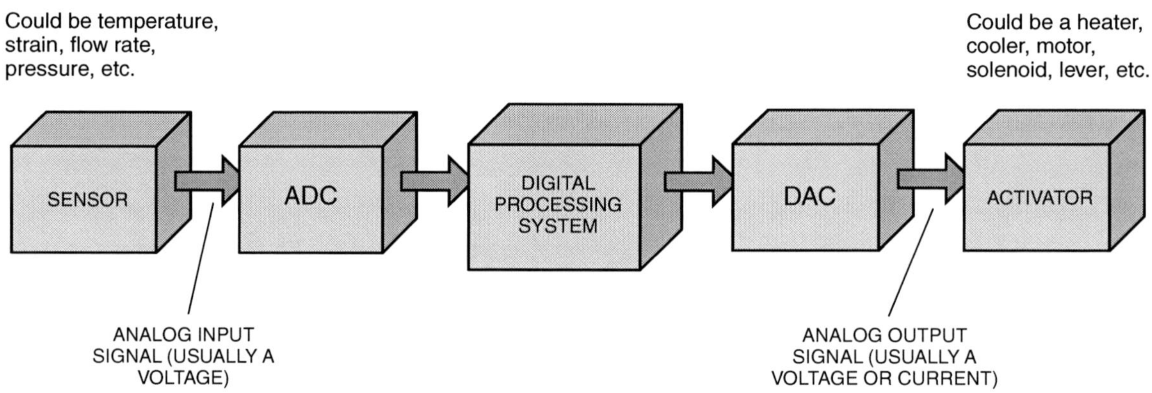

A data acquisition and control system using analog-to-digital and digital-to-analog converters.

3-39E3 Which of the following is not part of a MCU processor?

A. RAM
B. ROM
C. I/O
D. Voltage Regulator

Within the microprocessor control unit, we could find RAM, ROM, and I/O (input and output stages), but the voltage regulator would NOT be part of the insides of a microprocessor unit. Indeed, the equipment will have multiple voltage regulators, but not inside the MCU. **ANSWER D**

3-39E4 What portion of a microprocessor circuit is the pulse generator?

A. Clock
B. RAM
C. ROM
D. PLL

It is the CLOCK within the MCU that generates data pulses. **ANSWER A**

3-39E5 In a microprocessor, what is the meaning of the term "ALU"?

A. Automatic lock/unlock.
B. Arithmetical logic unit.
C. Auto latch undo.
D. Answer local unit.

The arithmetical logic unit (ALU) performs simple binary processes, shift processes, as well as addition, multiplication, subtraction and division. **ANSWER B**

3-39E6 What circuit interconnects the microprocessor with the memory and input/output system?
A. Control logic bus.
B. PLL line.
C. Data bus line.
D. Directional coupler.
Memory pulses and the input/output system are connected via the data bus line. This could include a physical conductor or an optical encoder/decoder. **ANSWER C**

Key Topic 40: Counters, Dividers, Converters

3-40E1 What is the purpose of a prescaler circuit?
A. Converts the output of a JK flip-flop to that of an RS flip-flop.
B. Multiplies an HF signal so a low-frequency counter can display the operating frequency.
C. Prevents oscillation in a low frequency counter circuit.
D. Divides an HF signal so that a low-frequency counter can display the operating frequency.
You will find a prescaler in frequency counters to divide down HF and VHF signals so they can be counted and displayed on a low-frequency counter. **ANSWER D**

3-40E2 What does the term "BCD" mean?
A. Binaural coded digit.
B. Bit count decimal.
C. Binary coded decimal.
D. Broad course digit.
Binary coded decimal, abbreviated BCD, results in a single digit of a decimal number represented by a binary equivalent. **ANSWER C**

3-40E3 What is the function of a decade counter digital IC?
A. Decode a decimal number for display on a seven-segment LED display.
B. Produce one output pulse for every ten input pulses.
C. Produce ten output pulses for every input pulse.
D. Add two decimal numbers.
A decade counter digital IC gives one output pulse for every 10 input pulses. **ANSWER B**

3-40E4 What integrated circuit device converts an analog signal to a digital signal?
A. DAC
B. DCC
C. ADC
D. CDC
The integrated circuit that converts analog signal from the RF stage to digital signals, within a radio, is called an analog-to-digital converter (ADC). **ANSWER C**

3-40E5 What integrated circuit device converts digital signals to analog signals?
A. ADC
B. DCC
C. CDC
D. DAC
And within that same radio, we need to come back out to a voice speaker system by using a digital to analog converter, DAC. **ANSWER D**

3-40E6 In binary numbers, how would you note the quantity TWO?
A. 0010
B. 0002
C. 2000
D. 0020
Look at this chart. It represents binary numbers that would represent the quantity TWO. **ANSWER A**

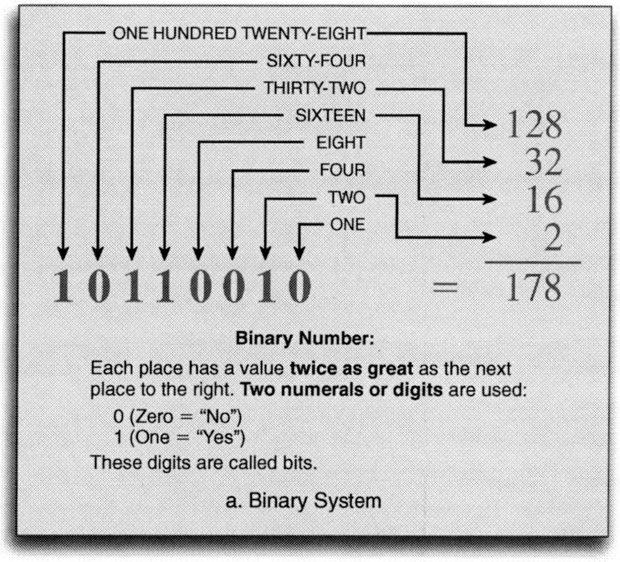

a. Binary System

In digital systems, binary nimbers are used to write and keep track of the many possible combinations of the two electrical states.

DIGITAL SIGNAL PROCESSING

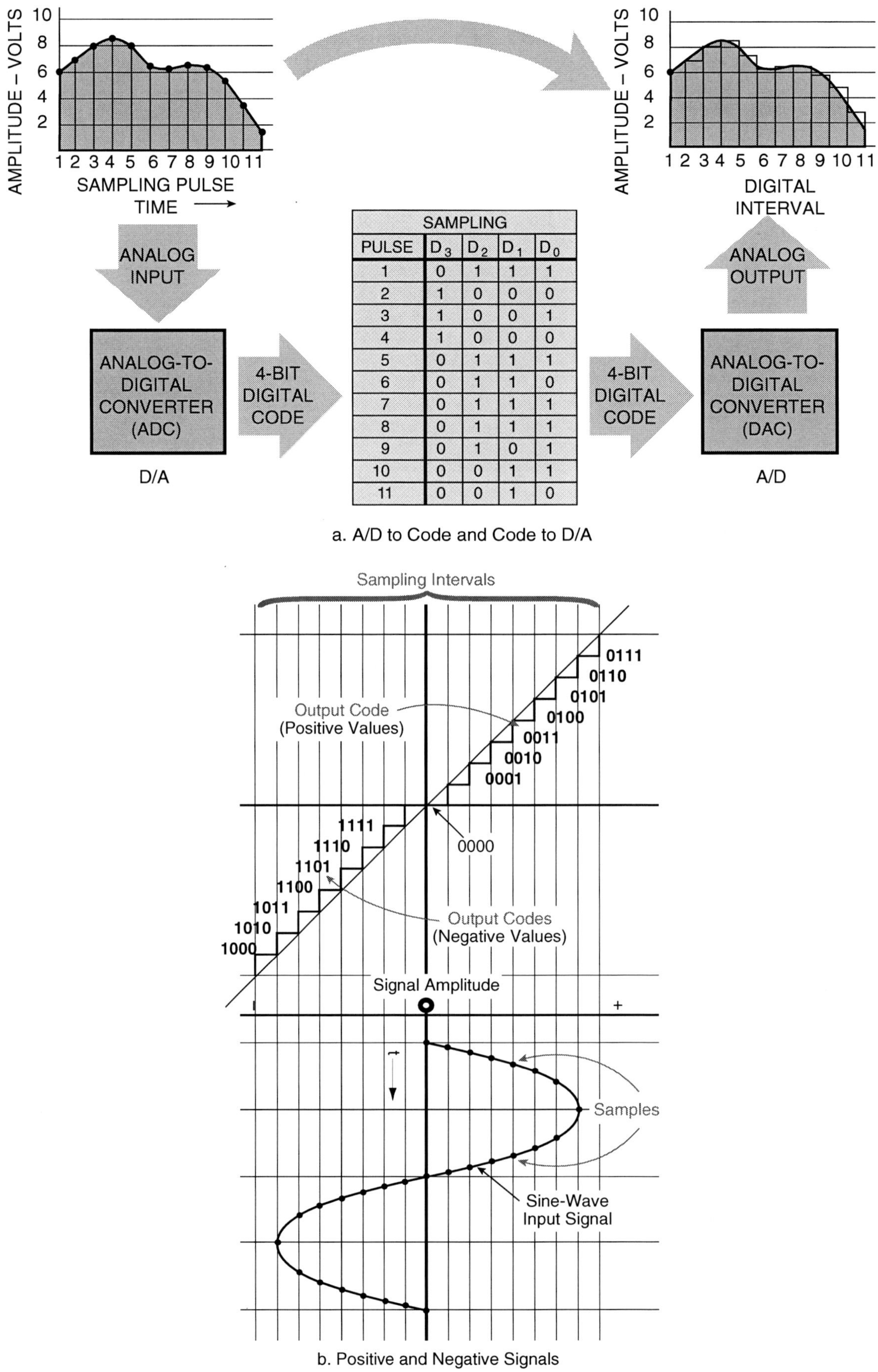

SAMPLING				
PULSE	D_3	D_2	D_1	D_0
1	0	1	1	1
2	1	0	0	0
3	1	0	0	1
4	1	0	0	0
5	0	1	1	1
6	0	1	1	0
7	0	1	1	1
8	0	1	1	1
9	0	1	0	1
10	0	0	1	1
11	0	0	1	0

a. A/D to Code and Code to D/A

b. Positive and Negative Signals

The A/D and D/A functions of digital signal processing.

Subelement F – Receivers

10 Key Topics, 10 Exam Questions, 2 Drawings

Key Topic 41: Receiver Theory

3-41F1 What is the limiting condition for sensitivity in a communications receiver?

A. The noise floor of the receiver.
B. The power supply output ripple.
C. The two-tone intermodulation distortion.
D. The input impedance to the detector.

Very weak signals get masked by the background noise of a receiver – the so-called noise floor of the receiver. The sensitivity of a receiver is limited by the noise generated by the devices inside, which contribute to the receiver's noise floor. **ANSWER A**

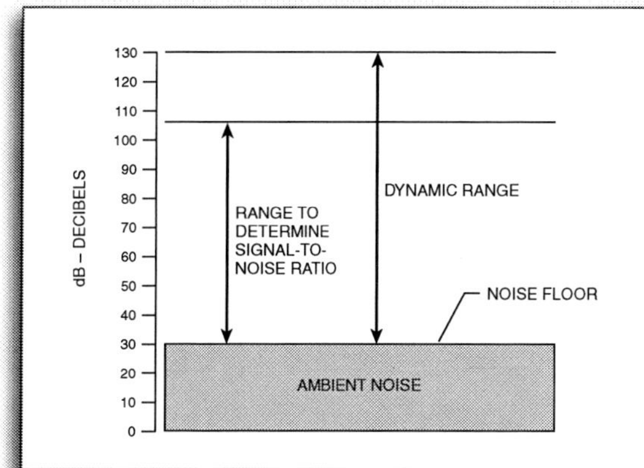

Noise Floor in a Receiver
Source: Installing and Maintaining Sound Systems, G. McComb, ©1996, Master Publishing, Inc.

3-41F2 What is the definition of the term "receiver desensitizing"?

A. A burst of noise when the squelch is set too low.
B. A reduction in receiver sensitivity because of a strong signal on a nearby frequency.
C. A burst of noise when the squelch is set too high.
D. A reduction in receiver sensitivity when the AF gain control is turned down.

You are up on a building rooftop servicing a base station when all of a sudden the incoming signal abruptly drops, and then pops up again at its normal level. Chances are the receiver of the equipment on which you are working is losing sensitivity because of the presence of a strong signal on an adjacent channel. Time to consult your spectrum analyzer, or a sensitive portable frequency counter, and determine what steps you are going to take to minimize receiver desensitization. **ANSWER B**

3-41F3 What is the term used to refer to a reduction in receiver sensitivity caused by unwanted high-level adjacent channel signals?

A. Desensitizing.
B. Intermodulation distortion.
C. Quieting.
D. Overloading.

If signals mysteriously come in strong, then abruptly get weak, and then come back strong again, chances are there is someone nearby transmitting on an adjacent frequency, desensitizing your receiver. The best way to check for adjacent signals is through the use of a frequency counter. **ANSWER A**

3-41F4 What is meant by the term noise figure of a communications receiver?

A. The level of noise entering the receiver from the antenna.
B. The relative strength of a received signal 3 kHz removed from the carrier frequency.
C. The level of noise generated in the front end and succeeding stages of a receiver.
D. The ability of a receiver to reject unwanted signals at frequencies close to the desired one.

For VHF and UHF satellite gear, always look for the receiver with the lowest noise figure and a hot front-end complement of transistors. **ANSWER C**

Noise figure, NF, is used to measure the noise performance of receivers. Noise figure is the ratio of the input S/N to the output S/N and is a measure of the noise introduced by the receiver. NF can be expressed as:

$$NF = 10 \log_{10} \times \frac{\text{input S/N}}{\text{output S/N}}$$

where NF = noise figure in **dB**
Input S/N = input signal-to-noise **ratio**
Output S/N = output signal-to-noise **ratio**
$\log_{10}$ = the logarithm to the **base 10**

As shown, NF, usually expressed in decibels, is 10 times the logarithm of the ratio of the *input-to-output signal-to-noise ratios*. A perfect receiver will have a NF of 0 dB, indicating that the receiver adds no noise to the received signal.

3-41F5 Which stage of a receiver primarily establishes its noise figure?

A. The audio stage.
B. The RF stage.
C. The IF strip.
D. The local oscillator.

It's the RF amplifier stage where we are most concerned about a low noise figure. **ANSWER B**

3-41F6 What is the term for the ratio between the largest tolerable receiver input signal and the minimum discernible signal?

A. Intermodulation distortion.
B. Noise floor.
C. Noise figure.
D. Dynamic range.

If your client's older two-way radio equipment sounds distorted when receiving extremely strong signals, chances are it has poor dynamic range. Good dynamic range allows the receiver to capture and send to the amplifier stage good, clean audio on extremely strong signal inputs all the way down to the minimum discernible signal. **ANSWER D**

Key Topic 42: RF Amplifiers

3-42F1 How can selectivity be achieved in the front-end circuitry of a communications receiver?

A. By using an audio filter.
B. By using an additional RF amplifier stage.
C. By using an additional IF amplifier stage.
D. By using a preselector.

A preselector achieves selectivity in the front-end circuitry of a communications receiver. When you change bands, different preselectors are automatically brought on line. **ANSWER D**

3-42F2 What is the primary purpose of an RF amplifier in a receiver?

A. To provide most of the receiver gain.
B. To vary the receiver image rejection by utilizing the AGC.
C. To improve the receiver's noise figure.
D. To develop the AGC voltage.

The RF amplifier, factory designed for your receiver, improves the receiver's noise figure. Low-noise transistors are used for an improved noise-figure rating. **ANSWER C**

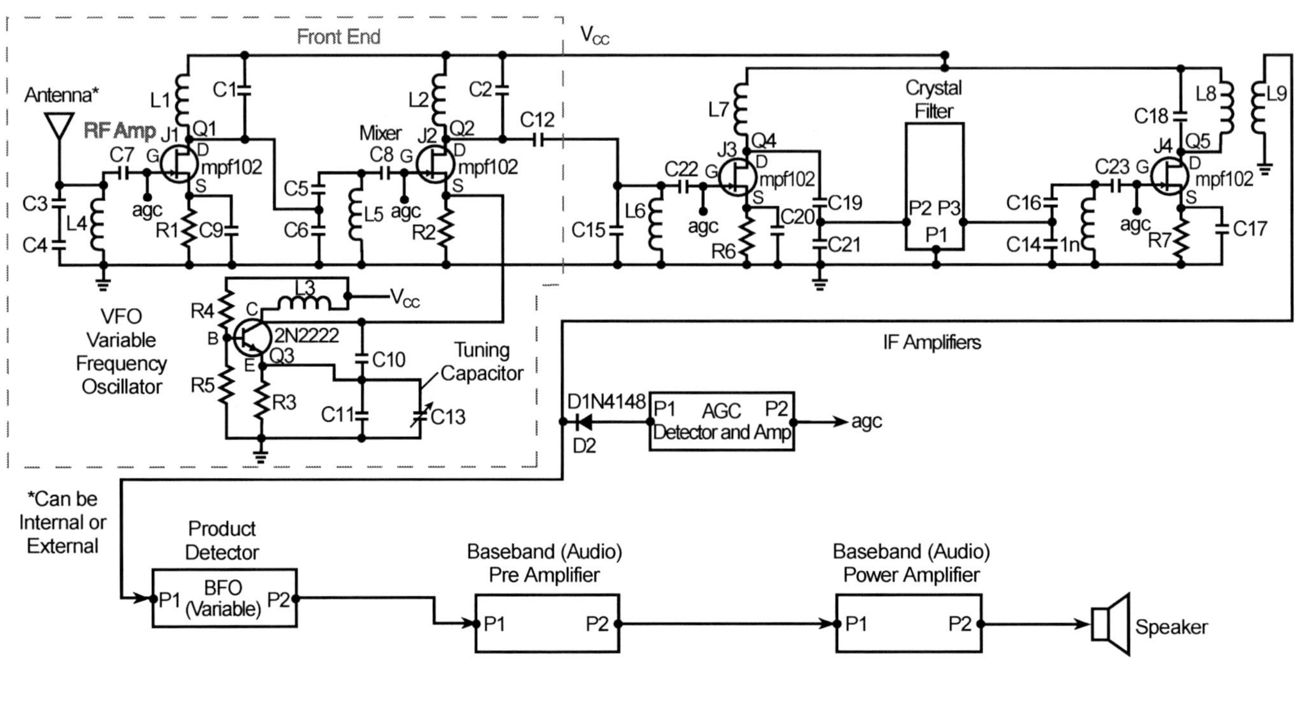

An SSB receiver

3-42F3 How much gain should be used in the RF amplifier stage of a receiver?

A. Sufficient gain to allow weak signals to overcome noise generated in the first mixer stage.
B. As much gain as possible short of self oscillation.
C. Sufficient gain to keep weak signals below the noise of the first mixer stage.
D. It depends on the amplification factor of the first IF stage.

Very sensitive receiver transistors now give the signal amplifier stages all the gain they need. If you try to add a pre-amplifier on a modern two-way transceiver, you could create more interference than the possible weak-signal amplification you would gain. You need enough gain to allow weak signals to overcome noise generated in the first mixer stage. **ANSWER A**

Receiver section in a communications transceiver.

3-42F4 Too much gain in a VHF receiver front end could result in this:

A. Local signals become weaker.
B. Difficult to match receiver impedances.
C. Dramatic increase in receiver current.
D. Susceptibility of intermodulation interference from nearby transmitters.

A VHF receiver, operating at high gain and off the desired curve for class A operation, may become non-linear. The result could be intermodulation distortion if a nearby strong signal comes on the air. The intermodulation interference may last only as long as the other powerful nearby frequency transmitter is UP, with greatest susceptibility to intermodulation in downtown harbor areas with strong 162 MHz continuous transmitting weather stations, and 152/157 MHz constant-on paging transmitters. **ANSWER D**

3-42F5 What is the advantage of a GaAsFET preamplifier in a modern VHF radio receiver?

A. Increased selectivity and flat gain.
B. Low gain but high selectivity.
C. High gain and low noise floor.
D. High gain with high noise floor.

The Gallium Arsenide Field Effect Transistor (GaAsFET) is widely used in UHF and microwave narrowband receivers. When marine band radio systems go narrowband, we will likely see increased use of the GaAsFET preamplifier, along with tighter filter sections in VHF marine band radios. **ANSWER C**

3-42F6 In what stage of a VHF receiver would a low noise amplifier be most advantageous?

A. IF stage.
B. Front end RF stage.
C. Audio stage.
D. Power supply.

A low noise front end is best employed at the first RF stage running Class A for minimum distortion, and a gain figure that will minimize intermodulation interference. **ANSWER B**

Key Topic 43: Oscillators

3-43F1 Why is the Colpitts oscillator circuit commonly used in a VFO (variable frequency oscillator)?

A. It can be phase locked.
B. It can be remotely tuned.
C. It is stable.
D. It has little or no effect on the crystal's stability.

The Colpitts oscillator is relatively stable. After you let things warm up, the Colpitts oscillator normally stays put on a single frequency. **ANSWER C**

3-43F2 What is the oscillator stage called in a frequency synthesizer?

A. VCO.
B. Divider.
C. Phase detector.
D. Reference standard.

In a PLL synthesizer, the output of the voltage controlled oscillator (VCO) feeds to a programmable divider. The divider is kept locked "rock on" by the phase comparator that is driven by a crystal-controlled reference oscillator. If the VCO drifts slightly, an error voltage from the phase comparator will travel down a feedback loop and apply the correct anti-drift DC error voltage, bringing the VCO back on to "rock" (crystal) solid frequency. Many times the reference oscillator crystal is encased within a "crystal oven" to further stabilize the entire PLL circuit. **ANSWER A**

3-43F3 What are three major oscillator circuits found in radio equipment?

A. Taft, Pierce, and negative feedback.
B. Colpitts, Hartley, and Taft.
C. Taft, Hartley, and Pierce.
D. Colpitts, Hartley, and Pierce.

Colpitts oscillators have a capacitor just like C in its name. Hartley is tapped, and a Pierce oscillator uses a crystal. **ANSWER D** *(See the illustrations at 3-28C6.)*

3-43F4 Which type of oscillator circuit is commonly used in a VFO (variable frequency oscillator)?

A. Colpitts.
B. Pierce.
C. Hartley.
D. Negative feedback.

Since the Colpitts oscillator uses a big capacitor in a variable frequency oscillator (VFO), it's the most common oscillator circuit for older VFO radios. Newer radios, even though they say they have a VFO, are really digitally controlled via an optical reader. They just look like they have a big capacitor behind that large tuning dial! **ANSWER A**

3-43F5 What condition must exist for a circuit to oscillate? It must:

A. Have a gain of less than 1.
B. Be neutralized.
C. Have sufficient negative feedback.
D. Have sufficient positive feedback.

The circuit must have just enough positive feedback to overcome losses in order to continue to properly oscillate. **ANSWER D**

3-43F6 In Figure 3F15, which block diagram symbol (labeled 1 through 4) is used to represent a local oscillator?

A. 1
B. 2
C. 3
D. 4

Working this diagram from left to right, the antenna feeds wide-band energy into the front end, which feeds #1, the mixer, which also has input from #2, the local oscillator. The output of the mixer is sent to the I.F. amplifiers. **ANSWER B**

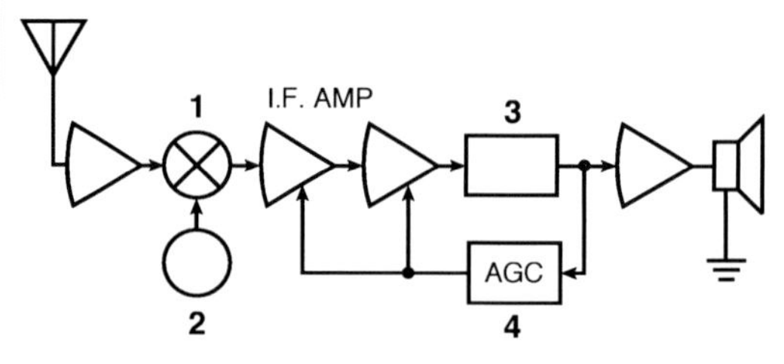

Fig. 3F15

Key Topic 44: Mixers

3-44F1 What is the image frequency if the normal channel is 151.000 MHz, the IF is operating at 11.000 MHz, and the LO is at 140.000 MHz?

A. 131.000 MHz.
B. 129.000 MHz.
C. 162.000 MHz.
D. 150.000 MHz.

Because mixing produces sum and difference frequencies of the received signals and the local oscillator signal, there is an opportunity for interfering signals that appear at the RF input to also produce intermediate frequency signals at the mixer output. In this problem, there is an image frequency that is greater than the station frequency by twice the intermediate frequency. The local oscillator is LOWER than the incoming frequency, so the image frequency equals the incoming frequency minus 2 times the intermediate frequency: 151 MHz -11 MHz -11 MHz = the correct answer at 129 MHz. **ANSWER B**

3-44F2 What is the mixing process in a radio receiver?

A. The elimination of noise in a wideband receiver by phase comparison.
B. The elimination of noise in a wideband receiver by phase differentiation.
C. Distortion caused by auroral propagation.
D. The combination of two signals to produce sum and difference frequencies.

Inside two-way radio equipment are several stages of mixers that combine two input signals to produce sum and different frequencies. These two inputs are the desired RF signal and the local oscillator (LO) signal. One of the output signals becomes the intermediate frequency (IF) signal inside the receiver section. **ANSWER D**

3-44F3 In what radio stage is the image frequency normally rejected?

A. RF.
B. IF.
C. LO.
D. Detector.

The radio frequency stage in any communications receiver has a double duty – offer sensitivity to weak stations on the desired frequency, and provide plenty of selectivity to strong stations above and below the desired station frequency. The RF stage must operate linear, Class A, minimizing intermodulation interference problems. In the modern VHF or UHF communications receiver, the RF stage may also include helical resonators in order to minimize intermodulation and the problems of out-of-band image frequency response. **ANSWER A**

3-44F4 What are the principal frequencies that appear at the output of a mixer circuit?

A. Two and four times the original frequency.
B. The sum, difference and square root of the input frequencies.
C. The original frequencies and the sum and difference frequencies.
D. 1.414 and 0.707 times the input frequency.

The output of a mixer circuit is your original two frequencies, and the sum and difference frequencies. The more elaborate the transceiver, the more mixing stages found in the sets. This helps filter out unwanted signals or phantom signals that could cause interference. The output of a mixer circuit is your original two frequencies, and the sum and difference frequencies. The more elaborate the transceiver, the more mixing stages found in the set. This helps filter out unwanted signals or phantom signals that could cause interference. There are also BALANCED and DOUBLE BALANCED mixers, which reject either one or both of the original frequencies, respectively. Double balanced mixers, DBMs, have become nearly standard in most modern receiver designs. However, a traditional mixer produces all four products, as well as many other minor products, of varying levels of importance. **ANSWER C**

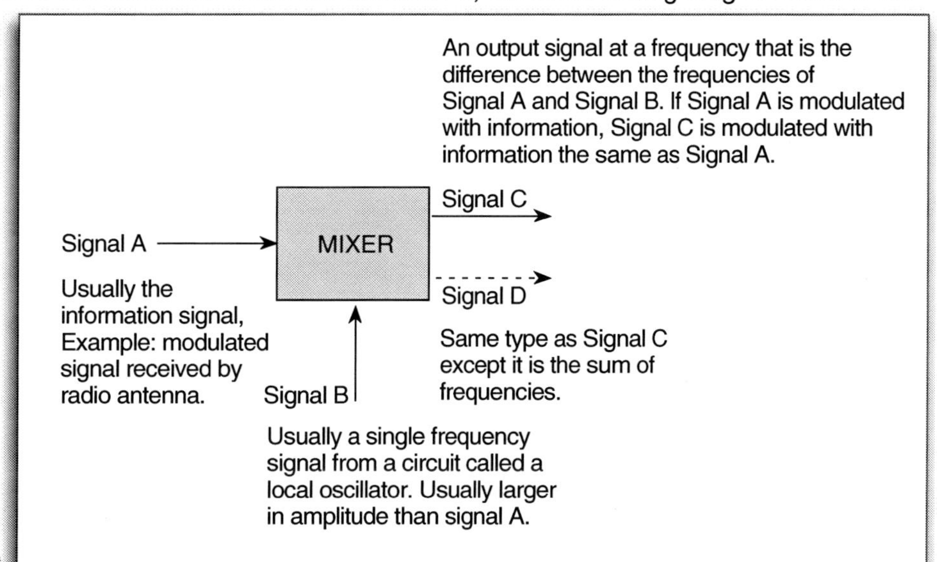

Block diagram of a mixer circuit

3-44F5 If a receiver mixes a 13.8 MHz VFO with a 14.255 MHz receive signal to produce a 455 kHz intermediate frequency signal, what type of interference will a 13.345 MHz signal produce in the receiver?

A. Local oscillator interference.
B. An image response.
C. Mixer interference.
D. Intermediate frequency interference.

When two signals are mixed, we get two primary products out: the sum of the two signals, and the difference between the two signals. In a superheterodyne receiver, where we use mixing to convert incoming signals to an I.F. (intermediate frequency) only one of these two products is desirable. The unwanted one is known as an image. In a superheterodyne receiver the image frequency is always twice the I.F. frequency away from the desired frequency. If our receiver has an I.F. frequency of 455 kHz, the image frequency is going to be 910 kHz away from our desired frequency. The image frequency for a radio signal at 14.255 MHZ is going to be 910 KHz (0.910 MHz) away from 14.255. 14.255 minus 0.910 = 13.345 MHz. **ANSWER B**

3-44F6 What might occur in a receiver if excessive amounts of signal energy overdrive the mixer circuit?

A. Automatic limiting occurs.
B. Mixer blanking occurs.
C. Spurious mixer products are generated.
D. The mixer circuit becomes unstable and drifts.

Any time we overdrive the RF input, or the mixer circuit, the overdriven mixer will produce serious spurious products. **ANSWER C**

Key Topic 45: IF Amplifiers

3-45F1 What degree of selectivity is desirable in the IF circuitry of a wideband FM phone receiver?

A. 1 kHz.
B. 2.4 kHz.
C. 4.2 kHz.
D. 15 kHz.

In your FM equipment, the filter must accommodate a minimum of +/- 5 kHz, a total of 10-kHz bandwidth. The 15-kHz filter is your best choice. **ANSWER D**

3-45F2 Which one of these filters can be used in micro-miniature electronic circuits?

A. High power transmitter cavity.
B. Receiver SAW IF filter.
C. Floppy disk controller.
D. Internet DSL to telephone line filter.

SAW stands for Surface-Acoustic-Wave, a filter using a quartz-type material called Lithium Niobate suitable for low loss radio receiver applications at microwave frequencies. The SAW filter network is ultramicro-miniature, so it fits on receiver printed circuit boards without taking up much real estate. **ANSWER B**

3-45F3 A receiver selectivity of 2.4 kHz in the IF circuitry is optimum for what type of signals?

A. CW.
B. Double-sideband AM voice.
C. SSB voice.
D. FSK RTTY.

A good 2.4-kHz filter in the IF circuitry is perfect for SSB voice. A 1.8 kHz filter would even be better, but might make the audio sound pinched. **ANSWER C**

3-45F4 A receiver selectivity of 10 kHz in the IF circuitry is optimum for what type of signals?

A. Double-sideband AM.
B. SSB voice.
C. CW.
D. FSK RTTY.

A very wide 10-kHz filter would be appropriate for tuning in shortwave broadcast stations on double-sideband AM. **ANSWER A**

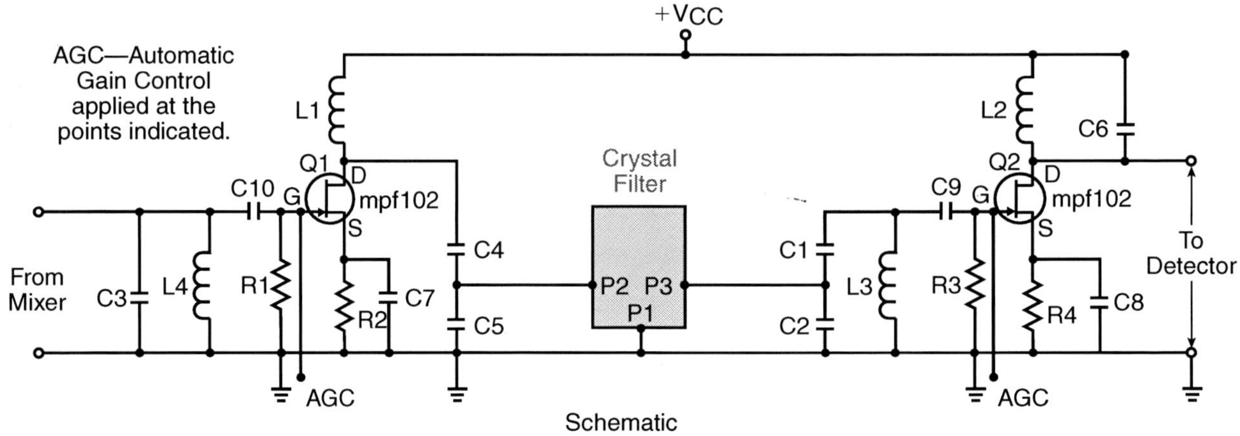

Cascaded IF amplifiers. The active devices are JFETs. The value of most components depends upon the frequency of operation.

3-45F5 What is an undesirable effect of using too wide a filter bandwidth in the IF section of a receiver?

A. Output-offset overshoot.
B. Undesired signals will reach the audio stage.
C. Thermal-noise distortion.
D. Filter ringing.

Using an AM 10-kHz IF filter for pulling in SSB signals would result in hearing undesired signals beyond the normal bandwidth of the desired signal. **ANSWER B**

3-45F6 How should the filter bandwidth of a receiver IF section compare with the bandwidth of a received signal?

A. Slightly greater than the received-signal bandwidth.
B. Approximately half the received-signal bandwidth.
C. Approximately two times the received-signal bandwidth.
D. Approximately four times the received-signal bandwidth.

For best fidelity, filter bandwidth should be slightly greater than the received signal bandwidth. If you are using one of those new SSB base stations, you have plenty of choices! **ANSWER A**

Key Topic 46: Filters and IF Amplifiers

3-46F1 What is the primary purpose of the final IF amplifier stage in a receiver?

A. Dynamic response.
B. Gain.
C. Noise figure performance.
D. Bypass undesired signals.

The purpose of the final IF amplifier stage in a receiver is both gain as well as selectivity. The answer for this question is gain. **ANSWER B**

3-46F2 What factors should be considered when selecting an intermediate frequency?

A. Cross-modulation distortion and interference.
B. Interference to other services.
C. Image rejection and selectivity.
D. Noise figure and distortion.

One of the key areas of receiver design is image rejection and selectivity. I prefer a selective receiver any day over one that is slightly more sensitive. **ANSWER C**

3-46F3 What is the primary purpose of the first IF amplifier stage in a receiver?

A. Noise figure performance.
B. Tune out cross-modulation distortion.
C. Dynamic response.
D. Selectivity.

The first intermediate-frequency stage in an amplifier provides selectivity and image rejection. **ANSWER D**

Mechanical filters in a communications receiver.

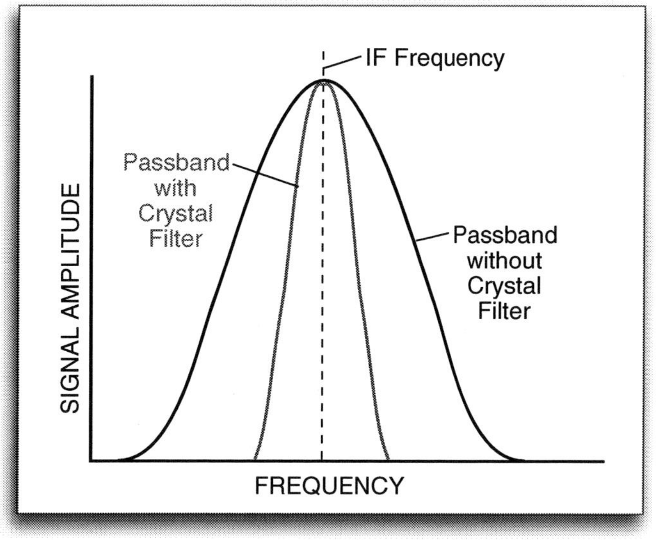

Effect of Crystal Filter on Passband.

3-46F4 What parameter must be selected when designing an audio filter using an op-amp?

A. Bandpass characteristics.
B. Desired current gain.
C. Temperature coefficient.
D. Output-offset overshoot.

What frequencies do you want an op-amp to amplify? This will help determine the bandpass characteristics required of the op amp. **ANSWER A**

3-46F5 What are the distinguishing features of a Chebyshev filter?

A. It has a maximally flat response over its passband.
B. It only requires inductors.
C. It allows ripple in the passband.
D. A filter whose product of the series- and shunt-element impedances is a constant for all frequencies.

This filter is named for the man who created the design. It sounds like "Chevy" connected to "Shev," but it's "Chebyshev." The Chebyshev filter may allow ripple to pass through on the pass band. **ANSWER C**

3-46F6 When would it be more desirable to use an m-derived filter over a constant-k filter?

A. When the response must be maximally flat at one frequency.
B. When the number of components must be minimized.
C. When high power levels must be filtered.
D. When you need more attenuation at a certain frequency that is too close to the cut-off frequency for a constant-k filter.

Occasionally, you need to knock down a close-in spur next to your transmitted signal. We would use an m-derived filter because it offers more attenuation at a certain frequency close to its cutoff. **ANSWER D**

Key Topic 47: Filters

3-47F1 A good crystal band-pass filter for a single-sideband phone would be?

A. 5 KHz.
B. 2.1 KHz.
C. 500 Hz.
D. 15 KHz.

A single-sideband emission is normally about 3 kHz wide. Without making the voice sound pinched, a 2.1-kHz filter is used on most worldwide SSB sets. **ANSWER B**

3-47F2 Which statement is true regarding the filter output characteristics shown in Figure 3F16?

A. C is a low pass curve and B is a band pass curve.
B. B is a high pass curve and D is a low pass curve.
C. A is a high pass curve and B is a low pass curve.
D. A is a low pass curve and D is a band stop curve.

Looking at Figure 3F16, we see frequency increases as we travel out on the baseline, and filter response as a function of power in the load. The vertical scale can be interpreted as "transmission" power, or filter pass response. Figure A is low pass. Figure B, bottom left, is band pass. Figure C, upper right, is high pass. And Figure D, lower right, is band stop. **ANSWER D**

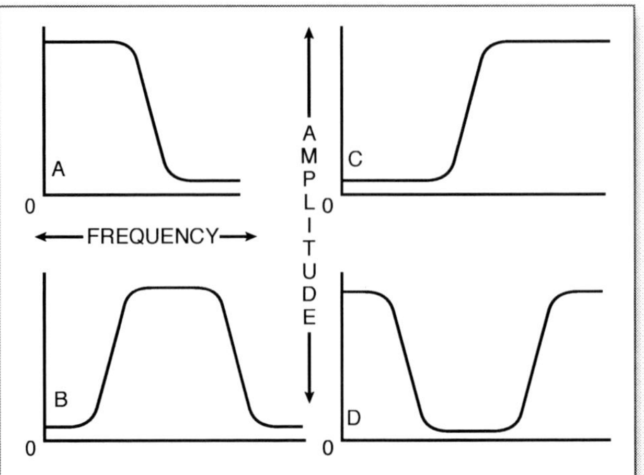

Fig. 3F16

3-47F3 What are the three general groupings of filters?

A. High-pass, low-pass and band-pass.
B. Inductive, capacitive and resistive.
C. Audio, radio and capacitive.
D. Hartley, Colpitts and Pierce.

High-pass filters are found in older TV sets. Low-pass filters are found in marine SSB sets. Band-pass filters may be found in VHF and UHF equipment. **ANSWER A**

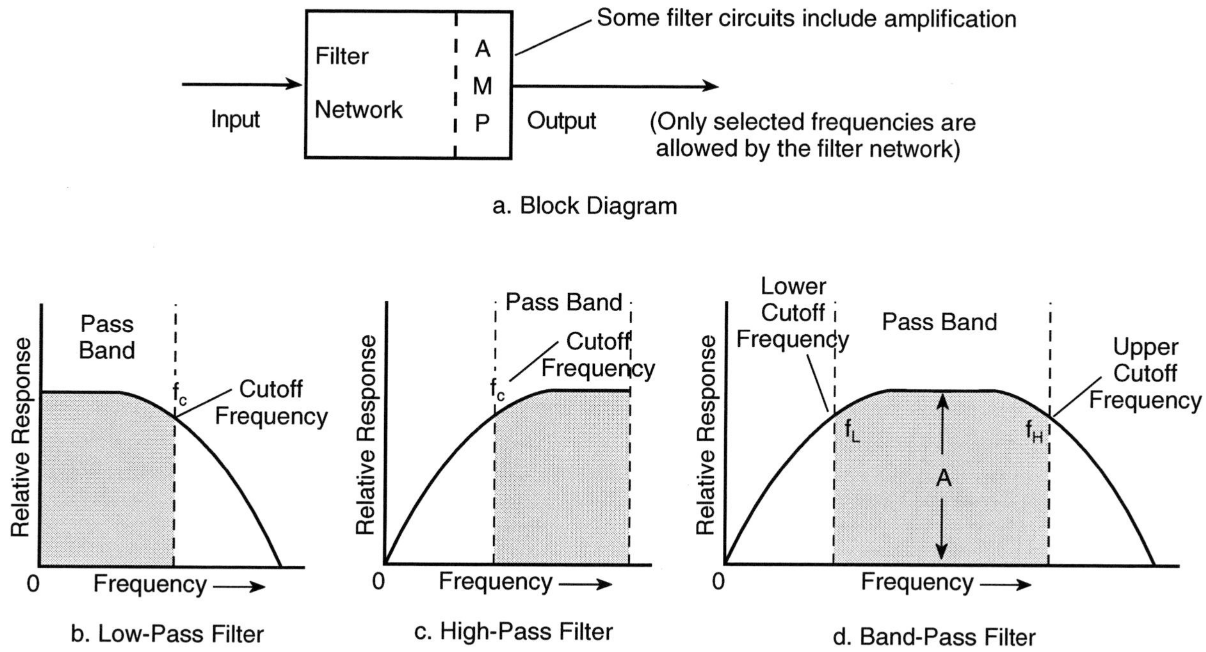

Filter networks are frequency selection circuits.

3-47F4 What is an m-derived filter?

A. A filter whose input impedance varies widely over the design bandwidth.
B. A filter whose product of the series- and shunt-element impedances is a constant for all frequencies.
C. A filter whose schematic shape is the letter "M".
D. A filter that uses a trap to attenuate undesired frequencies too near cutoff for a constant-k filter.

The m-derived filter has a tuned trap to attenuate undesired spurious emissions that are too near the normal cutoff for a constant-k filter. **ANSWER D**

3-47F5 What is an advantage of a constant-k filter?

A. It has high attenuation of signals at frequencies far removed from the pass band.
B. It can match impedances over a wide range of frequencies.
C. It uses elliptic functions.
D. The ratio of the cutoff frequency to the trap frequency can be varied.

The constant-k filter continues to suppress harmonics and spurs far removed from its pass band. **ANSWER A**

3-47F6 What are the distinguishing features of a Butterworth filter?

A. A filter whose product of the series- and shunt-element impedances is a constant for all frequencies.
B. It only requires capacitors.
C. It has a maximally flat response over its passband.
D. It requires only inductors.

The Butterworth filter can be counted on for its flat response over the entire pass band of frequencies. Unfortunately, its cutoff is not as sharp as other filters. **ANSWER C**

Key Topic 48: Detectors

3-48F1 What is a product detector?

A. It provides local oscillations for input to the mixer.
B. It amplifies and narrows the band-pass frequencies.
C. It uses a mixing process with a locally generated carrier.
D. It is used to detect cross-modulation products.

A product detector is found in SSB receivers. It mixes the incoming signal with a beat frequency oscillator signal. The BFO signal is a locally generated carrier that is mixed with the incoming signal. **ANSWER C**

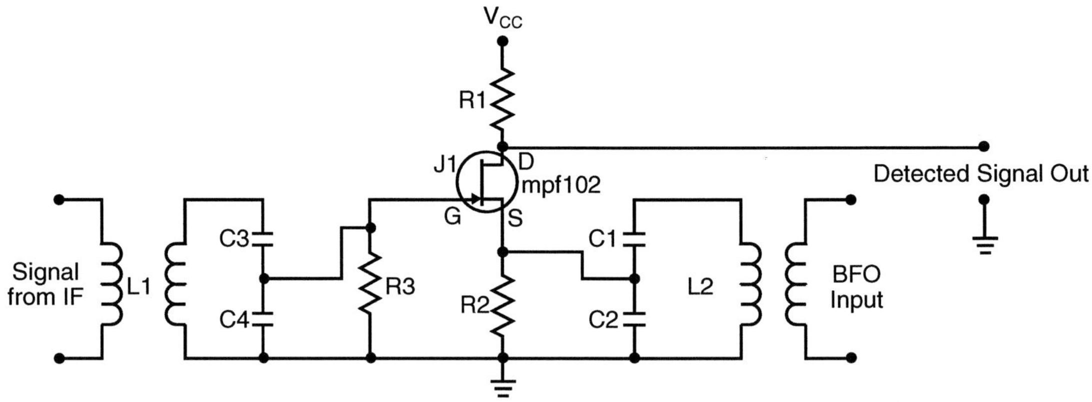

Product detector circuit used for SSB

3-48F2 Which circuit is used to detect FM-phone signals?

A. Balanced modulator.
B. Frequency discriminator.
C. Product detector.
D. Phase splitter.

FM signals must be detected differently than AM signals because frequency changes must be detected rather than amplitude changes. We use a reactance modulator to transmit FM. We use a frequency discriminator to detect an FM phone signal. **ANSWER B**

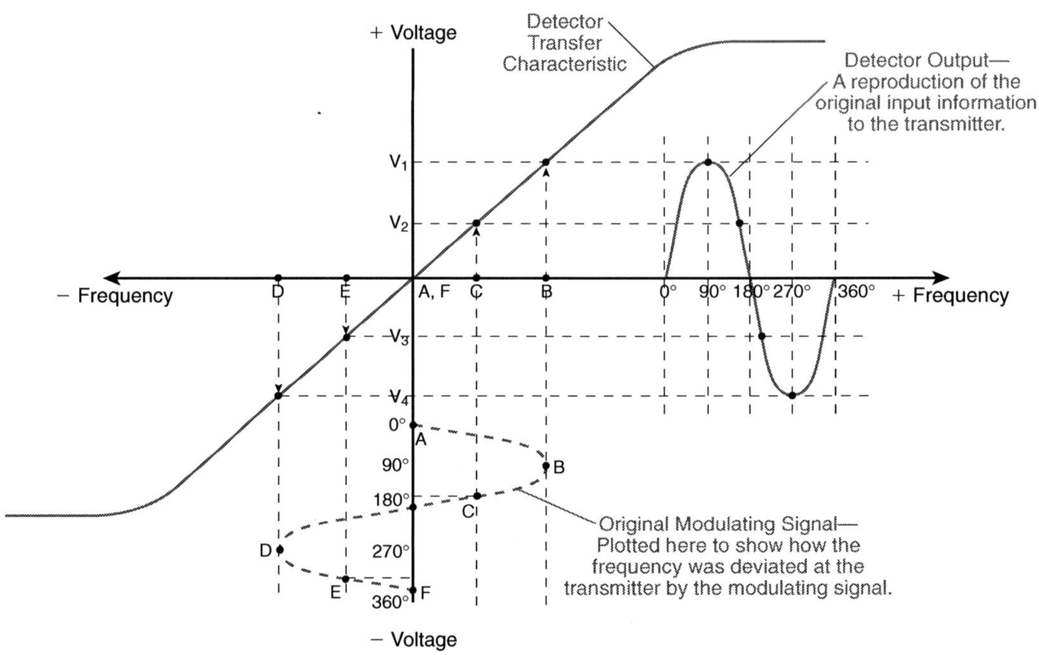

FM detection technique

3-48F3 What is the process of detection in a radio diode detector circuit?

A. Breakdown of the Zener voltage.
B. Mixing with noise in the transition region of the diode.
C. Rectification and filtering of RF.
D. The change of reactance in the diode with respect to frequency.

Since a diode only conducts for half of the AC signal, it may be used as a rectifier. By filtering out the radio frequency energy after rectification, detection is accomplished – the modulating signal is what remains. **ANSWER C**

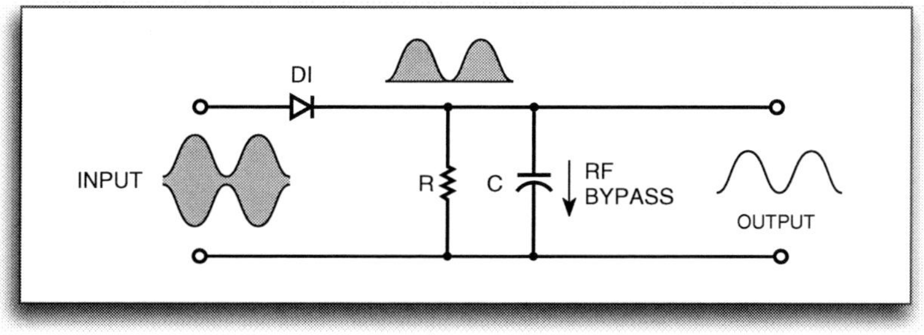

Diode Detector

3-48F4 What is a frequency discriminator in a radio receiver?

A. A circuit for detecting FM signals.
B. A circuit for filtering two closely adjacent signals.
C. An automatic band switching circuit.
D. An FM generator.

Anytime you see the words frequency discriminator you know that you are dealing with frequency modulation and an FM transceiver. **ANSWER A**

3-48F5 In a CTCSS controlled FM receiver, the CTCSS tone is filtered out after the:

A. IF stage but before the mixer.
B. Mixer but before the IF.
C. IF but before the discriminator.
D. Discriminator but before the audio section.

Continuous tone coded squelch system hum goes through most of the receiver sections, including the IF, mixer, and the discriminator, where it meets up with a CTCSS muting circuit to the audio stage, only allowing audio to pass if accompanied by the correct CTCSS tone. **ANSWER D**

3-48F6 What is the definition of detection in a radio receiver?

A. The process of masking out the intelligence on a received carrier to make an S-meter operational.
B. The recovery of intelligence from the modulated RF signal.
C. The modulation of a carrier.
D. The mixing of noise with the received signal.

Your radio's detector recovers intelligence from a modulated RF signal. You have modulation to put information on a carrier; you have detection to take information off a carrier. **ANSWER B**

Key Topic 49: Audio & Squelch Circuits

3-49F1 What is the digital signal processing term for noise subtraction circuitry?

A. Adaptive filtering and autocorrelation.
B. Noise blanking.
C. Noise limiting.
D. Auto squelch noise reduction.

Digital signal processing (DSP) may offer multiple functions to a communications receiver, including high-pass and low-pass filtering, bandpass and variable filtering, RF attenuation, notch filters, digital AGC, and adaptive filtering and autocorrelation. The latter DSP circuit literally "subtracts" noise that may be occluding that incoming waveform desired signal. The filter basically collapses against the radio signal, eliminating the noise between the desired audio information. The adaptive filtering is effective against a wide variety of "stationary" noises whose amplitude and pitch change slowly compared to the spectral variations that are characteristic of human speech. As the background noise changes, adaptive filtering and autocorrelation continue to track and subtract the noise, silencing noise from the desired signal. **ANSWER A**

3-49F2 What is the purpose of de-emphasis in the receiver audio stage?

A. When coupled with the transmitter pre-emphasis, flat audio is achieved.
B. When coupled with the transmitter pre-emphasis, flat audio and noise reduction is received.
C. No purpose is achieved.
D. To conserve bandwidth by squelching no-audio periods in the transmission.

In an FM transmitter, individual components in the modulator stage may introduce unwanted noise that rides along with the desired audio signal. As we approach UHF frequencies, individual components may introduce an unacceptable amount of circuit noise, and a transmitter pre-emphasis high-pass circuit that tracks modulation frequency increases is used to attenuate transmit circuit noise. In the companion FM receiver, de-emphasis incorporates a decreasing response characteristic as the modulating frequency goes higher, resulting in "flat audio" that is nearly absent of transmit noise. **ANSWER B**

3-49F3 What makes a Digital Coded Squelch work?

A. Noise.
B. Tones.
C. Absence of noise.
D. Digital codes.

Digital coded squelch, usually abbreviated DCS, incorporates digital encoding and decoding to recover only the signal from a specific DCS station or DCS groups. **ANSWER D**

3-49F4 What causes a squelch circuit to function?

A. Presence of noise.
B. Absence of noise.
C. Received tones.
D. Received digital codes.

The squelch circuit in a communications receiver is tied into the automatic gain control DC amplifier first transistor bias. When a carrier or SSB signal disappears into the noise, the inrush of constant noise may bias off the next transistor to keep the noise from reaching the audio stage and speaker. The squelch circuit itself creates an absence of noise, but it is actually pure noise and no AGC that engages the second transistor squelch circuit. Unfortunately, circuit hysteresis may cause weak signals to remain squelched out of the loud speaker system, so it is desirable to have a slow AGC circuit that may respond to a weak signal, and keep the squelch open as the signal fades in and out of the noise. **ANSWER A**

3-49F5 What makes a CTCSS squelch work?

A. Noise.
B. Tones.
C. Absence of noise.
D. Digital codes.

CTCSS is a continuous tone, about 0.5 kHz deviation, that travels on the transmitted audio waveform to the distant receiver. The receiver incorporates a CTCSS filter, and will only pass modulation when the correct CTCSS tone is continuously decoded. **ANSWER B**

EIA frequency codes with Motorola alphanumeric designators

Hz	Alpha Code	Hz	Alpha Code	Hz	Alpha Code	Hz	Alpha Code	Hz	Alpha Code
67.0	XZ	91.5	ZZ	127.3	3A	167.9	6Z	203.5	M1
69.3	WZ	94.8	ZA	131.8	3B	171.3	---	206.5	8Z
71.9	XA	97.4	ZB	136.5	4Z	173.8	6A	210.7	M2
74.4	WA	100.0	1Z	141.3	4A	177.3	---	218.1	M3
77.0	XB	103.5	1A	146.2	4B	179.9	6B	225.7	M4
79.7	WB	107.2	1B	151.4	5Z	183.5	---	229.1	9Z
82.5	YZ	110.9	2Z	156.7	5A	186.2	7Z	233.6	M5
85.4	YA	114.8	2A	159.8	---	189.9	---	241.8	M6
88.5	YB	118.8	2B	162.2	5B	192.8	7A	250.3	M7
		123.0	3Z	165.5	---	199.5	---	254.1	0Z

3-49F6 What radio circuit samples analog signals, records and processes them as numbers, then converts them back to analog signals?

A. The pre-emphasis audio stage.
B. The squelch gate circuit.
C. The digital signal processing circuit.
D. The voltage controlled oscillator circuit.

In a digital signal processing circuit, incoming analog audio signals are sampled, converted to numbers (a digital signal), and then converted back into analog signals at the DSP output. **ANSWER C**

Key Topic 50: Receiver Performance

3-50F1 Where would you normally find a low-pass filter in a radio receiver?

A. In the AVC circuit.
B. In the Oscillator stage.
C. In the Power Supply.
D. A and C, but not B.

There are countless areas of electronic circuits where low-pass filters might be found. One of the most common is in the power supply, after the rectifier. The filter capacitor and possibly a choke form a low-pass filter that removes all alternating currents, while allowing DC to pass. In an AVC (automatic volume control) circuit, a low-pass filter is used to determine the time constant or recovery time of the circuit. We don't want the circuit to filter out the desired modulation, but rather just the long-term average of the signal. We ordinarily would not use a low-pass filter in conjunction with an oscillator, because a properly working oscillator only has one frequency. The tank circuit of the oscillator itself generally functions as a bandpass filter. **ANSWER D**

3-50F2 How can ferrite beads be used to suppress ignition noise? Install them:

A. In the resistive high voltage cable every 2 years.
B. Between the starter solenoid and the starter motor.
C. Install them in the primary and secondary ignition leads.
D. In the antenna lead.

It is the primary and secondary ignition leads that radiate the strong spark pulses which create ignition noise. Ferrite beads strung on the leads may help reduce the higher-frequency ignition noises. **ANSWER C**

3-50F3 What is the term used to refer to the condition where the signals from a very strong station are superimposed on other signals being received?

A. Intermodulation distortion.
B. Cross-modulation interference.
C. Receiver quieting.
D. Capture effect.

If you hear more than one transmitted frequency in your receiver's signal, chances are it is cross-modulation interference. Remember, if they say the word received, it means "cross-mod." **ANSWER B**

3-50F4 What is cross-modulation interference?

A. Interference between two transmitters of different modulation type.
B. Interference caused by audio rectification in the receiver preamp.
C. Modulation from an unwanted signal heard in addition to the desired signal.
D. Harmonic distortion of the transmitted signal.

If you have driven your little handheld transceiver down to a big city, chances are you have heard the effects of cross-modulation interference. "Cross-mod" brings in modulation from an unwanted signal that rides on top of the signal you are presently listening to. **ANSWER C**

3-50F5 In Figure 3F15 at what point in the circuit (labeled 1 through 4) could a DC voltmeter be used to monitor signal strength?

A. 1
B. 2
C. 3
D. 4

On your upcoming GROL exam, the official figures to look over may either be incorporated within the test, or on a separate sheet at the end of the test. Be sure to check with your examiners if you cannot find the figure diagrams. In this figure, 3F15, we show a radio receiver and stage 4 is the automatic gain control (AGC) circuit. The AGC detector develops a DC voltage that is proportional to the incoming strength. This DC voltage is then fed back to the IF amplifier stages to control the overall receiver gain. The AGC keeps the receiver at a nearly-constant level, even though the RF signals may vary widely. This proportional DC voltage may also be sampled with a high impedance digital meter. The readings may then be compared to the service manual for incoming signal strengths to develop a specific AGC voltage. **ANSWER D**

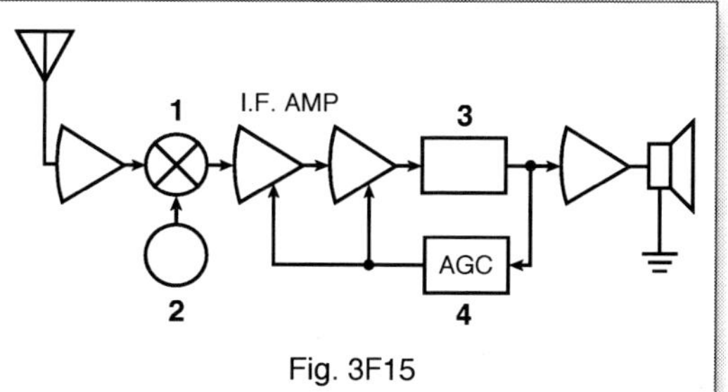

Fig. 3F15

3-50F6 Pulse type interference to automobile radio receivers that appears related to the speed of the engine can often be reduced by:

A. Installing resistances in series with spark plug wires.
B. Using heavy conductors between the starting battery and the starting motor.
C. Connecting resistances in series with the battery.
D. Grounding the negative side of the battery.

Factory-recommended resistance spark plugs may be necessary to reduce the interference in mobile two-way radio installations. There also are resistance spark-plug wires that help knock down the noise pulses. **ANSWER A**

Subelement G – Transmitters

6 Key Topics, 6 Exam Questions

Key Topic 51: Amplifiers-1

3-51G1 What class of amplifier is distinguished by the presence of output throughout the entire signal cycle and the input never goes into the cutoff region?

A. Class A. B. Class B C. Class C. D. Class D.

If it works over an entire signal cycle, it's a Class A amplifier. **ANSWER A**

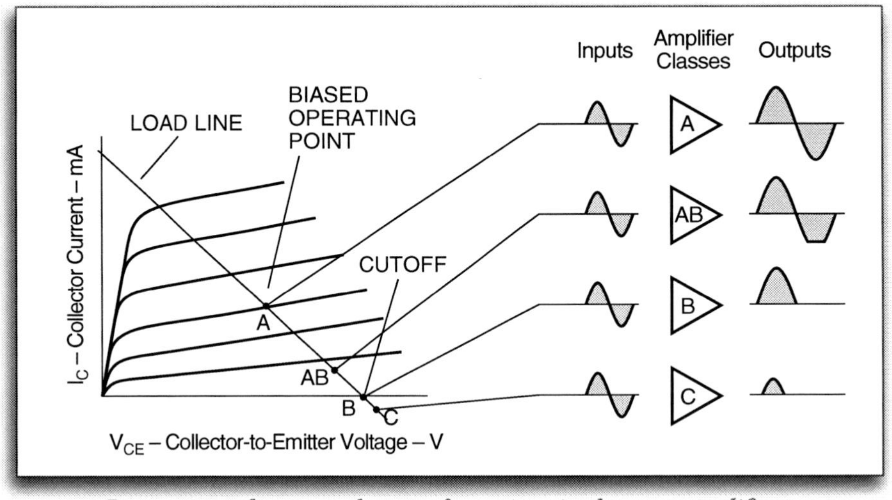

Bias points of various classes of transistorized power amplifiers.

Source: Advanced Class, 2nd Edition, G. West, © Copyright 1992, 1995 Master Publishing, Inc.

3-51G2 What is the distinguishing feature of a Class A amplifier?

A. Output for less than 180 degrees of the signal cycle.
B. Output for the entire 360 degrees of the signal cycle.
C. Output for more than 180 degrees and less than 360 degrees of the signal cycle.
D. Output for exactly 180 degrees of the input signal cycle.

The Class A amplifier works during the entire 360-degree signal cycle. It offers good linearity, but with all its constant work, may have low efficiency. **ANSWER B**

3-51G3 Which class of amplifier has the highest linearity and least distortion?

A. Class A. B. Class B. C. Class C. D. Class AB.

The best amplifier for a pure signal with the highest linearity and least distortion is a Class A amplifier. **ANSWER A**

3-51G4 Which class of amplifier provides the highest efficiency?

A. Class A. B. Class B. C. Class C. D. Class AB.

The highest efficiency comes out of a Class C amplifier. Since it operates less than 180 degrees of the cycle, it dissipates less power but, at the same time, produces narrow pulses which can cause loss of linearity. **ANSWER C**

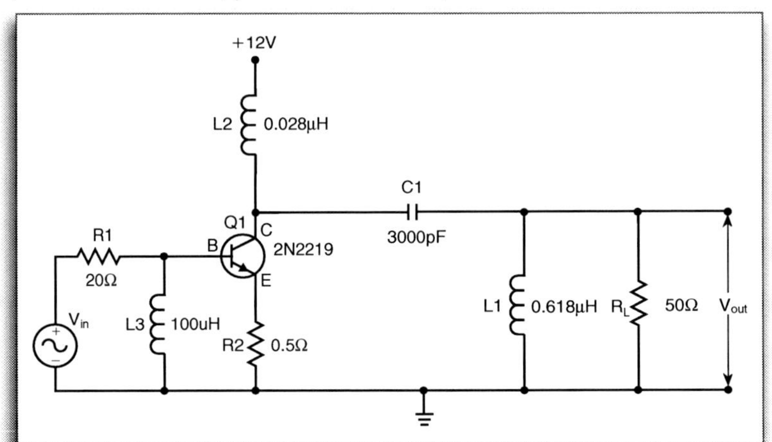

Bipolar Class-C tuned amplifier

3-51G5 What class of amplifier is distinguished by the bias being set well beyond cutoff?

A. Class A.
B. Class C.
C. Class B.
D. Class AB.

If the bias is set well beyond cutoff, the amplifier will only work less than 180 degrees of the cycle, and is a Class C amplifier. Class C amplifiers are very efficient, but not so great on linearity. **ANSWER B**

3-51G6 Which class of amplifier has an operating angle of more than 180 degrees but less than 360 degrees when driven by a sine wave signal?

A. Class A.
B. Class B.
C. Class C.
D. Class AB.

If it's more than 180, but less than 360, it must be a Class AB amplifier. **ANSWER D**

Key Topic 52: Amplifiers-2

3-52G1 The class B amplifier output is present for what portion of the input cycle?

A. 360 degrees.
B. Greater than 180 degrees and less than 360 degrees.
C. Less than 180 degrees.
D. 180 degrees.

The Class B amplifier only works during half of the input cycle, or a total of 180 degrees. We usually run Class B amplifiers in push/pull pairs. Their efficiency is higher than Class A amplifiers because each tube or transistor gets to rest for half of the cycle. **ANSWER D**

3-52G2 What input-amplitude parameter is most valuable in evaluating the signal-handling capability of a Class A amplifier?

A. Average voltage.
B. RMS voltage.
C. Peak voltage.
D. Resting voltage.

In a class A amplifier, peak voltage provides a double check on the capability and linearity of that amplifier. Many times, distortion occurs when the amplifier is handling a peak voltage. **ANSWER C**

3-52G3 The class C amplifier output is present for what portion of the input cycle?

A. Less than 180 degrees.
B. Exactly 180 degrees.
C. 360 degrees.
D. More than 180 but less than 360 degrees.

The Class C amplifier gets to rest a lot! It has great efficiency, but its linearity is not as good as a Class B or Class A amp. Class C amplifiers are best for digital modes. The Class C amplifier output is always less than 180 degrees of the input signal cycle. **ANSWER A**

3-52G4 What is the approximate DC input power to a Class AB RF power amplifier stage in an unmodulated carrier transmitter when the PEP output power is 500 watts?

A. 250 watts.
B. 600 watts.
C. 800 watts.
D. 1000 watts.

The Class AB amplifier is a real work horse, and is one of the most common amps found in worldwide radio shacks. It's 50% efficient. If you're putting out 500 watts, your input power is approximately 1000 watts. Remember, Class C amplifier efficiency is about 80%; Class B is approximately 60%; and Class AB is approximately 50%. These values will not be given on your test, so be sure to have them memorized before the big examination. **ANSWER D**

3-52G5 The class AB amplifier output is present for what portion of the input cycle?

A. Exactly 180 degrees.
B. 360 degrees
C. More than 180 but less than 360 degrees.
D. Less than 180 degrees.

The Class AB amplifier has some of the properties of both the Class A and the Class B amplifier. It has output for more than 180 degrees of the cycle, but less than 360 degrees of the cycle. **ANSWER C**

3-52G6 What class of amplifier is characterized by conduction for 180 degrees of the input wave?

A. Class A.
B. Class B.
C. Class C.
D. Class D.

If it's working for 180 degrees, it's a Class B amplifier. **ANSWER B**

Key Topic 53: Oscillators & Modulators

3-53G1 What is the modulation index in an FM phone signal having a maximum frequency deviation of 3,000 Hz on either side of the carrier frequency when the modulating frequency is 1,000 Hz?

A. 0.3
B. 3,000
C. 3
D. 1,000

Modulation index in an FM transceiver is calculated by dividing the frequency deviation by the modulation frequency. 3000Hz ÷ 1000Hz = a modulation index of 3. **ANSWER C**

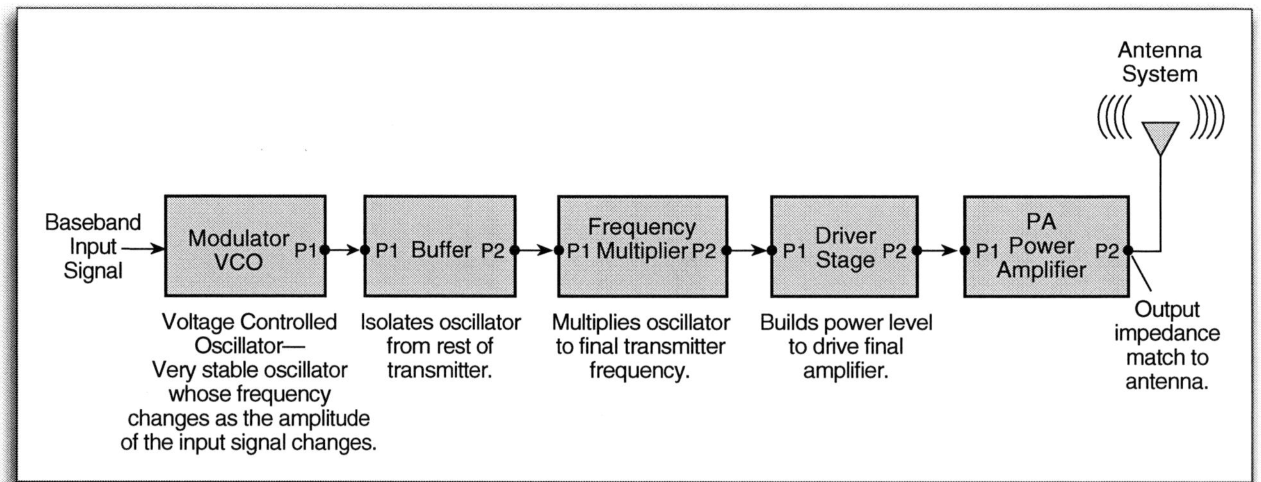

Block diagram of an FM transmitter

3-53G2 What is the modulation index of a FM phone transmitter producing a maximum carrier deviation of 6 kHz when modulated with a 2 kHz modulating frequency?

A. 3
B. 6,000
C. 2,000
D. 1

In this question, the carrier deviation is 6 kHz, with a modulating frequency of 2 kHz. 6 kHz ÷ 2 kHz = 3. **ANSWER A**

3-53G3 What is the total bandwidth of a FM phone transmission having a 5 kHz deviation and a 3 kHz modulating frequency?

A. 3 kHz.
B. 8 kHz.
C. 5 kHz.
D. 16 kHz.

The bandwidth of an FM transmission is determined by Carson's Rule where bandwidth equals 2 times the sum of the maximum carrier frequency deviation plus the maximum modulation frequency. 5 kHz + 3 kHz = 8 kHz, multiplied by 2 (the plus and minus swing of the FM signal) = 16 kHz. **ANSWER D**

3-53G4 How does the modulation index of a phase-modulated emission vary with RF carrier frequency?

A. It does not depend on the RF carrier frequency.
B. Modulation index increases as the RF carrier frequency increases.
C. It varies with the square root of the RF carrier frequency.
D. It decreases as the RF carrier frequency increases.

The modulation index of a phase-modulated emission is referenced to the frequency deviation divided by the highest modulating frequency. The modulation index does NOT depend on the actual RF carrier frequency itself. **ANSWER A**

3-53G5 How can a single-sideband phone signal be generated?

A. By driving a product detector with a DSB signal.
B. By using a reactance modulator followed by a mixer.
C. By using a loop modulator followed by a mixer.
D. By using a balanced modulator followed by a filter.

In a single-sideband transceiver, a balanced modulator generates the signal and a filter removes the unwanted lower or upper sideband. **ANSWER D**

3-53G6 What is a balanced modulator?

A. An FM modulator that produces a balanced deviation.
B. A modulator that produces a double sideband, suppressed carrier signal.
C. A modulator that produces a single sideband, suppressed carrier signal.
D. A modulator that produces a full carrier signal.

The balanced modulator produces an upper and lower sideband, with the carrier suppressed. Remember, the double sidebands are balanced into an upper and lower sideband around a suppressed carrier. **ANSWER B**

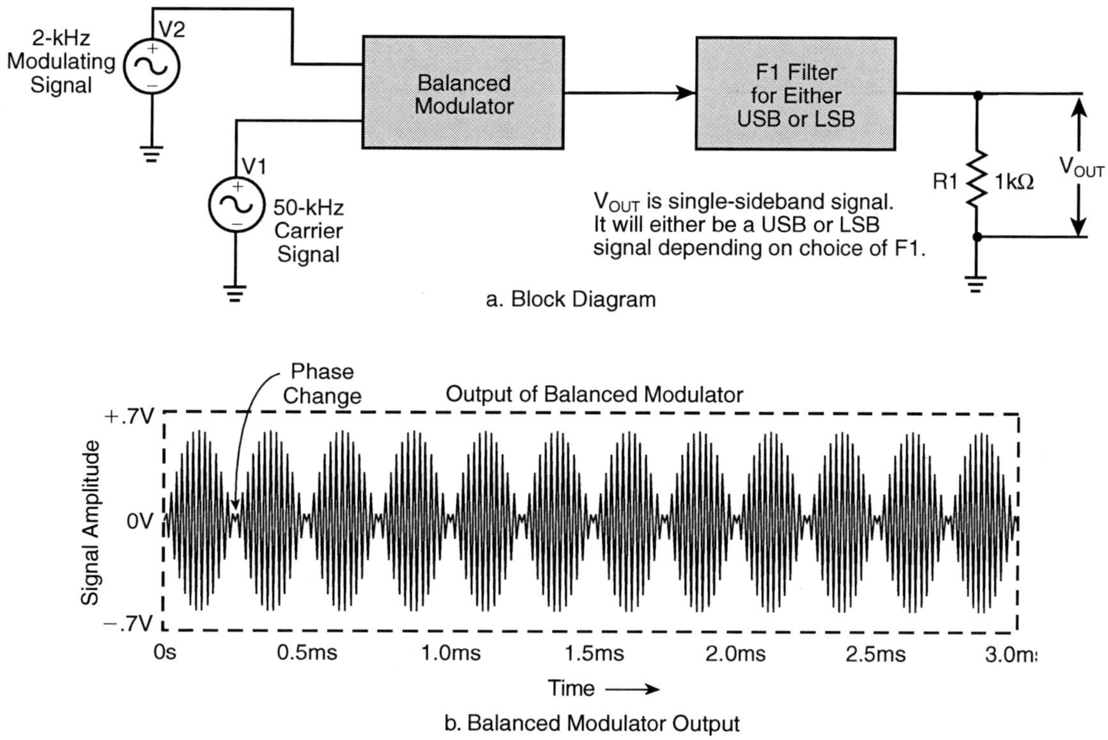

Generating an SSB signal

Key Topic 54: Resonance - Tuning Networks

3-54G1 What is an L-network?

A. A low power Wi-Fi RF network connection.
B. A network consisting of an inductor and a capacitor.
C. A "lossy" network.
D. A network formed by joining two low pass filters.

When troubleshooting two-way radio equipment, make absolutely sure you have the technical manual before going out to the job. This will help you spot different types of networks, such as the L-network that has an inductor and capacitor in it. **ANSWER B**

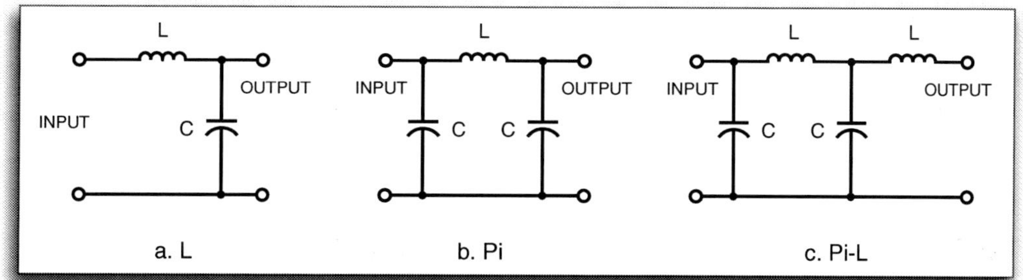

Matching Networks

3-54G2 What is a pi-network?

A. A network consisting of a capacitor, resistor and inductor.
B. The Phase inversion stage.
C. An enhanced token ring network.
D. A network consisting of one inductor and two capacitors or two inductors and one capacitor.

The pi-network consists of three components, looking like the Greek symbol pi (π). The legs could be either capacitors or coils, and the horizontal line at the top could be either a coil or a capacitor. **ANSWER D**

3-54G3 What is the resonant frequency in an electrical circuit?
- A. The frequency at which capacitive reactance equals inductive reactance.
- B. The highest frequency that will pass current.
- C. The lowest frequency that will pass current.
- D. The frequency at which power factor is at a minimum.

At resonance, X_L (inductive reactance) equals X_C (capacitive reactance). In a mobile series resonant whip antenna, there will be maximum current through the loading coil when the whip is tuned to resonance. On your base station vertical trap antenna or trap beam antenna, parallel resonant circuits offer high impedance, trapping out a specific length of the antenna for a certain frequency. At resonance, the resonant parallel circuit looks like a high impedance to any of your power trying to get through to the longer length of the antenna system. **ANSWER A**

3-54G4 Which three network types are commonly used to match an amplifying device to a transmission line?
- A. Pi-C network, pi network and T network.
- B. T network, M network and Z network.
- C. L network, pi network and pi-L network.
- D. L network, pi network and C network.

Worldwide and aeronautical long-range SSB equipment may have three different matching devices between the amplifier output and the transmission line – either pi, L, or pi-L networks. **ANSWER C**

3-54G5 What is a pi-L network?
- A. A Phase Inverter Load network.
- B. A network consisting of two inductors and two capacitors.
- C. A network with only three discrete parts.
- D. A matching network in which all components are isolated from ground.

The pi-L network consists of two coils and two capacitors. Depending on how they are arranged, it could be a high-pass network or a low-pass network. **ANSWER B**

3-54G6 Which network provides the greatest harmonic suppression?
- A. L network.
- B. Pi network.
- C. Pi-L network.
- D. Inverse L network.

The network with the most components will provide the greatest harmonic suppression. The pi-L-network is found in most worldwide marine SSB sets. **ANSWER C**

Key Topic 55: SSB Transmitters

3-55G1 What will occur when a non-linear amplifier is used with a single-sideband phone transmitter?
- A. Reduced amplifier efficiency.
- B. Increased intelligibility.
- C. Sideband inversion.
- D. Distortion.

Maritime mobile stations operating on marine SSB may legally employ single-sideband amplifiers for additional power output. The added power amplifier must be linear to prevent the signal from becoming distorted. Older tube-type amplifiers may not be rated for good linearity on an SSB emission. **ANSWER D**

3-55G2 To produce a single-sideband suppressed carrier transmission it is necessary to ____ the carrier and to ____ the unwanted sideband.
- A. Filter, filter.
- B. Cancel, filter.
- C. Filter, cancel.
- D. Cancel, cancel.

Marine and aviation SSB transceivers cancel the carrier for A3J operation and filter out the unwanted sideband. **ANSWER B**

Marine SSB tuned to upper sideband, on a marine channel.

3-55G3 In a single-sideband phone signal, what determines the PEP-to-average power ratio?

A. The frequency of the modulating signal.
B. The degree of carrier suppression.
C. The speech characteristics.
D. The amplifier power.

It's not all that easy to determine the ratio of peak envelope power (PEP) to average power in your worldwide SSB set because your speech characteristics may vary from those of another operator using the same equipment. **ANSWER C**

$PEP = P_{DC} \times E_{ff}$ Where: PEP = Peak Envelope Power in **watts**
P_{DC} = Input dc power in **watts**

3-55G4 What is the approximate ratio of peak envelope power to average power during normal voice modulation peak in a single-sideband phone signal?

A. 2.5 to 1.
B. 1 to 1.
C. 25 to 1.
D. 100 to 1.

Transmitting single sideband voice leads to a complex waveform with voice peaks of the transmitted power envelope at a ratio of about 2.5 to 1 to average power. 2.5 to 1 is an approximate figure, and none of the incorrect answers are close. **ANSWER A**

3-55G5 What is the output peak envelope power from a transmitter as measured on an oscilloscope showing 200 volts peak-to-peak across a 50-ohm load resistor?

A. 1,000 watts.
B. 100 watts.
C. 200 watts.
D. 400 watts.

The first step in solving this problem is to take the peak voltage and reduce it to RMS (root mean square). The 200 volts given in the problem is peak-to-peak, so the peak voltage is 100 volts. Now, multiply 100 volts by 0.707 = 70.7 volts RMS. RMS volts x average current = RMS power. So, RMS power (P)= $E^2 \div R$ = 99.96 watts (rounded to 100 watts.) **ANSWER B**

3-55G6 What would be the voltage across a 50-ohm dummy load dissipating 1,200 watts?

A. 245 volts.
B. 692 volts.
C. 346 volts.
D. 173 volts.

Your formula is the square root of power times resistance: $E = \sqrt{P \times R}$. Most calculators have a square root function, and your answer comes out 244.948 rounded to 245 volts. **ANSWER A**

Key Topic 56: Technology

3-56G1 How can intermodulation interference between two transmitters in close proximity often be reduced or eliminated?

A. By using a Class C final amplifier with high driving power.
B. By installing a terminated circulator or ferrite isolator in the feed line to the transmitter and duplexer.
C. By installing a band-pass filter in the antenna feed line.
D. By installing a low-pass filter in the antenna feed line.

Big ship radio systems may include several transceivers operating on and near VHF marine channels. Intermodulation may occur when two transmitters begin transmitting at the same time, like AIS and channel 13 bridge-to-bridge calls. A terminated circulator or ferrite isolator in the feedline to the transmitter on AIS or channel 13, may help minimize the problem of two transmitters creating additional signals within the VHF band. **ANSWER B**

3-56G2 How can parasitic oscillations be eliminated in a power amplifier?

A. By tuning for maximum SWR.
B. By tuning for maximum power output.
C. By neutralization.
D. By tuning the output.

We can reduce parasitic oscillations, and clean up our signal, through the steps of neutralization in a power amplifier. **ANSWER C**

3-56G3 What is the name of the condition that occurs when the signals of two transmitters in close proximity mix together in one or both of their final amplifiers, and unwanted signals at the sum and difference frequencies of the original transmissions are generated?

A. Amplifier desensitization.
B. Neutralization.
C. Adjacent channel interference.
D. Intermodulation interference.

It is common for a new AIS (Automatic Identification System) transmitter to come up on transmit at the same time as a marine VHF radio is transmitting, and when both stations are on the air simultaneously, additional low signal level emissions may be generated, called intermodulation products. **ANSWER D**

3-56G4 What term describes a wide-bandwidth communications system in which the RF carrier varies according to some pre-determined sequence?

A. Spread-spectrum communication.
B. AMTOR.
C. SITOR.
D. Time-domain frequency modulation.

Spread spectrum communications occupy a large swath of frequency domain, where the RF carrier waves may frequency-hop in a pre-determined sequence, which allows companion spread-spectrum stations to stay locked on. **ANSWER A**

3-56G5 How can even-order harmonics be reduced or prevented in transmitter amplifier design?

A. By using a push-push amplifier.
B. By operating class C.
C. By using a push-pull amplifier.
D. By operating class AB.

Today, most worldwide amplifiers are solid state. Even-order harmonics are reduced by using such an amplifier with a push-pull configuration. Don't speed read answer A! **ANSWER C**

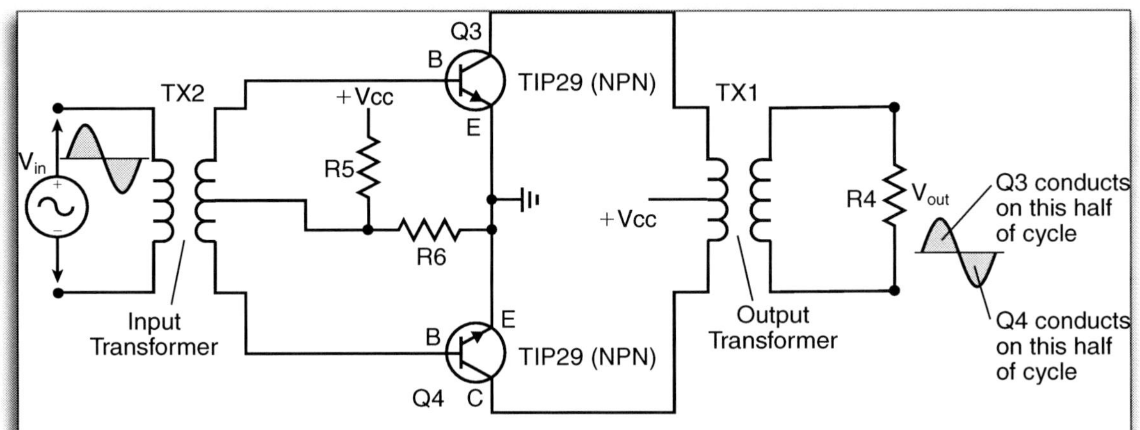

Push-Pull Amplifier

3-56G6 What is the modulation type that can be a frequency hopping of one carrier or multiple simultaneous carriers?

A. SSB.
B. FM.
C. OFSK.
D. Spread spectrum.

In frequency hopping, abbreviated FH, hundreds of individual frequency channels may be selected, one at a time, for random frequency hopping to an associated receiver that may error-correct in case one or two of the channels is occupied by another spread spectrum signal. Your little cordless phone, operating spread spectrum at 2.4 GHz, may occupy up to 80 frequencies, with each hop occurring every couple hundred milliseconds with a hop time of just micro-seconds. 802.11 describes the hopping sequences. 802.11 is spread spectrum communications, usually associated with Wireless Local Area Network, (WLAN), using a predetermined pseudo-random code for the sending and receiving of data on multiple frequencies, through frequency hopping techniques. **ANSWER D**

Subelement H – Modulation

3 Key Topics, 3 Exam Questions, 1 Drawing

Key Topic 57: Frequency Modulation

3-57H1 The deviation ratio is the:

A. Audio modulating frequency to the center carrier frequency.
B. Maximum carrier frequency deviation to the highest audio modulating frequency.
C. Carrier center frequency to the audio modulating frequency.
D. Highest audio modulating frequency to the average audio modulating frequency.

It's important to set the deviation ratio properly on your FM transceiver to prevent splatter, and a signal bandwidth that is too wide. Deviation ratio is the ratio of the maximum carrier swing to your highest audio modulating frequency when you speak into the microphone. **ANSWER B**

$$\text{Deviation Ratio} = \frac{\text{Maximum Carrier Frequency Deviation in kHz}}{\text{Maximum Modulation Frequency in kHz}}$$

3-57H2 What is the deviation ratio for an FM phone signal having a maximum frequency deviation of plus or minus 5 kHz and accepting a maximum modulation rate of 3 kHz?

A. 60 B. 0.16 C. 0.6 D. 1.66

Use formula at 3-57H1. Divide the maximum carrier frequency swing by the maximum modulation frequency. 3 kHz modulation frequency into 5 kHz frequency swing gives you a deviation ratio of 1.66. **ANSWER D**

3-57H3 What is the deviation ratio of an FM-phone signal having a maximum frequency swing of plus or minus 7.5 kHz and accepting a maximum modulation rate of 3.5 kHz?

A. 2.14 B. 0.214 C. 0.47 D. 47

Divide 3.5 (maximum modulation frequency) into 7.5 (maximum carrier frequency deviation), and you end up with 2.14, your deviation ratio. **ANSWER A**

3-57H4 How can an FM-phone signal be produced in a transmitter?

A. By modulating the supply voltage to a class-B amplifier.
B. By modulating the supply voltage to a class-C amplifier.
C. By using a balanced modulator.
D. By feeding the audio directly to the oscillator.

A frequency modulation phone signal is produced when the frequency of the carrier is changed instantaneously by the modulating signal. We can use a voltage controlled oscillator (VCO) to produce FM. When the signal at the input is a modulating voltage, the output frequency of the oscillator changes as the amplitude of the modulating voltage changes, producing frequency modulation. This frequency change is referred to as deviation, and the maximum change in the carrier frequency is the carrier's maximum deviation. **ANSWER D**

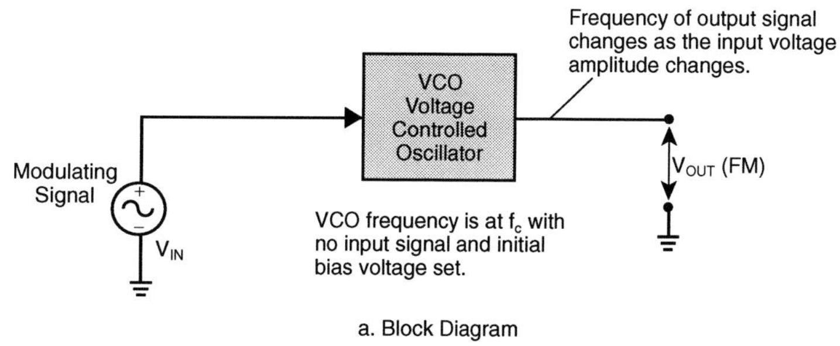

a. Block Diagram

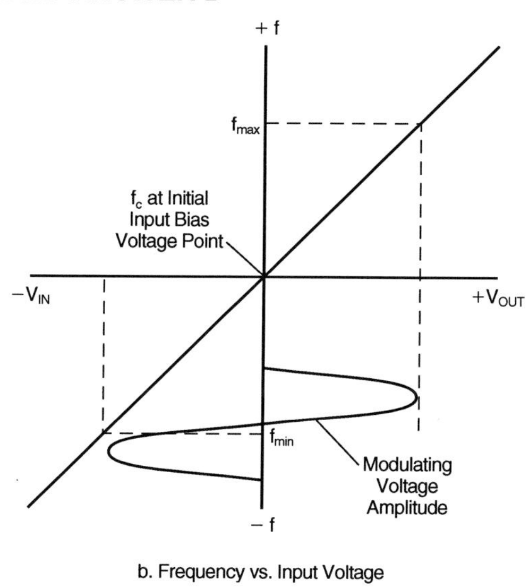

b. Frequency vs. Input Voltage

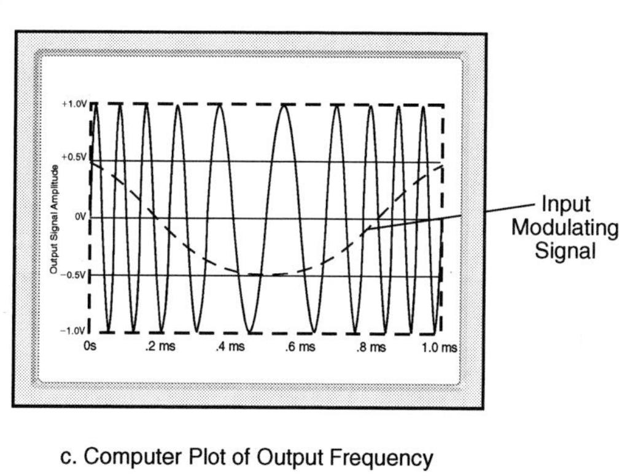

c. Computer Plot of Output Frequency

FM modulator using VCO.

3-57H5 What is meant by the term modulation index?

A. The ratio between the deviation of a frequency modulated signal and the modulating frequency.
B. The processor index.
C. The FM signal-to-noise ratio.
D. The ratio of the maximum carrier frequency deviation to the highest audio modulating frequency.

With modulation index, instead of looking at the highest audio frequency, we look at a specific modulating frequency. This is an important consideration for the amount of deviation on a packet station on FM. **ANSWER A**

3-57H6 In an FM-phone signal, what is the term for the maximum deviation from the carrier frequency divided by the maximum audio modulating frequency?

A. Deviation index.
B. Modulation index.
C. Deviation ratio.
D. Modulation ratio.

Simple division of the maximum audio frequency into the maximum deviation of the carrier frequency will lead you to the deviation ratio. Division produces a ratio. **ANSWER C**

Key Topic 58: SSB Modulation

3-58H1 In Figure 3H17, the block labeled 4 would indicate that this schematic is most likely a/an:

A. Audio amplifier.
B. Shipboard RADAR.
C. SSB radio transmitter.
D. Wireless LAN (local area network) computer.

Study Figure 3H17 and you will find that it is a block diagram of a single sideband (SSB) radio transmitter. **ANSWER C**

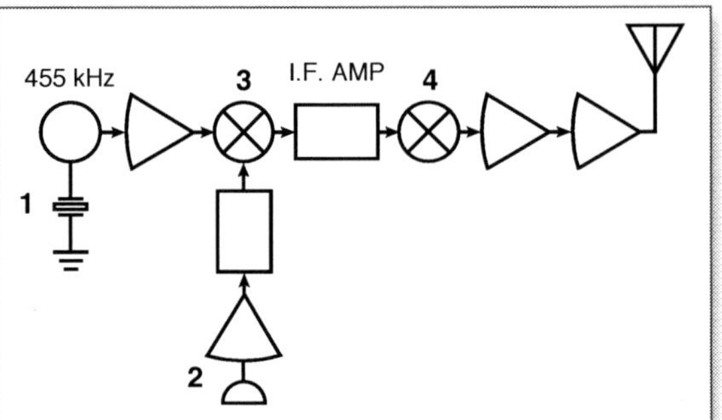

Fig. 3H17

3-58H2 In Figure 3H17, which block diagram symbol (labeled 1 through 4) represents where audio intelligence is inserted?

A. 1
B. 2
C. 3
D. 4

While we normally see a block diagram illustrating microphone input on the left side, Figure 3H17 the shows microphone input at symbol 2. **ANSWER B**

3-58H3 What kind of input signal could be used to test the amplitude linearity of a single-sideband phone transmitter while viewing the output on an oscilloscope?

A. Whistling in the microphone.
B. An audio frequency sine wave.
C. A two-tone audio-frequency sine wave.
D. An audio frequency square wave.

When viewing the transmit audio frequency sine wave on an oscilloscope, two non-harmonically related test tones are input on the microphone circuit, and the two tone audio frequency sine wave allows you to study closely the oscilloscope for amplitude linearity of the waveform. **ANSWER C**

3-58H4 What does a two-tone test illustrate on an oscilloscope?

A. Linearity of a SSB transmitter.
B. Frequency of the carrier phase shift.
C. Percentage of frequency modulation.
D. Sideband suppression.

Using an oscilloscope, the two tone test is your best way to examine the linearity of the transmit audio on a single sideband transmitter. **ANSWER A**

3-58H5 How can a double-sideband phone signal be produced?

A. By using a reactance modulator.
B. By varying the voltage to the varactor in an oscillator circuit.
C. By using a phase detector, oscillator, and filter in a feedback loop.
D. By modulating the supply voltage to a class C amplifier.

Here we are talking about double-sideband phone, amplitude modulation, with both sidebands present. Answer A is wrong because it's talking about FM. Answer D is correct because we modulate the supply voltage of the power amplifier stage. **ANSWER D**

3-58H6 What type of signals are used to conduct an SSB two-tone test?

A. Two audio signals of the same frequency, but shifted 90 degrees in phase.
B. Two non-harmonically related audio signals that are within the modulation band pass of the transmitter.
C. Two different audio frequency square wave signals of equal amplitude.
D. Any two audio frequencies as long as they are harmonically related.

The two non-harmonically related audio signals may be generated by an audio generator capable of an output of twin audio signals, within the modulation band pass of the transmitter. Keep these tones well below 2500 Hz. **ANSWER B**

Key Topic 59: Pulse Modulation

3-59H1 What is an important factor in pulse-code modulation using time-division multiplex?

A. Synchronization of transmit and receive clock pulse rates.
B. Frequency separation.
C. Overmodulation and undermodulation.
D. Slight variations in power supply voltage.

One of the most important factors in pulse-code modulation using time-division-multiplex is precise synchronization of TX and RX clock pulse rates, in addition to synchronization of the entire system to a national time standard. Our number one reference comes from the global positioning system (GPS) satellites, in addition to long range aid to navigation (LORAN) time bases. Even though the LORAN system is more than two decades old, it is our only backup system for time synchronization. **ANSWER A**

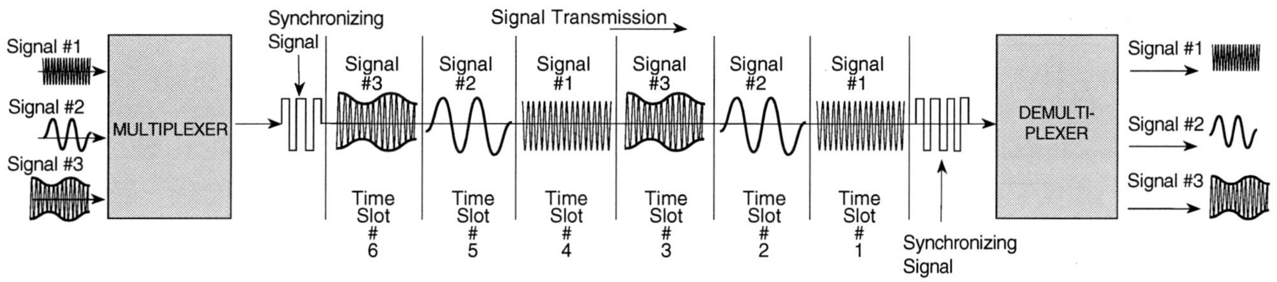

Time multiplexing and demultiplexing

3-59H2 In a pulse-width modulation system, what parameter does the modulating signal vary?

A. Pulse frequency.
B. Pulse duration.
C. Pulse amplitude.
D. Pulse intensity.

It is the duration of the pulse that the modulating signal varies. **ANSWER B**

3-59H3 What is the name of the type of modulation in which the modulating signal varies the duration of the transmitted pulse?

A. Amplitude modulation.
B. Frequency modulation.
C. Pulse-height modulation.
D. Pulse-width modulation.

When the modulating signal varies the duration of the transmitted pulse, it is called pulse-width modulation. **ANSWER D**

3-59H4 Which of the following best describes a pulse modulation system?

A. The peak transmitter power is normally much greater than the average power.
B. Pulse modulation is sometimes used in SSB voice transmitters.
C. The average power is normally only slightly below the peak power.
D. The peak power is normally twice as high as the average power.

One of the main advantages of pulse modulation is its lower power consumption compared with other modes. By operating with a low duty cycle because the transmitter is turned OFF much of the time, very high peak powers can be achieved while maintaining relatively low effective heating or average power. Different pulse modulation schemes use different average duty cycles, but all of them operate with considerably less than 100% duty cycle. **ANSWER A**

3-59H5 In a pulse-position modulation system, what parameter does the modulating signal vary?

A. The number of pulses per second.
B. The time at which each pulse occurs.
C. Both the frequency and amplitude of the pulses.
D. The duration of the pulses.

In a pulse-position modulation system, the position is the time at which each pulse occurs. The modulating signal varies the time when pulses occur. **ANSWER B**

3-59H6 What is one way that voice is transmitted in a pulse-width modulation system?

A. A standard pulse is varied in amplitude by an amount depending on the voice waveform at that instant.
B. The position of a standard pulse is varied by an amount depending on the voice waveform at that instant.
C. A standard pulse is varied in duration by an amount depending on the voice waveform at that instant.
D. The number of standard pulses per second varies depending on the voice waveform at that instant.

Remember pulse-width and duration. A standard pulse is varied in duration by an amount depending on the modulating waveform of the voice signal. **ANSWER C**

Subelement I – Power Sources

3 Key Topics, 3 Exam Questions

Key Topic 60: Batteries-1

3-60I1 When a lead-acid storage battery is being charged, a harmful effect to humans is:

A. Internal plate sulfation may occur under constant charging.
B. Emission of oxygen.
C. Emission of chlorine gas.
D. Emission of hydrogen gas.

If a lead acid storage battery is charged too rapidly, or is over-charged, it will begin to emit hydrogen gas, which is harmful to humans. **ANSWER D**

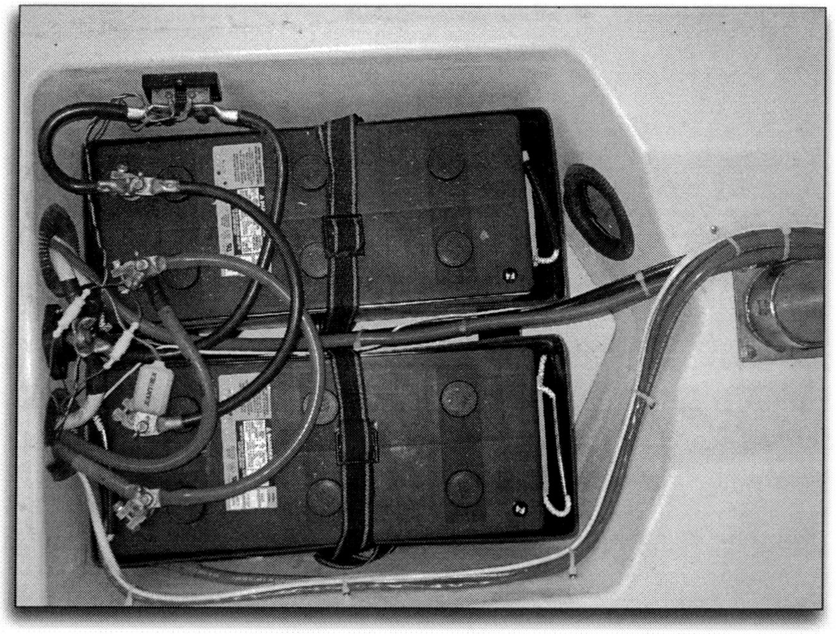

Marine storage batteries, in parallel

3-60I2 A battery with a terminal voltage of 12.5 volts is to be trickle-charged at a 0.5 A rate. What resistance should be connected in series with the battery to charge it from a 110-V DC line?

A. 95 ohms.
B. 300 ohms.
C. 195 ohms.
D. None of these.

First calculate the voltage drop (110 − 12.5 volts = 97.5) that must occur across the resistor. Then, apply Ohm's Law (R = E ÷ I). 97.5 ÷ 0.5 amps = 195 ohms. **ANSWER C**

3-60I3 What capacity in amperes does a storage battery need to be in order to operate a 50 watt transmitter for 6 hours? Assume a continuous transmitter load of 70% of the key-locked demand of 40 A, and an emergency light load of 1.5 A.

A. 100 ampere-hours.
B. 177 ampere-hours.
C. 249 ampere-hours.
D. None of these.

You don't need to worry about how many watts the transmitter puts out because they give you no voltage to work with, but they do give you 40 amps on a 70-percent duty cycle. This works out to be 28 amps, plus an additional 1.5 amps, for a total of 29.5 amps. Because the desired battery life is 6 hours, multiply 29.5 times 6, and you end up with 177 ampere-hours. Some of the questions provide information you don't really need to solve the problem. **ANSWER B**

3-60I4 What is the total voltage when 12 Nickel-Cadmium batteries are connected in series?
A. 12 volts.
B. 12.6 volts.
C. 15 volts.
D. 72 volts.

In this problem, you must remember that each nickel-cadmium battery delivers 1.25 volts. Batteries, when connected – to + (in series), add together. Multiply 1.25 volts by 12 batteries, and you get a series voltage of 15 volts. **ANSWER C**

Ni-Cad rechargeable 1.25 volt batteries in a marine hand held.

A new battery technology called Lithium Iron Phosphate (LiFePO4) offers a safe alternative to heavy lead acid and sensitive Lithium Ion battery systems. The Lithium Iron Phosphate batteries are half the weight, three times higher in capacity, and rechargeable beyond 2,000 cycles. See www.bioennopower.com for specifications. These batteries are great for your portable test equipment used in the field.

3-60I5 The average fully-charged voltage of a lead-acid storage cell is:
A. 1 volt.
B. 1.2 volts.
C. 1.56 volts.
D. 2.06 volts.

The fully charged voltage of a lead-acid storage cell is 2.06 volts. Keep in mind they are talking about an individual cell, not the entire series-configured battery. **ANSWER D**

3-60I6 A nickel-cadmium cell has an operating voltage of about:
A. 1.25 volts.
B. 1.4 volts.
C. 1.5 volts.
D. 2.1 volts.

The popular nickel-cadmium cell has an operating voltage of 1.25 volts; a similar-sized alkaline cell has a slightly higher voltage. It's important that all nickel-cadmium batteries be properly charged and exercised. Monitor both temperature and voltage during charging to ensure that the batteries will last as long as possible. **ANSWER A**

Key Topic 61: Batteries-2

3-61I1 When an emergency transmitter uses 325 watts and a receiver uses 50 watts, how many hours can a 12.6 volt, 55 ampere-hour battery supply full power to both units?
A. 1.8 hours.
B. 6 hours.
C. 3 hours.
D. 1.2 hours.

Operating power is 375 watts (325 + 50). Divide 375 by 12.6 volts to calculate the amount of current this emergency transmitter system is using (375 ÷ 12.6 = 29.76 amps at 12.6 VDC). Now, divide 29.76 amps into 55 ampere-hours to get the approximate operating time of 1.8 hours. Or you could multiply 12.6 × 55 = 693 watt hours and divide by 375 watts to get 1.85 hours. **ANSWER A**

3-61I2 What current will flow in a 6 volt storage battery with an internal resistance of 0.01 ohms, when a 3-watt, 6-volt lamp is connected?
A. 0.4885 amps.
B. 0.4995 amps.
C. 0.5566 amps.
D. 0.5795 amps.

Ohm's Law is P (Power) = E^2(Volts) ÷ R, but you must modify the formula to solve for resistance of the lamp. R = E^2 ÷ P. 36 ÷ 3 = 12 ohms. 12 + 0.01 = 12.01 ohms total resistance. 6V ÷ 12.01 = 0.4995 amps. **ANSWER B**

3-61I3 A ship RADAR unit uses 315 watts and a radio uses 50 watts. If the equipment is connected to a 50 ampere-hour battery rated at 12.6 volts, how long will the battery last?

A. 1 hour 43 minutes.
B. 28.97 hours.
C. 29 minutes.
D. 10 hours, 50 minutes.

Here is another easy problem. Divided the total watts used (315 + 50 = 365) by 12.6 volts to get 28.9 amps of constant current draw. Now, clear your calculator and divide 50 ampere-hours by 28.9 to get 1.7 hours. Finally, multiply 1.7 hours by 60 to get 103 minutes, which works out to be 1 hour and 43 minutes. **ANSWER A**

3-61I4 If a marine radiotelephone receiver uses 75 watts of power and a transmitter uses 325 watts, how long can they both operate before discharging a 50 ampere-hour 12 volt battery?

A. 40 minutes.
B. 1 hour.
C. 1 1/2 hours.
D. 6 hours.

Add 325 watts to 75 watts for a total power consumption of 400 watts. Divide that by 12.0 volts, to get the current draw of 33.33 amps. Divide this into 50-amp hours to find that the system will operate for exactly 1 1/2 hours. **ANSWER C**

3-61I5 A 6 volt battery with 1.2 ohms internal resistance is connected across two light bulbs in parallel whose resistance is 12 ohms each. What is the current flow?

A. 0.57 amps.
B. 0.83 amps.
C. 1.0 amps.
D. 6.0 amps.

The circuit has a battery in series with an internal resistance that feed current to two light bulbs in parallel. We will assume that the bulb resistance remains constant from cold to hot, which, in the real world, normally would not add up to the 12 ohms each as stated. Two 12 ohm resisters in parallel = 6 ohms. Thus, the total resistance across the battery is 1.2 ohms + 6 ohms = 7.2 ohms. To find the current, use $I = E \div R$
$6 \div 7.2 = 0.83$ amps. **ANSWER B**

3-61I6 A 12.6 volt, 8 ampere-hour battery is supplying power to a receiver that uses 50 watts and a RADAR system that uses 300 watts. How long will the battery last?

A. 100.8 hours.
B. 27.7 hours.
C. 1 hour.
D. 17 minutes or 0.3 hours.

In this question, first calculate the total power being consumed (300 watts + 50 watts = 350 watts). Then, divide 350 by 12.6 volts to determine the running current: a whopping 27 amps. Because 27 amps are being drawn from an emergency 8 ampere-hour battery, it should be obvious the system won't run for more than about a half hour. Divide 27 into 8, and you end up with 0.29 hours of operation. Multiply this by 60 to calculate minutes, and the answer is 17 minutes, the same as 0.3 hours. Not much time for a RADAR system! **ANSWER D**

Key Topic 62: Motors & Generators

3-62I1 What occurs if the load is removed from an operating series DC motor?

A. It will stop running.
B. Speed will increase slightly.
C. No change occurs.
D. It will accelerate until it falls apart.

A series DC motor must be matched to the load; the load keeps the motor spinning at a constant rate. If the load is removed from a series DC motor, the motor speeds up and could possibly fly apart. **ANSWER D**

Motor generator.

3-62I2 If a shunt motor running with a load has its shunt field opened, how would this affect the speed of the motor?

A. It will slow down.
B. It will stop suddenly.
C. It will speed up.
D. It will be unaffected.

A shunt motor running under load immediately speeds up if the field winding develops an open circuit. The field acts as a governor, and without a governor to create a counter EMF, the motor speeds up. **ANSWER C**

3-62I3 The expression "voltage regulation" as it applies to a shunt-wound DC generator operating at a constant frequency refers to:

A. Voltage fluctuations from load to no-load.
B. Voltage output efficiency.
C. Voltage in the secondary compared to the primary.
D. Rotor winding voltage ratio

In shipboard installations, voltage regulation is extremely important. The term "voltage regulation" applies to shunt-wound DC generators, operating at a fixed frequency, where no-load voltage is compared to voltage under load (load to no-load), just as we demonstrated in the previous questions. **ANSWER A**

3-62I4 What is the line current of a 7 horsepower motor operating on 120 volts at full load, a power factor of 0.8, and 95% efficient?

A. 4.72 amps.
B. 13.03 amps.
C. 56 amps.
D. 57.2 amps.

First convert the rated horsepower to output watts by multiplying 7 X 746 watts to get 5222 watts. Divide that by the power factor of 0.8 to get 6527.5 watts. Next, divide 6527.5 by 0.95, the efficiency, to get the actual power of 6871.05 volt amps. Finally, divide 6871.05 by the full load voltage (120 volts) to get the actual line current of 57.2 amps. **ANSWER D**

3-62I5 A 3 horsepower, 100 V DC motor is 85% efficient when developing its rated output. What is the current?

A. 8.545 amps.
B. 20.345 amps.
C. 26.300 amps.
D. 25.000 amps.

Here's the solution: The motor is rated at 3 horsepower. Therefore, since

$$1\text{ Hp} = 746\text{ watts}$$

$$3\text{ Hp} = 3 \text{ X } 746 = 2{,}238\text{ watts is } P_{OUT}$$

If the motor is only 85% efficient then the power input, P_{IN}, must be

$$P_{IN} \times 0.85 = P_{OUT}$$

$$P_{IN} = \frac{P_{OUT}}{0.85} = \frac{2238}{0.85} = 2632.9\text{ watts}$$

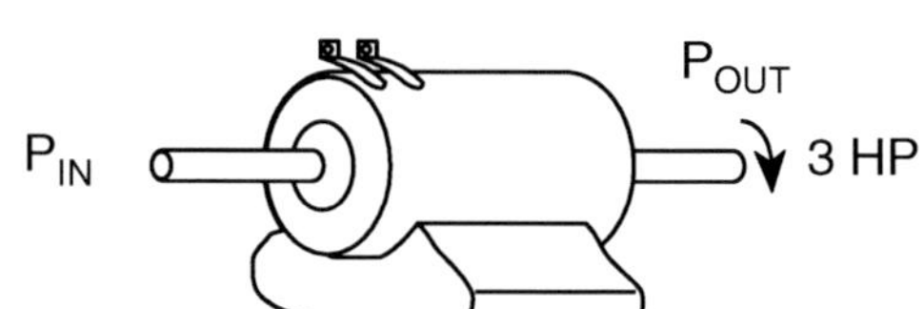

DC MOTOR

To find current with V_{IN} = 100 VDC

$$V_{IN} \times I_{IN} = P_{IN}$$

$$I_{IN} = \frac{P_{IN}}{V_{IN}} = \frac{2632.9}{100} = 26.3\text{ amps}$$

The current is 26.3 amps. **ANSWER C**

3-62I6 The output of a separately-excited AC generator running at a constant speed can be controlled by:

A. The armature.
B. The amount of field current.
C. The brushes.
D. The exciter.

We control the output voltage of a separately-excited AC generator by varying the field current. **ANSWER B**

Subelement J – Antennas

5 Key Topics, 5 Exam Questions

Key Topic 63: Antenna Theory

3-63J1 Which of the following could cause a high standing wave ratio on a transmission line?

A. Excessive modulation.
B. An increase in output power.
C. A detuned antenna coupler.
D. Low power from the transmitter.

Water inside an antenna mount will many times detune the antenna circuit. This causes high SWR, which can be detected with an SWR meter or analyzer. **ANSWER C**

3-63J2 Why is the value of the radiation resistance of an antenna important?

A. Knowing the radiation resistance makes it possible to match impedances for maximum power transfer.
B. Knowing the radiation resistance makes it possible to measure the near-field radiation density from transmitting antenna.
C. The value of the radiation resistance represents the front-to-side ratio of the antenna.
D. The value of the radiation resistance represents the front-to-back ratio of the antenna.

If we know the radiation resistance of a specific antenna, it will allow us to match impedances to that antenna for maximum power transfer. Most coaxial cables have a characteristic impedance of 50 to 52 ohms, and this is what is commonly used in land mobile, air, and marine installations. In an A.M. broadcast facility, it's also necessary to know the radiation resistance in order to calculate transmitted power. A radio frequency ammeter is inserted in the antenna at the feedpoint of every A.M. broadcast tower. Power is measured by taking the square of the antenna current and multiplying it by the radiation resistance. This radiation resistance is spelled out in the station license. Periodic verification of this value by measurement with an impedance bridge is an important part of broadcast facility maintenance. **ANSWER A**

3-63J3 A radio frequency device that allows RF energy to pass through in one direction with very little loss but absorbs RF power in the opposite direction is a:

A. Circulator.
B. Wave trap.
C. Multiplexer.
D. Isolator.

In microwave systems, the isolator keeps RF energy going in one direction, but opposes it from coming back in the opposite direction. **ANSWER D**

3-63J4 What is an advantage of using a trap antenna?

A. It may be used for multiband operation.
B. It has high directivity in the high-frequency bands.
C. It has high gain.
D. It minimizes harmonic radiation.

The trap antenna may be used for multiband operation, from a single feed line. There are trap dipoles, trap verticals, and trap beam antennas for multiband operation. **ANSWER A**

3-63J5 What is meant by the term radiation resistance of an antenna?

A. Losses in the antenna elements and feed line.
B. The specific impedance of the antenna.
C. The resistance in the trap coils to received signals.
D. An equivalent resistance that would dissipate the same amount of power as that radiated from an antenna.

Radiation resistance is the amount of power radiated from an antenna as electromagnetic waves. It is equal to the amount of DC power that would be radiated as heat if the antenna were replaced by a resistor of equal ohmic value. Radiation resistance is a positive measurement of an antenna's efficiency, as opposed to the DC losses of ohmic resistance. The feedpoint resistance is USUALLY, but not NECESSARILY the radiation resistance. **ANSWER D**

3-63J6 What is meant by the term antenna bandwidth?

A. Antenna length divided by the number of elements.
B. The frequency range over which an antenna can be expected to perform well.
C. The angle between the half-power radiation points.
D. The angle formed between two imaginary lines drawn through the ends of the elements.

As a ship station on worldwide frequencies switches to lower bands, such as the 4-MHz, 6-MHz, or 8-MHz band, antenna bandwidth becomes important. The associated automatic antenna tuner will probably seek a new LC setting with frequency excursions as little as 50 kHz. **ANSWER B**

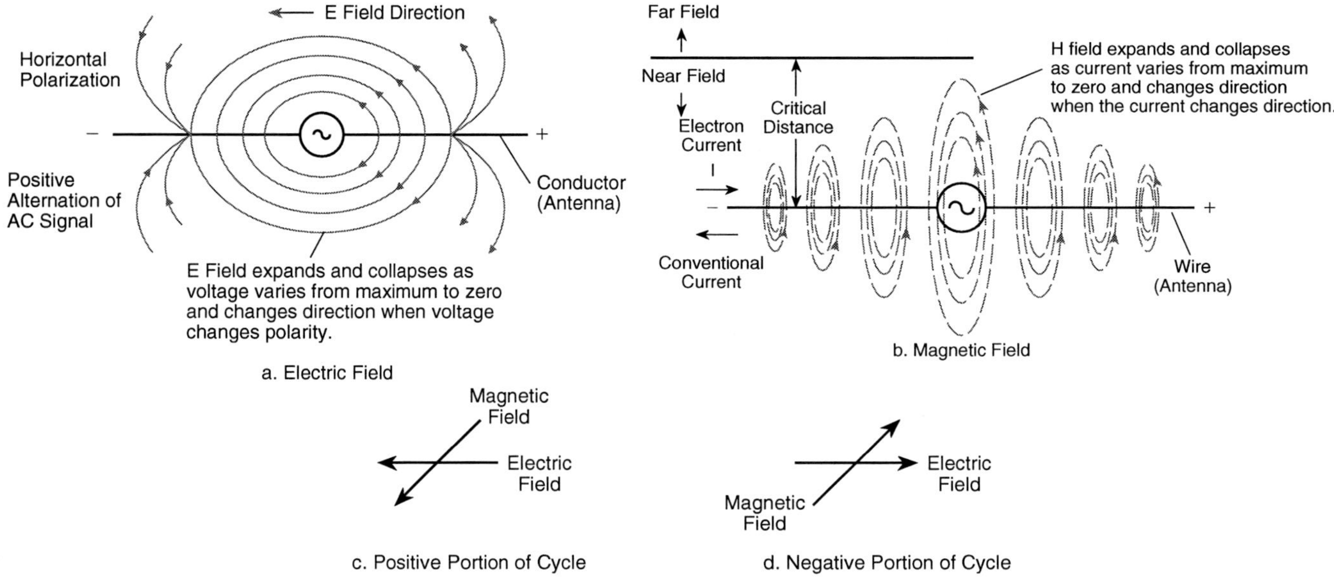

Electromagnetic antenna fields.

(*Source:* Antennas, A.J. Evans, K.E. Britain, ©1998, Master Publishing, Inc., Niles, IL)

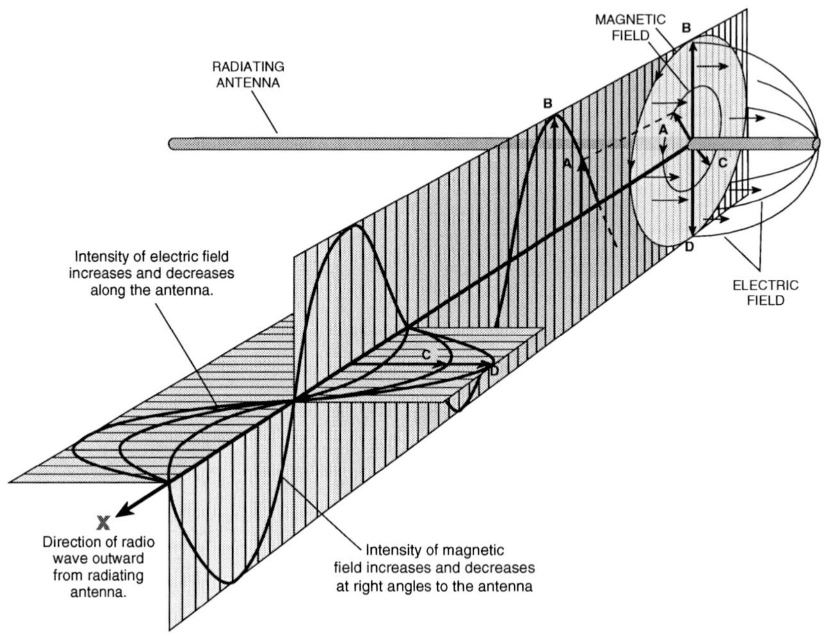

Electomagnetic wave travelling through space.

(*Source:* Basic Electronics, G. McWhorter, A.J. Evans, ©2000, Master Publishing, Inc., Niles, IL)

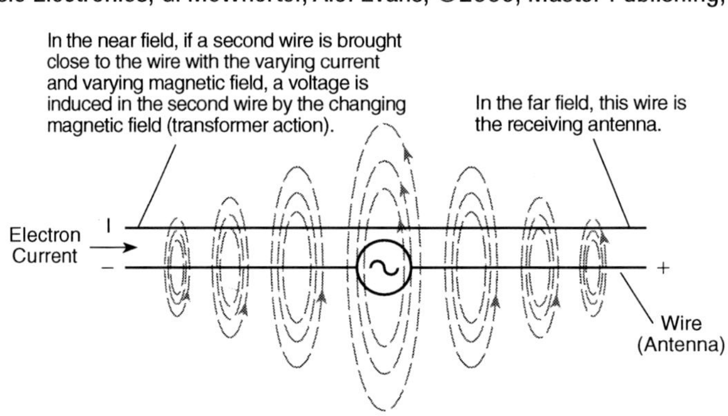

Magnetic field variations induce voltages in wires in the near field or in the far field. When the wire is in the far field it is a receiving antenna.

(*Source:* Basic Electronics, G. McWhorter, A.J. Evans, ©2000, Master Publishing, Inc., Niles, IL)

Why Characteristic Impedance is So Important

This figure shows why characteristic impedance, Z_0 of a transmission line is so important to the efficient transfer of power from the transmitter to the antenna. Standing-Wave Ratio (SWR) is a measure of how well the energy is being transferred over a transmission line to the load (the antenna). SWRs below 1.5 indicate quite efficient transfer of energy. An SWR of 1 is perfect.; SWRs above 2 begin to indicate poor matching.

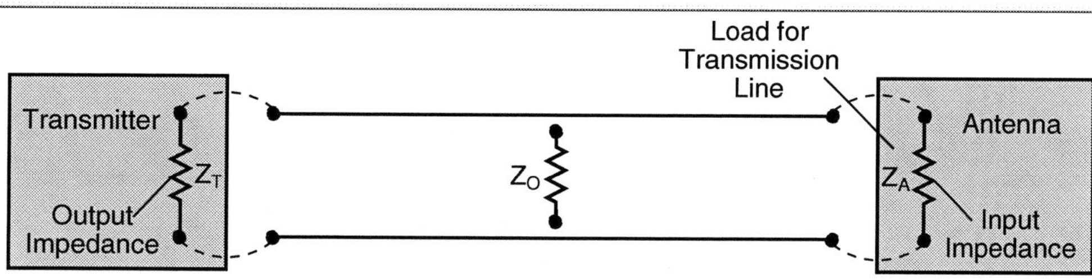

a. Typical Transmission Line for Coupling Transmitter Power to Antenna

Figure a shows a typical transmission line coupling transmitter to antenna. Transmitter output impedance is Z_T; Antenna input impedance is Z_A, and the transmission line impedance is Z_0.

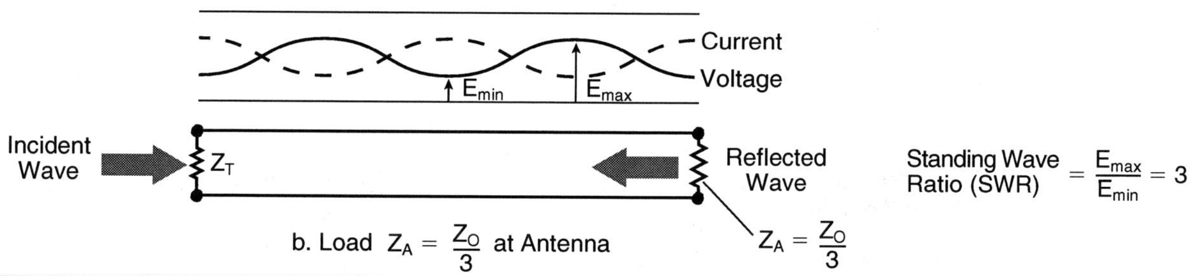

b. Load $Z_A = \frac{Z_O}{3}$ at Antenna

Figure b shows impedance mismatch illustrating that much of the signal power from the transmitter is reflected back down the line from the antenna, resulting an SWR of 3:1.

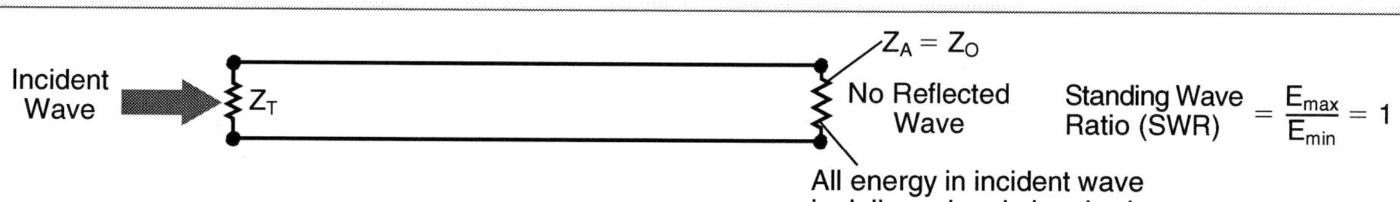

c. Load $Z_A = Z_O$ at Antenna

Figure c illustrates a perfect impedance match resulting in an SWR of 1:1, which provides maximum power output.

Key Topic 64: Voltage, Current and Power Relationships

3-64J1 What is the current flowing through a 52 ohm line with an input of 1,872 watts?
A. 0.06 amps.
B. 6 amps.
C. 28.7 amps.
D. 144 amps.
Recall the formula for calculating current when power and resistance are known:

$$I^2R = P$$

$$I_2 = \frac{P}{R} = \frac{1872}{52} = 36$$

$$I = \sqrt{36} = 6 \text{ amps}$$ **ANSWER B**

Where: I = Current in **amps**
R = resistance in **ohms**
P = power in **watts**

3-64J2 The voltage produced in a receiving antenna is:
A. Out of phase with the current if connected properly.
B. Out of phase with the current if cut to 1/3 wavelength.
C. Variable depending on the station's SWR.
D. Always proportional to the received field strength.
The voltage produced in a receiving antenna is always proportional to the received field strength at the antenna. Electromagnetic radio waves induce a voltage in the receiving antenna which produces a current in the antenna and input circuit to which the antenna is connected. **ANSWER D**

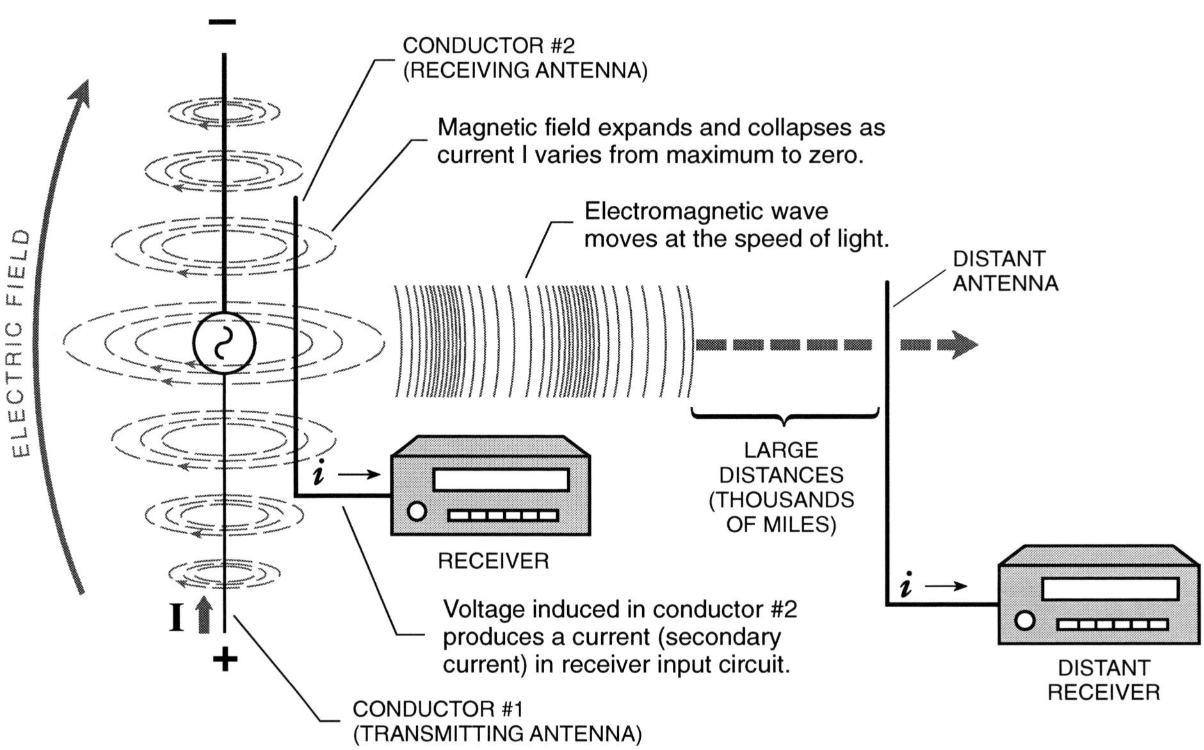

Transmitter to Receiver – Radio waves from transmitting antennas induce signals in receiving antennas as they pass by.
Source: Listening to Shortwave, Ken Winters, ©1993 Master Publishing, Inc., Niles, Illinois

---

3-64J3 Which of the following represents the best standing wave ratio (SWR)?
A. 1:1.
B. 1:1.5.
C. 1:3.
D. 1:4.
1:1 is the best SWR. **ANSWER A**

3-64J4 At the ends of a half-wave antenna, what values of current and voltage exist compared to the remainder of the antenna?

A. Equal voltage and current.
B. Minimum voltage and maximum current.
C. Maximum voltage and minimum current.
D. Minimum voltage and minimum current.

There is maximum voltage on the tip ends of a dipole, but minimum current. This is why you can sometimes spark a lead pencil (be careful) off of the tip ends of a dipole. **ANSWER C**

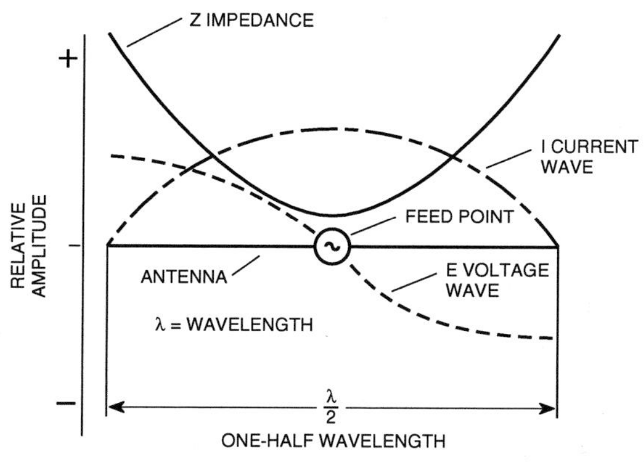

Voltage, Current and Impedance on a Half-wave Dipole

3-64J5 An antenna radiates a primary signal of 500 watts output. If there is a 2nd harmonic output of 0.5 watt, what attenuation of the 2nd harmonic has occurred?

A. 10 dB. B. 30 dB. C. 40 dB. D. 50 dB.

It's relatively simple to calculate a dB power change. Remember this table:

DB	Power Change
3 dB	2X Power change
6 dB	4X Power change
9 dB	8X Power change
10 dB	10X Power change
20 dB	100X Power change
30 dB	1000X Power change
40 dB	10,000X Power change

In this question, a signal of 500 watts output has a second harmonic at 0.5 watt. This is a 1000X power change, so the second harmonic is down 30 dB. **ANSWER B**

Derivation:

If $dB = 10 \log_{10} \frac{P_1}{P_2}$

then what power ratio is 20 dB?

$$20 = 10 \log_{10} \frac{P_1}{P_2}$$

$$\frac{20}{10} = \log_{10} \frac{P_1}{P_2}$$

$$2 = \log_{10} \frac{P_1}{P_2}$$

Remember: logarithm of a number is the exponent to which the base must be raised to get the number.

$$\therefore 10^2 = \frac{P_1}{P_2}$$

$$100 = \frac{P_1}{P_2}$$

Or $P_1 = 100\ P_2$

20 dB means P_1 is 100 times P_2

Decibels

3-64J6 There is an improper impedance match between a 30 watt transmitter and the antenna, with 5 watts reflected. How much power is actually radiated?

A. 35 watts. B. 30 watts. C. 25 watts. D. 20 watts.

This is a shore station, which is permitted higher output power than the 25 watts on a ship station. The 30 watt transmitter, with 5 watts reflected, will actually radiate 25 watts from the antenna. The reflected 5 watts is "lost" power. **ANSWER C**

Key Topic 65: Frequency and Bandwidth

3-65J1 A vertical 1/4 wave antenna receives signals:

A. In the microwave band.
B. In one vertical direction.
C. In one horizontal direction.
D. Equally from all horizontal directions.

While this answer sounds strange, the 1/4-wavelength vertical antenna receives its energy from other vertical antennas in all horizontal directions around the antenna. You have to think about this for a few seconds to get the picture – it receives in all directions – all horizontal directions. **ANSWER D**

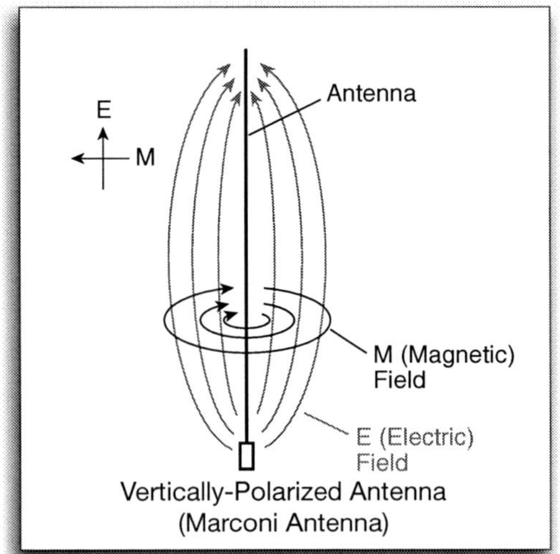

Polarization of antennas. An antenna's polarization is determined by the direction of the E (Electric) field.

3-65J2 The resonant frequency of a Hertz antenna can be lowered by:

A. Lowering the frequency of the transmitter.
B. Placing an inductance in series with the antenna.
C. Placing a condenser in series with the antenna.
D. Placing a resistor in series with the antenna.

The Hertz antenna is a half-wave dipole. If we add an inductance (coil) in series, it will lower its resonant frequency. **ANSWER B**

3-65J3 An excited 1/2 wavelength antenna produces:

A. Residual fields.
B. An electro-magnetic field only.
C. Both electro-magnetic and electro-static fields.
D. An electro-flux field sometimes.

All antennas radiate two fields—an electric field and a magnetic field. They are at right angles to each other. It is the electric field that is used when describing polarization. **ANSWER C**

3-65J4 To increase the resonant frequency of a 1/4 wavelength antenna:

A. Add a capacitor in series.
B. Lower capacitor value.
C. Cut antenna.
D. Add an inductor.

We add capacitance in series to raise the resonant frequency of a 1/4-wavelength antenna. Answer C would work, but is unrealistic atop a tall mast, so adding series capacitance within a matching device is the best solution. **ANSWER A**

3-65J5 What happens to the bandwidth of an antenna as it is shortened through the use of loading coils?

A. It is increased.
B. It is decreased.
C. No change occurs.
D. It becomes flat.

The shorter your mobile whip, the more loading you need, and bandwidth is decreased. Also, your efficiency usually goes down – so, for high efficiency and more bandwidth, go for a longer whip with less loading. **ANSWER B**

3-65J6 To lengthen an antenna electrically, add a:

A. Coil.
B. Resistor.
C. Battery.
D. Conduit.

To lengthen an antenna electrically, add an inductor, such as a coil, in series with the antenna at the base. **ANSWER A**

Key Topic 66: Transmission Lines

3-66J1 What is the meaning of the term velocity factor of a transmission line?
A. The ratio of the characteristic impedance of the line to the terminating impedance.
B. The velocity of the wave on the transmission line divided by the velocity of light in a vacuum.
C. The velocity of the wave on the transmission line multiplied by the velocity of light in a vacuum.
D. The index of shielding for coaxial cable.

Radio waves travel in free space at 300 million meters per second. But in coaxial cable transmission lines, and other types of feed lines, the radio waves move slower. Velocity factor is the velocity of the radio wave on the transmission line divided by the velocity in free space. **ANSWER B**

Large coax, with hollow center conductor, low loss.

3-66J2 What determines the velocity factor in a transmission line?
A. The termination impedance.
B. The line length.
C. Dielectrics in the line.
D. The center conductor resistivity.

There are some great low-loss coaxial cable types out there in radioland with different dielectrics separating the center conductor and the outside shield. It's the dielectric that determines the velocity factor. **ANSWER C**

3-66J3 Nitrogen is placed in transmission lines to:
A. Improve the "skin-effect" of microwaves.
B. Reduce arcing in the line.
C. Reduce the standing wave ratio of the line.
D. Prevent moisture from entering the line.

You can sometimes see large nitrogen tanks linked to radio tower transmission lines. The lines are filled with nitrogen to prevent moisture that could cause corrosion from getting into the transmission line. **ANSWER D**

3-66J4 A perfect (no loss) coaxial cable has 7 dB of reflected power when the input is 5 watts. What is the output of the transmission line?
A. 1 watt.
B. 1.25 watts.
C. 2.5 watts.
D. 5 watts.

This one you can do in your head. 6 dB reflection is a 4 times loss in this circuit. Since it's a little bit more than a 4 times loss at 5 watts, you would only have 1 watt getting out of the transmission line into the antenna circuit. **ANSWER A**

3-66J5 Referred to the fundamental frequency, a shorted stub line attached to the transmission line to absorb even harmonics could have a wavelength of:
A. 1.41 wavelength.
B. 1/2 wavelength.
C. 1/4 wavelength.
D. 1/6 wavelength.

The 1/4-wavelength shorted stub line is attached to a transmitter to absorb even harmonics on a single frequency antenna system. **ANSWER C**

3-66J6 If a transmission line has a power loss of 6 dB per 100 feet, what is the power at the feed point to the antenna at the end of a 200 foot transmission line fed by a 100 watt transmitter?

A. 70 watts.
B. 50 watts.
C. 25 watts.
D. 6 watts.

You can do this one in your head, too. If there is 6 dB loss for 100 feet of coax, there will be 12 dB loss for 200 feet. 12 dB loss is more than 10 dB which is easy to remember as a multiplier of 10. Knowing that your loss is greater than 10 times, resulting in 90 watts lost, and only 10 watts left, with 6 watts as the only answer that is close! Don't run small coax on VHF frequencies! **ANSWER D**

Decibels

It is important to know about decibels because they are used extensively in electronics. Look at the derivation for decibels and note it is a measure of the ratio of two powers, P_1 and P_2. Remember the power changes for different dB values. Also, since 6 dB is a four times change, 16 dB (6 dB + 10 dB) will be a 4 × 10 = 40 times change, and 26 dB (6 dB + 20 dB) will be a 4 × 100 = 400 times change. Thus, any dB value from 10 dB and above can be evaluted by using the power change of 1 dB to 9 dB and multiplying it by the 10 dB, 20 dB, 30 dB, 40 dB, 50 dB, 60 dB, etc. change. Thus, 57 dB (7 dB + 50 dB) will be a 5 × 100,000 = 500,000 times change.

dB	Power Change	$\frac{P_1}{P_2}$
1 dB	1¼X (1.26)	Power change
2 dB	1½X (1.58)	Power change
3 dB	2X (2.00)	Power change
4 dB	2½X (2.51)	Power change
5 dB	3X (3.16)	Power change
6 dB	4X (3.98)	Power change
7 dB	5X (5.01)	Power change
8 dB	6¼X (6.31)	Power change
9 dB	8X (7.94)	Power change

dB	Power Change	$\frac{P_1}{P_2}$
10 dB	10X	Power change
20 dB	100X	Power change
30 dB	1000X	Power change
40 dB	10,000X	Power change
50 dB	100,000X	Power change
60 dB	1,000,000X	Power change

Derivation:

If $\text{dB} = 10 \log_{10} \frac{P_1}{P_2}$

then what power ratio is 60 dB?

$$60 = 10 \log_{10} \frac{P_1}{P_2}$$

$$\frac{60}{10} = \log_{10} \frac{P_1}{P_2}$$

$$6 = \log_{10} \frac{P_1}{P_2}$$

Remember: logarithm of a number is the exponent to which the base must be raised to get the number.

$$\therefore 10^6 = \frac{P_1}{P_2}$$

$$1{,}000{,}000 = \frac{P_1}{P_2}$$

Or $P_1 = 1{,}000{,}000\ P_2$

60 dB means P_1 is one million times P_2

Key Topic 67: Effective Radiated Power

3-67J1 What is the effective radiated power of a repeater with 50 watts transmitter power output, 4 dB feedline loss, 3 dB duplexer and circulator loss, and 6 dB antenna gain?

A. 158 watts.
B. 39.7 watts.
C. 251 watts.
D. 69.9 watts.

You will have one of these six questions based on simple dB calculations at a repeater site. Don't worry about logarithms, you can do these in your head, and all you must remember is: 3 dB = a 2X increase; to twice the value; and −3 dB = a 2X loss to one half the value. To solve the problems, just add and subtract all the dB's, and if it ends up about +3 dB, then double your transmitter power output to end up with effective radiated power. If your answer ends up around −3 dB gain, cut your transmitter power in half for effective radiated power. In this question, we have −7 dB loss, offset by 6 dB gain for 1 dB net loss. If we started out with 50 watts, our effective radiated power (ERP) will be slightly less, so 39.7 is the only answer in the ballpark. **ANSWER B**

3-67J2 What is the effective radiated power of a repeater with 75 watts transmitter power output, 4 dB feedline loss, 3 dB duplexer and circulator loss, and 10 dB antenna gain?

A. 600 watts.
B. 75 watts.
C. 18.75 watts.
D. 150 watts.

We have 7 dB loss, and a great antenna system giving us 10 dB gain. This gives us a +3 dB gain, which is 2 times our power output. There it is, only one answer at 150 watts! **ANSWER D**

3-67J3 What is the effective radiated power of a repeater with 75 watts transmitter power output, 5 dB feedline loss, 4 dB duplexer and circulator loss, and 6 dB antenna gain?

A. 37.6 watts.
B. 237 watts.
C. 150 watts.
D. 23.7 watts.

9 dB loss, offset by 6 dB antenna gain gives a 3 dB net loss, or half power. Half of 75 watts is 37.5 watts, with answer A at 37.6 watts as the best answer. **ANSWER A**

3-67J4 What is the effective radiated power of a repeater with 100 watts transmitter power output, 4 dB feedline loss, 3 dB duplexer and circulator loss, and 7 dB antenna gain?

A. 631 watts.
B. 400 watts.
C. 25 watts.
D. 100 watts.

7 dB down the tubes, offset by 7 dB gain. No change on power output – still 100 watts ERP. **ANSWER D**

3-67J5 What is the effective radiated power of a repeater with 100 watts transmitter power output, 5 dB feedline loss, 4 dB duplexer and circulator loss, and 10 dB antenna gain?

A. 126 watts.
B. 800 watts.
C. 12.5 watts.
D. 1260 watts.

A total of 9 dB loss, offset by 10 dB gain, net +1 dB. Since we had 100 watts to start with and the net gain is 1 dB, the only close answer is 126 watts. **ANSWER A**

3-67J6 What is the effective radiated power of a repeater with 50 watts transmitter power output, 5 dB feedline loss, 4 dB duplexer and circulator loss, and 7 dB antenna gain?

A. 300 watts.
B. 315 watts.
C. 31.5 watts.
D. 69.9 watts.

Here we have 9 dB loss, and 7 dB gain. That's just about a half-power loss (-3 dB), and the only answer that is reasonably close is 31.5 watts. **ANSWER C**

Subelement K – Aircraft

6 Key Topics, 6 Exam Questions

Key Topic 68: Distance Measuring Equipment

3-68K1 What is the frequency range of the Distance Measuring Equipment (DME) used to indicate an aircraft's slant range distance to a selected ground-based navigation station?

A. 108.00 MHz to 117.95 MHz.
B. 108.10 MHz to 111.95 MHz.
C. 962 MHz to 1213 MHz.
D. 329.15 MHz to 335.00 MHz.

The distance measuring equipment (DME) that indicates an aircraft's slant range to a selected ground-based navigation station uses frequencies from 962 MHz to 1213 MHz. **ANSWER C**

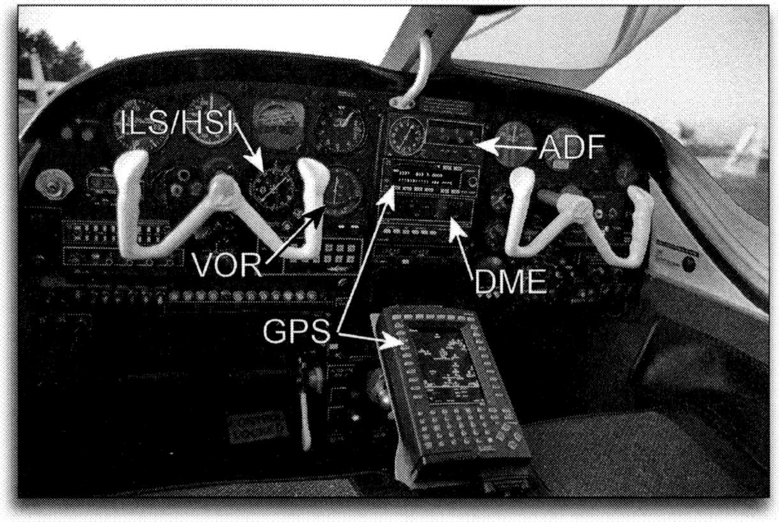

Aircraft instrumentation, with individual equipment callouts.

3-68K2 The Distance Measuring Equipment (DME) measures the distance from the aircraft to the DME ground station. This is referred to as:

A. DME bearing.
B. The slant range.
C. Glide Slope angle of approach.
D. Localizer course width.

The term "range," found in the correct answer, refers to distance to or distance from. The question asks the "distance from," and there is only one answer with the word "range" in it. Because the aeronautical station is well above the ground, the direct distance to the DME station is called the "slant range." **ANSWER B**

3-68K3 The Distance Measuring Equipment (DME) ground station has a built-in delay between reception of an interrogation and transmission of the reply to allow:

A. Someone to answer the call.
B. The VOR to make a mechanical hook-up.
C. Operation at close range.
D. Clear other traffic for a reply.

The DME retransmission delay is to avoid uncoordinated operation when the aircraft is very close to the DME ground station. At UHF frequencies, the transmission speed of 12.36 μs/nm is too fast to avoid possible sync failures at close range. The calibrated delays can be either 50, 56, or 62 microseconds, depending on the DME type/channel. The aircraft receiver is programmed to recognize different channels, designated by X, Y, W, or Z. This delay can also be adjusted to calibrate the aircraft DME, as well as to adjust the distance readout for ILS. Typically, the calibrated distance should be zero at touchdown on an ILS associated DME, otherwise distance adjustments can be referenced on the various approach plates used by the pilot. This transponder delay is not be to confused with the randomly selected delay between pulse pairs that the interrogating DME uses to identify its unique signal. **ANSWER C**

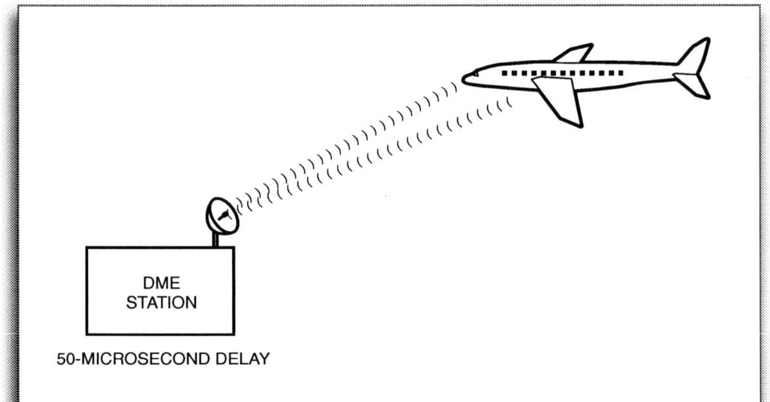

Measuring Aircraft's Distance to Station.

3-68K4 What is the main underlying operating principle of an aircraft's Distance Measuring Equipment (DME)?

A. A measurable amount of time is required to send and receive a radio signal through the Earth's atmosphere.
B. The difference between the peak values of two DC voltages may be used to determine an aircraft's distance to another aircraft.
C. A measurable frequency compression of an AC signal may be used to determine an aircraft's altitude above the earth.
D. A phase inversion between two AC voltages may be used to determine an aircraft's distance to the exit ramp of an airport's runway.

DME equipment uses the principle of RADAR and time delays between a transmit and received signal. DME equipment measures the amount of time required to send and receive a radio signal through the Earth's atmosphere and converts that time to distance. The key word is time. **ANSWER A**

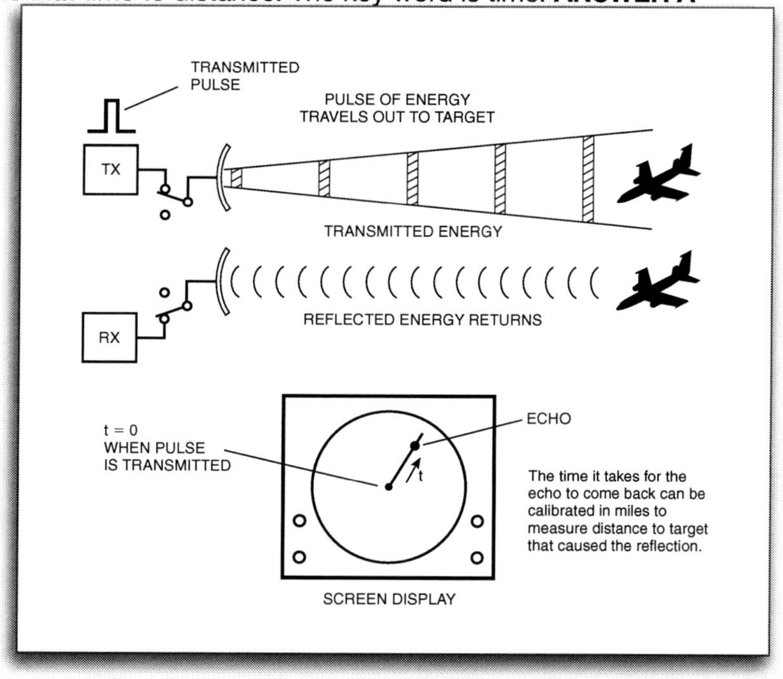

RADAR Echo
Source: *Technology Dictionary*, ©1987, Master Publishing, Inc.

3-68K5 What radio navigation aid determines the distance from an aircraft to a selected VORTAC station by measuring the length of time the radio signal takes to travel to and from the station?

A. RADAR.
B. Loran C.
C. Distance Marking (DM).
D. Distance Measuring Equipment (DME).

It is the distance measuring equipment, abbreviated DME, that measures the distance from an aircraft to a selected VORTAC station. The DME determines the distance based on the time delay of the radio signal. **ANSWER D**

3-68K6 The majority of airborne Distance Measuring Equipment systems automatically tune their transmitter and receiver frequencies to the paired __ / __ channel.

A. VOR/marker beacon.
B. VOR/LOC.
C. Marker beacon/glideslope.
D. LOC/glideslope.

The distance measuring equipment (DME) receives the VHF omnidirectional range (VOR) NAVAID and localizer (LOC) at the same time. The LOC is the left-right information portion of an instrument landing system. **ANSWER B**

Key Topic 69: VHF Omnidirectional Range (VOR)

3-69K1 All directions associated with a VOR station are related to:

A. Magnetic north.
B. North pole.
C. North star.
D. None of these.

All VOR bearings are related to magnetic North, not true North. **ANSWER A**

3-69K2 The rate that the transmitted VOR variable signal rotates is equivalent to how many revolutions per second?

A. 60.
B. 30.
C. 2400.
D. 1800.

A VOR station transmits two signals simultaneously. One signal is omnidirectional, and the other signal is rotated to create a phase shift. The horizontal dipole VOR transmitting antenna rotates 30 times per second, creating an in-phase condition at the 360 degree radial, and an out-of-phase condition at the 180 degree radial. This signal is interpreted by the airborne receiver, and the pilot knows which radial the aircraft is flying on. **ANSWER B**

3-69K3 What is the frequency range of the ground-based Very-high-frequency Omnidirectional Range (VOR) stations used for aircraft navigation?

A. 108.00 kHz to 117.95 kHz.
B. 329.15 MHz to 335.00 MHz.
C. 329.15 kHz to 335.00 kHz.
D. 108.00 MHz to 117.95 MHz.

The frequency range for the Very-high-frequency Omnidirectional Range stations is from 108 MHz to 117.95 MHz. Watch out for answer A – it is incorrect because it is in kHz; VHF is in MHz. **ANSWER D**

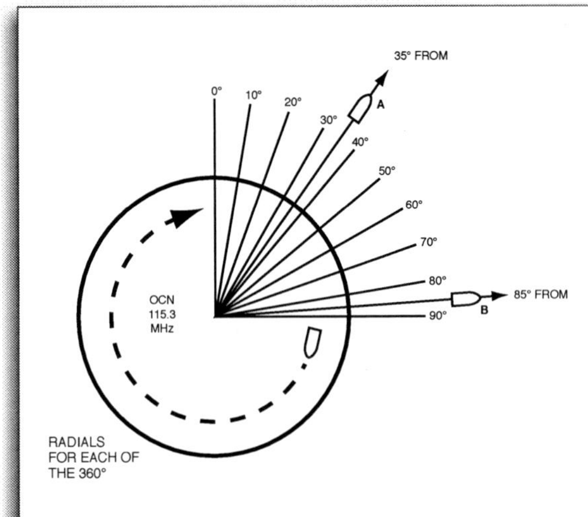

VOR signals may be received up to 100 miles away, when the aircraft is above 3,000 feet.

3-69K4 Lines drawn from the VOR station in a particular magnetic direction are:

A. Radials.
B. Quadrants.
C. Bearings.
D. Headings.

The lines from a VOR station are called radials, and all VOR stations transmit 360 distinct radial signals. The signals are magnetically created by phase shifts between the VOR sub-carriers. **ANSWER A**

VOR signals can even be received on air band hand held transceivers, in case of total cockpit instrumentation black out!

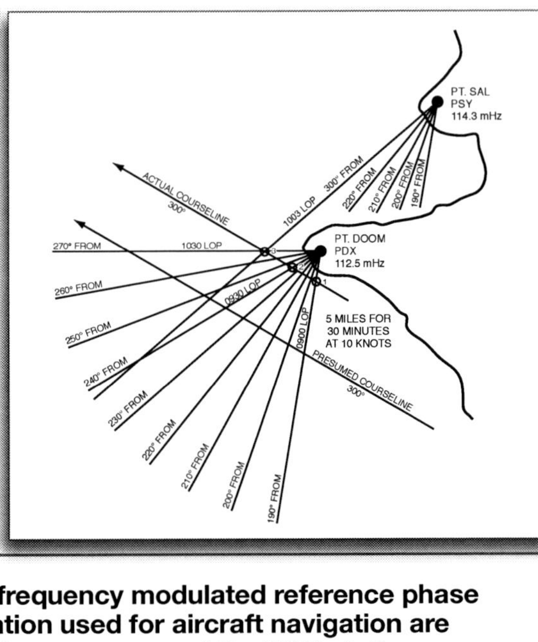

3-69K5 The amplitude modulated variable phase signal and the frequency modulated reference phase signal of a Very-high-frequency Omnidirectional Range (VOR) station used for aircraft navigation are synchronized so that both signals are in phase with each other at ____________ of the VOR station.

A. 180 degrees South, true bearing position.
B. 360 degrees North, magnetic bearing position.
C. 180 degrees South, magnetic bearing position.
D. 0 degrees North, true bearing position.

When both VOR signals are in phase, they indicate exactly 360 degrees magnetic North. Not true North, but magnetic North. **ANSWER B**

3-69K6 What is the main underlying operating principle of the Very-high-frequency Omnidirectional Range (VOR) aircraft navigational system?

A. A definite amount of time is required to send and receive a radio signal.
B. The difference between the peak values of two DC voltages may be used to determine an aircraft's altitude above a selected VOR station.
C. A phase difference between two AC voltages may be used to determine an aircraft's azimuth position in relation to a selected VOR station.
D. A phase difference between two AC voltages may be used to determine an aircraft's distance from a selected VOR station.

The VOR navigation receiver in an aircraft compares the phase differences between two AC voltages to determine an aircraft's azimuth position in relation to a selected VOR station. **ANSWER C**

Key Topic 70: Instrument Landing System (ILS)

3-70K1 What is the frequency range of the localizer beam system used by aircraft to find the centerline of a runway during an Instrument Landing System (ILS) approach to an airport?

A. 108.10 kHz to 111.95 kHz.
B. 329.15 MHz to 335.00 MHz.
C. 329.15 kHz to 335.00 kHz.
D. 108.10 MHz to 111.95 MHz.

The VHF frequency range for the localizer beam system for an instrument landing system approach is 108.1 MHz to 111.95 MHz, just above the FM radio band. If you're near an airport, crank your FM receiver dial all the way to the right, and you can usually hear the localizer. Watch out for answer A; these are the right numbers, but answer A is in kHz instead of MHz. **ANSWER D**

3-70K2 What is the frequency range of the marker beacon system used to indicate an aircraft's position during an Instrument Landing System (ILS) approach to an airport's runway?

A. The outer, middle, and inner marker beacons' UHF frequencies are unique for each ILS equipped airport to provide unambiguous frequency-protected reception areas in the 329.15 to 335.00 MHz range.
B. The outer marker beacon's carrier frequency is 400 MHz, the middle marker beacon's carrier frequency is 1300 MHz, and the inner marker beacon's carrier frequency is 3000 MHz.
C. The outer, the middle, and the inner marker beacon's carrier frequencies are all 75 MHz but the marker beacons are 95% tone-modulated at 400 Hz (outer), 1300 Hz (middle), and 3000 Hz (inner).
D. The outer, marker beacon's carrier frequency is 3000 kHz, the middle marker beacon's carrier frequency is 1300 kHz, and the inner marker beacon's carrier frequency is 400 kHz.

The marker beacon system that illustrates an aircraft's position during an ILS approach uses an outer marker, middle marker, and inner marker, all on 75 MHz. A rising tone indicates the aircraft is coming in for a landing. The outer marker is modulated at 400 Hz, the middle marker at 1300 Hz, and the inner marker at 3000 Hz. Watch out for those incorrect answers that list the tone-modulated audio MHz or kHz. There is only one answer with the correct 75-MHz carrier frequency. **ANSWER C**

3-70K3 Which of the following is a required component of an Instrument Landing System (ILS)?

A. Altimeter: shows aircraft height above sea-level.
B. Localizer: shows aircraft deviation horizontally from center of runway.
C. VHF Communications: provide communications to aircraft.
D. Distance Measuring Equipment: shows aircraft distance to VORTAC station.

The localizer is part of the instrument landing system (ILS). The altimeter, VHF COM radio, and DME equipment are not part of the instrument landing system, although they have a vital roll in air safety. **ANSWER B**

3-70K4 What type of antenna is used in an aircraft's Instrument Landing System (ILS) glideslope installation?

A. A vertically polarized antenna that radiates an omnidirectional antenna pattern.
B. A balanced loop reception antenna.
C. A folded dipole reception antenna.
D. An electronically steerable phased-array antenna that radiates a directional antenna pattern.

The type of antenna used in an aircraft instrument landing system glideslope installation is a folded dipole reception antenna. **ANSWER C**

3-70K5 Choose the only correct statement about the localizer beam system used by aircraft to find the centerline of a runway during an Instrument Landing System (ILS) approach to an airport. The localizer beam system:

A. Operates within the assigned frequency range of 108.10 to 111.95 GHz.
B. Produces two amplitude modulated antenna patterns; one pattern above and one pattern below the normal 2.5 degree approach glide path of the aircraft.
C. Frequencies are automatically tuned-in when the proper glide slope frequency is selected on the aircraft's Navigation and Communication (NAV/COMM) transceiver.
D. Produces two amplitude modulated antenna patterns; one pattern with an audio frequency of 90 Hz and one pattern with an audio frequency of 150 Hz, one left of the runway centerline and one right of the runway centerline.

The two audio frequencies that are amplitude modulated for the localizer beam are 90 Hz and 150 Hz. No other answer has 90 and 150 in it. **ANSWER D**

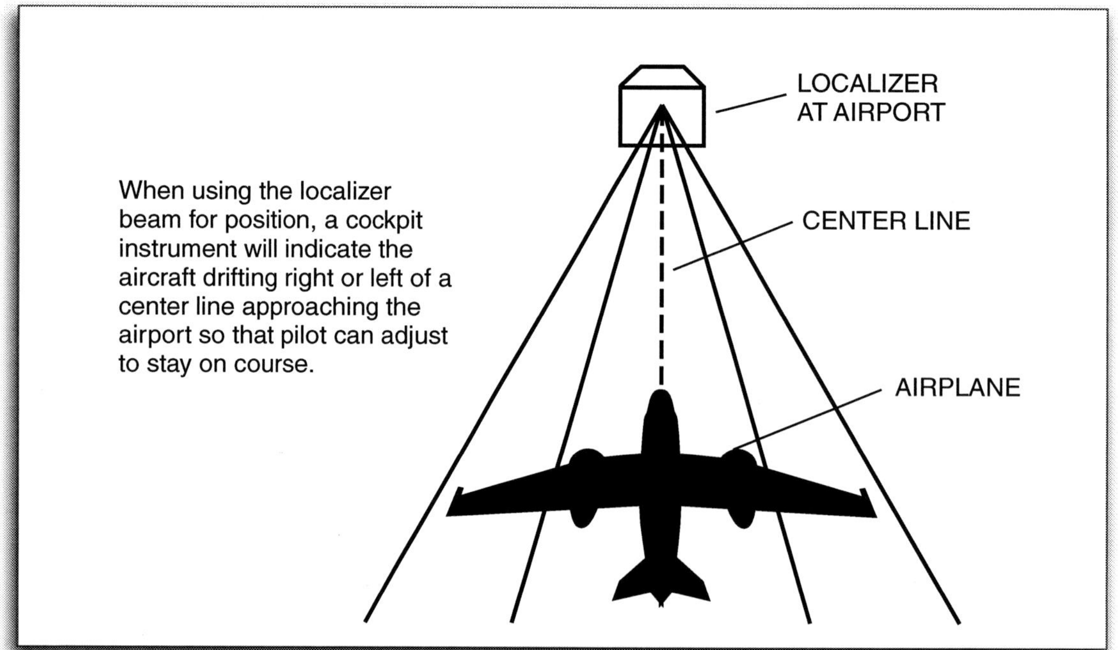

Flight Control with Localizer

3-70K6 On runway approach, an ILS Localizer shows:

A. Deviation left or right of runway center line.
B. Deviation up and down from ground speed.
C. Deviation percentage from authorized ground speed.
D. Wind speed along runway.

The localizer in an instrument landing system shows deviation to the left or right of the runway center line. The localizer approach is a non-precision approach. **ANSWER A**

The Global Navigation Satellite Systems (GNSS) are fast becoming an important part in aviation navigation, including runway approach and landings. New systems recently added to our own GPS constellation are Galileo from Europe, GLONASS from Russia, and our own new MEOSAR satellite constellation. Will this mean that ILS will go away soon? Not for a long time! Google GNSS and read up on all it has to offer.

Key Topic 71: Automatic Direction Finding Equipment (ADF) & Transponders

3-71K1 What is the frequency range of an aircraft's Automatic Direction Finding (ADF) equipment?

A. 190 kHz to 1750 kHz.
B. 190 MHz to 1750 MHz.
C. 108.10 MHz to 111.95 MHz.
D. 108.00 MHz to 117.95 MHz.

Automatic direction finders work on low- and medium-frequency bands from 190 kHz to 1750 kHz. Watch out for answer B – it is incorrectly stated in MHz. **ANSWER A**

3-71K2 What is meant by the term "night effect" when using an aircraft's Automatic Direction Finding (ADF) equipment? Night effect refers to the fact that:

A. All Non Directional Beacon (NDB) transmitters are turned-off at dusk and turned-on at dawn.
B. Non Directional Beacon (NDB) transmissions can bounce-off the Earth's ionosphere at night and be received at almost any direction.
C. An aircraft's ADF transmissions will be slowed at night due to the increased density of the Earth's atmosphere after sunset.
D. An aircraft's ADF antennas usually collect dew moisture after sunset which decreases their effective reception distance from an NDB transmitter.

At nighttime, the ionosphere can refract and bounce NDB signals, and through back scatter could be received from a direction different than where the beacon is actually located. **ANSWER B**

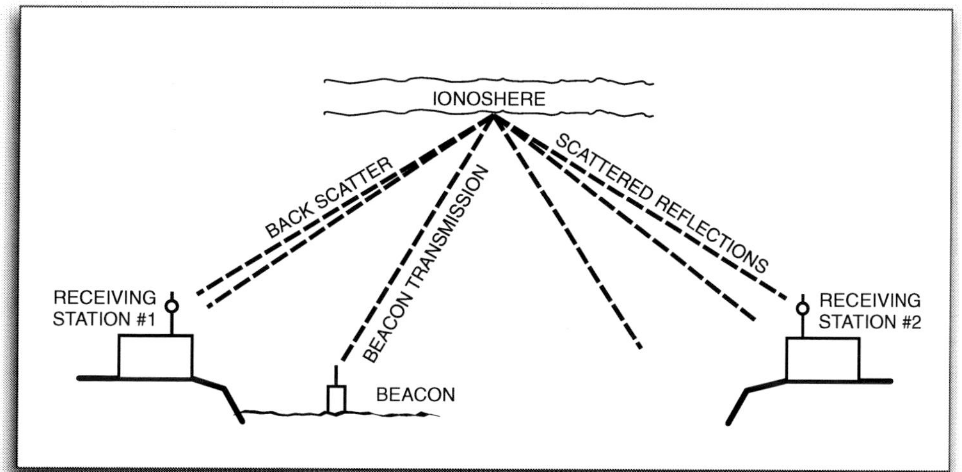

"Night Effect" Using ADF Equipment

3-71K3 What are the transmit and receive frequencies of an aircraft's mode C transponder operating in the Air Traffic Control RADAR Beacon System (ATCRBS)?

A. Transmit at 1090 MHz, and receive at 1030 MHz
B. Transmit at 1030 kHz, and receive at 1090 kHz
C. Transmit at 1090 kHz, and receive at 1030 kHz
D. Transmit at 1030 MHz, and receive at 1090 MHz

The mode-C transponder transmits at 1090 MHz, and receives 60 MHz lower at 1030 MHz. Watch out for incorrect answers that have this backwards, and incorrect answers in kHz instead of MHz! **ANSWER A**

3-71K4 In addition to duplicating the functions of a mode C transponder, an aircraft's mode S transponder can also provide:

A. Primary RADAR surveillance capabilities.
B. Long range lightning detection.
C. Mid-Air collision avoidance capabilities.
D. Backup VHF voice communication abilities.

The mode-S transponder has all the functions of a mode-C transponder; however, mode-S transponders also have mid-air collision avoidance capabilities. **ANSWER C**

3-71K5 What type of encoding is used in an aircraft's mode C transponder transmission to a ground station of the Air Traffic Control RADAR Beacon System (ATCRBS)?

A. Differential phase shift keying.
B. Pulse position modulation.
C. Doppler effect compressional encryption.
D. Amplitude modulation at 95%.

The encoding for a mode-C transponder is PPM. This stands for pulse position modulation, not parts per million! **ANSWER B**

3-71K6 Choose the only correct statement about an aircraft's Automatic Direction Finding (ADF) equipment.

A. An aircraft's ADF transmission exhibits primarily a line-of-sight range to the ground-based target station and will not follow the curvature of the Earth.
B. Only a single omnidirectional sense antenna is required to receive an NDB transmission and process the signal to calculate the aircraft's bearing to the selected ground station.
C. All frequencies in the ADF's operating range except the commercial standard broadcast stations (550 to 1660 kHz) can be utilized as a navigational Non Directional Beacon (NDB) signal.
D. An aircraft's ADF antennas can receive transmissions that are over the Earth's horizon (sometimes several hundred miles away) since these signals will follow the curvature of the Earth.

An aircraft's reception of older, non-directional beacons on low frequency and medium frequency sometimes propagates well beyond the Earth's horizon because these lower frequencies tend to follow the curvature of the Earth and are sometimes refracted by the ionosphere. **ANSWER D**

Key Topic 72: Aircraft Antenna Systems and Frequencies

3-72K1 What type of antenna pattern is radiated from a ground station phased-array directional antenna when transmitting the PPM pulses in a Mode S interrogation signal of an aircraft's Traffic alert and Collision Avoidance System (TCAS) installation?

A. 1090 MHz directional pattern.
B. 1030 MHz omnidirectional pattern.
C. 1090 MHz omnidirectional pattern.
D. 1030 MHz directional pattern.

A 1030 MHz directional pattern is radiated from a phased-array directional antenna when P1 or P3 signals are transmitted. **ANSWER D**

3-72K2 What type of antenna is used in an aircraft's Instrument Landing System (ILS) marker beacon installation?

A. An electronically steerable phased-array antenna that radiates a directional antenna pattern.
B. A folded dipole reception antenna.
C. A balanced loop reception antenna.
D. A horizontally polarized antenna that radiates an omnidirectional antenna pattern.

For the instrument landing system marker beacon, a balanced loop reception antenna is used. **ANSWER C**

3-72K3 What is the frequency range of an aircraft's Very High Frequency (VHF) communications?

A. 118.000 MHz to 136.975 MHz (worldwide up to 151.975 MHz).
B. 108.00 MHz to 117.95 MHz.
C. 329.15 MHz to 335.00 MHz.
D. 2.000 MHz to 29.999 MHz.

Aeronautical VHF channels are found from 118.000 to 136.975 MHz (and sometimes higher) throughout the world. **ANSWER A**

3-72K4 Aircraft Emergency Locator Transmitters (ELT) operate on what frequencies?

A. 121.5 MHz.
B. 243 MHz.
C. 121.5 and 243 MHz.
D. 121.5, 243 and 406 MHz.

The modern aircraft Emergency Locater Transmitter will send a data burst every 50 seconds on 406 MHz, which will be received by both GEOSAR and LEOSAR satellites. This data burst contains ELT user information and may also include an imbedded GPS derived latitude and longitude. The ELT will also continuously sweep a homing signal on 121.5 MHz, with a second harmonic on 243 MHz. The homing signal is intended for ground stations to track down the exact location of the activated ELT. Low Earth Orbit SARSAT satellites no longer Doppler shift compute this 121.5 MHz signal. **ANSWER D**

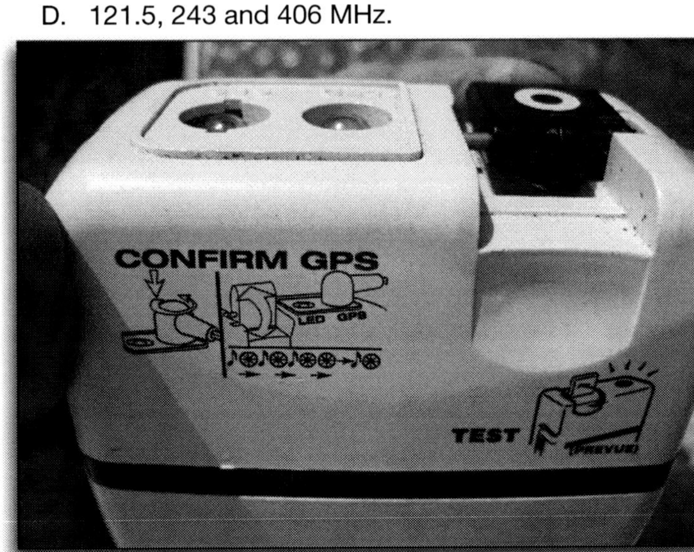

Portable Emergency Locator Transmitter, from ACR.

3-72K5 What is the frequency range of an aircraft's radio altimeter?
A. 962 MHz to 1213 MHz.
B. 329.15 MHz to 335.00 MHz.
C. 4250 MHz to 4350 MHz.
D. 108.00 MHz to 117.95 MHz.

Here is one more frequency to file in memory before the big test – radio altimeter from 4250 MHz to 4350 MHz. The answers are the only four-digit numbers (up in the RADAR range). The radio altimeter signal is a frequency-modulated continuous wave. **ANSWER C**

Aircraft cockpit instruments are now going to glass panel displays.

3-72K6 What type of antenna is attached to an aircraft's Mode C transponder installation and used to receive 1030 MHz interrogation signals from the Air Traffic Control Radar Beacon System (ATCRBS)?
A. An electronically steerable phased-array directional antenna.
B. An L-band monopole blade-type omnidirectional antenna.
C. A folded dipole reception antenna.
D. An internally mounted, mechanically rotatable loop antenna.

An L-band monopole blade-type omnidirectional antenna is used to receive 1030-MHz mode-C transponder interrogations. **ANSWER B**

Key Topic 73: Equipment Functions

3-73K1 Some aircraft and avionics equipment operates with a prime power line frequency of 400 Hz. What is the principle advantage of a higher line frequency?
A. 400 Hz power supplies draw less current than 60 Hz supplies allowing more current available for other systems on the aircraft.
B. A 400 Hz power supply generates less heat and operates much more efficiently than a 60 Hz power supply.
C. The magnetic devices in a 400 Hz power supply such as transformers, chokes and filters are smaller and lighter than those used in 60 Hz power supplies.
D. 400 Hz power supplies are much less expensive to produce than power supplies with lower line frequencies.

Many high-voltage aircraft avionics operate with a power line frequency of 400-Hz, rather than lower AC frequencies. This allows the magnetic devices like filters, chokes, and transformers to be much smaller and lighter than those in 60-Hz devices. Smaller and lighter components are necessary for the basic aerodynamic designs of most aircraft. **ANSWER C**

3-73K2 Aviation services use predominantly ____ microphones.
A. Dynamic
B. Carbon
C. Condenser
D. Piezoelectric crystal

The aviation service normally uses dynamic microphones because of their rugged characteristics and their ability to withstand vibration and heat. Dynamic microphones require impedance-matching amplifiers to match their very low output impedance to the high-impedance audio input stage on the aviation radio set. **ANSWER A**

3-73K3 Typical airborne HF transmitters usually provide a nominal RF power output to the antenna of ____ watts, compared with ____ watts RF output from a typical VHF transmitter.

A. 10, 50
B. 50, 10
C. 20, 100
D. 100, 20

Both aviation and marine high frequency 2 MHz – 30 MHz SSB transceivers typically output 100 watts peak envelope power. The aviation and marine VHF channelized transceivers generally run between 20 and 25 watts, maximum output. Aviation VHF equipment runs AM, and marine VHF equipment runs FM. **ANSWER D**

3-73K4 Before ground testing an aircraft RADAR, the operator should:

A. Ensure that the area in front of the antenna is clear of other maintenance personnel to avoid radiation hazards.
B. Be sure the receiver has been properly shielded and grounded.
C. First test the transmitter connected to a matched load.
D. Measure power supply voltages to prevent circuit damage.

RADAR energy is at dangerous levels near the transmitting antenna. Before testing an aircraft radar on the ground, be sure that the area in front of the antenna is clear of other maintenance personnel so you don't bombard them with radiation from the radar. **ANSWER A**

3-73K5 What type of antenna is used in an aircraft's Very High Frequency Omnidirectional Range (VOR) and Localizer (LOC) installations?

A. Vertically polarized antenna that radiates an omnidirectional antenna pattern.
B. Horizontally polarized omnidirection reception antenna.
C. Balanced loop transmission antenna.
D. Folded dipole reception antenna.

The antenna for VOR and LOC installations is a horizontally polarized omnidirection reception antenna.
ANSWER B

3-73K6 What is the function of a commercial aircraft's SELCAL installation? SELCAL is a type of aircraft communications _________.

A. Device that allows an aircraft's receiver to be continuously calibrated for signal selectivity.
B. System where a ground-based transmitter can call a selected aircraft or group of aircraft without the flight crew monitoring the ground-station frequency.
C. Transmission that uses sequential logic algorithm encryption to prevent public "eavesdropping" of crucial aircraft flight data.
D. System where an airborne transmitter can selectively calculate the line-of-sight distance to several ground-station receivers.

Selective calling that allows base stations to contact an individual aircraft is preceded by a series of tones that trigger a specific aircraft's SELCAL decoder. The decoder normally remains silent so the flight crew does not need to audibly monitor for the ground-station call. **ANSWER B**

Subelement L – Installation, Maintenance & Repair
8 Key Topics, 8 Exam Questions

Key Topic 74: Indicating Meters

3-74L1 What is a 1/2 digit on a DMM?
A. Smaller physical readout on the left side of the display.
B. Partial extended accuracy on lower part of the range.
C. Smaller physical readout on the right side.
D. Does not apply to DMMs.

Often, when you're looking for a new digital multimeter (DMM or DVM), you'll see it specified as having "3-1/2 digit" accuracy, or similar. Does this mean that half of one of the numbers is missing? Actually, yes it does! The FIRST digit only includes 4 segments: a MINUS sign, a number 1 (made up of two segments), and a decimal point. Since this is a SIGNIFICANT digit, you can actually double the range of a 3 digit meter. With three digits, you can display from −999 to +999 (or decimal parts thereof). With that extra 1, or minus 1, you can measure from −1999 to +1999. That's a lot of extra "brainpower" for a few extra segments! **ANSWER B**

Digital Multimeter.

3-74L2 A 50 microampere meter movement has an internal resistance of 2,000 ohms. What applied voltage is required to indicate half-scale deflection?
A. 0.01 volts
B. 0.10 volts
C. 0.005 volts.
D. 0.05 volts.

In this Ohm's Law problem, you multiply current by resistance, yielding full-scale deflection. Then, divide your answer by 2 to get the final answer of 0.05 volts. Remember that 50 microamperes is 6 places to the left for amps (0.000050 amps), and 2 kilohms is 3 decimal places to the right for ohms (2000 ohms).
Check with: $50 \times 10^{-6} \times 2 \times 10^{3} = 100 \times 10^{-3} = 50 \times 10^{-3} = 0.05$ V. **ANSWER D**

3-74L3 What is the purpose of a series multiplier resistor used with a voltmeter?
A. It is used to increase the voltage-indicating range of the voltmeter.
B. A multiplier resistor is not used with a voltmeter.
C. It is used to decrease the voltage-indicating range of the voltmeter.
D. It is used to increase the current-indicating range of an ammeter, not a voltmeter.

The old Simpson 260 multimeter uses multiplier resistors to increase the voltage-indicating range of the voltmeter part of the multimeter. **ANSWER A**

3-74L4 What is the purpose of a shunt resistor used with an ammeter?
A. A shunt resistor is not used with an ammeter.
B. It is used to decrease the ampere indicating range of the ammeter.
C. It is used to increase the ampere indicating range of the ammeter.
D. It is used to increase the voltage indicating range of the voltmeter, not the ammeter.

In the older Simpson meter, a shunt resistor increases the ampere indicating range of the ammeter by shunting some of the current around the meter movement, rather than through it. **ANSWER C**

The range of this ammeter may be expanded by using a shunt resistor.

3-74L5 What instrument is used to indicate high and low digital voltage states?

A. Ohmmeter.
B. Logic probe.
C. Megger.
D. Signal strength meter.

It is the job of the logic probe to indicate high and low voltage states in a digital circuit. **ANSWER B**

Logic probe to trace digital circuits.

3-74L6 What instrument may be used to verify proper radio antenna functioning?

A. Digital ohm meter.
B. Hewlett-Packard frequency meter.
C. An SWR meter.
D. Different radio.

The instrument used to measure transmitter power output, as well as antenna forward and reflected power levels, is the Standing Wave Ratio (SWR) meter. **ANSWER C**

Key Topic 75: Test Equipment

3-75L1 How is a frequency counter used?

A. To provide reference points on an analog receiver dial thereby aiding in the alignment of the receiver.
B. To heterodyne the frequency being measured with a known variable frequency oscillator until zero beat is achieved, thereby indicating the unknown frequency.
C. To measure the deviation in an FM transmitter in order to determine the percentage of modulation.
D. To measure the time between events, or the frequency, which is the reciprocal of the time.

You must never directly couple a frequency counter to your transmitter output. Most portable counters have a small rubber antenna that will pick up the signal from a nearby antenna quite nicely. Most counters can read out a signal within 50 feet. The more expensive the counter, the further away it can read a transmitting signal. If you plan to use the counter outside, make sure you purchase a counter with an LCD display. Stay away from LED displays for outside use because you can't see their readout in the bright sunlight! **ANSWER D**

3-75L2 What is a frequency standard?

A. A well-known (standard) frequency used for transmitting certain messages.
B. A device used to produce a highly accurate reference frequency.
C. A device for accurately measuring frequency to within 1 Hz.
D. A device used to generate wide-band random frequencies.

The time standard stations WWV and WWVH may allow field checks of frequencies at 5, 10, 15, and 20 MHz. The most accurate frequency standard is the 10 MHz output of a GPS receiver. The Rubidium Standard is a secondary frequency reference source. **ANSWER B**

3-75L3 What equipment may be useful to track down EMI aboard a ship or aircraft?

A. Fluke multimeter.
B. An oscilloscope.
C. Portable AM receiver.
D. A logic probe.

EMI stands for ElectroMagnetic Interference. EMI can cause interference problems on frequencies below 30 MHz. A small portable AM receiver – including an AM portable broadcast receiver – is a handy tool to localize the source of EMI aboard a ship or aircraft. **ANSWER C**

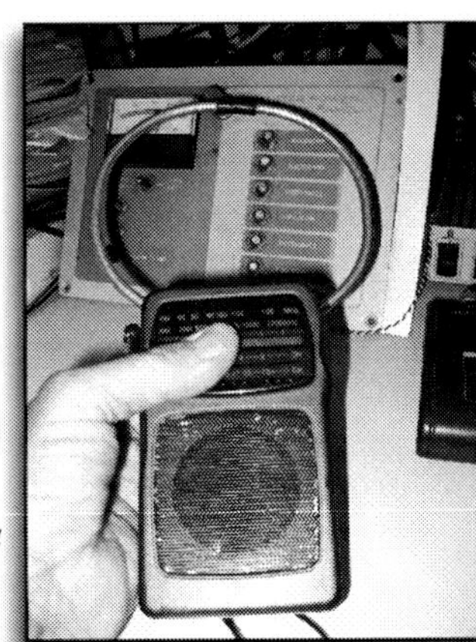

Simple AM radio with added loop antenna for sniffing RFI from a voltage panel.

3-75L4 On an analog wattmeter, what part of the scale is most accurate and how much does that accuracy extend to the rest of the reading scale?

A. The accuracy is only at full scale, and that absolute number reading is carried through to the rest of the range. The upper 1/3 of the meter is the only truly calibrated part.
B. The accuracy is constant throughout the entire range of the meter.
C. The accuracy is only there at the upper 5% of the meter, and is not carried through at any other reading.
D. The accuracy cannot be determined at any reading.

The accuracy is only at full scale, and that absolute number reading is carried through to the rest of the range. Many analog watt meters will use individual slugs for certain power scales. If you expect to accurately measure a 100 watt output transmitter, use a 100 watt slug, not a 500 watt or one kilowatt slug! Also, only depend on the upper 1/3 of the meter to be well calibrated. **ANSWER A**

Professional wattmeter.

3-75L5 Which of the following frequency standards is used as a time base standard by field technicians?

A. Quartz Crystal.
B. Rubidium Standard.
C. Cesium Beam Standard.
D. LC Tank Oscillator.

The rubidium oscillator is a dramatic improvement over the common quartz crystal. The rubidium oscillator takes about 3 minutes to lock up, and provides a much improved time standard. A 10 MHz sine wave rubidium oscillator may sometimes be found on internet swap sites for under $500! The rubidium oscillator requires a heat sink, and they take +20 to +36 volts input, at about 1.7 amps on power on, and then settling in to about ½ an amp. **ANSWER B**

3-75L6 Which of the following contains a multirange AF voltmeter calibrated in dB and a sharp, internal 1000 Hz bandstop filter, both used in conjunction with each other to perform quieting tests?

A. SINAD meter.
B. Reflectometer.
C. Dip meter.
D. Vector-impedance meter.

The term SINAD stands for signal plus noise and distortion. We would use a SINAD meter along with a multi-range audio frequency volt meter to calibrate an analog receiver. **ANSWER A**

Key Topic 76: Oscilloscopes

3-76L1 What is used to decrease circuit loading when using an oscilloscope?

A. Dual input amplifiers. B. 10:1 divider probe. C. Inductive probe. D. Resistive probe.

An oscilloscope 10:1 divider probe would allow us to work with integrated circuits, minimizing any circuit loading by the test probe. **ANSWER B**

3-76L2 How does a spectrum analyzer differ from a conventional oscilloscope?

A. The oscilloscope is used to display electrical signals while the spectrum analyzer is used to measure ionospheric reflection.
B. The oscilloscope is used to display electrical signals in the frequency domain while the spectrum analyzer is used to display electrical signals in the time domain.
C. The oscilloscope is used to display electrical signals in the time domain while the spectrum analyzer is used to display electrical signals in the frequency domain.
D. The oscilloscope is used for displaying audio frequencies and the spectrum analyzer is used for displaying radio frequencies.

The spectrum analyzer displays the strength of signals at particular frequencies according to frequency (in a frequency domain) along a horizontal axis. A signal can be locked in and centered in the middle of the screen with 25 kHz to the left and 25 kHz to the right for a close examination of any off-frequency spurs it might have. A simple oscilloscope cannot display the signal in the frequency domain – it displays the signals on a time axis (in a time domain) going from left to right. **ANSWER C**

3-76L3 What stage determines the maximum frequency response of an oscilloscope?

A. Time base.
B. Horizontal sweep.
C. Power supply.
D. Vertical amplifier.

When choosing an oscilloscope, select one with the greatest bandwidth within your price range. It is the vertical amplifier that determines the maximum frequency response of an oscilloscope. **ANSWER D**

Oscilloscope.

3-76L4 What factors limit the accuracy, frequency response, and stability of an oscilloscope?

A. Sweep oscillator quality and deflection amplifier bandwidth.
B. Tube face voltage increments and deflection amplifier voltage.
C. Sweep oscillator quality and tube face voltage increments.
D. Deflection amplifier output impedance and tube face frequency increments.

Every service shop should have a full-featured oscilloscope. Many new marine and aeronautical transceivers employ digital signal processing (DSP). This demands a stable, high-quality sweep oscillator circuit inside the scope, and the scope must also have enough bandwidth to handle high-frequency applications. **ANSWER A**

3-76L5 An oscilloscope can be used to accomplish all of the following except:

A. Measure electron flow with the aid of a resistor.
B. Measure phase difference between two signals.
C. Measure velocity of light with the aid of a light emitting diode.
D. Measure electrical voltage.

There is plenty that you can do with an oscilloscope, but one thing you cannot do is measure the velocity of light with the aid of a light emitting diode. You can measure voltage, phase, and electron flow with an oscilloscope, but not the velocity of light. **ANSWER C**

3-76L6 What instrument is used to check the signal quality of a single-sideband radio transmission?

A. Field strength meter.
B. Signal level meter.
C. Sidetone monitor.
D. Oscilloscope.

When servicing single sideband transmitters, it is the two tone test, observed on an oscilloscope, that lets you examine linearity of the transmitted signal. I also like to bring along a companion SSB receiver, first removing the antenna on the receiver, and then listening for a good quality single sideband transmission. **ANSWER D**

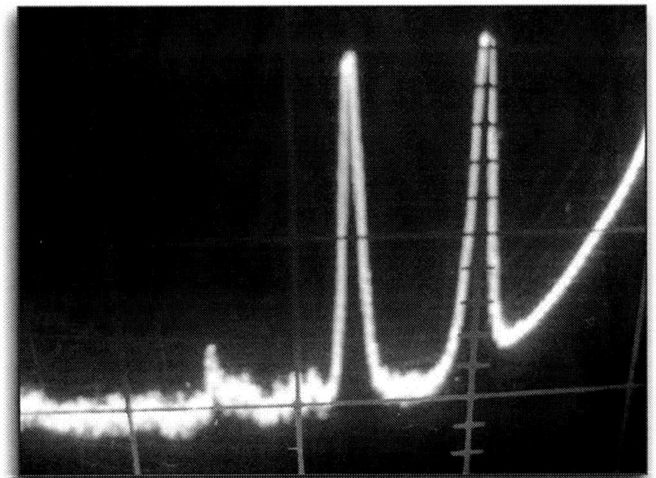

Spectrum analyzer measuring a two-tone test on an aircraft SSB.

Key Topic 77: Specialized Instruments

3-77L1 A(n) ____ and ____ can be combined to measure the characteristics of transmission lines. Such an arrangement is known as a time-domain reflectometer (TDR).

A. Frequency spectrum analyzer, RF generator.
B. Oscilloscope, pulse generator.
C. AC millivolt meter, AF generator.
D. Frequency counter, linear detector.

You might be sent out to troubleshoot a marine communications system aboard a big ship with hundreds of feet of coaxial cable feedline. An oscilloscope and a pulse generator could help determine how far up the line there is a pinch, crack, or break. This device is known as a time-domain reflectometer and is very handy when troubleshooting long runs of coaxial cable. **ANSWER B**

3-77L2 What does the horizontal axis of a spectrum analyzer display?

A. Amplitude.
B. Voltage.
C. Resonance.
D. Frequency.

The horizontal axis of a spectrum analyzer displays the frequency of the signal. **ANSWER D**

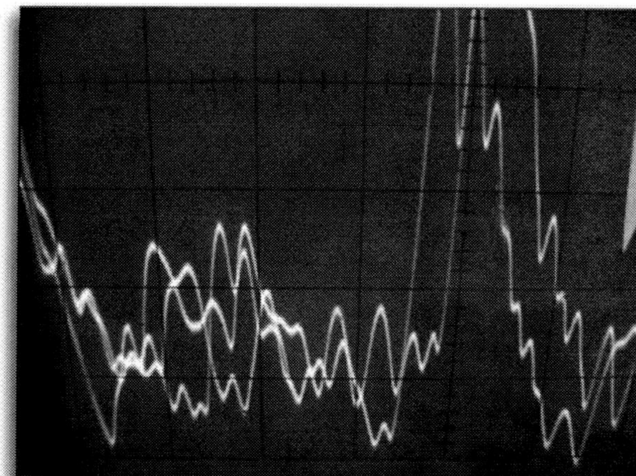

Spectrum Scope shows lots of birdies.

3-77L3 What does the vertical axis of a spectrum analyzer display?

A. Amplitude.
B. Duration.
C. Frequency.
D. Time.

The vertical axis of the spectrum analyzer displays the amplitude of the signal. **ANSWER A**

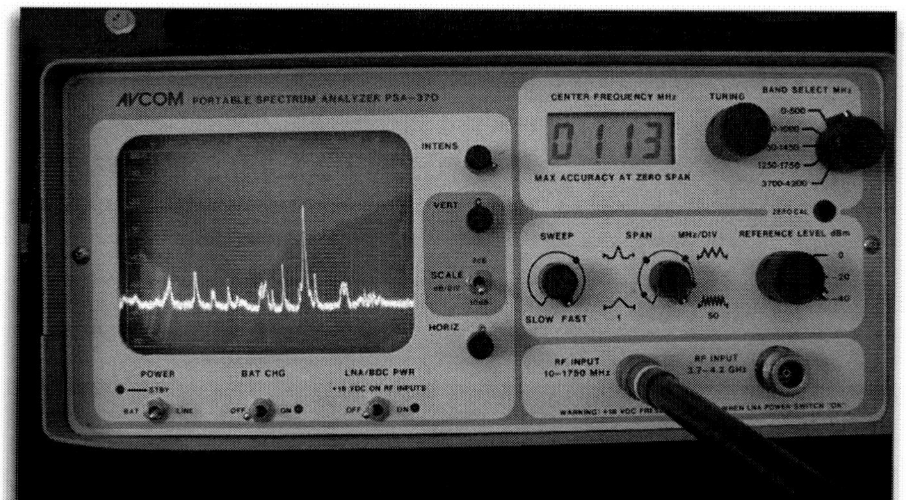

The spectrum analyzer.

3-77L4 What instrument is most accurate when checking antennas and transmission lines at the operating frequency of the antenna?

A. Time domain reflectometer.
B. Wattmeter.
C. DMM.
D. Frequency domain reflectometer.

The frequency domain reflectometer is a test instrument capable of measuring both antenna resonance as well as feedline impedance. **ANSWER D**

3-77L5 What test instrument can be used to display spurious signals in the output of a radio transmitter?

A. A spectrum analyzer.
B. A wattmeter.
C. A logic analyzer.
D. A time domain reflectometer.

You can check for spurious signals with a spectrum analyzer. **ANSWER A**

3-77L6 What instrument is commonly used by radio service technicians to monitor frequency, modulation, check receiver sensitivity, distortion, and to generate audio tones?

A. Oscilloscope.
B. Spectrum analyzer.
C. Service monitor.
D. DMM.

The service monitor is usually a minimum $10,000 piece of equipment. Some technicians may own one, and likely the service shop has several portable or fixed service monitors. Service monitors contain an accurate frequency time base for measuring transmitter frequency, deviation capabilities to measure FM bandwidth, a signal generator to measure receiver sensitivity, and many more features that replace multiple pieces of individual equipment with one nice, neat package. **ANSWER C**

Service monitor.
Courtesy of Hewlett-Packard Company

Key Topic 78: Measurement Procedures

3-78L1 Can a P25 radio system be monitored with a scanner?

A. Yes , regardless if it has P25 decoding or not.
B. No.
C. Yes, if the scanner has P25 decoding.
D. Yes, but it must also have P26 decoding.

P25 is also referred to APCO – 25, and Project 25, a standard for digital radio communications. This standard allows interoperability between many different agencies using the P25 protocol. The agencies could be Federal, state, city, and business land mobile radio users. Many P25 radios operate at 4800 baud digital modulation, requiring a digital receiver capable of quadrature phase shift keying. It is important to remember that relatively inexpensive programmable P25 scanners can easily decode this open architecture digital signal. These scanners, plus some P25 transceivers, may also support analog modulation. P25 with added layers of encryption is a proprietary safeguard against conventional P25 scanner reception. With encryption, not even the P25 scanner can decode the transmission. **ANSWER C**

Uniden P25 Trunk Tracker IV scanner for digital reception.

3-78L2 Which of the following answers is true?

A. The RF Power reading on a CDMA (code division multiple access) radio will be very accurate on an analog power meter.
B. The RF Power reading on a CDMA radio is not accurate on an analog power meter.
C. Power cannot be measured using CDMA modulation.
D. None of the above.

The conventional analog power meter would not be suitable to measure CDMA power output because the spread spectrum digital code has momentary periods of zero power output during pulses. **ANSWER B**

3-78L3 What is a common method used to program radios without using a "wired" connection?

A. Banding.
B. Using the ultraviolet from a programmed radio to repeat the programming in another.
C. Infra-red communication.
D. Having the radio maker send down a programming signal via satellite.

Land mobile radios are generally not permitted to offer keyboard frequency selection by the operator. Programming must be accomplished before the radio is ready for field use, and a handy way of programming multiple land mobile radios to a specific set of channels is through an infrared link between the master radio and the other radios for cloning, if the radios are so equipped. **ANSWER C**

3-78L4 What is the common method for determining the exact sensitivity specification of a receiver?

A. Measure the recovered audio for 12 dB of SINAD.
B. Measure the recovered audio for 10 dB of quieting.
C. Measure the recovered audio for 10 dB of SINAD.
D. Measure the recovered audio for 25 dB of quieting.

The term SINAD stands for Signal + Noise And Distortion. SINAD equals signal + noise + distortion divided by noise + distortion. SINAD is always expressed in dB, with 12 dB corresponding to a 4:1 signal to noise ratio. **ANSWER A**

3-78L5 A communications technician would perform a modulation-acceptance bandwidth test in order to:

A. Ascertain the audio frequency response of the receiver.
B. Determine whether the CTCSS in the receiver is operating correctly.
C. Verify the results from a 12 dB SINAD test.
D. Determine the effective bandwidth of a communications receiver.

As we continue to pursue narrow band modulation techniques, a modulation-acceptance bandwidth test may determine whether a communications receiver is designed for 12.5 kHz bandwidth, or even twice as narrow – 6 kHz bandwidth. On FM, this test is best performed with a service monitor, but the sharp technician can quickly spot an incompatible transmitter and receiver where clipping occurs causing the signal to sound syllable-interrupted. **ANSWER D**

3-78L6 What is the maximum FM deviation for voice operation of a normal wideband channel on VHF and UHF?

A. 2.5 kHz
B. 5.0 kHz
C. 7.5 kHz
D. 10 kHz

We consider a wideband FM signal to have a deviation of 5 kHz. Deviation is proportional to the amplitude of the modulating signal, and is the frequency swing AWAY from an unmodulated frequency. Narrowband VHF channels have 2.5 kHz deviation, but this question asks about wideband deviation, 5.0 kHz. **ANSWER B**

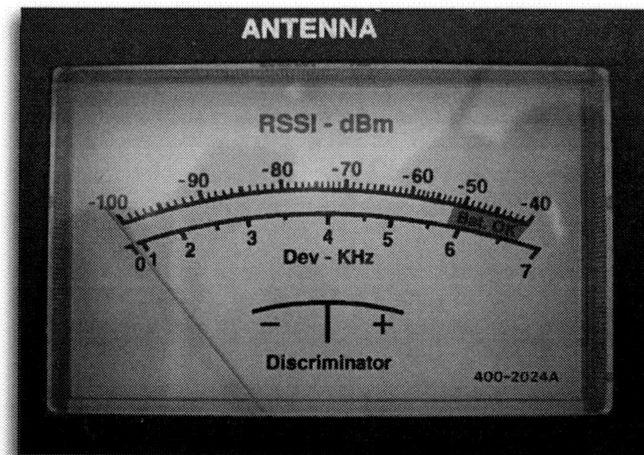

Deviation meter.

Key Topic 79: Repair Procedures

3-79L1 When soldering or working with CMOS electronics products or equipment, a wrist strap:

A. Must have less than 100,000 ohms of resistance to prevent static electricity.
B. Cannot be used when repairing TTL devices.
C. Must be grounded to a water pipe.
D. Does not work well in conjunction with anti-static floor mats.

Metal oxide semiconductors are extremely sensitive to static electricity device failure. Even CMOS devices with "gate protection" (a protective diode) may fail permanently during circuit tests if the technician accidentally creates a static discharge from his body to the metal instrument and then to the CMOS device. The chassis of the turned-off equipment needs a low resistance ground before servicing. A ground mat may help discharge the service person's static buildup. If soldering is required, the soldering iron tip should also have a ground lead firmly attached to the chassis ground. Finally, the technician's grounding wrist strap is also helpful in protecting the delicate CMOS circuit. The wrist strap typically has very low resistance – always less then 100,000 ohms. **ANSWER A**

3-79L2 Which of the following is the preferred method of cleaning solder from plated-through circuit-board holes?

A. Use a dental pick.
B. Use a vacuum device.
C. Use a soldering iron tip that has a temperature above 900 degrees F.
D. Use an air jet device.

Removing solder from a circuit board, if the board is only one-sided, might be accomplished with a solder wick. However, now that circuit boards are two-sided and use plated-through circuit board holes, solder from both sides is removed in a single operation using a solder vacuum pump. In the field, a bulb syringe (as a last resort) may sometimes work. **ANSWER B**

3-79L3 What is the proper way to cut plastic wire ties?
A. With scissors.
B. With a knife.
C. With semi-flush diagonal pliers.
D. With flush-cut diagonal pliers and cut flush.

Plastic wire ties can be dangerous if sharp edges are left exposed. Plastic wire ties, uncut, at eye level are extremely dangerous! Always cut your wire ties with flush-cut diagonal pliers, and cut the extra tie material as close to the clamp mechanism as possible. Using diagonal pliers may also be a good way to cinch tight the wire tie, before cutting it flush. **ANSWER D**

Cut plastic ties carefully!

3-79L4 The ideal method of removing insulation from wire is:
A. The thermal stripper.
B. The pocket knife.
C. A mechanical wire stripper.
D. The scissor action stripping tool.

The best way to remove insulation from old or new wire is with a thermal stripper, which literally melts the insulation for easy removal. Removing insulation with a simple pocket knife, diagonals, or scissors can sometimes cut into the wire, making it weaker, or can remove an anti-corrosion coating from the wire. The thermal stripper is your best tool. **ANSWER A**

3-79L5 A "hot gas bonder" is used:
A. To apply solder to the iron tip while it is heating the component.
B. For non-contact melting of solder.
C. To allow soldering both sides of the PC board simultaneously.
D. To cure LCA adhesives.

When removing a large scale integrated circuit chip, the multiple contacts present a de-soldering challenge. The best tool for this work is a hot gas bonder that heats up the solder nearly instantly, allowing all LSI leads to free up from the molten solder, and making it easy to lift the device out without distortion of the solder pads. This process is reversed to put in the new LSI chip. **ANSWER B**

3-79L6 When repairing circuit board assemblies it is most important to:
A. Use a dental pick to clear plated-through holes.
B. Bridge broken copper traces with solder.
C. Wear safety glasses.
D. Use a holding fixture.

Any time you are working around hot solder or sharp components, or on a tower, or on the ground below the tower, or around antenna elements, WEAR YOUR SAFETY GLASSES! Prevent permanent blindness by always wearing your eye protection gear. **ANSWER C**

Key Topic 80: Installation Codes & Procedures

3-80L1 What color is the binder for pairs 51-75 in a 100-pair cable?

A. Red
B. Blue
C. Black
D. Green

Have you ever picked up a chunk of surplus telephone cable with hundreds of color coded wires and wondered how they ever kept them all straight? Well, back in the dark ages, Ma Bell (the telephone company for you young whippersnappers) had this curious habit of doing everything right. The hierarchical color coding scheme was just one such innovation. It could easily be memorized for hundreds or even thousands of wires! You've probably used a small segment of this hierarchy if you've ever worked with ethernet cables. You probably have used:

BLUE/WHITE-BLUE
ORANGE/WHITE-ORANGE
GREEN/WHITE-GREEN
BROWN/WHITE-BROWN
SLATE/WHITE-SLATE

Each of these form a "twisted pair," one half of each being the dominant "group" color, and the other half being the "pair" color. This entire group is the WHITE group. Most home networking hobbyists never move beyond the white group. In fact, with ethernet, you never even get to the SLATE pair.
We have four more GROUP colors though:

RED
BLACK
YELLOW
VIOLET

Five pair colors, each contained within five group colors, gives us a total of 25 pairs. A 25 pair cable is your "basic" fat cable.

Now, what if we have more than 25 pairs? Well, we group our entire 25 pair cable into a group binding color. Oh no! Does this mean we have to memorize a whole new set of colors for the bindings? Relax! Guess what? We just re-use the same pair colors from the original lowest level of the hierarchy: the binding colors are: BLUE, ORANGE, GREEN, BROWN, SLATE.

So, you see with three layers of hierarchy, we can quickly build a 125 pair cable. Well, what happens when we use up all 125 pairs? Are we at the end of our rope? Nope! We go back to our "BINDING BINDING" color, which is the same as, you guessed it, the GROUP COLORS: WHITE, RED, BLACK, YELLOW, VIOLET. We can now have a 625 pair cable, and we've only needed to memorize TEN COLORS! Pretty slick, eh?

We can expand this scheme forever (theoretically) alternating between the pair colors and the group colors at each successive layer of the hierarchy. The biggest cable I've worked on is a 125 pair cable, which I did with NO CHARTS whatsoever. If I can do it, so can you!

Now, going to our original question, we're working with a 100 pair cable. (We aren't OBLIGATED to use all five bindings, but we ARE obligated to start out with the BLUE binding.) Starting at pair 51, puts us in our third BINDING GROUP (each having 25 pairs, remember) starting with BLUE. So... Blue, Orange, GREEN! **ANSWER D**

Thanks to Ma Bell, the color coding on 100-pair cables actually makes it easy to install complex phone and computer wiring.

3-80L2 What is most important when routing cables in a mobile unit?

A. That cables be cut to the exact length.
B. Assuring accessibility of the radio for servicing from outside the vehicle.
C. Assuring radio or electronics cables do not interfere with the normal operation of the vehicle.
D. Assuring cables are concealed under floor mats or carpeting.

You've been instructed to run power and data control cables in a mobile unit – this could be a vehicle, an aircraft, or a ship. An important consideration is to keep large DC voltage cables away from any magnetic sensors. Remember, a current-carrying conductor can generate a magnetic field. Also make sure that both ends of your DC power cable are well fused. Make sure that there is no chance that any type of cabling might interfere with anything mechanical at the installation site. If you are installing any type of two-way radio equipment, perform transmitter tests to insure the transmit signal does not cause any erratic behavior on the electronics within the mobile unit. **ANSWER C**

3-80L3 Why should you not use white or translucent plastic tie wraps on a radio tower?

A. White tie wraps are not FAA approved.
B. UV radiation from the Sun deteriorates the plastic very quickly.
C. The white color attracts wasps
D. The black tie wraps may cause electrolysis.

White and translucent tie wraps are subject to cracking and deterioration caused by ultraviolet rays from the Sun. Any tie wraps exposed to the Sun should be rated UV safe and are usually black in color. **ANSWER B**

3-80L4 What is the 6th pair color code in a 25 pair switchboard cable as is found in building telecommunications interconnections?

A. Blue/Green, Green/Blue.
B. Red/Blue, White/Violet.
C. Red/Blue, Blue/Red.
D. White/Slate, Slate/White.

If you show up for a job interview at your local phone company central office (CO) and show the boss that you know the HIERARCHICAL COLOR CODE by heart, you'll probably get the job! The hierarchy is simple. Pairs, groups, group bindings, binding bindings. If you get the pairs and groups right, it's all you'll need to know.

Pairs are:	Groups are:
BLUE	WHITE
ORANGE	RED
GREEN	BLACK
BROWN	YELLOW
SLATE	VIOLET

The 6TH pair will put us in the second group, the RED group. We're the first pair in the red group, the BLUE pair. **ANSWER C**

3-80L5 What tolerance off of plumb should a single base station radio rack be installed?

A. No tolerance allowed.
B. Just outside the bubble on a level.
C. All the way to one end.
D. Just inside the bubble on a level.

When installing a radio rack in a transmitter facility, use a level to insure the rack is plumb, and make sure that any variance is just inside the bubble on a level. **ANSWER D**

3-80L6 What type of wire would connect an SSB automatic tuner to an insulated backstay?

A. GTO-15 high-voltage cable.
B. RG8U.
C. RG213.
D. 16-gauge two-conductor.

The high voltage cable going to the antenna from an automatic antenna tuner is rated to 15,000 volts, and is called GTO-15. This is similar to neon sign cable wire employed in high voltage installations to prevent arcing. Never substitute coaxial cable in a high voltage antenna tuner output circuit. **ANSWER A**

GTO-15 high voltage single wire cable , used for an antenna lead in , as well as to the tuner's circuit board RF output.

Key Topic 81: Troubleshooting

3-81L1 On a 150 watt marine SSB HF transceiver, what would be indicated by a steady output of 75 watts when keying the transmitter on?

A. There is probably a defect in the system causing the carrier to be transmitted.
B. One of the sidebands is missing.
C. Both sidebands are being transmitting.
D. The operation is normal.

A marine SSB will show almost no power output when unmodulated in the J3E single sideband mode. If you detect a steady output carrier when transmitting in the SSB mode, you may first suspect that you accidentally put the radio in the AM mode, full carrier. But if you are in SSB running a terminal node controller tied into a computer, you might suspect either RF feedback or the terminal node controller sending out an unwanted carrier. Unplug the TNC and if the carrier disappears, you have isolated the problem. However, if the carrier remains in the J3E mode, you may have a defect within the transmitter modulator section. **ANSWER A**

3-81L2 The tachometer of a building's elevator circuit experiences interference caused by the radio system nearby. What is a common potential "fix" for the problem?

A. Replace the tachometer of the elevator.
B. Add a .01 μF capacitor across the motor/tachometer leads.
C. Add a 200 μF capacity across the motor/tachometer leads.
D. Add an isolating resistor in series with the motor leads.

Any micro-amp meter movement may give a false reading in the presence of a strong RF signal. Wind and speed equipment, elevator equipment, and engine tachometers using a micro-amp meter will likely react to strong RF fields. Any fix you do is temporary; it will take the certified elevator repair technician to install a permanent fix. **ANSWER B**

3-81L3 A common method of programming portable or mobile radios is to use a:

A. A laptop computer.
B Dummy load.
C. A wattmeter.
D. A signal generator.

Radio manufacturers may provide their authorized service shops with restricted programming software, specific to individual radios, to be loaded onto shop computers as well as technician LAPTOP computers for field work. At no time should these restricted computer "loads" be distributed to the general public or to non-technicians. Keep your programming software secure! **ANSWER A**

NOTE: The programming software is protected by copyright, and almost every manufacturer *will prosecute* for illegal use of their software. It is cheaper to buy their software than to deal with a lawsuit and attorneys' fees when you get caught using pirated, non-authorized software. The radio shop that did not get the sale or service of the radio is quick to identify improper use of proprietary software.

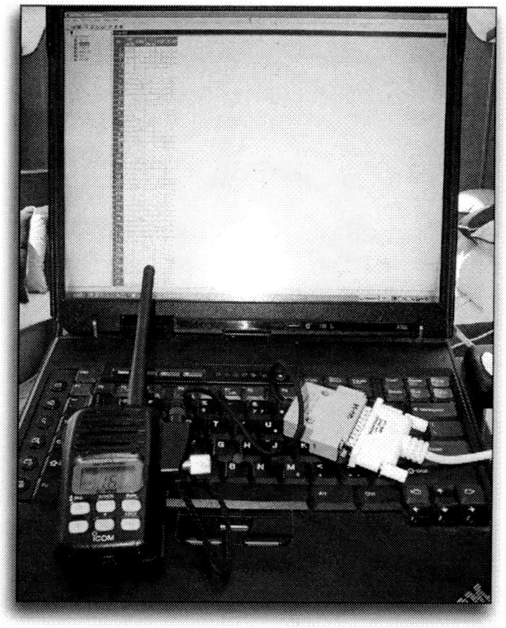

A laptop for programming hand-held transceivers in the field.

3-81L4 In a software-defined transceiver, what would be the best way for a technician to make a quick overall evaluation of the radio's operational condition?

A. Set up a spectrum analyzer and service monitor and manually verify the manufacturer's specifications.
B. Use another radio on the same frequency to check the transmitter.
C. Use the built-in self-test feature.
D. Using on-board self-test routines are strictly prohibited by the FCC in commercial transmitters. Amateur Radio is the only service currently authorized to use them.

The software defined transceiver most always contains its own diagnostic and self-test functions. This allows the field technician to review operating parameters without necessarily bringing along a complete service monitor. Although spectral output is still best checked by an external service monitor, letting the SDR operate a self-test on its own internal program is a good way to quickly check how the radio is working. **ANSWER C**

3-81L5 How might an installer verify correct GPS sentence to marine DSC VHF radio?

A. Press and hold the red distress button.
B. Look for latitude and longitude on the display.
C. Look for GPS confirmation readout.
D. Ask for VHF radio check position report.

The marine electronics industry has not standardized plugs and jacks for interconnecting an external GPS output to the radio's GPS input. A simple way to observe a marine VHF properly processing external GPS information is to look on the radio's screen and see if the latitude, longitude, plus UTC time show up on the display. Keep in mind that GPS time may be a few seconds off from UTC because the satellites do not compensate for Leap seconds. **ANSWER B**

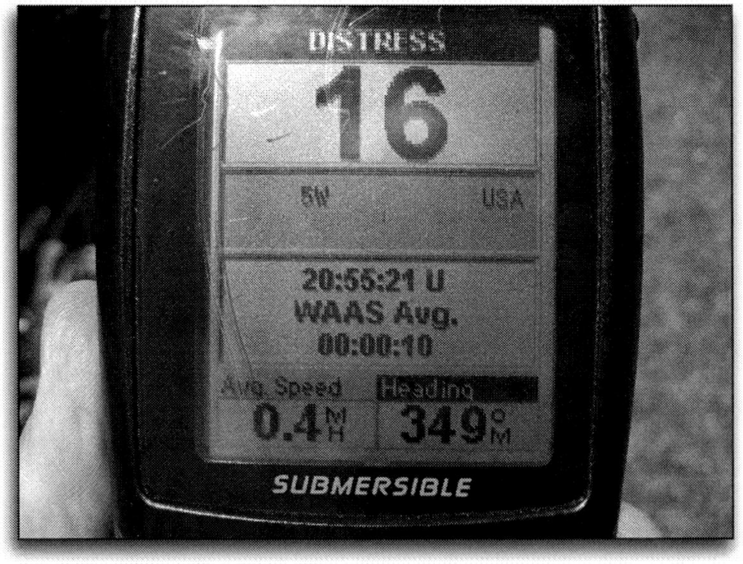

The speed and course readout indicates this VHF is receiving GPS information correctly.

3-81L6 What steps must be taken to activate the DSC emergency signaling function on a marine VHF?

A. Separate 12 volts to the switch.
B. Secondary DSC transmit antenna.
C. GPS position input.
D. Input of registered 9-digit MMSI.

The marine electronics technician installing shipboard two-way radio equipment should assist the mariner in programming the nine digit MMSI number into each transceiver to comply with GMDSS protocol. Without the imbedded MMSI number that is input by the user or technician, that particular piece of equipment WILL NOT send out an emergency DSC distress call. **ANSWER D**

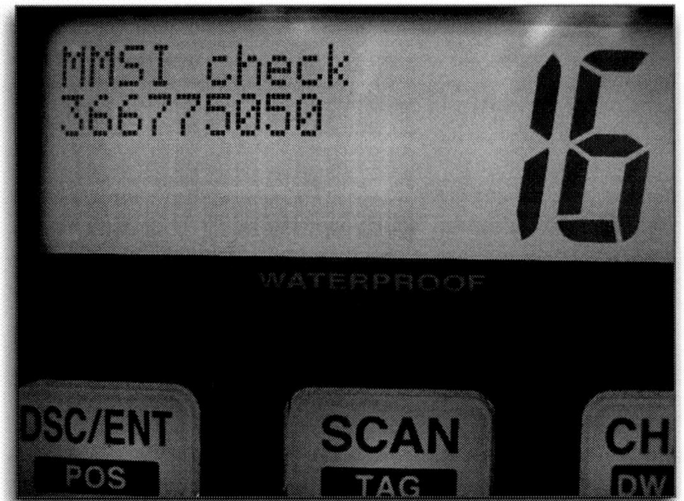

MMSI number assigned to this radio.

Subelement M – Communications Technology

3 Key Topics, 3 Exam Questions

Key Topic 82: Types of Transmissions

3-82M1 What term describes a wide-bandwidth communications system in which the RF carrier frequency varies according to some predetermined sequence?

A. Amplitude compandored single sideband.
B. SITOR.
C. Time-domain frequency modulation.
D. Spread spectrum communication.

Spread spectrum communications are now very popular. Virtually all cordless phones use spread spectrum communications. The frequency hops in a pre-arranged sequence, eliminating interference and eavesdropping. The global positioning system (GPS) at 1500 MHz uses spread-spectrum techniques. **ANSWER D**

3-82M2 Name two types of spread spectrum systems used in most RF communications applications?

A. AM and FM.
B. QPSK or QAM.
C. Direct Sequence and Frequency Hopping.
D. Frequency Hopping and APSK.

Spread spectrum communications may use either direct sequence or frequency hopping for its signaling technique. **ANSWER C**

3-82M3 What is the term used to describe a spread spectrum communications system where the center frequency of a conventional carrier is altered many times per second in accordance with a pseudo-random list of channels?

A. Frequency hopping.
B. Direct sequence.
C. Time-domain frequency modulation.
D. Frequency compandored spread spectrum.

Spread spectrum communications alters the center frequency of an RF carrier in pseudo-random manner, causing the frequency to "hop" around from channel to channel. This is termed frequency hopping. **ANSWER A**

3-82M4 A TDMA radio uses what to carry the multiple conversations sequentially?

A. Separate frequencies.
B. Separate pilot tones.
C. Separate power levels.
D. Separate time slots.

TDMA stands for Time Division Multiple Access, and allows the transmission and reception of multiple conversations sequentially, separated by specific time slots. **ANSWER D**

3-82M5 Which of the following statements about SSB voice transmissions is correct?

A. They use A3E emission which are produced by modulating the final amplifier.
B. They use F3E emission which is produced by phase shifting the carrier.
C. They normally use J3E emissions, which consists of one sideband and a suppressed carrier.
D. They may use A1A emission to suppress the carrier.

Single Sideband, found in both aeronautical as well as marine equipment, usually employs upper sideband with the carrier suppressed. Some fixed land services using SSB may employ either lower or upper sideband, again with a suppressed carrier. Years ago, high frequency public coast stations operated SSB with a noticeable carrier to assist receive stations to lock on to the "pilot carrier" for frequency stability. This type of transmission has been all but eliminated in favor of suppressed carrier SSB. **ANSWER C**

3-82M6 What are the two most-used PCS (Personal Communications Systems) coding techniques used to separate different calls?

A. QPSK and QAM.
B. CDMA and GSM.
C. ABCD and SYZ.
D. AM and Frequency Hopping.

Cellular phones may use code division multiple access (CDMA) and, for worldwide operation, may also offer GSM (Global System for Mobile Communications). **ANSWER B**

Key Topic 83: Coding and Multiplexing

3-83M1 What is a CODEC?

A. A device to read Morse code.
B. A computer operated digital encoding compandor.
C. A coder/decoder IC or circuitry that converts a voice signal into a predetermined digital format for encrypted transmission.
D. A voice amplitude compression chip.

The proprietary CODEC may be a small integrated circuit chip that encodes a digital format to a proprietary encrypted data signal, and then have that data signal decoded by the other CODEC, for secure communications. **ANSWER C**

3-83M2 The GSM (Global System for Mobile Communications) uses what type of CODEC for digital mobile radio system communications?

A. Regular-Pulse Excited (RPE).
B. Code-Excited Linear Predictive (CLEP).
C. Multi-Pulse Excited (MPE).
D. Linear Excited Code (LEC).

GSM (Global System for Mobile Communications) is an international standard for digital radio communications that includes cell phones. A CODEC (Compressor/Decompressor) is a device or computer program that allows voice data to be compressed into a limited bandwidth channel for transmission and reassembled for reception with minimum distortion. Various schemes for compression and decompression have evolved, but the current state-of-the-art technique is the Regular Pulse Excited (RPE) algorithm. **ANSWER A**

3-83M3 Which of the following codes has gained the widest acceptance for exchange of data from one computer to another?

A. Gray.
B. Baudot.
C. Morse.
D. ASCII.

ASCII stands for American Standard Code for Information Interchange, and it has gained the widest acceptance for exchanging data from one computer to another. **ANSWER D**

					0	1	0	1	1	0	0	1
					0	0	1	1	1	1	0	0
1					1	1	1	1	0	0	0	0
0	0	0	0		@	P	‘	p	0	sp	NUL	DLE
1	0	0	0		A	Q	a	q	1	!	SOH	DC1
0	1	0	0		B	R	b	r	2	"	STX	DC2
1	1	0	0		C	S	c	s	3	#	ETX	DC3
0	0	1	0		D	T	d	t	4	$	EOT	DC4
1	0	1	0		E	U	e	u	5	%	ENQ	NAK
0	1	1	0		F	Z	f	v	6	&	ACK	SYN
1	1	1	0		G	W	g	w	7	’	BEL	ETB
0	0	0	1		H	X	h	x	8	(	BS	CAN
1	0	0	1		I	Y	i	y	9	)	HT	EM
0	1	0	1		J	Z	j	z	:	*	LF	SUB
1	1	0	1		K	[	k	{	;	+	VT	ESC
0	0	1	1		L	\	l	l	<	,	FF	FS
1	0	1	1		M	]	m	}	=	-	CR	GS
0	1	1	1		N	^	n	~	>	.	SO	RS
1	1	1	1		O	_	o	DEL	?	/	SI	US

American Standard Code for Information Interchange

3-83M4 The International Organization for Standardization has developed a seven-level reference model for a packet-radio communications structure. What level is responsible for the actual transmission of data and handshaking signals?

A. The physical layer.
B. The transport layer.
C. The communications layer.
D. The synchronization layer.

The data and handshaking signals are in the physical layer. **ANSWER A**

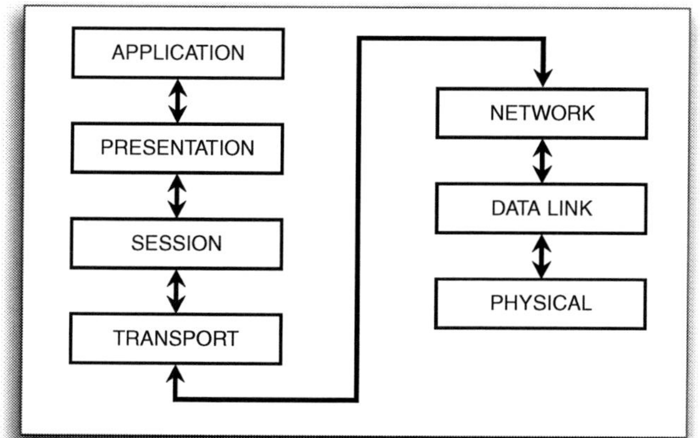

ISO Model for Open Systems Interconnection

3-83M5 What CODEC is used in Phase 2 P25 radios?

A. IWCE
B. IMBC
C. IMMM
D. AMBE

The phase 2 APCO 25 radio may use AMBE CODEC for encryption. **ANSWER D**

3-83M6 The International Organization for Standardization has developed a seven-level reference model for a packet-radio communications structure. The _______ level arranges the bits into frames and controls data flow.

A. Transport layer.
B. Link layer.
D. Synchronization layer.
C. Communications layer.

The link layer arranges the bits into frames and also controls data flow. **ANSWER B**

Key Topic 84: Signal Processing, Software and Codes

3-84M1 What is a SDR?

A. Software Deviation Ratio.
B. Software Defined Radio.
C. SWR Meter.
D. Static Dynamic Ram.

The software defined radio (SDR) allows the computer to handle most functions that were originally relegated to discrete transmitter and receiver components. True, transmitter final amplifier "bricks" are not part of the computer program, but most other functions take discrete radio components and, through digital signal processing, put them in the digital domain. The SDR is not confined to specific modulation techniques, nor is it subject to long waits in performance improvement. By running new software, the SDR can always be kept technically up-to-date. **ANSWER B**

3-84M2 What does the DSP not do in a modern DSP radio?

A. Control frequency.
B. Control modulation.
C. Control detection.
D. Control SWR.

This question asks about digital signal processing, and what it does NOT do inside a radio. DSP may indeed control frequency, modulation, and offer receiver detection, but DSP does NOT correct nor control the antenna system Standing Wave Ratio. **ANSWER D**

DSP speaker added to a non-DSP radio system.

3-84M3 Which statement best describes the code used for GMDSS-DSC transmissions?
A. A 10 bit error correcting code starting with bits of data followed by a 3 bit error correcting code.
B. A 10 bit error correcting code starting with a 3 bit error correcting code followed by 7 bits of data.
C. An 8 bit code with 7 bits of data followed by a single parity bit.
D. A 7 bit code that is transmitted twice for error correction.
Marine radio global maritime distress safety system digital selective calling transmissions are made up of 10 bit error correcting code, starting with bits of data followed by 3 bit error correcting code. The error correcting code is useful in high frequency installations where signal to noise ratios are constantly changing. **ANSWER A**

3-84M4 Which is the code used for SITOR-A and -B transmissions?
A. The 5 bit baudot telex code.
B. Each character consists of 7 bits with 3 "zeros" and 4 "ones".
C. Each character consists of 7 bits with 4 "zeros" and 3 "ones".
D. Each character has 7 bits of data and 3 bits for error correction.
Simplex telex over radio (SITOR) A and B transmissions have each character consisting of 7 bits with 4 zeros and 3 ones. This has proven to be one of the best digital signaling methods over high frequency with intense signal fades. **ANSWER C**

3-84M5 Which of the following statements is true?
A. The Signal Repetition character (1001100) is used as a control signal in SITOR-ARQ.
B. The Idle Signal (a) (0000111) is used for FEC Phasing Signal 1.
C. The Idle Signal (b) (0011001) is used for FEC Phasing Signal 2.
D. The Control Signal 1 (0101100) is used to determine the time displacement in SITOR-B.
With forward-error-correction phasing signal 1, the idle signal (a) (0000111) is employed for FEC. **ANSWER B**

3-84M6 What principle allows multiple conversations to be able to share one radio channel on a GSM channel?
A. Frequency Division Multiplex.
B. Double sideband.
C. Time Division Multiplex.
D. None of the above.
On the modern GSM phone, multiple conversations may share a single radio channel through TIME division multiplex. (TDMA) **ANSWER C**

Subelement N – Marine

5 Key Topics, 5 Exam Questions

Key Topic 85: VHF

3-85N1 What is the channel spacing used for VHF marine radio?

A. 10 kHz.
B. 12.5 kHz.
C. 20 kHz.
D. 25 kHz.

Presently, the marine VHF channel spacing is 25 kHz. About 30 years ago, marine VHF spacing was cut down from 50 kHz to 25 kHz, creating twice as many ship-to-ship and ship-to-shore marine VHF channels. **ANSWER D**

3-85N2 What VHF channel is assigned for distress and calling?

A. 70
B. 16
C. 21A
D. 68

VHF Channel 16 (156.800 MHz) is the distress and calling channel for voice. **ANSWER B**

3-85N3 What VHF Channel is used for Digital Selective Calling and acknowledgement?

A. 16
B. 21A
C. 70
D. 68

Digital Selective Calling and acknowledgement is normally carried out on VHF Channel 70, 156.525 MHz. **ANSWER C**

3-85N4 Maximum allowable frequency deviation for VHF marine radios is:

A. +/- 5 kHz.
B. +/- 15 kHz.
C. +/- 2.5 kHz.
D. +/- 25 kHz.

The allowable frequency deviation for modulation in the VHF marine radio is +/- 5 kHz. Most VHF radio services have switched over to narrow band operation, but the marine radio service remains wide band, with only marine Coast Guard Auxiliary channels going narrow band. **ANSWER A**

3-85N5 What is the reason for the USA-INT control or function?

A. It changes channels that are normally simplex channels into duplex channels.
B. It changes some channels that are normally duplex channels into simplex channels.
C. When the control is set to "INT" the range is increased.
D. None of the above.

Duplex operation is the means by which simultaneous two-way radio communications may take place. This is most simply achieved by transmitting on one frequency and receiving on another. Originally, this allowed a marine radio to interface directly with coastal telephone exchanges in a rather "seamless" manner, allowing the ship's radio to work just like a telephone. The USA-INT (USA-INTERNATIONAL) function on many marine radios quickly changes the channel assignments to simplex (single channel) frequencies. All of the "A" channels are basically duplex channels used in Europe and elsewhere, and therefore are INT channels. We did not need that many duplex channels here in the USA, so we designated them as simplex, or "A" Channels. **ANSWER B**
The "A" simplex channels will soon drop the "A" suffix, and use the new ITU "10" prefix. See page 51 for details.

3-85N6 How might an installer verify correct GPS sentence to marine DSC VHF radio?

A. Look for latitude and longitude, plus speed, on VHF display.
B. Press and hold the red distress button.
C. Look for GPS confirmation readout.
D. Ask for VHF radio check position report.

Digital Selective Calling (DSC) is a system whereby a VHF radio can report the location of a vessel based on GPS (Global Positioning System) information. GPS information (latitude, longitude, altitude, and time) are transmitted in a data stream or "sentence" that is interpreted by the DSC VHF radio and continuously displayed on a readout on the radio. If the GPS and the DSC VHF radio are "talking" to each other, the coordinates and the ship's speed should be updating. **ANSWER A**

Key Topic 86: MF-HF, SSB-SITOR

3-86N1 What is a common occurrence when voice-testing an SSB aboard a boat?

A. Ammeter fluctuates down with each spoken word.
B. Voltage panel indicator lamps may glow with each syllable.
C. Automatic tuner cycles on each syllable.
D. Minimal voltage drop seen at power source.

When testing a marine single sideband transceiver into an active antenna, such as a marine whip or insulated stay, it is quite common to see instrument panel lamps glow with each syllable. Different bands will cause different instrumentation lamps to illuminate during normal speech. Radio frequency emissions from the antenna system are coupling into the instrument panel wiring, causing the small incandescent and neon lamps to glow slightly during modulation. **ANSWER B**

At this modern navigation station, the wood and fiberglass will not shield electrical wiring behind the panel, and SSB energy may cause meters and lamps to deflect on modulation peaks

3-86N2 What might contribute to apparent low voltage on marine SSB transmitting?

A. Blown red fuse.
B. Too much grounding.
C. Blown black negative fuse.
D. Antenna mismatch.

The marine single sideband carries red and black wiring to battery positive and battery negative. The sideband is also grounded to the ship's grounding and bonding system through the use of copper foil. This ultimately leads back to battery negative, but with much higher resistance than the direct battery negative connection with the black wire. Occasionally, something onboard, like a refrigeration or anchor windlass motor, may pull excessive amounts of current out of both the red and black wires, as well as the common ground circuit. Pulling current through the ground circuit may cause other electronic devices to blow their own battery negative fuses, preventing the errant refrigerator or anchor windlass motor from pulling current through the equipment's ground system. Once the black fuse blows, the SSB is now pulling battery current through the ground foil system, incurring significant voltage drops on transmit, yet looking normal on receive. **ANSWER C**

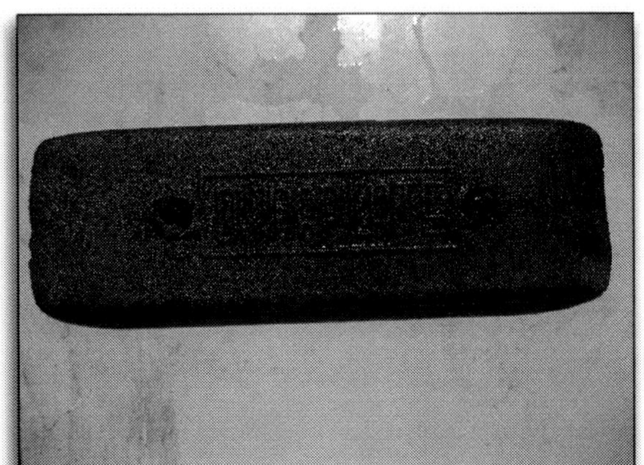

The ground shoe, left, is connected to the SSB tuner's ground system using 3″ wide copper foil. See http://www.metal-cable.com.

3-86N3 What type of wire connects an SSB automatic tuner to an insulated backstay?

A. RG8U.
B. RG213.
C. 16-gauge two-conductor.
D. GTO-15 high-voltage cable.

The single plastic insulated, stranded wire going to the insulated backstay is called GTO-15, the 15 standing for 15,000 volts, similar to neon sign wire. This wire is typically 12 gauge stranded, and silver tinned, surrounded by 2 layers of high voltage insulation. **ANSWER D**

3-86N4 Which of the following statements concerning SITOR communications is true?

A. ARQ message transmissions are made in data groups consisting of three-character blocks.
B. ARQ transmissions are acknowledged by the Information Receiving Station only at the end of the message.
C. ARQ communications rely upon error correction by time diversity transmission and reception.
D. Forward error correction is an interactive mode.

ARQ stands for Automatic Repeat reQuest. Each character contains four mark and three space bits. In mode A, ARQ, the sending station transmits three characters to the receiving station, and the receiving station acknowledges correct receipt of these characters or asks for a repeat. With good propagation, SITOR offers near perfect copy, but under poor band conditions, it may take several minutes to receive several lines of perfect copy text. **ANSWER A**

3-86N5 The sequence ARQ, FEC, SFEC best corresponds to which of the following sequences?

A. One-way communications to a single station, one-way communications to all stations, two-way communications.
B. One-way communications to all stations, two-way communications, one-way communications to a single station.
C. Two way communications, one-way communications to all stations, one-way communications to a single station.
D. Two way communications, one-way communications to a single station, one-way communications to all stations.

ARQ would be the exchange of digital communications between specific transmitting and receiving stations. FEC, Forward Error Correction, sends each character twice, and is a one way communication to all stations. SFEC is a one way communication with forward error correction to a specific, single station. 2-1-1. **ANSWER C**

3-86N6 Which of the following statements concerning SITOR communications is true?

A. Communication is established on the working channel and answerbacks are exchanged before FEC broadcasts can be received.
B. In the ARQ mode each character is transmitted twice.
C. Weather broadcasts cannot be made in FEC because sending each character twice would cause the broadcast to be prohibitively long.
D. Two-way communication with the coast radio station using FEC is not necessary to be able to receive the broadcasts.

Many times maritime weather reports are transmitted using SITOR and, using Forward Error Correction, the associated receiving stations do not need to first establish any communications with the transmitting station. **ANSWER D**

Key Topic 87: Survival Craft Equipment: VHF, SARTs & EPIRBs

3-87N1 What causes the SART to begin a transmission?

A. When activated manually, it begins radiating immediately.
B. After being activated the SART responds to RADAR interrogation.
C. It is either manually or water activated before radiating.
D. It begins radiating only when keyed by the operator.

To intensify a small target return to a distant or nearby RADAR system, a SART (Search and Rescue Transponder) will show up as 12 equally spaced dots in a straight line, appearing on the RADAR screen of any vessel using RADAR equipment on SART bands. The 12 equally spaced dots will point in the direction to the activated SART. Extremely small targets, such as a small lifeboat in heavy seas, now shows up dramatically on a nearby RADAR screen. RADAR operators must insure that any interference filter on their equipment is turned off in order to be able to detect a SART signal. The SART begins transmitting the dashes as soon as it receives a RADAR signal on 9 GHz. **ANSWER B**

3-87N2 How should the signal from a Search And Rescue Radar Transponder appear on a RADAR display?

A. A series of dashes.
B. A series of spirals all originating from the range and bearing of the SART.
C. A series of twenty dashes.
D. A series of 12 equally spaced dots.

The 12 equally spaced dots will point to the direction of the activated SART, with the ship's own position being in the center of the RADAR scope. **ANSWER D**

3-87N3 In which frequency band does a search and rescue transponder operate?

A. 9 GHz
B. 3 GHz
C. S-band
D. 406 MHz

SART transponders operate on the 9 GHz X band frequencies. They will not show up on other bands of RADAR equipment, so make sure that any search and rescue attempts look for an activated SART on the common X band 9 GHz frequencies. **ANSWER A**

> The International Maritime Organization (IMO) has amended the GMDSS regulations to permit Search and Rescue Locating Device Transponders on marine VHF AIS channels as an alternative to 9 GHz SART transponders.
> Also, Part 95 Satellite Emergency Notification Devices (SENDs) must receive FCC certification before they are sold to the general public. The U.S. Coast Guard will also be involved in this effort to insure these non-406 MHz devices on separate satellite systems meet certain specifications for reliability during an emergency activation.
> And expect to see new innovations in man-overboard alerting devices, too, called Maritime Survivor Locating Devices.

ACR Search and Rescue Transponder, SART.

3-87N4 Which piece of required GMDSS equipment is the primary source of transmitting locating signals?

A. Radio Direction Finder (RDF).
B. A SART transmitting on 406 MHz.
C. Survival Craft Transceiver.
D. An EPIRB transmitting on 406 MHz.

The 406 MHz EPIRB is a beacon that broadcasts a distress signal to the COSPAS-SARSAT satellite system, which pinpoints the beacon's location. Through Doppler shift analysis, it takes about one hour for this pinpointing process to take place. But, good news, the 406 MHz beacon may also carry an imbedded GPS allowing a data burst every 50 seconds of GPS-derived latitude and longitude, received by the geosynchronous GEOSAR satellite, and instantly relayed to ground stations. A GPS-enabled beacon eliminates the waiting time required for the traditional Low Earth Orbit LEOSAR satellites to obtain a doppler position fix. The GPS position is updated every 20 minutes. **ANSWER D**

3-87N5 Which of the following statements concerning satellite EPIRBs is true?

A. Once activated, these EPIRBs transmit a signal for use in identifying the vessel and for determining the position of the beacon.
B. The coded signal identifies the nature of the distress situation.
C. The coded signal only identifies the vessel's name and port of registry.
D. If the GMDSS Radio Operator does not program the EPIRB, it will transmit default information such as the follow-on communications frequency and mode.

The 406 MHz EPIRB transmits the beacon's registration information, not only alerting that there is an emergency, but allowing rescue agencies to make phone calls to verify that the beacon is not a false alarm at home. **ANSWER A**

3-87N6 What statement is true regarding 406 MHz EPIRB transmissions?

A. Allows immediate voice communications with the RCC.
B. Coding permits the SAR authorities to know if manually or automatically activated.
C. Transmits a unique hexadecimal identification number.
D. Radio Operator programs an I.D. into the SART immediately prior to activation.

Each 406 MHz EPIRB transmits a unique hexadecimal identification number, along with GPS information, if GPS is part of the EPIRB device. Some GPS receivers are built into the actual equipment where other gear is constantly updated by ship's GPS position until the EPIRB is pulled out of the holder and activated. Either way, an EPIRB with a GPS signal will dramatically speed up the rescue over an EPIRB with just a 406 MHz signal. Each activated 406 MHz also transmits an extremely low power 121.5 MHz/243 MHz homing tone, providing rescue agencies one additional method to track down locally the activated EPIRB. The COSPAS-SARSAT search and rescue system no longer monitors for 121.5 MHz signals. **ANSWER C**

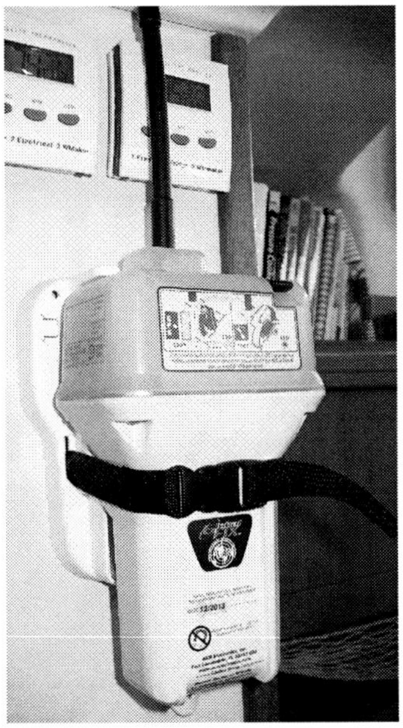

406 MHz ACR EPIRB, positioned for immediate deployment in a companionway.

Key Topic 88: FAX, NAVTEX

3-88N1 What is facsimile?

A. The transmission of still pictures by slow-scan television.
B. The transmission of characters by radioteletype that form a picture when printed.
C. The transmission of printed pictures for permanent display on paper.
D. The transmission of video by television.

In the marine and aviation radio services, the most common use of radio facsimile is for the printed pictures of weather charts. This is called "WEFAX", and is the term used to refer to the broadcast of facsimile images containing weather charts and satellite imagery. In the United States, these broadcasts are typically prepared by the National Weather Service (NWS), a branch of the National Oceanographic and Atmospheric Administration (NOAA). **ANSWER C**

Weather facsimile receiver with paper charts.

Photo courtesy of Furuno Electric Co., Ltd.

3-88N2 What is the standard scan rate for high-frequency 3 MHz - 23 MHz weather facsimile reception from shore stations?

A. 240 lines per minute.
B. 120 lines per minute.
C. 150 lines per second.
D. 60 lines per second.

High frequency short-wave weather facsimile transmissions are sent at 120 lines per minute, roughly half the resolution of polar orbiting satellites. HF weather facsimile reception takes 10 minutes for each chart product, and weather forecasters give you all forms of synopsis and forecasts with the appropriate weather symbols. **ANSWER B**

3-88N3 What would be the bandwidth of a good crystal lattice band-pass filter for weather facsimile HF (high frequency) reception?

A. 500 Hz at -6 dB.
B. 6 kHz at -6 dB.
C. 1 kHz at -6 dB.
D. 15 kHz at -6 dB.

Attempting to recover a weather facsimile signal on a traditional single sideband transceiver without a narrow band filter will lead to excessive noise, seen as a light gray caste to the received image. If the single sideband transceiver has the capability to switch in a 1 kHz at -6dB filter, you will likely see an immediate improvement to the received picture. **ANSWER C**

3-88N4 Which of the following statements about NAVTEX is true?

A. Receives MSI broadcasts using SITOR-B or FEC mode.
B. The ship station transmits on 518 kHz.
C. The ship receives MSI broadcasts using SITOR-A or ARQ mode.
D. NAVTEX is received on 2182 kHz using SSB.

NAVTEX are one-way shore-to-ship broadcasts using SITOR-B or FEC modes of important local Coast Guard information, including warnings and discrepancies to aids to navigation, distress call relays, and important weather information. Aboard ships, these NAVTEX digital broadcasts are usually received on a dedicated receiver with its own dedicated antenna, to continuously receive MSI messages when approaching port. **ANSWER A**

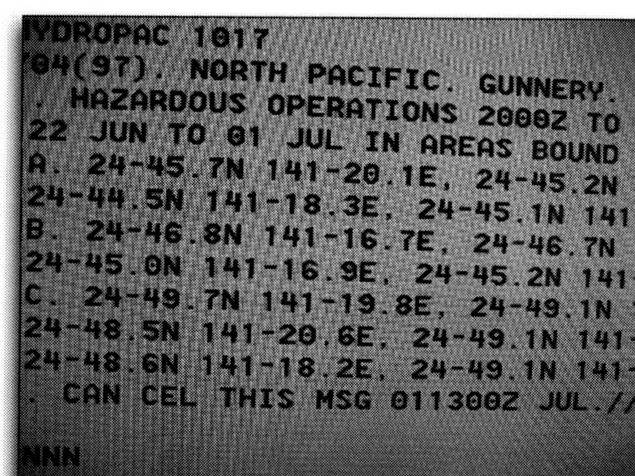

NAVTEX decoded with software.

3-88N5 Which of the following is the primary frequency that is used exclusively for NAVTEX broadcasts internationally?

A. 2187.5 kHz.
B. 4209.5 kHz.
C. VHF channel 16.
D. 518 kHz.

All NAVTEX broadcasts are made on 518 kHz, using narrow-band direct printing 7-unit forward error correcting (FEC or SITOR-B) transmission. Broadcasts use 100 baud FSK modulation, with a frequency shift of 170 Hz. The center frequency of the audio spectrum applied to a single sideband transmitter is 1700 Hz. The receiver 6 dB bandwidth should be between 270-340 Hz. **ANSWER D**

3-88N6 What determines whether a NAVTEX receiver does not print a particular type of message content?

A. The message does not concern your vessel.
B. The subject indicator matches that programmed for rejection by the operator.
C. The transmitting station ID covering your area has not been programmed for rejection by the operator.
D. All messages sent during each broadcast are printed.

The dedicated NAVTEX receiver may be programmed to ignore duplicate messages, and also programmed to ignore routine traffic that might be received via a different system. This could include routine Notice to Mariners. Emergency traffic will always pass through, as long as the NAVTEX receiver is turned on. **ANSWER B**

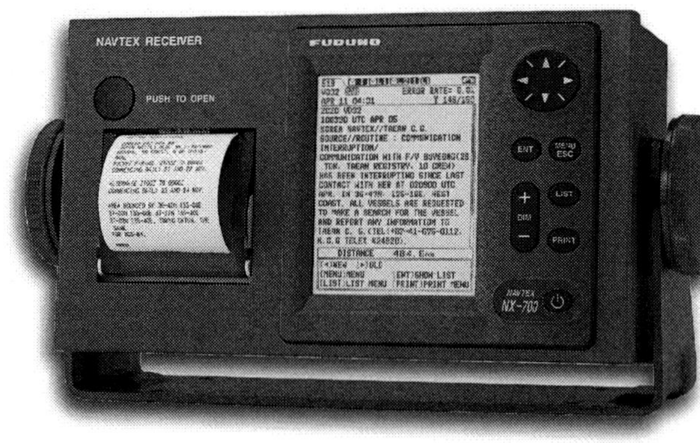

NAVTEX receiver with built in LCD screen and paper log sheets.

Photo courtesy of Furuno Electric Co., Ltd.

Key Topic 89: NMEA Data

3-89N1 What data language is bi-directional, multi-transmitter, multi-receiver network?

A. NMEA 2000.
B. NMEA 0181.
C. NMEA 0182.
D. NMEA 0183.

NMEA stands for National Marine Electronics Association, an industry organization whose mission is to strengthen marine electronics products and services through standards, dealer education, communication, training, and certification. NMEA 2000 is a low-cost, moderate capacity, bi-directional multi-transmitter/multi-receiver network topology for interconnecting marine electronic devices. These devices are connected in a strict bus topology, based on a continuous backbone. Under the NMEA standard, each device is connected to the backbone via individual taps and drop cables. **ANSWER A**

3-89N2 How should shielding be grounded on an NMEA 0183 data line?

A. Unterminated at both ends.
B. Terminated to ground at the talker and unterminated at the listener.
C. Unterminated at the talker and terminated at the listener.
D. Terminated at both the talker and listener.

Original NMEA 0183 data language was one-way , from "talker" to "listener" – from a GPS to an automatic pilot, for example. Shielded cable is not required for the data transfer, but recommended in noisy environments. The shield would be grounded at the NMEA device SENDING the data (like a GPS output), and the shield left "floating" (un-terminated) at the RECEIVING (listener) device (like the autopilot). **ANSWER B**

New installations of marine electronics will likely use the new NMEA 2000 network, where all connections between cables, and between cables and equipment, are made via a drop cable with approved NMEA cable and connectors. The NMEA 2000 cable is two sets of twisted pair of wires, and a shield. The two twisted pairs consist of power and signal. The power pair is primarily used to power the CAN transceiver with the product. The 120-ohm terminating resistors are installed on the signal pair at BOTH ends of the NEMA 2000 backbone. The terminating resistors help minimize noise on the network.

3-89N3 What might occur in NMEA 2000 network topology if one device in line should fail?
A. The system shuts down until the device is removed.
B. Other electronics after the failed device will be inoperable.
C. The main fuse on the backbone may open.
D. There will be no interruption to all other devices.
The NMEA recommends network planning to verify that the network cabling and devices can be connected in the intended manner, while still meeting all the applicable requirements. In order to accommodate a wide variety of maritime applications, the specifications outlined by the NMEA provide a similarly broad variety of options for cabling and powering NMEA 2000 devices, with remaining devices continuing to work even if a device in a single location should fail. **ANSWER D**

3-89N4 In an NMEA 2000 device, a load equivalence number (LEN) of 1 is equivalent to how much current consumption?
A. 50 mA
B. 10 mA
C. 25 mA
D. 5 mA
Marine electronic manufacturers offering NMEA 2000 equipment will specify the amount of current drawn from the NMEA backbone wiring using a load equivalent number (LEN). The LEN number represents a specific range of current use, with 0-50 mA equivalent to LEN 1. A device pulling more than 75 mA has a LEN 2 (in the range of 50mA -100mA). **ANSWER A**

3-89N5 An NMEA 2000 system with devices in a single location may be powered using this method:
A. Dual mid-powered network.
B. End-powered network.
C. Individual devices individually powered.
D. No 12 volts needed for NMEA 2000 devices.
NMEA 2000 products are interconnected in a strict bus topology over a continuous backbone which feeds end-powered voltage up the line to operate each piece of equipment's computer. However, the equipment itself, such as a high current automatic pilot, will have its own voltage source that operates the rest of the equipment. **ANSWER B**

3-89N6 What voltage drop at the end of the last segment will satisfy NMEA 2000 network cabling plans?
A. 0.5 volts
B. 2.0 volts
C. 1.5 volts
D. 3.0 volts
Marine electronics equipment with no need for external electrical power connections may draw no more than 1 amp of current from the backbone. A voltage drop of less than 1.5 volts at the end of the last product will satisfy NMEA 2000 network cabling plans. "In order to meet the requirements of Section 8.3.3.1, the maximum network distribution voltage drop shall be equal to or less than the voltage given in table 18 for networks powered from the indicated source," showing 1.5 volts permissible drop from a battery, and 5 volts to an isolated power supply. **ANSWER C**

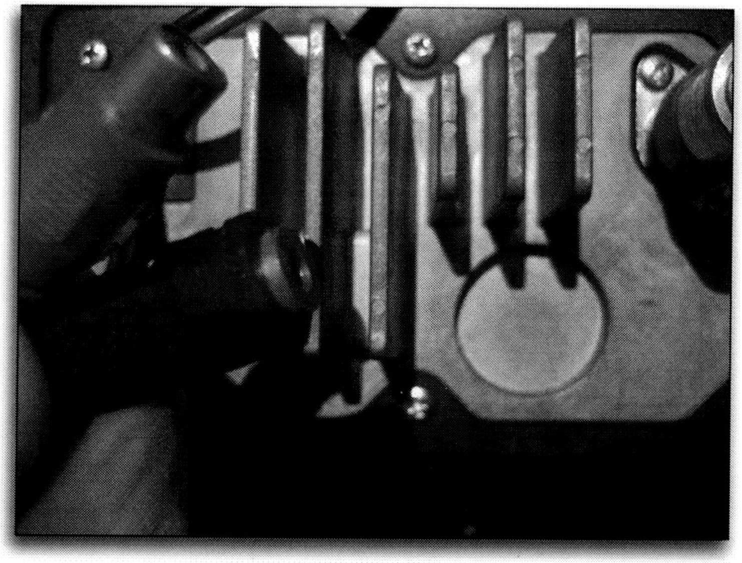

NMEA Input and Output Data Jacks.

If you are considering a job in the marine electronics field, the NMEA offers several levels of marine electronics installer programs. Learn more about the NMEA by visiting www.NMEA.org. If you are actively seeking employment in the marine electronics industry, the NMEA is your one and only marine electronics association to join.

Subelement O – RADAR

5 Key Topics, 5 Exam Questions

Key Topic 90: RADAR Theory

3-90O1 What is the normal range of pulse repetition rates?

A. 2,000 to 4,000 pps.
B. 1,000 to 3,000 pps.
C. 500 to 1,000 pps.
D. 500 to 2,000 pps.

Okay, RADAR techs, here are a few common letters to remember: PPS = pulses per second; PRR = pulse repetition rate, and PRF = pulse repetition frequency. They all mean the same thing! The normal range of pulse repetition rates on most marine RADARs is between 500 to 2,000 pulses per second. The time in between each pulse is called the pulse repetition interval, and this is equal to 1 divided by the pulse repetition rate, with an answer in micro-seconds. Pulse repetition rate may also be called pulse repetition frequency. Pulse repetition interval is sometimes called pulse repetition time. Pulse repetition rate is the number of pulses per second sent to the target. It also is the number of pulses per second sent to the magnetron. Small targets many miles away from the operating radar are best detected with a lower pulse repetition rate, which is usually an automatic function as the operator switches to a longer range that will generate a longer pulse width. **ANSWER D**

3-90O2 The RADAR range in nautical miles to an object can be found by measuring the elapsed time during a RADAR pulse and dividing this quantity by:

A. 0.87 seconds.
B. 1.15 μs.
C. 12.346 μs.
D. 1.073 μs.

We know that it takes 6.173 (previously rounded to 6.2) microseconds for a RADAR wave to travel 1 nautical mile. Therefore, it will take twice that, or 12.35 microseconds, to make the trip from the RADAR antenna to the target 1 nautical mile away and then back to the antenna. **ANSWER C**

Radar antenna safely above everyone on deck.

3-90O3 What is the normal range of pulse widths?

A. .05 μs to 0.1 μs.
B. .05 μs to 1.0 μs.
C. 1.0 μs to 3.5 μs.
D. 2.5 μs to 5.0 μs.

The normal range of RADAR pulse widths is .05 μS to 1.0 μS (micro-seconds). If we were transmitting 2,400 pulses-per-second, with each pulse 0.12 micro-seconds in LENGTH, this would yield 288 micro-seconds of energy transmitted in that second. If we had 800 pulses, each 0.4 micro-seconds in length, this would amount to 320 micro-seconds of energy within one second, but each second of time has 1,000,000 micro-seconds – too long a pulse on short range would not allow enough time for the receiver to respond to the echo. Too short a pulse at long range may not illuminate the target adequately. The modern RADAR will usually develop the correct algorithm of pulses-per-second and pulse length. **ANSWER B**

3-90O4 Shipboard RADAR is most commonly operated in what band?

A. VHF.
B. UHF.
C. SHF.
D. EHF.

Marine RADARs operate on super high frequency bands – 3000 MHz to 30,000MHz. **ANSWER C**

3-9O5 The pulse repetition rate (prr) of a RADAR refers to the:

A. Reciprocal of the duty cycle.
B. Pulse rate of the local oscillator.
C. Pulse rate of the klystron.
D. Pulse rate of the magnetron.

The pulse repetition rate (PRR) is the number of pulses per second sent to the target. It also is the number of pulses per second sent from the magnetron. **ANSWER D**

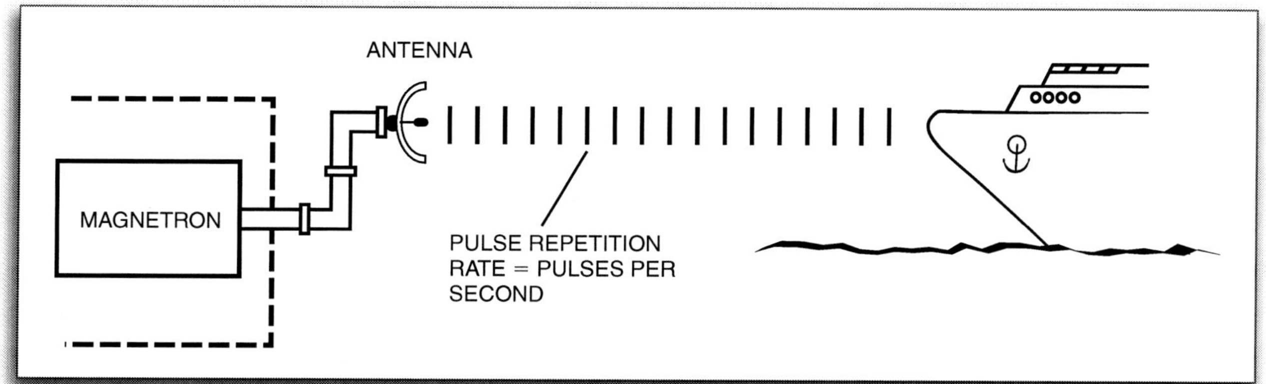

Pulse Repetition Rate from Antenna is Same as from Magnetron

3-9O6 If the elapsed time for a RADAR echo is 62 microseconds, what is the distance in nautical miles to the object?

A. 5 nautical miles.
B. 87 nautical miles.
C. 37 nautical miles.
D. 11.5 nautical miles.

It takes a radar signal 6.2 microseconds to travel 1 nautical mile. In 62 microseconds, the radar signal will travel 10 total miles – 5 miles to the target and 5 miles back. $62 \div (6.2 \times 2) = 5$. **ANSWER A**

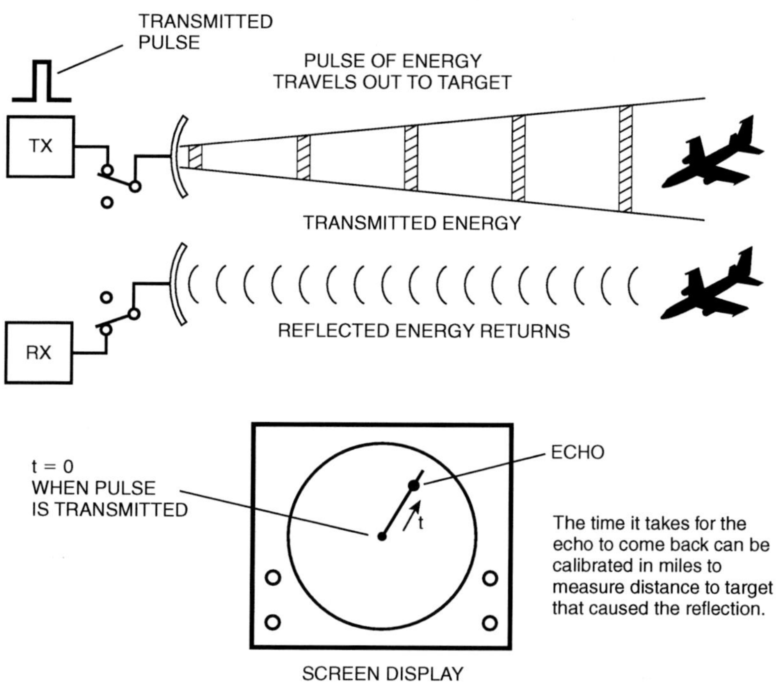

RADAR Echo
Source: Technology Dictionary, ©1987, Master Publishing, Inc.

Key Topic 91: Components

3-9101 The ATR box:

A. Prevents the received signal from entering the transmitter.
B. Protects the receiver from strong RADAR signals.
C. Turns off the receiver when the transmitter is on.
D. All of the above.

The ATR circuit keeps received echoes out of the transmitter. **ANSWER A**

3-9102 What is the purpose or function of the RADAR duplexer/circulator? It is a/an:

A. Coupling device that is used in the transition from a rectangular waveguide to a circular waveguide.
B. Electronic switch that allows the use of one antenna for both transmission and reception.
C. Modified length of waveguide that is used to sample a portion of the transmitted energy for testing purposes.
D. Dual section coupling device that allows the use of a magnetron as a transmitter.

The duplexer/circulator is like an electronic transmit/receive switch, toggling outward bound pulses from the magnetron to the antenna waveguide, while also passing incoming echo pulses from the antenna's same waveguide to the receiver. This allows one RADAR antenna to both transmit energy and receive the weak echoes. The duplexer/circulator prevents transmit pulses from entering the receiver directly. The rare failure of the duplexer/circulator could easily damage the RADAR receiver circuitry. **ANSWER B**

3-9103 What device can be used to determine the performance of a RADAR system at sea?

A. Echo box.
B. Klystron.
C. Circulator.
D. Digital signal processor.

An echo box is a good way to test RADAR performance when far out to sea with no local targets to return an echo. The echo box is a resonant cavity constructed in such a way as to slightly delay the return echo when it is reflected back to the receiver. **ANSWER A**

3-9104 What is the purpose of a synchro transmitter and receiver?

A. Synchronizes the transmitted and received pulse trains.
B. Prevents the receiver from operating during the period of the transmitted pulse.
C. Transmits the angular position of the antenna to the indicator unit.
D. Keeps the speed of the motor generator constant.

The RADAR synchro keeps the PPI sweep line in the same angular position as the rotating antenna. When servicing a marine RADAR, it is advised to go out for a sea trial. Point the bow of the ship to a distant target, such as a buoy and make sure that the bow, buoy, and heading marker are in perfect angular alignment, dead ahead. Next, repeat this test off the stern. **ANSWER C**

3-9105 Digital signal processing (DSP) of RADAR signals (compared with analog) causes:

A. Improved display graphics.
B. Improved weak signal or target enhancement.
C. Less interference with SONAR systems.
D. Less interference with other radio communications equipment.

Digital signal processing (DSP) allows for the subtraction of both internal and external noise, enhancing weak echo response. **ANSWER B**

Learn about CHIRP technology, the latest digital improvement to RADAR, on page 285.

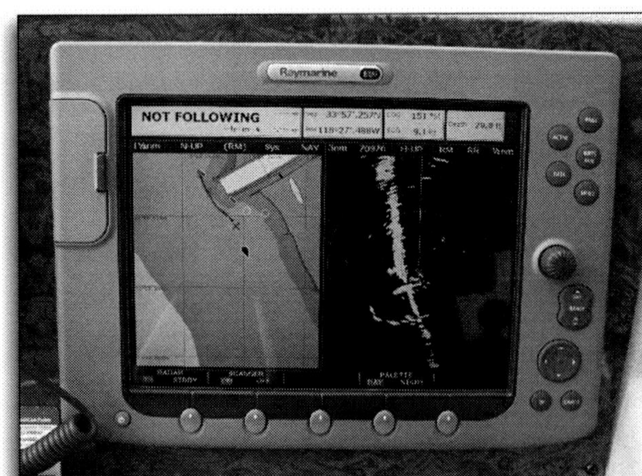

Here the modern RADAR also displays charts, either to the side, or overlaid.

3-9106 The component or circuit providing the transmitter output power for a RADAR system is the:

A. Thyratron.
B. SCR.
C. Klystron.
D. Magnetron.

Transmitter power output is delivered by the magnetron. **ANSWER D**

Key Topic 92: Range, Pulse Width & Repetition Rate

3-92O1 When a RADAR is being operated on the 48 mile range setting, what is the most appropriate pulse width (PW) and pulse repetition rate (pps)?

A. 1.0 µs PW and 2,000 pps.
B. 0.05 µs PW and 2,000 pps.
C. 2.5 µs PW and 2,500 pps.
D. 1.0 µs PW and 500 pps.

The absolute upper limit for a PRR is determined by the fact that we must never send out more than one pulse for every one that returns. Otherwise, we don't know which pulse we're looking at. At 48 miles, the one-way trip time for the pulse is 48 x (1 ÷ 186.292) = 0.000258 seconds. We double that to get our round trip time of 0.000515 seconds. The time between our pulses must be greater than 0.000515 seconds. To find the PPS, we simply take the reciprocal of our period, or 1 ÷ 0.000515, which gives us 1942 PPS. Our PRR must be LESS than 1942 PPS. This eliminates answers A, B, and C. **ANSWER D**

PULSE LENGTHS...

LENGTH OF TIME OF EACH PULSE	DISTANCE EACH PULSE TRAVELS	CLOSEST DISTANCE FOR PRESENTATION
1.0 MICROSECONDS	984 FEET	492 FEET
0.5 MICROSECONDS	492 FEET	246 FEET
0.25 MICROSECONDS	246 FEET	123 FEET
0.1 MICROSECONDS	98 FEET	49 FEET
0.05 MICROSECONDS	49 FEET	25 FEET

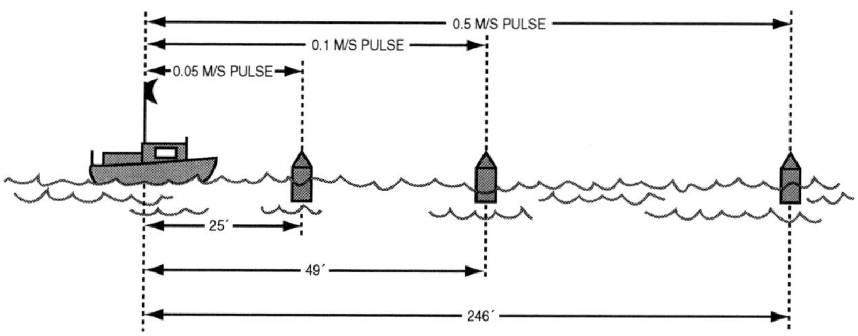

REPETITION FREQUENCIES...

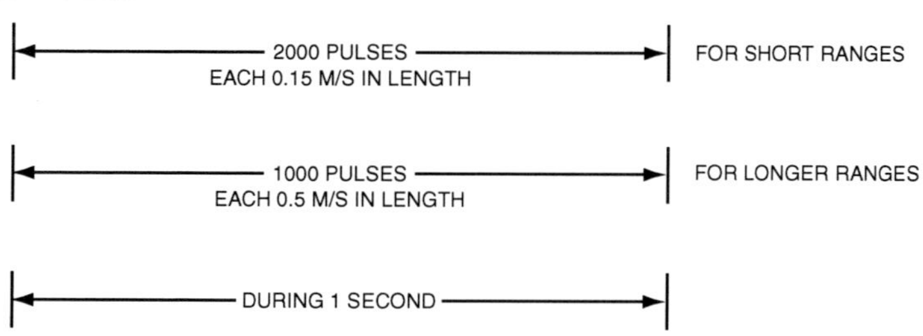

3-92O2 When a RADAR is being operated on the 6 mile range setting what is the most appropriate pulse width and pulse repetition rate?

A. 1.0 µs PW and 500 pps.
B. 2.0 µs PW and 3,000 pps.
C. 0.25 µs PW and 1,000 pps.
D. 0.01 µs PW and 500 pps.

Borrowing from problem 3-92O1 above, we can see that our maximum PRF for unambiguous returns is around 16,000 PPS. (If the maximum PRF for 48 miles is 1,942 we can increase our PRF by a factor of eight for an object only 6 miles away). We are in no danger of "overclocking" even at 3,000 PPS, as in Answer B. However, answer B suggests a 2.0 µs pulse width, which is WAY overkill for such a short distance. We can give our radar a bit of a rest by using a tiny 0.25µs pulse width, which will still give us a comfortably strong signal with good resolution for fast moving boats, or even your odd low-flying airplane! We still want to use a moderately fast PRR, however, so 1000 PPS is a good compromise. **ANSWER C**

The big black domes house microwave antennas, for RADAR, INMARSAT, and satellite TV direct reception. Twin open radar arms are also painted black, on this big motor yacht.

3-9203 We are looking at a target 25 miles away. When a RADAR is being operated on the 25 mile range setting what is the most appropriate pulse width and pulse repetition rate?

A. 1.0 μs PW and 500 pps.
B. 0.25 μs PW and 1,000 pps.
C. 0.01 μs PW and 500 pps.
D. 0.05 μs PW and 2,000 pps.

As in question 3-9201, the maximum PRR is non-negotiable. We MUST have a period between pulses that is greater than the round trip time. Remember, period is the inverse of the PRR. First, let's figure out the round trip time of a target 25 miles away. It is 2 x (25 ÷ 186,000) = 0.000269 seconds. This is equal to a PRR of 1 ÷ 0.000269, or 3717 pulses per second (PPS), our maximum acceptable PRR. All four answers fall comfortably within that limit, so all of them would "work," but not all of them would work equally well. 25 miles is still a pretty distant target, so we want a lot of energy to get a strong return. Greater pulse width (PW) means more energy, so we have only one BEST answer. **ANSWER A**

3-9204 What pulse width and repetition rate should you use at long ranges?

A. Narrow pulse width and slow repetition rate.
B. Narrow pulse width and fast repetition rate.
C. Wide pulse width and fast repetition rate.
D. Wide pulse width and slow repetition rate.

On long-range radar detection, the RADAR will automatically switch to a wider pulse width at a slower repetition rate. **ANSWER D**

3-9205 What pulse width and repetition rate should you use at short ranges?

A. Wide pulse width and fast repetition rate.
B. Narrow pulse width and slow repetition rate.
C. Narrow pulse width and fast repetition rates.
D. Wide pulse width and slow repetition rates.

On short-range target detection, the RADAR will automatically select narrow pulse width and a faster repetition rate. **ANSWER C**

3-9206 When a RADAR is being operated on the 1.5 mile range setting, what is the most appropriate pulse width and pulse repetition rate?

A. 0.25 μs PW and 1,000 pps.
B. 0.05 μs PW and 2,000 pps.
C. 1.0 μs PW and 500 pps.
D. 2.5 μs PW and 2,500 pps.

The modern ship RADAR has no operator functions to manually adjust pulse width or pulse repetition rates – this is accomplished automatically when the operator changes range scales. The two functions are interrelated, and are strictly controlled by the range selected on the RADAR controls. Veteran RADAR operators indeed had more options with older RADAR gear, but now range adjustments automatically select the optimum pulse width and pulse repetition rate. **ANSWER B**

Key Topic 93: Antennas & Waveguides

3-9301 How does the gain of a parabolic dish antenna change when the operating frequency is doubled?

A. Gain does not change.
B. Gain is multiplied by 0.707.
C. Gain increases 6 dB.
D. Gain increases 3 dB.

From 1270 MHz on up, you may wish to use a parabolic dish antenna for microwave work. If you double the frequency, the gain of the dish increases by 6 dB, a 4 times increase. **ANSWER C**

Parabolic Antenna

3-9302 What type of antenna or pickup device is used to extract the RADAR signal from the wave guide?

A. J-hook.
B. K-hook.
C. Folded dipole.
D. Circulator.

Some open-array RADAR antennas terminate the waveguide run with a tuned J hook flare, capped with a thin fiberglass window, to allow RADAR energy to pass from and into the rigid waveguide. The waveguide system may also contain a tuned J hook probe for extracting microwave energy from within the waveguide. **ANSWER A**

3-9303 What happens to the beamwidth of an antenna as the gain is increased? The beamwidth:

A. Increases geometrically as the gain is increased.
B. Increases arithmetically as the gain is increased.
C. Is essentially unaffected by the gain of the antenna.
D. Decreases as the gain is increased.

The larger the RADAR parabola, the greater will be the energy concentration, leading to an increase in gain. The decreased beamwidth allows more energy to be concentrated into a narrow plane for better target discrimination. **ANSWER D**

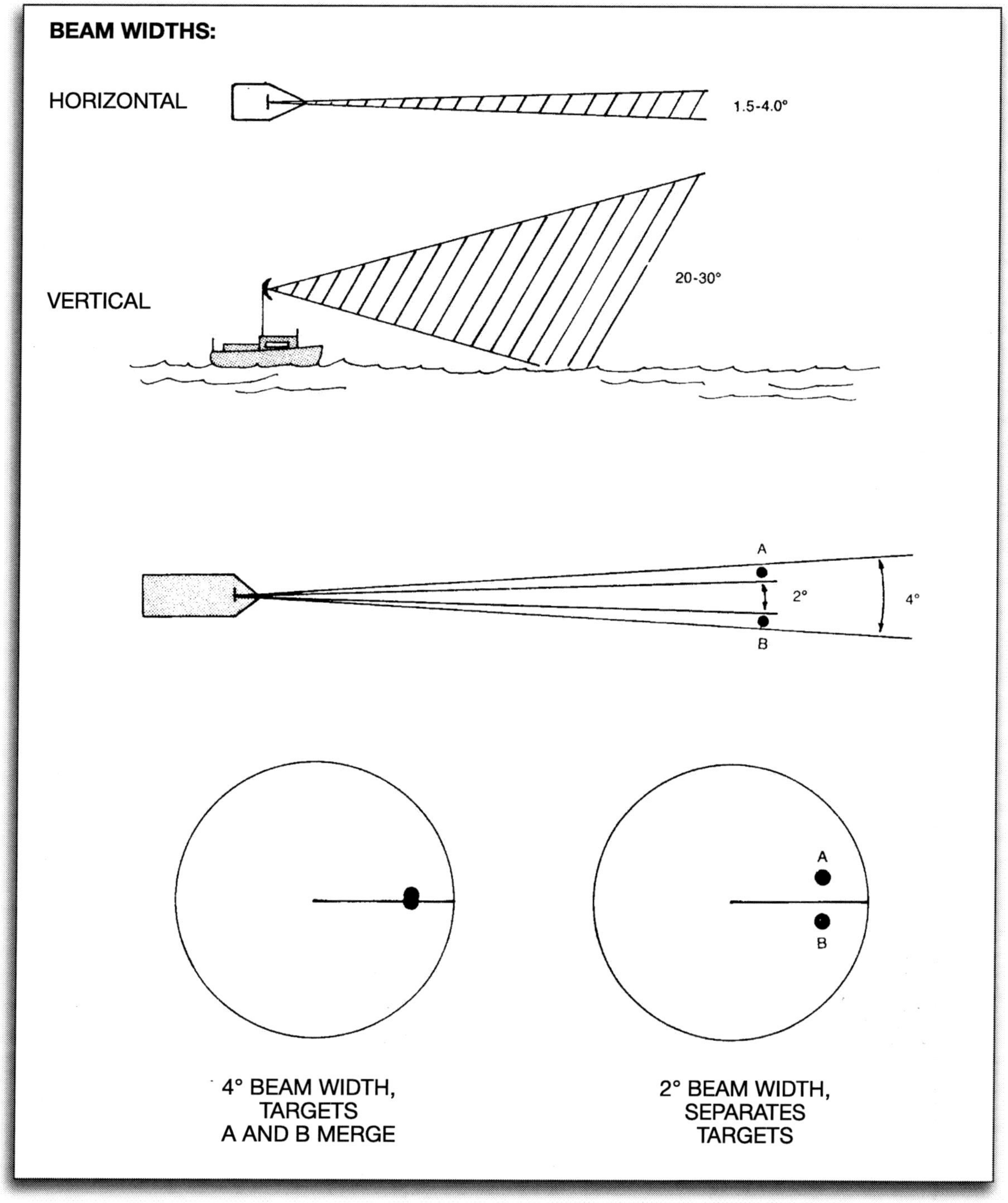

3-93O4 A common shipboard RADAR antenna is the:

A. Slotted array.
B. Dipole.
C. Stacked Yagi.
D. Vertical Marconi.

An interesting RADAR antenna is the slotted array, which uses a number of in-phase radiating openings. This acts much like a co-linear array, but the slotted array is horizontal, allowing for the horizontal lobe to be sharpened. A parabolic trough to the rear of the slotted array also allows unwanted vertical beamwidth to be reduced, too. **ANSWER A**

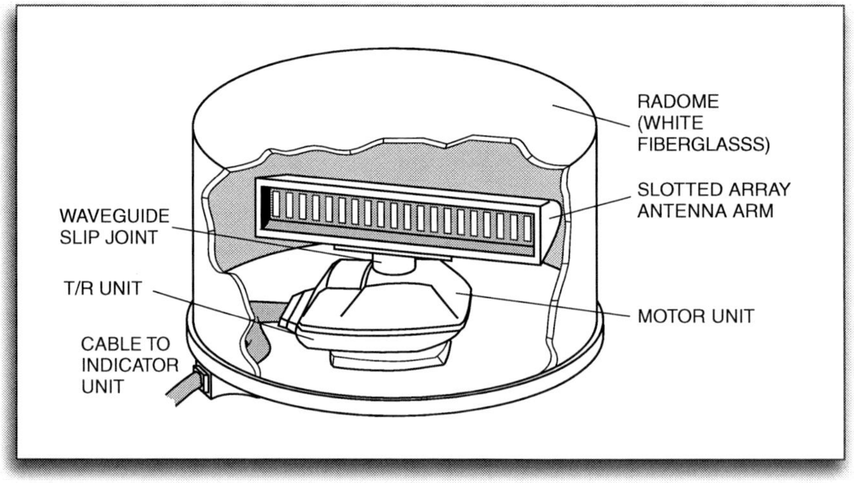

Slotted Array Antenna.

3-93O5 Conductance takes place in a waveguide:

A. By interelectron delay.
B. Through electrostatic field reluctance.
C. In the same manner as a transmission line.
D. Through electromagnetic and electrostatic fields in the walls of the waveguide.

In a waveguide, the conductance takes place in the walls of the waveguide through magnetic and electric fields. **ANSWER D**

Close-up photo of a waveguide.

3-93O6 To couple energy into and out of a waveguide use:

A. Wide copper sheeting.
B. A thin piece of wire as an antenna.
C. An LC circuit.
D. Capacitive coupling.

With the transmitter turned off, and the waveguide feed assembly out of circuit, take a look inside and spot the tiny thin piece of wire that is acting as the antenna element. This couples energy into and out of a waveguide. **ANSWER B**

Key Topic 94: RADAR Equipment

3-94O1 The permanent magnetic field that surrounds a traveling-wave tube (TWT) is intended to:
A. Provide a means of coupling.
B. Prevent the electron beam from spreading.
C. Prevent oscillations.
D. Prevent spurious oscillations.
The permanent magnetic field that surrounds a TWT helps focus the electron beam. **ANSWER B**

3-94O2 Prior to testing any RADAR system, the operator should first:
A. Check the system grounds.
B. Assure the display unit is operating normally.
C. Inform the airport control tower or ship's master.
D. Assure no personnel are in front of the antenna.
RADAR systems use microwave signals to detect targets. These are the same frequencies used to heat food in microwave ovens. Remember the Amana Radar Range? You don't want to cook your fellow sailors when testing your RADAR, so make sure no one is standing in front of the antenna! **ANSWER D**

3-94O3 In the term "ARPA RADAR," ARPA is the acronym for which of the following?
A. Automatic RADAR Plotting Aid.
B. Automatic RADAR Positioning Angle.
C. American RADAR Programmers Association.
D. Authorized RADAR Programmer and Administrator.
The ARPA RADAR takes the mystery out of distant target movements. Unlike a conventional display showing only the position of target echoes, the ARPA RADAR shows target history, target direction, and closest point of approach. There are even RADAR alarms that sound if the ARPA RADAR detects that you may be on a collision course. **ANSWER A**

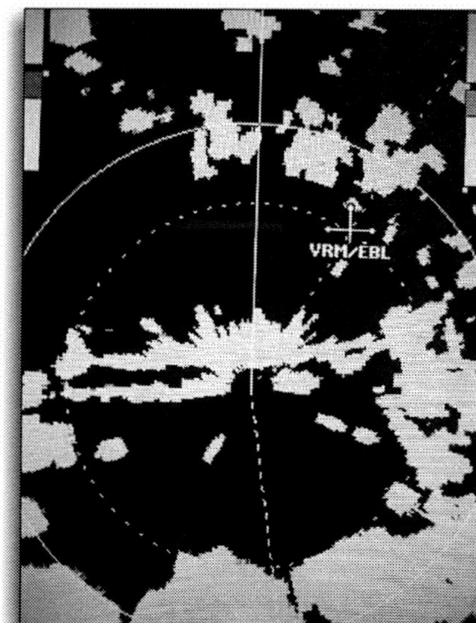

An ARPA RADAR display.

3-94O4 Which of the following is NOT a precaution that should be taken to ensure the magnetron is not weakened:
A. Keep metal tools away from the magnet.
B. Do not subject it to excessive heat.
C. Keep the TR properly tuned.
D. Do not subject it to shocks and blows.
The RADAR magnetron is a delicate network and may be damaged by a metal tool flying into the magnet. This question asks which of the following is NOT a precaution to insure the magnetron is not damaged. Keeping the TR properly tuned may have little influence on damage to the magnetron. **ANSWER C**

3-94O5 Exposure to microwave energy from RADAR or other electronics devices is limited by U.S. Health Department regulations to _______ mW/centimeter.
A. 0.005
B. 5.0
C. 0.05
D. 0.5
Microwave energy from a RADAR unit to your body may not exceed 5 mW per centimeter. The higher we go in frequency, the more critical the consideration to stay out of the main lobe of a marine RADAR. **ANSWER B**

3-94O6 RADAR collision avoidance systems utilize inputs from each of the following except your ship's:
A. Gyrocompass.
B. Navigation position receiver.
C. Anemometer.
D. Speed indicator.
The ship's anemometer, which measures wind speed, will not add to your RADAR collision avoidance input. However, your ship speed will, and same with navigation position from a GPS, as well as your gyro compass. But the anemometer will have no input. **ANSWER C**

Subelement P – Satellite

4 Key Topics, 4 Exam Questions

Key Topic 95: Low Earth Orbit Systems

3-95P1 What is the orbiting altitude of the Iridium satellite communications system?

A. 22,184 miles.
B. 11,492 miles.
C. 4,686 miles.
D. 485 miles.

The Iridium satellite communications system offers worldwide voice and data, with a constellation of satellites in low Earth orbit, 485 miles up. If you are in a remote area of the world, not near a mutual downlink Earth station, an Iridium satellite will automatically pass your message through to another Iridium satellite, providing good voice quality from just a small handheld satellite transceiver. **ANSWER D**

Iridium Phone

> The Iridium Low Earth Orbit satellite service may soon be a partner with GMDSS. It has been so reliable, employing 10 new satellites with worldwide satellite relay capabilities, that RTCM is carefully weighing all its possibilities in a distress situation. Learn more at www.RTCM.ORG. Also, visit WWW.NAVCEN.USCG.GOV/?PageName=MaritimeTelecomms (then click GMDSS/TF) to learn more.

3-95P2 What frequency band is used by the Iridium system for telephone and messaging?

A. 965 - 985 MHz.
B. 1616 -1626 MHz.
C. 1855 -1895 MHz.
D. 2415 - 2435 MHz.

The Iridium satellite system operates on the 1600 MHz band. **ANSWER B**

3-95P3 What services are provided by the Iridium system?

A. Analog voice and Data at 4.8 kbps.
B. Digital voice and Data at 9.6 kbps.
C. Digital voice and Data at 2.4 kbps.
D. Analog voice and Data at 9.6 kbps.

Iridium provides worldwide voice and data, at 2.4 kbps. **ANSWER C**

3-95P4 Which of the following statements about the Iridium system is true?

A. There are 48 spot beams per satellite with a footprint of 30 miles in diameter.
B. There are 48 satellites in orbit in 4 orbital planes.
C. The inclination of the orbital planes is 55 degrees.
D. The orbital period is approximately 85 minutes.

Each Iridium satellite contains 48 spot beams, illuminating the Earth up to 30 miles in diameter per beam. If the satellite does not see a mutual downlink station, it will relay your voice to an alternate Iridium satellite, providing worldwide coverage. **ANSWER A**

3-95P5 What is the main function of the COSPAS-SARSAT satellite system?

A. Monitor 121.5 MHz for voice distress calls.
B. Monitor 406 MHz for distress calls from EPIRBs.
C. Monitor 1635 MHz for coded distress calls.
D. Monitor 2197.5 kHz for hexadecimal coded DSC distress messages.

The main function of COSPAS-SARSAT satellite system is to monitor 406 MHz distress signals from activated EPIRBs, PLBs, and ELTs. No longer is there a satellite guard of 121.5 MHz distress alerts, although most emergency beacons transmit a low power 121.5 MHz signal for ground station direction finding. **ANSWER B**

3-95P6 How does the COSPAS-SARSAT satellite system determine the position of a ship in distress?
A. By measuring the Doppler shift of the 406 MHz signal taken at several different points in its orbit.
B. The EPIRB always transmits its position which is relayed by the satellite to the Local User Terminal.
C. It takes two different satellites to establish an accurate position.
D. None of the above.

The COSPAS-SARSAT satellite system determines the location of ship-to-shore activated 406 MHz distress signals by measuring the Doppler shift, taken at several points in its orbit. The Doppler shift estimated position will lead to a search area of 2.3 nautical miles radius, 12.5 square mile search area, with an average 1 hour search and rescue notification. In addition to Doppler shift measurements, a geosynchronous satellite called SARSAT listens for the 406 MHz built in GPS position data burst, providing search and rescue teams a 0.05 nautical mile radius search area (110 yards!), or a 0.008 square nautical mile search area, with a 5 minute notification to search and rescue agencies. It is always advised that ocean going vessels purchase the more expensive EPIRB with built in GPS capability. **ANSWER A**

Key Topic 96: INMARSAT Communications Systems-1

3-96P1 What is the orbital altitude of INMARSAT Satellites?
A. 16, 436 miles.
B. 22,177 miles.
C. 10, 450 miles.
D. 26,435 miles.

The orbital altitude of INMARSAT satellites is 22,177 miles. This puts them in a geosynchronous orbit with Earth, so they appear motionless in the sky as the Earth rotates. **ANSWER B**

3-96P2 Which of the following describes the INMARSAT Satellite system?
A. AOR at 35° W, POR-E at 165° W, POR-W at 155° E and IOR at 56.5° E.
B. AOR-E at 25° W, AOR-W at 85° W, POR at 175° W and IOR at 56.5° E.
C. AOR-E at 15.5° W, AOR-W at 54° W, POR at 178° E and IOR at 64.5° E.
D. AOR at 40° W, POR at 178° W, IOR-E at 109° E and IOR-W at 46° E.

The Earth is covered by INMARSAT satellites for both voice and data communications. AOR-East is located at 15.5° west, AOR-West is located at 54° west, POR at 178° east, and IOR at 64.5° east. All major shipping areas are covered with INMARSAT satellite signals, other than the extreme North and South poles. **ANSWER C**

3-96P3 What are the directional characteristics of the INMARSAT-C SES antenna?
A. Highly directional parabolic antenna requiring stabilization.
B. Wide beam width in a cardioid pattern off the front of the antenna.
C. Very narrow beam width straight-up from the top of the antenna.
D. Omnidirectional.

The INMARSAT-C SES antenna provides excellent functionality for computer capabilities aboard a ship. Think of INMARSAT-C for computer. The INMARSAT-C antenna is not much larger than a football and is omnidirectional. **ANSWER D**

The INMARSAT-C SES antenna is about the size of a football, and is dwarfed by these larger TV and Mini-M black domes.

3-96P4 When engaging in voice communications via an INMARSAT-B terminal, what techniques are used?

A. CODECs are used to digitize the voice signal.
B. Noise-blanking must be selected by the operator.
C. The voice signal must be compressed to fit into the allowed bandwidth.
D. The voice signal will be expanded at the receiving terminal.

An INMARSAT-B terminal is what you would find aboard large cruise ships, using a large tracking antenna, offering digital and voice capability with a CODEC used to digitize the voice signal. The INMARSAT-B phone call sounds a little metallic, but provides passengers on cruise ships nearly worldwide phone and computer access – at a price. **ANSWER A**

3-96P5 Which of the following statements concerning INMARSAT geostationary satellites is true?

A. They are in a polar orbit, in order to provide true global coverage.
B. They are in an equatorial orbit, in order to provide true global coverage.
C. They provide coverage to vessels in nearly all of the world's navigable waters.
D. Vessels sailing in equatorial waters are able to use only one satellite, whereas other vessels are able to choose between at least two satellites.

The INMARSAT geostationary satellites provide amazing clarity for data and voice transmissions (with the right service) and cover nearly all of the world's navigable waters, other than the North and South poles. **ANSWER C**

3-96P6 Which of the following conditions can render INMARSAT -B communications impossible?

A. An obstruction, such as a mast, causing disruption of the signal between the satellite and the SES antenna when the vessel is steering a certain course.
B. A satellite whose signal is on a low elevation, below the horizon.
C. Travel beyond the effective radius of the satellite.
D. All of these.

An INMARSAT-B satellite antenna is usually located high atop a cruise ship super structure to minimize any blocking of the satellite by any metal on the ship. If the satellite is low on the horizon – when cruising in Alaska or Antarctica, for example – you have likely traveled beyond the effective coverage of the satellite and this, along with obstructions from the ship super structure, will not allow the signal to get through. **ANSWER D**

Satellite and RADAR antennas always mount as high as possible for range and safety.

Key Topic 97: INMARSAT Communications Systems-2

3-97P1 What is the best description for the INMARSAT-C system?

A. It provides slow speed telex and voice service.
B. It is a store-and-forward system that provides routine and distress communications.
C. It is a real-time telex system.
D. It provides world-wide coverage.

INMARSAT-C is for data only. Think of "C" for computer. INMARSAT-C may also provide distress communications.
ANSWER B

The football-sized INMARSAT-C antenna is much smaller when compared to these larger communications satellite antennas.

3-97P2 The INMARSAT mini-M system is a:

A. Marine SONAR system.
B. Marine global satellite system.
C. Marine depth finder.
D. Satellite system utilizing spot beams to provide for small craft communications.

The Inmarsat mini-M is a dependable voice, data, and fax system. The 3-axis, fully stabilized antenna system weighs only about 11 lbs, and is only 11.5 inches high and 10 inches in diameter, making it small enough for even modest sized boats. **ANSWER D**

3-97P3 What statement best describes the INMARSAT-B services?

A. Voice at 16 kbps, Fax at 14.4 kbps and high-speed Data at 64/54.
B. Store and forward high speed data at 36/48 kbps.
C. Voice at 3 kHz, Fax at 9.6 kbps and Data at 4.8 kbps.
D. Service is available only in areas served by highly directional spot beam antennas.

INMARSAT-B offers clear but hollow-sounding voice at 16 kbps, fax at a modest 14.4 kbps, and high speed data at 64/54 kbps. **ANSWER A**

3-97P4 Which INMARSAT systems offer High Speed Data at 64/54 kbps?

A. C.
B. B and C.
C. Mini-M.
D. B, M4 and Fleet.

High speed data at 64/54 kbps is available with INMARSAT-B, Inmarsat-M4, and Fleet. Fleet broadband may also be available up to 432 kbps with an antenna as small as 10 inches in diameter, and low monthly airtime packages for voice, fax, data, and internet. There is also an affordable, fixed-rate, unlimited access plan, and a flexible per-megabyte airtime plan, too, for INMARSAT Fleet broadband service. **ANSWER D**

3-97P5 When INMARSAT-B and INMARSAT-C terminals are compared:

A. INMARSAT-C antennas are small and omni-directional, while INMARSAT-B antennas are larger and directional.
B. INMARSAT-B antennas are bulkier but omni-directional, while INMARSAT-C antennas are smaller and parabolic, for aiming at the satellite.
C. INMARSAT-B antennas are parabolic and smaller for higher gain, while INMARSAT-C antennas are larger but omni-directional.
D. INMARSAT-C antennas are smaller but omni-directional, while INMARSAT-B antennas are parabolic for lower gain.

The INMARSAT-B antenna is what you see aboard passenger ships, under that white dome that stands about 15 feet tall. INMARSAT-C antennas are omni-directional, sized about like a football, and can go nearly anywhere on a small ship. **ANSWER A**

3-97P6 What services are provided by the INMARSAT-M service?

A. Data and Fax at 4.8 kbps plus e-mail.
B. Voice at 3 kHz, Fax at 9.6 kbps and Data at 4.8 kbps.
C. Voice at 6.2 kbps, Data at 2.4 kbps, Fax at 2.4 kbps and e-mail.
D. Data at 4.8 kbps and Fax at 9.6 kbps plus e-mail.

INMARSAT-M offers digital voice, fax, and data communications for a reasonable per-minute fee, and works with a relatively small motorized tracking antenna unit, offering voice at 6.2 kbps, data at 2.4 kbps, fax at 2.4 kbps, and e-mail. **ANSWER C**

Key Topic 98: GPS

3-98P1 Global Positioning Service (GPS) satellite orbiting altitude is:

A. 4,686 miles.
B. 24,184 miles.
C. 12,554 miles.
D. 247 miles.

The Department of Defense gives us our global positioning service (GPS), based on a constellation of 24 satellites up 12,554 miles. These satellites orbit the Earth twice a day, are arrayed in 6 orbital planes, with 4 satellites per plane. Our position is calculated from distance measurements to all satellites in view. Just 4 distance measurements are required for a position fix within a radius of approximately 15 yards! **ANSWER C**

3-98P2 The GPS transmitted frequencies are:

A. 1626.5 MHz and 1644.5 MHz.
B. 1227.6 MHz and 1575.4 MHz.
C. 2245.4 and 2635.4 MHz.
D. 946.2 MHz and 1226.6 MHz.

GPS transmits in L-band frequencies, with 1575.42 MHz for civilians, in addition to 1227.6 MHz for the military. Additional frequencies are coming on the air soon! **ANSWER B**

3-98P3 How many GPS satellites are normally in operation?

A. 8
B. 18
C. 24
D. 36

GPS satellites total 24, plus a few spares, in 6 orbital planes, inclined 55 degrees with 4 satellites in each plane. **ANSWER C**

3-98P4 What best describes the GPS Satellites orbits?

A. They are in six orbital planes equally spaced and inclined about 55 degrees to the equator.
B. They are in four orbital planes spaced 90 degrees in a polar orbit.
C. They are in a geosynchronous orbit equally spaced around the equator.
D. They are in eight orbital planes at an altitude of approximately 1,000 miles.

The 24 GPS satellites are in 6 orbital planes, equally spaced, and inclined about 55 degrees to the Equator, 4 satellites per orbital plane. **ANSWER A**

3-98P5 How many satellites must be received to provide complete position and time?

A. 1
B. 2
C. 3
D. 4

It takes 4 satellites to provide a position fix and time, including altitude. While you could get by with 3 by discounting any position deep in space, your best answer is four. **ANSWER D**

3-98P6 What is DGPS?

A. Digital Ground Position System.
B. A system to provide additional correction factors to improve position accuracy.
C. Correction signals transmitted by satellite.
D. A system for providing altitude corrections for aircraft.

DGPS is Differential Global Positioning. Correction signals provide additional factors to improve position accuracy, taking into account atmospheric and ionospheric disturbances as measured by a reference station. The DGPS signals are sent out on our old radio beacon band, 285 to 325 kHz, and provide excellent signal strengths near harbor entrances. The Wide Area Augmentation System (WAAS) is yet another form of position error correction, and is used in aeronautical applications. **ANSWER B**

Subelement Q – SAFETY

2 Key Topics, 2 Exam Questions

Key Topic 99: Radiation Exposure

3-99Q1 Compliance with MPE, or Maximum Permissible Exposure to RF levels (as defined in FCC Part 1, OET Bulletin 65) for "controlled" environments, are averaged over _______ minutes, while "uncontrolled" RF environments are averaged over ______ minutes.

A. 6, 30.
B. 30, 6.
C. 1, 15.
D. 15, 1.

OET bulletin 65 describes maximum permissible exposure levels to radio frequency (RF) levels. The station operator is considered in the "controlled environment" of the emissions from his or her station, and time averaging for controlled exposure is 6 minutes. The general population is considered to be in the "uncontrolled environment," and time averaging for the general public is over a much longer 30 minute period of time, to give them additional protection. **ANSWER A**

3-99Q2 Sites having multiple transmitting antennas must include antennas with more than _______% of the maximum permissible power density exposure limit when evaluating RF site exposure.

A. Any
B. 5
C. 1
D. 12.5

If you add a transmitting antenna at a site that has multiple transmitters, and if you increase the maximum power density exposure by more than 5%, you will need to conduct a maximum permissible power density exposure evaluation. **ANSWER B**

3-99Q3 RF exposure from portable radio transceivers may be harmful to the eyes because:

A. Magnetic fields blur vision.
B. RF heating polarizes the eye lens.
C. The magnetic field may attract metal particles to the eye.
D. RF heating may cause cataracts.

Your eyes are the most sensitive part of your body to damage from RF heating inside the eyeball through concentrations of (generally) microwave radiation. Never stand in front of an operating RADAR antenna system! **ANSWER D**

3-99Q4 At what aggregate power level is an MPE (Maximum Permissible Exposure) study required?

A. 1000 Watts ERP.
B. 500 Watts ERP.
C. 100 Watts ERP.
D. Not required.

With any installation that has over 1000 watts effective radiated power, you are required to conduct a maximum permissible exposure study. **ANSWER A**

3-99Q5 Why must you never look directly into a fiber optic cable?

A. High power light waves can burn the skin surrounding the eye.
B. An active fiber signal may burn the retina and infra-red light cannot be seen.
C. The end is easy to break.
D. The signal is red and you can see it.

Never assume that a fiber optic cable is not energized. Just because you can't see any light coming out of it, it might be operated at infrared light levels which cannot be seen, and could damage your corneas or retinas. **ANSWER B**

3-99Q6 If the MPE (Maximum Permissible Exposure) power is present, how often must the personnel accessing the affected area be trained and certified?

A. Weekly.
B. Monthly.
C. Yearly.
D. Not at all.

Any personnel who have access to an area where high-power RF radiation is present must be trained and certified at least once a year. **ANSWER C**

Key Topic 100: Safety Steps

3-100Q1 What device can protect a transmitting station from a direct lightning hit?

A. Lightning protector.
B. Grounded cabinet.
C. Short lead in.
D. There is no device to protect a station from a direct hit from lightning.

Unfortunately, there is nothing to absolutely guarantee a tower, mast, or superstructure may not be hit by a direct lightning strike. Although there are static dissipation devices for the tops of towers, there is no device to protect a station and tower from the direct lightning strike. **ANSWER D**

A lightning static dissipater mounted atop a sailboat mast.

3-100Q2 What is the purpose of not putting sharp corners on the ground leads within a building?

A. No reason.
B. It is easier to install.
C. Lightning will jump off of the ground lead because it is not able to make sharp bends.
D. Ground leads should always be made to look good in an installation, including the use of sharp bends in the corners.

All of your grounding at a radio station should take gradual bends and not sharp corners. A gradual bend will keep lightning flowing to a safe ground exit point. Sharp corners may cause lightning to jump off the ground lead, zapping other sensitive electronics nearby, including YOU! **ANSWER C**

3-100Q3 Should you use a power drill without eye protection?

A. Yes.
B. No.
C. It's okay as long as you keep your face away from the drill bit.
D. Only in an extreme emergency.

Always wear eye protection when you are working around power tools. This includes the simple power drill that could launch tiny pieces of metal or wood in the direction of your eyes. Wear eye protection at all times! **ANSWER B**

3-100Q4 What class of fire is one that is caused by an electrical short circuit and what is the preferred substance used to extinguish that type of fire?

A. FE28.
B. FE29.
C. FE30.
D. FE31.

When extinguishing a fire caused by an electrical short circuit, you want to make sure that the fire suppression chemical is NON-conductive, FE-30 content. **ANSWER C**

3-100Q5 Do shorted-stub lightning protectors work at all frequencies?

A. Yes.
B. No, the short also kills the radio signals.
C. No, the short enhances the radio signal at the tuned band.
D. No, only at the tuned frequency band.

A shorted-stub lightning protector might work at a specific frequency, but it will not work to protect equipment at all frequencies. **ANSWER D**

3-100Q6 What is a GFI electrical socket used for?

A. To prevent electrical shock by sensing ground path current and shutting the circuit down.
B. As a gold plated socket.
C. To prevent children from sticking objects in the socket.
D. To increase the current capacity of the socket.

A ground fault interrupter (GFI)is a circuit to prevent electrical shock by sensing ground path current and shutting the circuit down by opening up the electrical contacts. **ANSWER A**

CHAPTER F

RADAR – RAdio Detection And Ranging

Today's ship, shore, and aeronautical RADAR systems remain unmatched by any other technology, given their elegant, simple capabilities. Send an electromagnetic wave out and measure the time delay for the return echo to determine the distance to a target. Synchronize an antenna unit with a 360 degree display and you have a bearing to the target. Tilt the antenna upwards and you can calculate the altitude of a target. Continuously transmit and receive the radio signal and you can determine the speed of a target. Among the many advantages of RADAR is its ability to operate long- or close-range; its accuracy, and its ability to penetrate darkness, fog and adverse weather.

No other aeronautical or shipboard radio system can do what RADAR does, with only one signal needed. No GPS required. No "IDENT" squawk needed. No digital packet required from the other station. One RADAR and its returned echo are all that is required.

Photo courtesy of Furuno Electric Co., Ltd.

HISTORY OF RADAR

Echo navigation has been around for centuries. Early ships would blow a whistle, or ring a bell, and listen for an echo. Sound travels at the rate of 1,100 feet per second, and if the echo took 6 seconds to return, the entire trip would be 6,600 feet, with a one-way distance to the headlands of 3,300 feet. In water, an acoustic SONAR "ping" travels at 4,900 feet per second. If it bounces off the side of a metal hull at a depth of 400 feet, the interval between the send and the return echo would be about 1/6 of a second. The electromagnetic waves used by RADAR to create an echo travel at the speed of light, 186,282 statute miles per second, or about 982 feet in one microsecond. The RADAR mile is measured in nautical miles, which are 15% longer than statute miles, and RADAR range is usually measured in microseconds.

No single inventor, scientist, or engineer can claim to be the "inventor" of RADAR. The concept of radio signals reflecting off of objects was known by many, and they all contributed to the development of RADAR.

The first to use radio waves to detect "the presence of distant metallic objects" was achieved in 1904 by German scientist Christian Hülsmeyer, who demonstrated the feasibility of detecting the presence of a ship in dense fog, but not its distance. In August 1917 Nikola Tesla first established principles regarding frequency and power level for the first primitive RADARs, noting that, "...we may determine the relative position or course of a moving object, such as a vessel at sea, the distance traversed by the same, or its speed."

In 1922, scientists working for the U.S. Naval Research Laboratory noted a difference in received signals caused by reflections from a wooden vessel on the Potomac River. They reasoned that these different echoes could be used to detect and track enemy ships. A few years later, physicists at Washington's Carnegie Institution measured the distance to the ionosphere by "pulse ranging." By beaming a series of short signals of slightly different frequencies to the ionosphere, they were able to determine its height by measuring the time that each pulse took to return to Earth.

The first practical use of RADAR was during the 1930's for military use in detecting aircraft and surface vessels. France, Germany and the United States all developed experimental early warning RADAR systems. RADAR was later installed in airplanes to detect enemy surface vessels.

RADAR came into its own during World War II when it became a vital weapon. Anti-aircraft guns were controlled by RADAR. Airborne RADAR played a large part in strategic bombing. In 1941, RADAR-directed gunfire sank many enemy ships. Using RADAR, naval battleships could maneuver close to enemy coasts during low visibility. And the British made extensive use of RADAR defending themselves from German air raids.

Today, RADAR has become an indispensible tool in many aspects of daily life including civil aviation, maritime shipping and recreational boating, national defense systems, and weather forecasting, to name only a few.

HOW DOES RADAR WORK?

Modern RADAR systems use an extremely wide range of frequencies, from the HF to infrared range. Microwave frequencies are the most commonly used, since they can travel through clouds and bad weather without absorption and noise. RADAR operates on two principles: first, objects illuminated by a radio frequency (RF) beam reflect radiation in all directions and, second, the speed of RF is constant.

While the reflected energy is scattered in all directions, a detectable portion is also reflected back to the place where it originated. To detect and locate objects, a RADAR transmits radio frequency (RF) energy and picks up the reflections with a radio receiver. The receiver is so sensitive that power levels of less than a millionth of a millionth of a watt can be detected and amplified.

Radio signals, like all electromagnetic waves, travel at the speed of light. That is 186,282 statute miles per second or 0.186 miles (982 feet) in a microsecond. By dividing the time it takes for the return echo to be received by two (the reflected signals must make a two-way trip) one can accurately gauge the distance to an object. Therefore, an echo returned in one microsecond indicates that a target is 491 feet away from the RADAR set. A "RADAR mile" refers to the round trip distance in nautical miles, which are 15 percent longer than statute miles. 1.15 statute miles equals 1 nautical mile; therefore, radio waves travel 0.162 nautical miles in a microsecond.

Like light, RF can be focused in a certain direction. An object is located by scanning a region and the beam position that yields the strongest echoes indicates the existence of a reflecting surface of the object.

TYPES OF RADAR

There are two general types of RADAR – pulsed and continuous wave (CW). A pulsed RADAR transmits its RF energy in short bursts or pulses. A CW RADAR operates constantly instead of in short bursts. Magnetrons are the vacuum tube oscillators that generate high-power microwave pulses. The energy can be focused into a narrow beam with the so-called RADAR "dish" – an antenna with a parabolic shape.

Pulse RADAR

Simple pulse RADARs transmit a beam of powerful on-off bursts of microwave radiation. The RADAR burst lasts only a few millionths of a second. During the interval between the pulses, the return echo is received and interpreted as shown in *Figure 6-1*. A high-speed electronic switch called a duplexer turns the transmitter on and off, and changes the antenna from "transmit" to "receive." This permits the same antenna to be used for transmitting and receiving. The pulse duration is determined by the minimum range of the remote object and must be short enough so as not to mask the return echo. Pulse RADARs are used primarily to determine distance and direction.

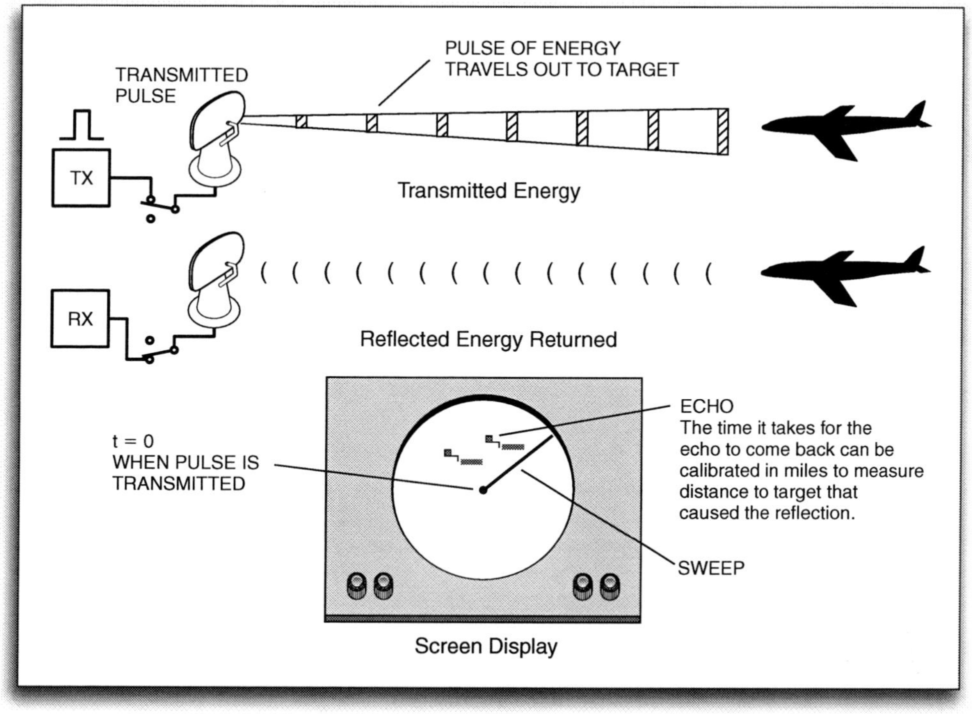

Figure 6-1

Distance (range) can be determined by measuring the time between the transmission of the pulse and the reception of the return echo. A pulse RADAR set automatically converts the round trip time into distance to the object. Direction is determined by synchronizing a scan on a display screen to a sweeping antenna, which focuses a narrow beam of energy out ahead of it. An object in the path of the highly directional beam reflects an echo that appears in position on the display. Pulse RADAR can track a moving object by continuously measuring distance and direction at regular intervals as the antenna sweeps an area.

RADAR maps can be produced by scanning many pulses over an area and plotting the echoes from each direction on a RADAR screen.

Continuous-Wave RADAR

Pulse RADAR provides no information about target velocity. Continuous wave (CW) RADAR transmits and receives at the same time. There are no pulses. The problem of interference between the transmitter and receiver is overcome by isolating them as much as possible and by designing the receiving system so that it can distinguish differences between the transmitted and reflected signals. Systems that can do this are of two types: Doppler RADAR, and frequency-modulated (FM) RADAR.

Doppler RADAR

If the object is moving, the transmitted and reflected signals will have different frequencies. This is known as a Doppler shift. When the outgoing wave strikes an object that is approaching the RADAR antenna, the wave is reflected at a slightly higher frequency. The reflection is bounced back at a lower frequency when the target moves away from the antenna. The greater the frequency difference between the transmitted and reflected waves, the faster the object is moving. Doppler RADAR does not determine distance, but rate of change of distance. Doppler is used by police to detect speeding motorists, by the military to direct weapons fire at moving targets, and by weather stations to track thunderstorms and tornadoes.

Frequency-Modulated RADAR

If the target is stationary, differences may be created by continuously varying the transmitter's frequency, as with frequency modulation. FM RADAR, unlike Doppler RADAR, can determine distance. The return echo will have a slightly different frequency than the outbound burst. The difference between the transmitter frequency and the reflection is converted into distance. The farther away the object is, the greater the frequency difference.

Because of the difficulty in isolating the transmitter and receiver, and the ease of measuring distance with pulses, CW RADAR is not as widely used as pulsed RADAR.

RADAR Displays

RADAR sets come in all sizes, from small, four-pound, hand-held police version to those with giant parabolic dish antennas weighing tons and a thousand feet in diameter! But they all have somewhat similar parts – a transmitter, a receiver, an antenna, and a display. If it is a pulsed RADAR, a TR electronic switch is required.

The received RADAR information can be displayed in a number of ways. Some sets have analog meters and digital readouts. Others have a circular cathode-ray tube (CRT) screen, commonly called a PPI — for Plan Position Indicator.

The most common CRT PPI display is a circular map-like picture of the area scanned by the RADAR beam. The observing station is at the center of the display. Each pulse appears as an electron beam from the center of the tube to the outer edge. RADAR echoes appear as target echoes on the screen. Concentric rings indicate distance. Land based RADARs are calibrated in statute miles; ship RADAR sets in nautical miles. Speed is determined by noting how fast the target echoes move across the screen.

Aircraft RADAR Display

Today, PPI phosphorous cathode ray tubes are being replaced by digital technology with computer-driven raster scan displays. As a result, all types of digital display techniques can be used.

AVIATION AND MARITIME RADAR

The largest non-military users of RADAR are police departments for speed-detection, weather-forecasting (rain reflects RADAR signals), civil aviation, and ship navigation. Air controllers are able to determine the position of planes in the air at least 50 miles from an airport. Planes map the areas over which they fly. RADAR altimeters show how high a plane is flying. Weather RADAR detects storms in the area.

The modern ship RADAR uses a color Raster scan display, and may be a blend of RADAR imaging, electronic charting, overlaid with Doppler RADAR-derived weather, and overlaid with canned aerial photography, and then overlaid again with collision avoidance. Sensory overload at its best – many a mariner has gone aground by overly depending on GPS to drive an electronic chart program to faithfully show where they are, down a narrow ship channel. RADAR, by itself, lets you determine accurately everything on the water around you. NO charts – no aerial photography – just the straight RADAR facts.

Aviation, ground, and marine RADARs need no other radio signals to do their job. All those add-ons to the RADAR display are additional information beyond the raw RADAR display that is most important on the screen.

Even though aviation has thousands of aircraft squawking their identity and sending collision avoidance data, the straight search RADAR gives the pilots a true heads up on everything around them. This could include a flock of geese, easily spotted by airborne RADAR, dead ahead. Or, down at the harbor, RADAR can see a cluster of kayakers dangerously close to a restricted vessel traffic zone.

RADAR is used by vessels of all sizes; from small pleasure boats to huge cargo ships and ocean liners. RADAR provides visibility in bad weather, spots other vessels, and highlights possible obstacles to navigation. A ship's position can be determined by RADAR echoes from reflector buoys.

Ship RADAR is one of the most important pieces of navigation equipment in the wheelhouse. RADAR, combined with depth sounding and a sharp lookout, is an absolute requirement when entering congested shipping lanes. Although the Global Positioning System (GPS) receiver and associated electronic chart display is important to stay on track in the vessel traffic system (VTS), only RADAR will show targets all around the vessel not necessarily printed on paper charts or stored in electronic chart displays.

Harbor masters keep track of ships using RADAR, and the Coast Guard uses RADAR for search and rescue. RADAR beacons (RACONS) operate somewhat like a RADAR set in reverse. A RADAR beacon is triggered by pulses from a RADAR set and sends back a distinctive reply that provides precise location and bearing information.

Coast RADAR Station in France

CAUTION

A word of caution! RADAR waves are DANGEROUS! Never expose yourself to the output of the antenna which radiates RADAR microwaves. Be extremely cautious of thousands of volts inside the RADAR transceiver system. Never allow a turned-on RADAR to sweep you with radiated transmit energy. Never stand close or look into a transmitting RADAR system. These are microwaves, and they are DANGEROUS.

RADAR ENDORSEMENT FOR THE COMMERCIAL LICENSES

The importance of shipboard RADAR is so great that the Federal Communications Commission requires passing the commercial radio operator Element 8 Ship RADAR Endorsement examination as part of the commercial radio licensing process.

The Ship RADAR Endorsement is not a stand-alone license. It must be placed on the General Radiotelephone Operator License (GROL), the Global Maritime Distress and Safety System Radio Maintainer's License (GMDSS/DM), or the Radiotelegraph Operator's License (T). Only those technicians whose commercial radio operator license bears this RADAR endorsement may open up and repair, maintain or internally adjust ship RADAR equipment. Refer to Chapter B for the eligibility requirements to obtain a Ship RADAR Endorsement. A Ship RADAR Endorsement is not needed to routinely operate an aircraft or ship RADAR set, but the operation must be limited to manipulation of external controls.

SECONDARY RADAR (IFF)

Although not directly related to maritime radar, the Secondary Surveillance RADAR (SSR), also known as the IFF or Identification Friend or Foe technology, is widely used in our airspaces and should be well understood by any RADAR technician. An IFF RADAR relies on an on-board transponder aboard an aircraft which will transmit a coded message "squawk" back to a ground-based interrogating RADAR. While a standard Primary Surveillance RADAR (PSR) can derive a lot of information from passive targets, an IFF system will give a positive identification of an aircraft within the RADAR's range. Military and civilian aviation RADAR systems use both SSR reflections and PSR "squawks" and correlates the two signals for extreme reliability.

SSR RADARs use far less power than their associated PSRs because they effectively regenerate the interrogating signal by means of the onboard transponder. For a passively reflected RADAR signal, the free space attenuation goes up as the fourth power of the distance, which is why long distance primary surveillance RADARs use extremely high effective radiated power. The free space attenuation of an SSR is only the square of the distance between the RADAR and the target.

A PSR/SSR combination RADAR is largely immune to many counter-measures, such as jamming, and can also compensate for unfavorable weather conditions.

CHAPTER G

Marine RADAR Endorsement

Element 8 Question Pool

The FCC Commercial Element 8 written examination covers ship RADAR techniques – the specialized theory and practice applicable to the proper installation, servicing, and maintenance of aircraft and ship radar equipment in general use for aviation and marine navigation purposes. The Element 8 examination is used to prove that the examinee possesses the qualifications required to adjust, repair and maintain marine and aircraft RADARs that can only be serviced by a person holding a RADAR Endorsement.

The RADAR Endorsement is not a stand-alone license. It may be placed only on General Radiotelephone Operator Licenses (PG), GMDSS Radio Maintainer's License (DM or DB), or Radiotelegraph Operator license (T). Only persons whose commercial radio operator license that bears this endorsement may repair, maintain or internally adjust ship and aircraft RADAR equipment. The Ship RADAR Endorsement is valid for the lifetime of the holder.

The Element 8 question pool contains a total of 300 questions covering 6 Subelement topic areas. There are 50 Key Topics in the question pool. Each contains 6 questions. One question is taken at random from each Key Topic to create an examination with 50 questions. To pass, you must correctly answer at least 38 of the 50 questions on the written exam. *Table G-1* summarizes the Subelement topics and number of questions in each topic area.

Table G-1. Summary of Element 8 Question Pool

Subelement Topic	Key Topics	No. of Questions	Examination Questions
RADAR Principles	10	60	10
Transmitting Systems	8	48	8
Receiving Systems	10	60	10
Display & Control Systems	10	60	10
Antenna Systems	5	30	5
Installation, Maintenance & Repair	7	42	7
Total	**50**	**300**	**50**

Each Element 8 examination is administered by a Commercial Operator License Examination Manger (COLEM). The COLEM must construct the examination by selecting 50 questions from the Element 8 question pool contained in this book, as explained above. Each question in the pool contains the question, and 4 multiple-choice answers – one correct answer and three wrong answers (distracters). The COLEM may change the order of the four answer choices, but cannot change any of the question or answer content.

In this book, official Element 8 questions and 4 answer choices are followed by an explanation of why the correct answer is right, and the identification of the correct answer.

Subelement A – RADAR Principles

10 Key Topics, 10 Exam Questions

Key Topic 1 – Marine RADAR Systems

8-1A1 Choose the most correct statement containing the parameters which control the size of the target echo.

A. Transmitted power, antenna effective area, transmit and receive losses, RADAR cross section of the target, range to target.
B. Height of antenna, power radiated, size of target, receiver gain, pulse width.
C. Power radiated, antenna gain, size of target, shape of target, pulse width, receiver gain.
D. Magnetron gain, antenna gain, size of target, range to target, wave-guide loss.

The size and signal strength of a RADAR target echo may be determined by the target cross sectional area, range to the target, transmit and receive path losses, RADAR antenna effective area, and your transmitter power output. Some of the other answers look pretty good, but go with answer A, and look at the drawings to see why distant small targets appear and disappear on the display! Antenna effective area are the key words in the correct answer! **ANSWER A**

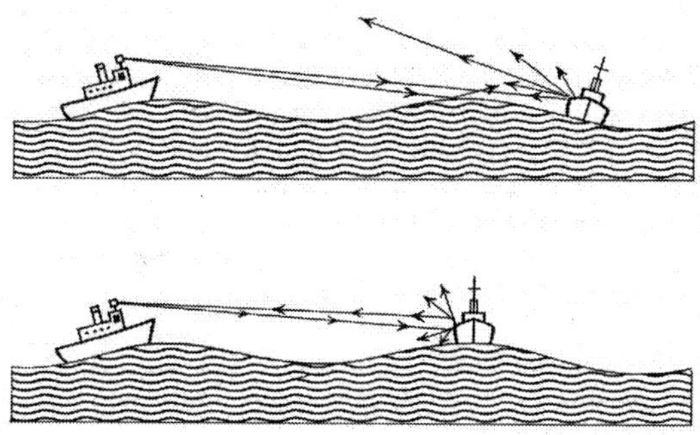

Heavy seas can cause targets to fade in and out on the RADAR display.

8-1A2 Which of the following has NO effect on the maximum range capability?

A. Carrier frequency.
B. Recovery time.
C. Pulse repetition frequency.
D. Receiver sensitivity.

Maximum range of a RADAR system IS influenced by peak power output, receiver sensitivity, pulse repetition frequency, and the RADAR band frequency of operation. But the question asks what has NO effect on maximum range, and that would be RECOVERY TIME. Recovery time is associated with MINIMUM range calculations as a period of time where the receiver is unable to receive the echo, a function of the TR tube to respond to within 6 dB of normal sensitivity at the conclusion of the transmitter pulse. Recovery time would not factor in maximum range capability. **ANSWER B**

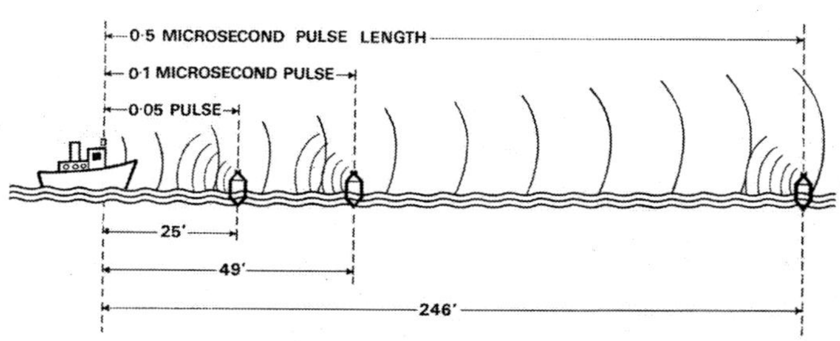

Longer range, longer pulse lengths.

8-1A3 What type of transmitter power is measured over a period of time?

A. Average.
B. Peak.
C. Reciprocal.
D. Return.

Pulse RADARs transmit energy in variable pulse lengths and variable pulse repetition time. A RADAR pulse has a peak power much greater than the average transmitted power, since the transmitter is OFF most of the time. The AVERAGE power is the peak power times the DUTY CYCLE of the pulse. But the RADAR pulse width is so short that conventional power measuring devices may not react fast enough. Because of this, RADAR power output also factors in pulse-repetition TIME resulting in AVERAGE power. Average power is the total energy content of the complete pulse cycle, as illustrated in this figure. Average power = peak power × pulse width ÷ pulse repetition time. **ANSWER A**

8-1A4 What RADAR component controls timing throughout the system?

A. Power supply.
B. Indicator.
C. Synchronizer.
D. Receiver.

The synchronizer is a timing circuit which feeds multiple RADAR circuits, such as transmitter, receiver, duplexer, indicator, and any other RADAR circuit needing timing pulses. If the synchronizer circuit should fail, the entire RADAR will cease to operate. **ANSWER C**

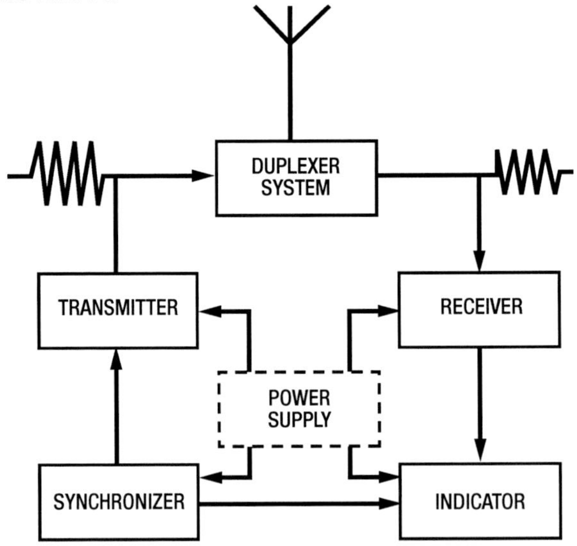

Functional block diagram of a basic RADAR system.

RADAR COMPONENTS

Pulse radar systems can be functionally divided into the six essential components shown here. These components are briefly described in the following paragraphs and will be explained in detail after that:

- The SYNCHRONIZER (also referred to as the TIMER or KEYER) supplies the synchronizing signals that time the transmitted pulses, the indicator, and other associated circuits.
- The TRANSMITTER generates electromagnetic energy in the form of short, powerful pulses.
- The DUPLEXER allows the same antenna to be used for transmitting and receiving.
- The ANTENNA SYSTEM routes the electromagnetic energy from the transmitter, radiates it in a highly directional beam, receives any returning echoes, and routes those echoes to the receiver.
- The RECEIVER amplifies the weak, electromagnetic pulses returned from the reflecting object and reproduces them as video pulses that are sent to the indicator.
- The INDICATOR produces a visual indication of the echo pulses in a manner that, at a minimum, furnishes range and bearing information.

8-1A5 Which of the following components allows the use of a single antenna for both transmitting and receiving?

A. Mixer.
B. Duplexer.
C. Synchronizer.
D. Modulator.

The RADAR duplexer allows one antenna to toggle between transmit and receive. The duplexer also provides isolation to any transmitter pulses working back into the RADAR receiver, with fast receiver time recovery for close-in target detection. Further, the duplexer must not load the receiver in any fashion, as the receiver must receive extremely weak target echoes. **ANSWER B**

8-1A6 The sweep frequency of a RADAR indicator is determined by what parameter?

A. Carrier frequency.
B. Pulse width.
C. Duty cycle.
D. Pulse repetition frequency.

RADAR sweep frequency always applies to a particular type of RADAR called a CHIRP or FMCW RADAR. In a CHIRP RADAR, frequency sweep is always synchronous with the Pulse Repetition Frequency. The sweep frequency of the RADAR indicator may be automatically pre-selected by the RADAR operator adjusting the range scale selector. The pulse repetition frequency, sometimes called pulse repetition rate, may be pre-determined depending on RADAR range settings. "Sweep" should not be confused with "scan rate" as you would have on a PPI display. This is synchronized with the rotation speed of the RADAR beam. **ANSWER D**

Key Topic 2 – Distance and Time

8-2A1 A radio wave will travel a distance of three nautical miles in:

A. 6.17 microseconds.
B. 37.0 microseconds.
C. 22.76 microseconds.
D. 18.51 microseconds.

RADAR and radio waves travel at approximately the speed of light; it takes 6.17 μs for a RADAR wave to travel one nautical mile from the transmitter. To travel three nautical miles, it would take 18.51 (6.17 × 3) microseconds to travel this distance, and your correct answer is 18.51 microseconds. The way you can remember the number 6 in microseconds per nautical mile is never to get "seasix." **ANSWER D**

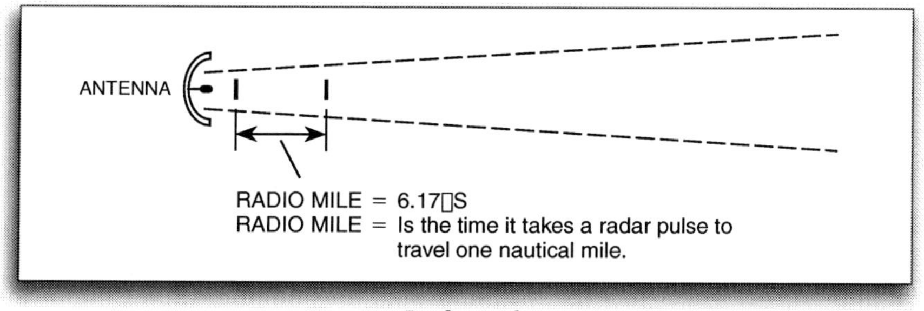

Radio Mile

8-2A2 One RADAR mile is how many microseconds?

A. 6.2.
B. 528.0.
C. 12.34.
D. 0.186.

It takes a RADAR pulse 6.17 microseconds to travel out one nautical mile. One "RADAR mile" considers the echo return, so 2 X 6.17 = 12.34 microseconds. Be careful of A, it is only the time out, not out and back. **ANSWER C**

RADAR Mile = 2 × 6.17 μs Where: RADAR Mile = Time out and back to target at one nautical mile in μs.

8-2A3 RADAR range is measured by the constant:

A. 150 meters per microsecond.
B. 150 yards per microsecond.
C. 300 yards per microsecond.
D. 18.6 miles per microsecond.

150 meters is the round-trip distance that a microwave travels in one microsecond. **ANSWER A**

8-2A4 If a target is 5 miles away, how long does it take for the RADAR echo to be received back at the antenna?

A. 51.4 microseconds.
B. 123 microseconds.
C. 30.75 microseconds.
D. 61.7 microseconds.

If the target is 5 miles away, the RADAR wave must travel a total of 10 miles – 5 miles to the target, and 5 miles return as an echo. Ten nautical miles times 6.17 microseconds per mile gives us about 61.7 microseconds. Watch out for C; this is not a one-way trip! **ANSWER D**

8-2A5 How long would it take for a RADAR pulse to travel to a target 10 nautical miles away and return to the RADAR receiver?

A. 12.34 microseconds.
B. 1.234 microseconds.
C. 123.4 microseconds.
D. 10 microseconds.

Be sure to read these types of questions carefully to see whether or not they want the range to the target, or the range to the target and back again as a time delay. It takes 6.17 microseconds for the RADAR wave to travel 1 nautical mile, or 10 nautical miles to the target in 61.7 microseconds, and the return echo would take another 61.7 microseconds, or a total time delay of 123.4 microseconds. Remember, up and back, a RADAR mile, means times two of the target distance. **ANSWER C**

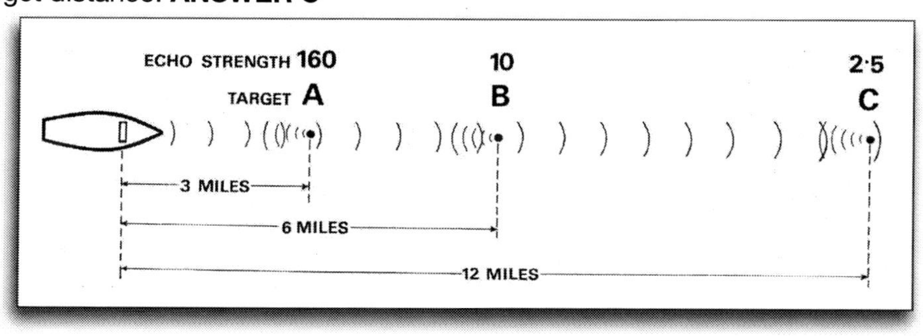

Echo Strength

8-2A6 What is the distance in nautical miles to a target if it takes 308.5 microseconds for the RADAR pulse to travel from the RADAR antenna to the target and back.

A. 12.5 nautical miles.
B. 25 nautical miles.
C. 50 nautical miles.
D. 2.5 nautical miles.

Divide 308.5 microseconds by 6.17 to find the total distance the signal travels (to the target and back to the RADAR receiver). Then, divide the answer by 2 to find the one-way distance to the target. 308.5 ÷ 6.17 = 50 ÷ 2 = 25. **ANSWER B**

Key Topic 3 – Frequency and Wavelength

8-3A1 Frequencies generally used for marine RADAR are in the ___ part of the radio spectrum.

A. UHF.
B. EHF.
C. SHF.
D. VHF.

VHF = 30 - 300 MHz
UHF = 300 - 3,000 MHz
SHF = 3,000 - 30,000 MHz
EHF = 30,000 - 300,000 MHz

The three marine RADAR bands are 2,900 - 3,100 MHz, 9,300 - 9,500 MHz, and 14,000 - 14,050 MHz. All three marine RADAR bands fall into the super-high frequency range of 3,000 - 30,000 MHz. The most popular, X-band (9,300 - 9,500 MHz), is sometimes noted in gigahertz (9.3 - 9.5 GHz). **ANSWER C**

8-3A2 Practical RADAR operation requires the use of microwave frequencies so that:

A. Stronger target echoes will be produced.
B. Ground clutter interference will be minimized.
C. Interference to other communication systems will be eliminated.
D. Non-directional antennas can be used for both transmitting and receiving.

RADARs operate line-of-sight. RADAR echoes depend on the cross-sectional area of the target, relative to the wavelength. Microwave frequencies faithfully reflect off of metal and land, and are partially reflected off fiberglass. Lower frequencies penetrate these materials and do not produce the return echoes necessary for RADAR operation. **ANSWER A**

8-3A3 An S-band RADAR operates in which frequency band?

A. 1 - 2 GHz.
B. 4 - 8 GHz.
C. 8 - 12 GHz.
D. 2 - 4 GHz.

An S-band RADAR operates on marine assigned frequencies 2,900 - 3,100 MHz. The S-band range is sometimes abbreviated 2 - 4 GHz. **ANSWER D**

8-3A4 A RADAR operating at a frequency of 3 GHz has a wavelength of approximately:

A. 1 centimeter.
B. 10 centimeters.
C. 3 centimeters.
D. 30 centimeters.

Wavelength in cm = 30,000/frequency in MHz (30,000 ÷ 3,000 = 10.0 cm). Remember, for this formula the frequency is in MHz. **ANSWER B**

8-3A5 The major advantage of an S-band RADAR over an X-band RADAR is:

A. It is less affected by weather conditions.
B. It has greater bearing resolution.
C. It is mechanically less complex.
D. It has greater power output.

Lower-frequency RADARs on the S-band exhibit poor reflectivity from nearby thunderstorms, and thus have the ability to “see” better through light snow and rain. **ANSWER A**

8-3A6 An X band RADAR operates in which frequency band?

A. 1 - 2 GHz.
B. 2 - 4 GHz.
C. 4 - 8 GHz.
D. 8 - 12 GHz.

An X-band RADAR operates from 9,300 to 9,500 MHz. **ANSWER D**

Key Topic 4 – Power, Pulse Width, PRR

8-4A1 A pulse RADAR has a pulse repetition frequency (PRF) of 400 Hz, a pulse width of 1 microsecond, and a peak power of 100 kilowatts. The average power of the RADAR transmitter is:

A. 25 watts.
B. 40 watts.
C. 250 watts.
D. 400 watts.

Duty cycle = pulse width ÷ pulse repetition interval. Since the pulse repetition interval equals 1/PRF, then the duty cycle is equal to the pulse width times PRF duty cycle = PW X PRF, Then, $1 \times 10^{-6} \times 400 = 4 \times 10^{-4} =$ 0.0004. Average power is equal to peak power times duty cycle, or 0.0004 × 100,000 = 40 watts. **ANSWER B**

8-4A2 A shipboard RADAR transmitter has a pulse repetition frequency (PRF) of 1,000 Hz, a pulse width of 0.5 microseconds, peak power of 150 KW, and a minimum range of 75 meters. Its duty cycle is:

A. 0.5.
B. 0.05.
C. 0.005.
D. 0.0005.

Duty cycle = pulse width ÷ pulse repetition interval. Because the pulse repetition interval equals 1/PRF. Then the duty cycle is equal to the pulse width times PRF. (0.5×10^{-6}) X (1000) = 0.5×10^{-3} = 0.0005. 150 KW and 75 meters are not needed for the calculation. **ANSWER D**

8-4A3 A pulse RADAR transmits a 0.5 microsecond RF pulse with a peak power of 100 kilowatts every 1600 microseconds. This RADAR has:

A. An average power of 31.25 watts.
B. A PRF of 3200.
C. A maximum range of 480 kilometers.
D. A duty cycle of 3.125 percent.

Duty cycle is equal to ($0.5 \times 10^{-6} \div 1600 \times 10^{-6} = 0.5 \times 10^{-6} \div 1.6 \times 10^{-3} = 0.0003125$). Average power equals 100 KW × 0.0003125 = 31.25 watts. **ANSWER A**

8-4A4 If a RADAR transmitter has a pulse repetition frequency (PRF) of 900 Hz, a pulse width of 0.5 microseconds and a peak power of 15 kilowatts, what is its average power output?

A. 15 kilowatts.
B. 13.5 watts.
C. 6.75 watts.
D. 166.67 watts.

Average power output = peak power X pulse repetition frequency X pulse width.

$15{,}000 \times 900 \times (0.5 \times 10^{-6}) = 1.5 \times 10^{4} \times 9 \times 10^{2} \times 0.5 \times 10^{-6} = 6.75$ watts. Everything gets multiplied together, but remember 0.5 microseconds is 0.5×10^{-6}. **ANSWER C**

8-4A5 What is the average power if the RADAR set has a PRF of 1000 Hz, a pulse width of 1 microsecond, and a peak power rating of 100 kilowatts?

A. 10 watts.
B. 100 watts.
C. 1000 watts.
D. None of these.

Multiply 100,000 watts × 1000 Hz X (1.0×10^{-6} seconds) = 100 watts. See the previous question for more details. **ANSWER B**

8-4A6 A search RADAR has a pulse width of 1.0 microsecond, a pulse repetition frequency (PRF) of 900 Hz, and an average power of 18 watts. The unit's peak power is:

A. 200 kilowatts.
B. 180 kilowatts.
C. 20 kilowatts.
D. 2 kilowatts.

Peak power = average power ÷ duty cycle. The duty cycle is $900 \times (1 \times 10^{-6}) = 0.0009$ seconds.
18 watts ÷ 0.0009 = 20,000 watts, or 20 KW. **ANSWER C**

Key Topic 5 – Range, Pulse Width, PRF

8-5A1 For a range of 5 nautical miles, the RADAR pulse repetition frequency should be:

A. 16.2 Hz or more.
B. 16.2 MHz or less.
C. 1.62 kHz or more.
D. 16.2 kHz or less.

5 × 2 × 6.17 = 61.7 microseconds. 1 ÷ 0.0000618 seconds = 16181 Hz = 16.2 kHz, or less. **ANSWER D**

8-5A2 For a range of 100 nautical miles, the RADAR pulse repetition frequency should be:

A. 8.1 kHz or less.
B. 810 Hz or less.
C. 8.1 kHz or more.
D. 81 kHz or more.

100 × 2 X 6.17 = 1234 microseconds. 1 ÷ 0.001234 seconds = 810 Hz, or less. **ANSWER B**

8-5A3 The minimum range of a RADAR is determined by:
A. The frequency of the RADAR transmitter.
B. The pulse repetition rate.
C. The transmitted pulse width.
D. The pulse repetition frequency.

The minimum range at which a RADAR can detect a target is determined by the pulse duration, sometimes called pulse width or pulse length. As the RADAR operator switches down to a close-in scale, the RADAR automatically shortens its pulse width to as little as 0.1 microsecond. A 0.1 microsecond pulse covers approximately 160 yards of range on the display. Small-boat marine RADARs use even shorter pulses for target detection as close as 75 feet. **ANSWER C**

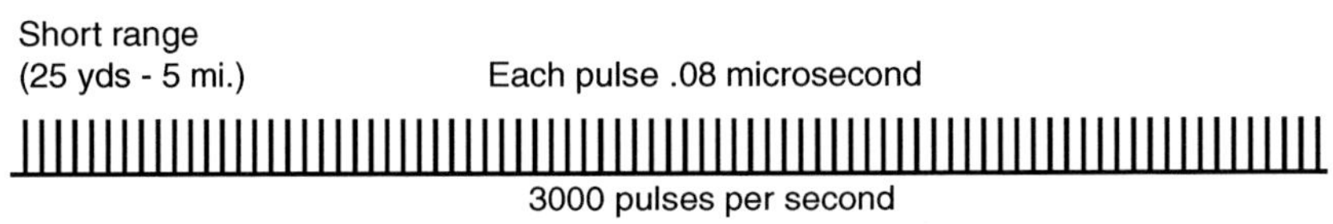

On minimum range, 3,000 pulses per second are sent.

8-5A4 Short range RADARs would most likely transmit:
A. Narrow pulses at a fast rate.
B. Narrow pulses at a slow rate.
C. Wide pulses at a fast rate.
D. Wide pulses at a slow rate.

To detect an echo when the target is close by, you need a narrow pulse with a fast pulse repetition rate. A wide, slow pulse would actually cover up the return echo. **ANSWER A**

8-5A5 For a range of 30 nautical miles, the RADAR pulse repetition frequency should be:
A. 0.27 kHz or less.
B. 2.7 kHz or less.
C. 27 kHz or more.
D. 2.7 Hz or more.

First calculate the amount of time for a RADAR wave to travel 30 nautical miles out, and 30 nautical miles back: $30 \times 2 \times 6.17 = 370.2$ microseconds. The maximum pulse repetition frequency required to allow that signal to return is 1 divided by the pulse time in seconds, or $1 \div 0.0003702$ seconds $= 2701$ Hz, which converted to kHz is 2.7 kHz, or less. Watch out for answers that say "or more." They would not work because the pulse repetition rate would not have enough time for the pulse to return to the receiver before another pulse was sent. As a target gets closer, you want to be able to detect progressively smaller amounts of movement, which span a greater subtended angle for a given speed. High PRFs are also useable where ambiguity might be a problem if you use a RANGE GATE to eliminate "impossible solutions." Why would you want to do this? High PRFs give you more effective power if you use averaging techniques; sometimes useful for very weak targets. RANGE GATES have to be used with caution however, so as to not inadvertently blank the real thing. **ANSWER B**

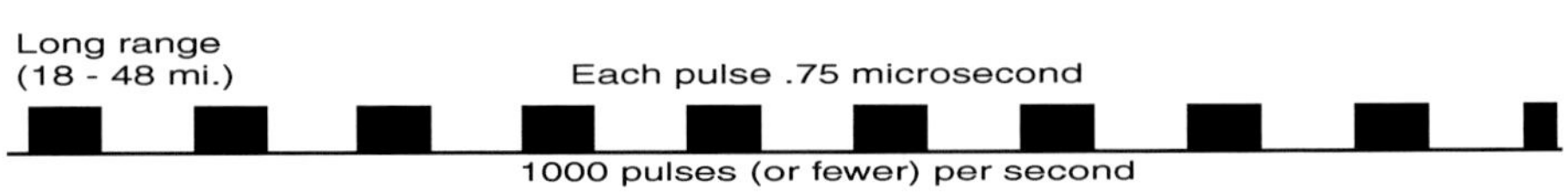

On long range, 1000 pulses per second are sent.

8-5A6 For a range of 10 nautical miles, the RADAR pulse repetition frequency (PRF) should be:
A. Approximately 8.1 kHz or less.
B. 900 Hz.
C. 18.1 kHz or more.
D. 120.3 microseconds.

Let's first calculate the amount of time for a RADAR wave to travel 10 miles out and 10 miles back: $10 \times 2 \times 6.17 = 123.4$ microseconds. The maximum pulse repetition frequency required to allow that signal to return is 1 divided by the time it takes the pulse to travel (in seconds), or $1 \div 0.0001234$ seconds $= 8.104$ kHz. 18.1 kHz "or more" isn't right because a higher pulse repetition rate would not have enough time for the pulse to return to the receiver before another pulse was sent. **ANSWER A**

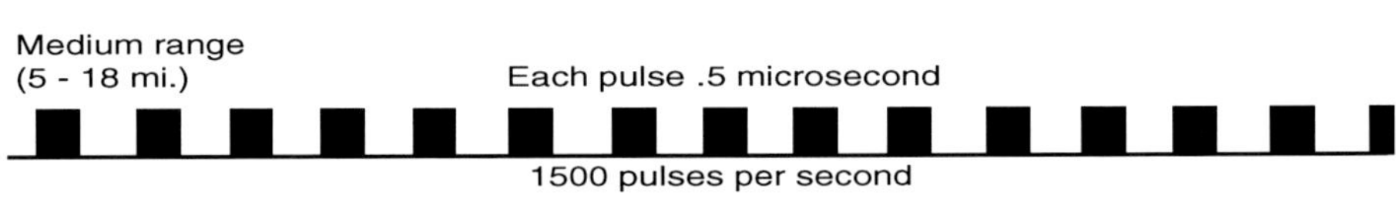

On medium range, 1500 pulses per second are sent.

Key Topic 6: Pulse Width - Pulse Repetition Rates

8-6A1 If the PRF is 2500 Hz, what is the PRI?

A. 40 microseconds.
B. 400 microseconds.
C. 250 microseconds.
D. 800 microseconds.

Pulse repetition interval (PRI) = 1 ÷ pulse repetition rate = 1 ÷ 2,500 = 400 μ seconds. PRF, pulse repetition frequency, and PRR, pulse repetition rate, are the same thing — the number of pulses per second. Pulse repetition interval (PRI), sometimes called pulse repetition time (PRT), is 1 ÷ PRF, or the period of the pulse repetition frequency. PRI is the time from the leading edge of a pulse to the leading edge of the next pulse. PRI = 1 ÷ PRF = 1 ÷ 2500 = 0.4×10^{-3} = 0.0004 = 400×10^{-6} The period of the pulses is 400 microseconds. **ANSWER B**

8-6A2 If the pulse repetition frequency (PRF) is 2000 Hz, what is the pulse repetition interval (PRI)?

A. 0.05 seconds.
B. 0.005 seconds.
C. 0.0005 seconds.
D. 0.00005 seconds.

To calculate pulse repetition interval (PRI), divide 1 second by the pulse repetition frequency: 1 ÷ 2000 cycles per second = 0.0005 seconds. **ANSWER C**

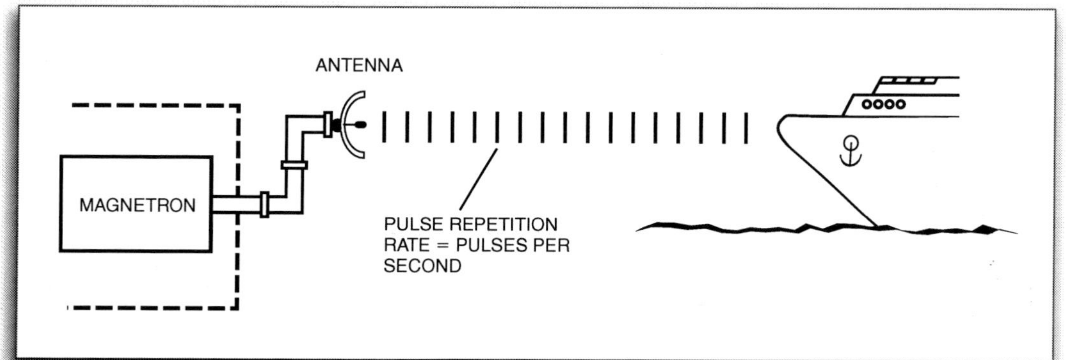

Pulse Repetition Rate

8-6A3 The pulse repetition rate (PRR) refers to:

A. The reciprocal of the duty cycle.
B. The pulse rate of the local oscillator tube.
C. The pulse rate of the klystron.
D. The pulse rate of the magnetron.

The pulse repetition rate (PRR) is the number of pulses per second sent to the target. It is also the number of pulses per second sent by the magnetron. **ANSWER D**

8-6A4 If the RADAR unit has a pulse repetition frequency (PRF) of 2000 Hz and a pulse width of 0.05 microseconds, what is the duty cycle?

A. 0.0001.
B. 0.0005.
C. 0.05.
D. 0.001.

Here the RADAR has a pulse repetition rate of 2,000 Hz, and a pulse width of 0.05 microseconds. The duty cycle is calculated by dividing the pulse width (0.05×10^{-6} seconds) by the pulse repetition interval (1 ÷ 2000): $0.05 \times 10^{-6} \div (1 \div 2000) = 0.0001$. **ANSWER A**

8-6A5 Small targets are best detected by:

A. Short pulses transmitted at a fast rate.
B. Using J band frequencies.
C. Using a long pulse width with high output power.
D. All of these answers are correct.

Small targets many miles away from the operating RADAR are best detected with a lower pulse repetition rate, which should automatically generate a longer pulse width. Higher power RADARs will many times display small target echoes that otherwise might be missed with lower power output RADARs. **ANSWER C**

8-6A6 What is the relationship between pulse repetition rate and pulse width?
A. Higher PRR with wider pulse width.
B. The pulse repetition rate does not change with the pulse width.
C. The pulse width does not change with the pulse repetition rate.
D. Lower PRR with wider pulse width.

As the operator switches to a longer range scale, the modern RADAR will automatically select a lower pulse repetition rate along with a wider pulse width. This will allow the detection of distant small targets. To give you an idea of range and pulse width variables, ponder this: 2,400 pulses, each pulse 0.12 microseconds in length, will yield 288 microseconds of energy transmitted during each second. If we had 800 pulses, each 0.4 microseconds in length, this would amount to 320 microseconds of energy in 1 second. But each second of time has 1,000,000 microseconds. Too long a pulse on short range would not allow enough time for the receiver to respond to the echo. Too short a pulse at long range may not illuminate the target as a received echo. When the operator selects the desired range, the modern RADAR uses automatic circuits that yield the best combination of all transmitting and receiving RADAR parameters. **ANSWER D**

Key Topic 7 – Components-1

8-7A1 What component of a RADAR receiver is represented by block 46 in Fig. 8A1?
A. The ATR box.
B. The TR box.
C. The RF Attenuator.
D. The Crystal Detector.

Block 46 in Diagram 8A1 is the TR box. **ANSWER B**

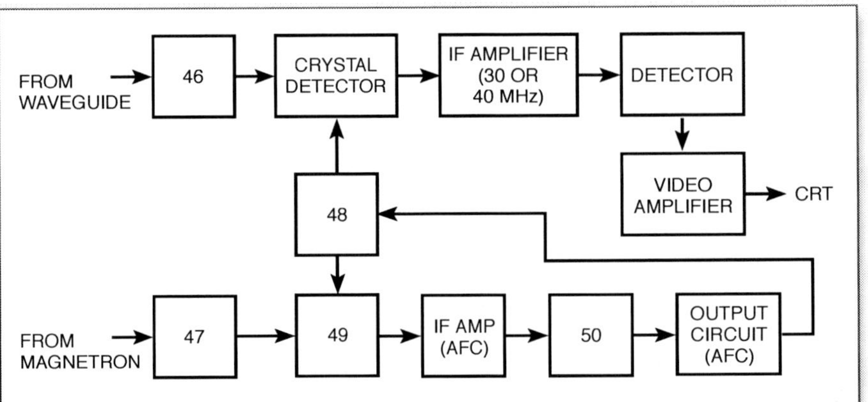

Figure 8A1

8-7A2 A basic sample-and-hold circuit contains:
A. An analog switch and an amplifier.
B. An analog switch, a capacitor, and an amplifier.
C. An analog multiplexer and a capacitor.
D. An analog switch, a capacitor, amplifiers and input and output buffers.

A capacitor samples the input voltage until the amplifier is placed in the hold mode by an analog switch. The input and outputs are buffered. **ANSWER D**

8-7A3 When comparing a TTL and a CMOS NAND gate:
A. Both have active pull-up characteristics.
B. Both have three output states.
C. Both have comparable input power sourcing.
D. Both employ Schmitt diodes for increased speed capabilities.

Both the TTL and the CMOS NAND gate have an active pull-up characteristic. **ANSWER A**

8-7A4 Silicon crystals:
A. Are very sensitive to static electric charges.
B. Should be wrapped in lead foil for storage.
C. Tolerate very low currents.
D. All of these.

Working with silicon crystals requires you to take great precautions not to accidentally zap them with a static discharge. They can tolerate only an extremely low current, and for protection when shipping, they are wrapped in lead foil to protect against a static discharge. **ANSWER D**

8-7A5 Which is typical current for a silicon crystal used in a RADAR mixer or detector circuit?
A. 3 mA.
B. 15 mA.
C. 50 mA.
D. 100 mA.
The typical current for a silicon crystal in a RADAR mixer or detector is no more than 3 or 4 milliamps, maximum. Any more would burn it out. **ANSWER A**

8-7A6 What component of a RADAR receiver is represented by block 47 in Fig. 8A1?
A. The ATR box.
B. The TR box.
C. The RF Attenuator.
D. The Crystal Detector.
Refer to the diagram in question 8-7A1. Block 47 in Diagram 8A1 is the RF attenuator. **ANSWER C**

Key Topic 8 – Components-2

8-8A1 The basic frequency determining element in a Gunn oscillator is:
A. The power supply voltage.
B. The type of semiconductor used.
C. The resonant cavity.
D. The loading of the oscillator by the mixer.
A varactor can be used to vary the frequency for narrow bandwidth applications. But for wide bandwidths applications, such as RADAR, YIG tuning may be employed. In YIG tuning, the resonance of a cavity is changed by using a magnetic coil. **ANSWER C**

8-8A2 Which of the following is not a method of analog-to-digital conversion?
A. Delta-sigma conversion.
B. Dynamic-range conversion.
C. Switched-capacitor conversion.
D. Dual-slope integration.
Dynamic range is specifically related to the amplitude levels of multiple signals that a receiver can accommodate. This is not an indication of an analog-to-digital converter. **ANSWER B**

8-8A3 When comparing TTL and CMOS logic families, which of the following is true:
A. CMOS logic requires a supply voltage of 5 volts ±20%, whereas TTL logic requires 5 volts ±5%.
B. Unused inputs should be tied high or low as necessary only in the CMOS family.
C. At higher operating frequencies, CMOS circuits consume almost as much power as TTL circuits.
D. When a CMOS input is held low, it sources current into whatever it drives.
One of the features of CMOS is the low power consumption at low frequencies compared to TTL. However, due to increased capacitance at high frequencies, the power consumption is about the same. **ANSWER C**

8-8A4 The primary operating frequency of a reflex klystron is controlled by the:
A. Dimensions of the resonant cavity.
B. Level of voltage on the control grid.
C. Voltage applied to the cavity grids.
D. Voltage applied to the repeller plate.
The larger the resonant cavity of the klystron, the lower the frequency. The resonant cavity determines the primary operating frequency of a reflex klystron transmitter stage. **ANSWER A**

8-8A5 A Gunn diode oscillator takes advantage of what effect?
A. Negative resistance.
B. Avalanche transit time.
C. Bulk-effect.
D. Negative resistance and bulk-effect.
Gunn diodes are not really diodes in the normal rectification sense. They are used as oscillators and only retain the diode name because they have two elements. The Gunn diode operates because of the bulk semiconductor material (bulk effect) and capacitance and negative resistance. **ANSWER D**

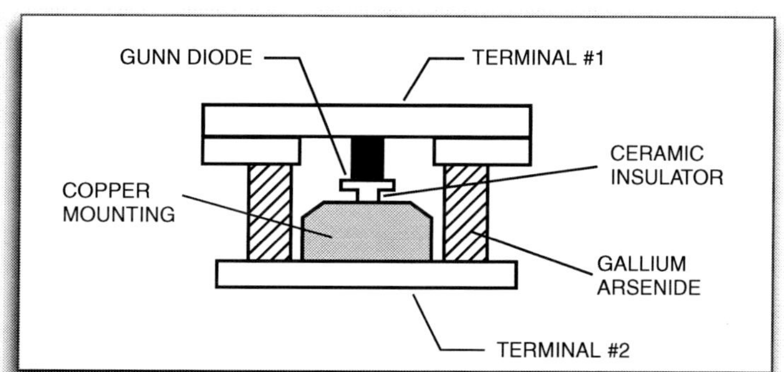

Gunn Diode

8-8A6 Fine adjustments of a reflex klystron are accomplished by:

A. Adjusting the flexible wall of the cavity.
B. Varying the repeller voltage.
C. Adjusting the AFC control system.
D. Varying the cavity grid potential.

Fine adjustments of a reflex klystron are accomplished using small changes in the control voltage on the repeller plates. Large adjustment are made by changing the cavity volume. **ANSWER B**

Key Topic 9 – Circuits-1

8-9A1 Blocking oscillators operate on the formula of:

A. T = R × C.
B. I = E/R.
C. By using the receiver's AGC.
D. None of the above are correct.

The blocking oscillator produces equally-spaced timing trigger pulses, with a pulse repetition rate determined by the operating frequency of the master oscillator. The wave-shaping circuit is based on the formula T = R × C. **ANSWER A**

8-9A2 The block diagram of a typical RADAR system microprocessor is shown in Fig. 8A2. Choose the most correct statement regarding this system.

A. The ALU is used for address decoding.
B. The Memory and I/O communicate with peripherals.
C. The control unit executes arithmetic manipulations.
D. The internal bus is used simultaneously by all units.

The two memory systems, along with the CPU (Central Processing Unit), provide the operating instructions to the RADAR system. The I/O (Input/Output) device is primarily responsible for data transfers between the CPU and all other supporting circuits outside the CPU. It can be seen from the diagram that the memory and I/O buss will receive input from GPS, LORAN, GYN and boat speed in a marine system. **ANSWER B**

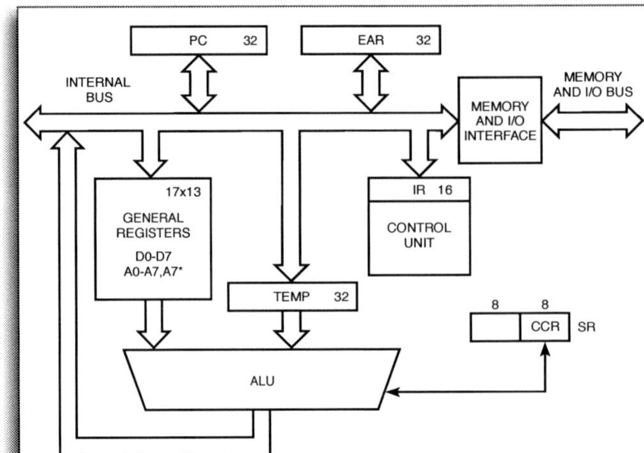

Figure 8A2

8-9A3 The phantastron circuit is capable of:

A. Stabilizing the magnetron.
B. Preventing saturation of the RADAR receiver.
C. Being used to control repeller voltage in the AFC system.
D. Developing a linear ramp voltage when triggered by an external source.

The phantastron circuit is a variable-length sawtooth generator which develops a linear ramp voltage when triggered by an external source. The phantastron is found in the sweep circuit, feeding a steep sawtooth to the sweep amplifier, dependent on the range selected by the RADAR operator. **ANSWER D**

8-9A4 The block diagram of a typical RADAR system microprocessor is shown in Fig. 8A2. Choose the most correct statement regarding this system.

A. The ALU executes arithmetic manipulations.
B. The ALU is used for address decoding.
C. General registers are used for arithmetic manipulations.
D. Address pointers are contained in the control unit.

The ALU (Arithmetic Logic Unit) is the computer's "brain" capable of adding and subtracting binary numbers that a microprocessor executes according to its program. The ALU can be thought of as the "thinking" part of a computer system. **ANSWER A**

8-9A5 In the Line-Driver/Coax/Line-receiver circuit shown in Fig. 8A3, what component is represented by the blank box marked "X"?

A. 25-ohm resistor.
B. 51-ohm resistor.
C. 10-microhm inductor.
D. 20-microhm inductor.

A 51-ohm resistor is required to match the impedance of the cable. This allows efficient signal transfer and maintains a clean waveform. **ANSWER B**

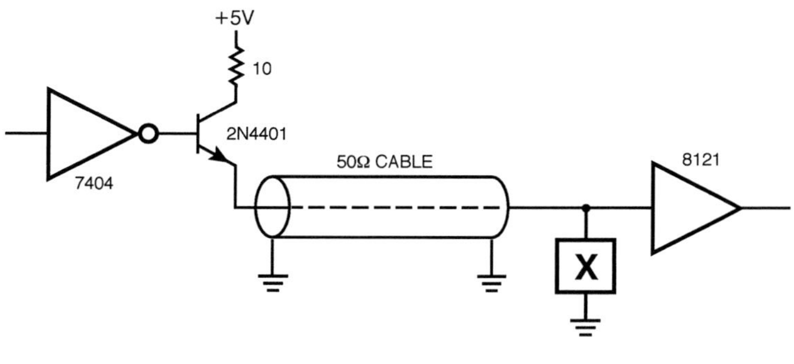

Figure 8A3

8-9A6 Choose the most correct statement:

A. The magnetron anode is a low voltage circuit.
B. The anode of the magnetron carries high voltage.
C. The filament of the magnetron carries dangerous voltages.
D. The magnetron filament is a low voltage circuit.

The filament is bombarded by electrons from the output of a magnetron, which travels in a loop. Plate voltage and field strength must be monitored to prevent destructive filament bombardment. The filament of the magnetron will carry a dangerous voltage. **ANSWER C**

Key Topic 10 – Circuits-2

8-10A1 In the circuit shown in Fig. 8A4, U5 pins 1 and 4 are high and both are in the reset state. Assume one clock cycle occurs of Clk A followed by one cycle of Clk B. What are the output states of the two D-type flip flops?

A. Pin 5 low, Pin 9 low.
B. Pin 5 high, Pin 9 low.
C. Pin 5 low, Pin 9 high.
D. Pin 5 high, Pin 9 high.

Pins 1 and 4 do not have an effect on the logic condition. At the start, pins 5 and 9 are low, and pins 6 and 8 are high. One clock cycle of Clk A causes pins 5 and 12 to take the state of the D input on pin 2, a high. Pins 6 and 2 go low with no effect. On the clock cycle at Clk B, pin 9 takes the state of 12, which is high. **ANSWER D**

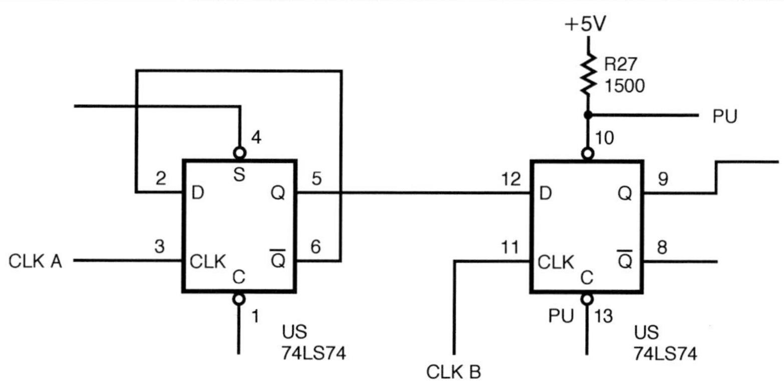

Figure 8A4

8-10A2 If more light strikes the photodiode in Fig. 8A5, there will be:

A. Less diode current.
B. No change in diode current.
C. More diode current.
D. There is wrong polarity on the diode.

More diode current passes when photons (light) strike the photodiode.
ANSWER C

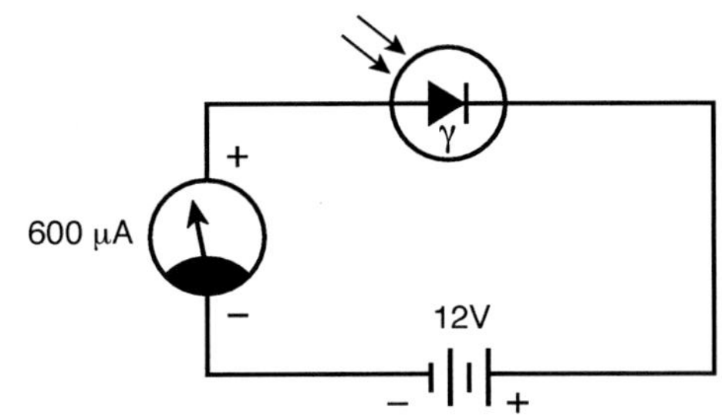

Figure 8A5

8-10A3 In the circuit shown in Fig. 8A6, which of the following is true?

A. With A and B high, Q1 is saturated and Q2 is off.
B. With either A or B low, Q1 is saturated and Q2 is off.
C. With A and B low, Q2 is on and Q4 is off.
D. With either A or B low, Q1 is off and Q2 is on.

When either A or B is low, Q1 is turned on and saturates, which pulls the base of Q2 low, turning off Q2. **ANSWER B**

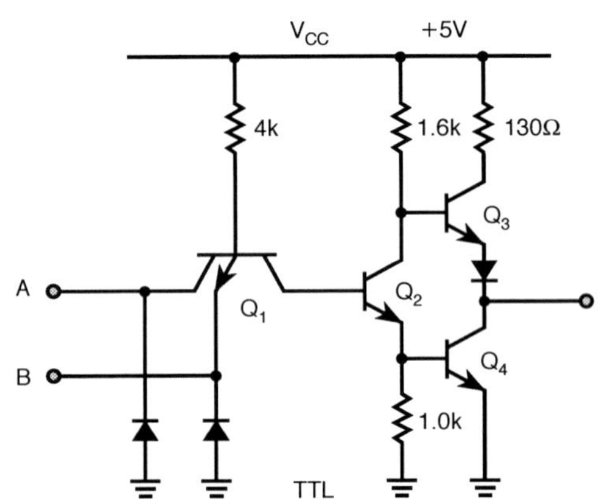

Figure 8A6

8-10A4 What is the correct value of R_s in Fig. 8A7, if the voltage across the LED is 1.9 Volts with 5 Volts applied and I_f max equals 40 milliamps?

A. 4700 ohms.
B. 155 ohms.
C. 77 ohms.
D. 10000 ohms.

If 1.9 volts is applied across the diode, then 3.1 volts is applied across R_s in Figure 8A7. Even though the forward bias current is typically between 10 and 20 mA, they expect you to use the stated maximum current of 40 milliamps. So 3.1 ÷ 0.04 = 77.5 ohms. **ANSWER C**

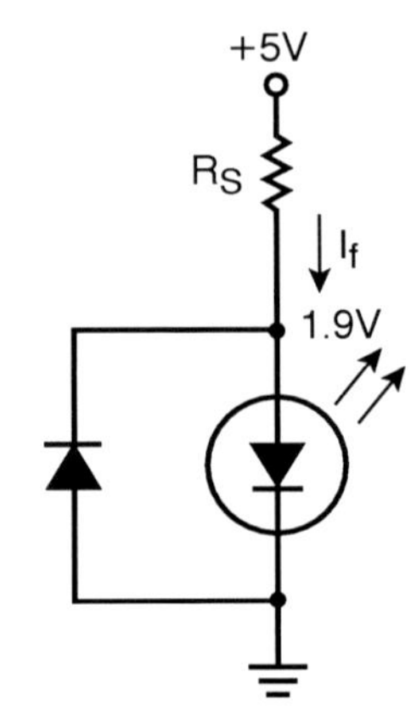

Figure 8A7

Photo diode within the silver can.

8-10A5 The block diagram of a typical RADAR system microprocessor is shown in Fig. 8A2. Choose the most correct statement regarding this system.

A. The ALU is used for address decoding.
B. General registers are used for arithmetic manipulations.
C. The control unit executes arithmetic manipulations.
D. Address pointers are contained in the general registers.

RADAR system microprocessors are very similar to personal computers, or PCs. A microprocessor, by definition, is an integrated circuit, or a circuit that contains all the computational and control circuitry for a small computer. It can both input and output information on data pins which can be "bussed" to other circuitry or devices. Moving through a series of steps, information is sent through the data pins and then either inputted or bussed to other areas to perform certain tasks. The information is encoded to a unique pattern of "1's" and "0's" and are arranged in either 4, 8, 16, or 32-bit groups depending on whether it is a 4-bit, 8-bit, 16-bit, or 32-bit microprocessor. The encoded information is then sent along these grouped data lines. The signal lines are grouped into 3 categories: data bus lines, address bus lines, and control bus lines. The address lines are controlled by a program counter in the general registers which "point" or sequence the counting of the "1's" and "0's" and provide the instructions to the RADAR unit. Addressing, then, is program controlled and it is in the general registers where the RADAR's instructions are generated. It should be noted, however, that this sequence is normally altered by information brought into the microprocessor, so that the RADAR can be "programmed" to meet specific needs. **ANSWER D**

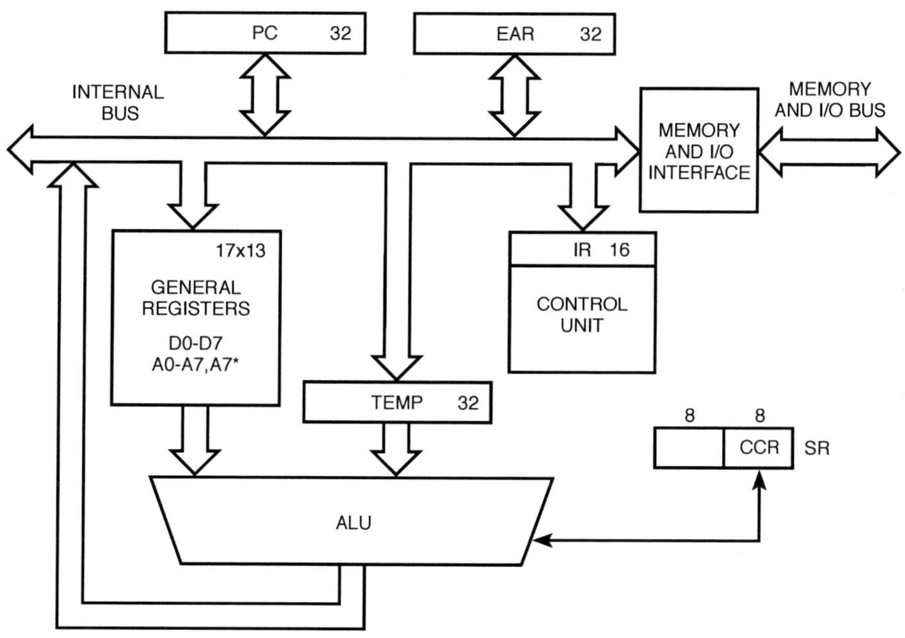

Figure 8A2

8-10A6 You are troubleshooting a component on a printed circuit board in a RADAR system while referencing the Truth Table in Fig. 8A8. What kind of integrated circuit is the component?

A. D-type Flip-Flop, 3-State, Inverting.
B. Q-type Flip-Flop, Non-Inverting.
C. Q-type Directional Shift Register, Dual.
D. D to Q Convertor, 2-State.

The Dn column indicates this is a D-type flip-flop. OE is enable. **ANSWER A**

TRUTH TABLE

INPUTS			OUTPUT
$\overline{OE}$	CP	Dn	$\overline{Qn}$
L	_/‾	H	L
L	_/‾	L	H
L	L	X	No Change
H	X	X	Z

Note: X = Don't care
Z = High impedance state
_/‾ = Low-to-High transition

Figure 8A8

Subelement B – Transmitting Systems

8 Key Topics, 8 Exam Questions

Key Topic 11 – Transmitting Systems

8-11B1 The magnetron is used to:

A. Generate the output signal at the proper operating frequency.
B. Determine the shape and width of the transmitted pulses.
C. Modulate the pulse signal.
D. Determine the pulse repetition rate.

The magnetron is a thermionic vacuum tube with a large, permanent magnet surrounding the tube, and is found in the power output stage of a RADAR. When properly assembled, the magnetron will generate a stable output signal at the proper operating frequency. **ANSWER A**

8-11B2 The purpose of the modulator is to:

A. Transmit the high voltage pulses to the antenna.
B. Provide high voltage pulses of the proper shape and width to the magnetron.
C. Adjust the pulse repetition rate.
D. Tune the Magnetron to the proper frequency.

The modulator is found in the transmitter to produce properly-timed, high-amplitude, rectangular pulses which are then fed to the magnetron. **ANSWER B**

8-11B3 Which of the following statements about most modern RADAR transmitter power supplies is false?

A. High voltage supplies may produce voltages in excess of 5,000 volts AC.
B. There are usually separate low voltage and high voltage supplies.
C. Low voltage supplies use switching circuits to deliver multiple voltages.
D. Low voltage supplies may supply both AC and DC voltages.

Here they ask which statement is FALSE. Most RADARs may feed the modulator with a high-voltage, DC (NOT AC) power supply, using half wave, full wave, or bridge rectification. These power supplies MUST be cooled by an airflow system. **ANSWER A**

8-11B4 The purpose of the Pulse Forming Network is to:

A. Act as a low pass filter.
B. Act as a high pass filter.
C. Produce a pulse of the correct width.
D. Regulate the pulse repetition rate.

The pulse-forming network is found in the RADAR modulator, and produces a pulse with a specific correct width. **ANSWER C**

8-11B5 The purpose of the Synchronizer is to:

A. Generate the modulating pulse to the magnetron.
B. Generate a timing signal that establishes the pulse repetition rate.
C. Insure that the TR tube conducts at the proper time.
D. Control the pulse width.

The synchronizer is a timer circuit to calculate the interval between transmitted pulses and to insure the interval is of the proper length. These timing pulses insure synchronized operation as related to the pulse repetition rate. **ANSWER B**

8-11B6 Which of the following is not part of the transmitting system?

A. Magnetron.
B. Modulator.
C. Pulse Forming Network.
D. Klystron.

The klystron may be found in the RADAR receiver local oscillator, using 30 MHz or 60 MHz intermediate frequency. **ANSWER D**

Key Topic 12 – Magnetrons

8-12B1 High voltage is applied to what element of the magnetron?

A. The waveguide.
B. The anode.
C. The plate cap.
D. The cathode.

High voltage is found within the heart of the magnetron on the sealed cathode. Cathode current is several amps, and the cathode is large so that it can withstand the tremendous amount of heat that develops. Keep metal tools well away from the magnetron because of its large permanent magnet. The exposed anode is at ground potential, and a large negative voltage of several kilovolts is applied to the cathode during the magnetron's oscillating state. **ANSWER D**

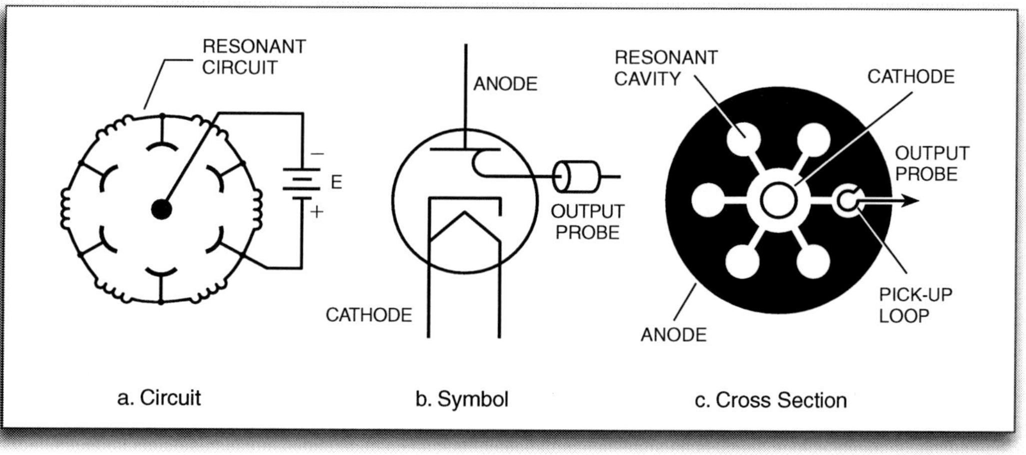

A Magnetron Generates Transmitter Power

8-12B2 The characteristic of the magnetron output pulse that relates to accurate range measurement is its:

A. Amplitude.
B. Decay time.
C. Rise time.
D. Duration.

The rise time of the magnetron output must be as short as possible to produce the desired rapid pulse rates. If the rise time is too long, the measured target distance is inaccurate. **ANSWER C**

8-12B3 What device is used as a transmitter in a marine RADAR system?

A. Magnetron.
B. Klystron.
C. Beam-powered pentode.
D. Thyratron.

The magnetron produces high-power outputs for marine RADARs. The klystron (incorrect Answer B) is generally used as the oscillator for the system. **ANSWER A**

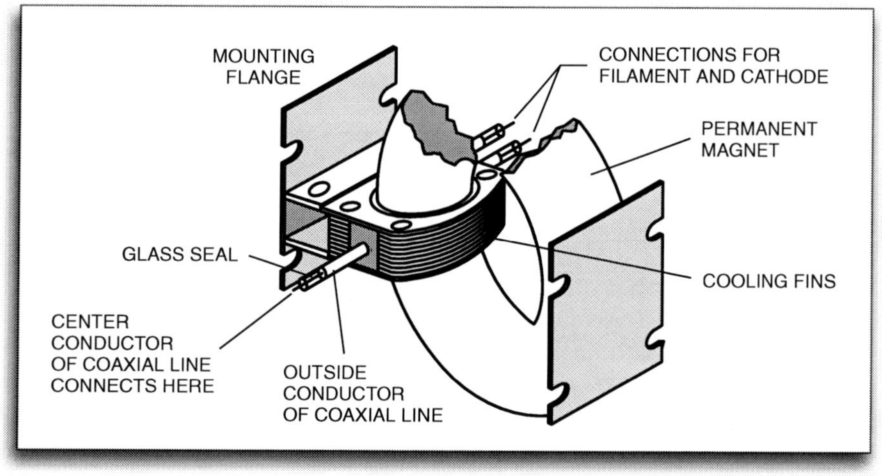

Magnetron

8-12B4 The magnetron is:

A. A type of diode that requires an internal magnetic field.
B. A triode that requires an external magnetic field.
C. Used as the local oscillator in the RADAR unit.
D. A type of diode that requires an external magnetic field.

The magnetron is a diode with a magnetic field between the cathode and anode; the magnetic field controls the tube. When the magnet gets weak, the magnetron must be replaced. **ANSWER D**

8-12B5 A negative voltage is commonly applied to the magnetron cathode rather than a positive voltage to the magnetron anode because:

A. The cathode must be made negative to force electronics into the drift area.
B. A positive voltage would tend to nullify or weaken the magnetic field.
C. The anode can be operated at ground potential for safety reasons.
D. The cavities might not be shock-excited into oscillation by a positive voltage.

The magnetron anode is normally operated at ground potential to reduce the likelihood of an accidental shock when technicians are working on a turned-on RADAR TR (transmitter / receiver section) unit. **ANSWER C**

8-12B6 The anode of a magnetron is normally maintained at ground potential:

A. Because it operates more efficiently that way.
B. For safety purposes.
C. Never. It must be highly positive to attract the electrons.
D. Because greater peak-power ratings can be achieved.

The anode is at ground potential to protect operations personnel from shocks and to eliminate the problem of insulating the magnetron from the chassis. **ANSWER B**

Magnetron

Key Topic 13 – Modulation

8-13B1 In a solid-state RADAR modulator, the duration of the transmitted pulse is determined by:

A. The thyratron.
B. The magnetron voltage.
C. The pulse forming network.
D. The trigger pulse.

The pulse forming network determines the pulse width, which in turn determines the duration of the transmitted pulse. **ANSWER C**

8-13B2 The modulation frequency of most RADAR systems is between:

A. 60 and 500 Hz.
B. 3000 and 6000 Hz.
C. 1500 and 7500 Hz.
D. 1000 and 3000 Hz,

A new type of RADAR called FMCW (Frequency Modulated Continuous Wave) RADAR is sometimes known as a CHIRP RADAR. In a chirp RADAR, very long pulses are transmitted, rather than our more familiar short pulses. However, during the period of the pulse, the frequency of the signal is swept across a wide range of frequencies. If you could HEAR this radio signal, it would sound like a rapidly sliding calliope, or perhaps a bird being strangled. We know from earlier that long signals have lots of energy, but they also have very poor resolution and even worse close-in response. Using a very tricky technique called PULSE COMPRESSION, of which CHIRP is just one style, we can "fool" our entire RADAR system into seeing the very long pulse as being a very high amplitude pulse. Using this technique, we really gain the best of both worlds, within limits. In the receiver, we use a device called a transverse filter, which has a delay time that is proportional to the frequency. When the frequency is low, there is a shorter time delay; when the frequency is high, it has a longer delay. When a chirped signal encounters this transverse filter, it tends to "pile up" upon itself, converting that long drawn out energy pulse into a short, high amplitude pulse in the receiver I.F. The end result is, we have the resolution of a RADAR having a much shorter pulse, but the long distance potency of a long-pulsed RADAR. Now, there is no such thing as a free lunch. We have sacrificed the capability of detecting very close targets, because we really have no way of separating our long drawn out pulse from our long, drawn out reflection. It's also not really great at detecting fast-moving targets without some additional fancy footwork. The one thing pulse compression does do for us, is that it gives us a much more powerful RADAR system for a given antenna size and transmitter peak power. Most marine RADARs use pulse modulation with a PRR of 500 to 2,000 Hz. This eliminates distractors B, C, and D since the minimums are 1,000 Hz and higher and the maximums are all above 2,000 Hz. This leaves 60 to 500 Hz as the correct answer. This frequency range probably refers to continuous wave (CW) RADAR. **ANSWER A**

8-13B3 A shipboard RADAR uses a PFN driving a magnetron cathode through a step-up transformer. This results in which type of modulation?

A. Frequency modulation.
B. Amplitude modulation.
C. Continuous Wave (CW) modulation.
D. Pulse modulation.

The term "PFN" stands for Pulse-Forming Network, which is a type of pulse modulation. **ANSWER D**

8-13B4 In a pulse modulated magnetron what device determines the shape and width of the pulse?

A. Pulse Forming Network.
B. Thyratron.
C. LC parallel circuit.
D. Dimensions of the magnetron cavity.

A pulse-forming network is a rectangular wave generator based on an LC network that alternately stores, and then releases, the energy. **ANSWER A**

8-13B5 What device(s) may act as the modulator of a RADAR system?

A. Magnetron.
B. Klystron.
C. Echo box.
D. Thyratron or a silicon-controlled rectifier (SCR).

A Thyratron, or an SCR, is used to discharge the pulse-forming network to supply pulsed power to the magnetron. **ANSWER D**

8-13B6 The purpose of a modulator in the transmitter section of a RADAR is to:

A. Improve bearing resolution.
B. Provide the correct wave form to the transmitter.
C. Prevent sea return.
D. Control magnetron power output.

The modulator is found in the RADAR transmitter, and controls RADAR transmit pulse width by means of a rectangular DC pulse of the required duration and amplitude. **ANSWER B**

Key Topic 14 – Pulse-Forming Networks Modulation

8-14B1 The pulse developed by the modulator may have an amplitude greater than the supply voltage. This is possible by:

A. Using a voltage multiplier circuit.
B. Employing a resonant charging choke.
C. Discharging a capacitor through an inductor.
D. Discharging two capacitors in series and combining their charges.

At resonance, the amplitude of the pulse can be well above the supply voltage. This is made possible by employing a resonant charging choke. **ANSWER B**

8-14B2 Pulse transformers and pulse-forming networks are commonly used to shape the microwave energy burst RADAR transmitter. The switching devices most often used in such pulse-forming circuits are:

A. Power MOSFETS and Triacs.
B. Diacs and SCR's.
C. Thyratrons and BJT's.
D. SCR's and Thyratrons.

The switching devices found in the pulse-forming circuit of most RADARs are silicon-controlled rectifiers (SCR's) and Thyratrons. **ANSWER D**

8-14B3 The purpose of the pulse-forming network is to:

A. Determine the width of the modulating pulses.
B. Determine the pulse repetition rate.
C. Act as a high pass filter.
D. Act as a log pass filter.

The PFN alternately stores and releases energy in approximately rectangular waves, which determines the width of the modulating pulses. **ANSWER A**

See page 285 for a discussion of new Pulse Compression methods.

8-14B4 The shape and duration of the high-voltage pulse delivered to the magnetron is established by:

A. An RC network in the keyer stage.
B. The duration of the modulator input trigger.
C. An artificial delay line.
D. The time required to saturate the pulse transformer.

It is the delay line that shapes and determines the duration of the high-voltage pulse delivered to the magnetron. **ANSWER C**

8-14B5 Pulse-forming networks are usually composed of the following:

A. Series capacitors and shunt inductors.
B. Series inductors and shunt capacitors.
C. Resonant circuit with an inductor and capacitor.
D. None of the above.

The pulse-forming network controls the duration of the modular pulses, dependent on the value of the inductors and capacitors in each LC section. **ANSWER B**

8-14B6 An artificial transmission line is used for:

A. The transmission of RADAR pulses.
B. Testing the RADAR unit, when actual targets are not available.
C. Determining the shape and duration of pulses.
D. Testing the delay time for artificial targets.

The artificial transmission line in the RADAR modulator is actually a storage element. The modulator switch controls the pulse-repetition rate, and when the modulator switch is open, the transmission line charges. When the switch is closed, the transmission line discharges through the series circuit. The transmission line simulates the actual lines and allows determination of the shapes and duration of the pulses. **ANSWER C**

Key Topic 15 – TR - ATR - Circulators - Directional Couplers-1

8-15B1 The ferrite material in a circulator is used as a(an):

A. Electric switch.
B. Saturated reactor.
C. Loading element.
D. Phase shifter.

A circulator is a 3- or 4-port device that passes energy from one port to the next port, but in only one direction. A ferrite material within the circulator has low-forward loss, and high-reverse loss. The effect is a phase shift from port to port. **ANSWER D**

8-15B2 In a circular resonant cavity with flat ends, the E-field and the H-field form with specific relationships. The:

A. E-lines are parallel to the top and bottom walls.
B. E-lines are perpendicular to the end walls.
C. H-lines are perpendicular to the side walls.
D. H-lines are circular to the end walls.

The E-field and the H-field are perpendicular to each other in the resonant cavity. To remember this, think which way it would be easier to bend the waveguide. The E plane is the “easy” way, and the H plane is the “hard” way. **ANSWER B**

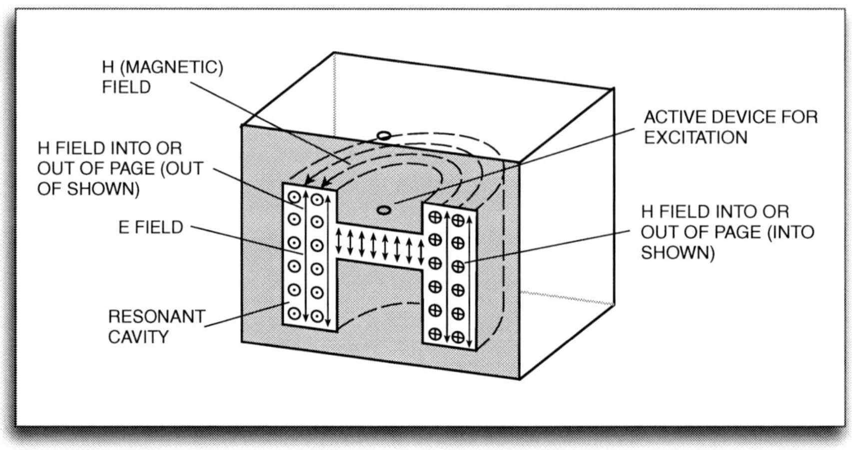

Cross Section of a Resonant Cavity

8-15B3 A ferrite circulator is most commonly used in what portion of a RADAR system?

A. The antenna.
B. The modulator.
C. The duplexer.
D. The receiver.

The duplexer action is similar to that of a TR switch on a transceiver. There is enough isolation in the duplexer to keep transmit power from feeding back into, and destroying, the receiver. The duplexer circuit is accomplished by a ferrite circulator for simultaneous transmit and receive capabilities without the need of any mechanical switching. **ANSWER C**

8-15B4 A circulator provides what function in the RF section of a RADAR system?

A. It replaces the TR cell and functions as a duplexer.
B. It cools the magnetron by forcing a flow of circulating air.
C. It permits tests to be made to the thyristors while in use.
D. It transmits antenna position to the indicator during operation.

A circulator eliminates the need for the older TR cell, and the circulator will then perform as a duplexer. **ANSWER A**

8-15B5 A directional coupler has an attenuation of -30 db. A measurement of 100 milliwatts at the coupler indicates the power of the line is:

A. 10 watts.
B. 100 watts.
C. 1,000 watts.
D. 10,000 watts.

Minus 30 dB is 1,000 times down in power level. 1,000 times 0.1 watts is 100 watts. When you convert everything to dB, your system losses and gains all become simple addition and subtraction problems. Decibels were invented to make things EASY! **ANSWER B**

10 dB = 10 times
20 dB = 100 times
30 dB = 1,000 times

8-15B6 What is the purpose or function of the RADAR duplexer/circulator?

A. An electronic switch that allows the use of one antenna for both transmission and reception.
B. A coupling device that is used in the transition from a rectangular waveguide to a circular waveguide.
C. A modified length of waveguide used to sample a portion of the transmitted energy for testing purposes.
D. A dual section coupling device that allows the use of a magnetron as a transmitter.

You can consider the duplexer and circulator as an electronic switch that allows the outgoing energy and incoming echo to switch between the antenna and the receiver without outgoing energy directly coupling into the receiver causing burnout. **ANSWER A**

Key Topic 16 – TR - ATR - Circulators - Directional Couplers-2

8-16B1 The ATR box:

A. Protects the receiver from strong RADAR signals.
B. Prevents the received signal from entering the transmitter.
C. Turns off the receiver when the transmitter is on.
D. All of the above.

ATR stands for Antitransmit-Receive Tube. The ATR tube disconnects the RADAR transmitter during the period of echo receive. The ATR "box" keeps the weak echo from going the wrong way into the transmitter, rather than where it should go, to the receiver. Thus, the ATR "box" is a transmitter isolator to the received signals coming from the antenna unit. If this isolation does not occur, a signal entering the transmitter port will be re-reflected, causing a misleading reflection to occur in the receiver. Conversely, the TR tube has the primary job of disconnecting the receiver during pulse transmit. **ANSWER B**

8-16B2 When a pulse RADAR is radiating, which elements in the TR box are energized?

A. The TR tube only.
B. The ATR tube only.
C. Both the TR and ATR tubes.
D. Neither the TR nor ATR tubes.

As the RADAR is transmitting a pulse, gas switches in both the TR and the ATR tubes becomes ionized; therefore, both elements are energized. **ANSWER C**

8-16B3 The TR box:

A. Prevents the received signal from entering the transmitter.
B. Protects the receiver from the strong RADAR pulses.
C. Turns off the receiver when the transmitter is on.
D. Protects the receiver from the strong RADAR pulses and turns off the receiver when the transmitter is on.

In the TR box, a solid-state circuit acts as a mechanical TR relay. The transmit/receive box protects the receiver from the high-powered signals produced during transmit. **ANSWER D**

8-16B4 What device is located between the magnetron and the mixer and prevents received signals from entering the magnetron?

A. The ATR tube.
B. The TR tube.
C. The RF Attenuator.
D. A resonant cavity.

The ATR circuit (tube) keeps received echoes out of the transmitter. **ANSWER A**

8-16B5 A keep-alive voltage is applied to:

A. The crystal detector.
B. The ATR tube.
C. The TR tube.
D. The magnetron.

A keep-alive voltage is always kept on the TR tube to insure it has the fastest possible response time for the changeover. **ANSWER C**

8-16B6 A DC keep-alive potential:

A. Is applied to a TR tube to make it more sensitive.
B. Partially ionizes the gas in a TR tube, making it very sensitive to transmitter pulses.
C. Fully ionizes the gas in a TR tube.
D. Is applied to a TR tube to make it more sensitive and partially ionizes the gas in a TR tube, making it very sensitive to transmitter pulses.

The TR tube is a gas switch. When the gas is not ionized, there is a high impedance to the signal. When the gas is ionized, it tends to short circuit the line by presenting a low-impedance path. An additional electrode is inserted to provide a keep-alive voltage to keep the gas partially ionized. This allows the gas to be more quickly and easily fully ionized by the transmit pulse. **ANSWER D**

Key Topic 17 – Timer - Trigger - Synchronizer Circuits

8-17B1 What RADAR circuit determines the pulse repetition rate (PRR)?

A. Discriminator.
B. Timer (synchronizer circuit).
C. Artificial transmission line.
D. Pulse-rate-indicator circuit.

The pulse repetition rate is determined by the RADAR timer synchronizer circuit. **ANSWER B**

8-17B2 The triggering section is also known as the:

A. PFN.
B. Timer circuit.
C. Blocking oscillator.
D. Synchronizer.

The RADAR synchronizer is sometimes referred to as a timer or keyer, supplying synchronizing signals that time transmit pulses, indictor display, and many other circuits in the receiver. **ANSWER D**

8-17B3 Operation of any RADAR system begins in the:

A. Triggering section.
B. Magnetron.
C. AFC.
D. PFN.

The RADAR system MUST operate in a specific timed relationship, including the interval between transmitted pulses. Timing pulses relate to pulse repetition frequency, with additional RADAR components timed by the output of the synchronizer or by timing signals from the transmitter as it is turned on. This is the heart of a RADAR system, the triggering section synchronizer. **ANSWER A**

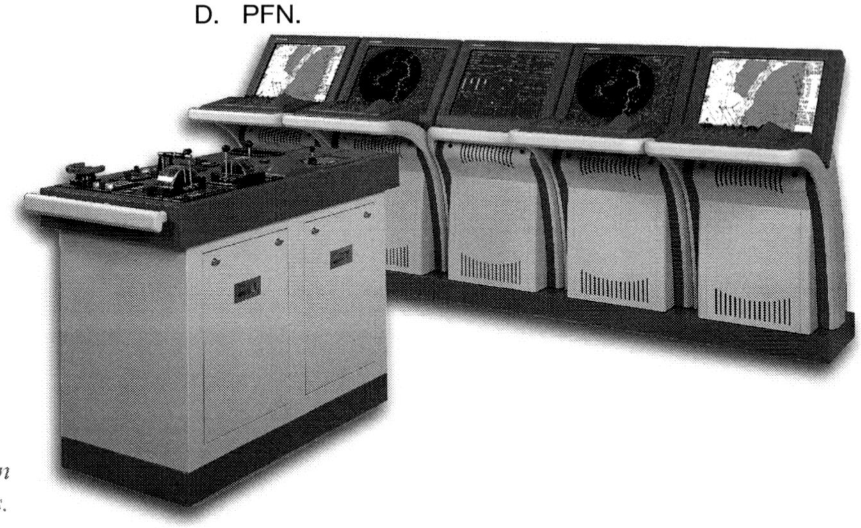

This big ship integrated navigation station has multiple RADAR displays.

Photo courtesy of Furuno Electric Co., Ltd.

8-17B4 The timer circuit:

A. Determines the pulse repetition rate (PRR).
B. Determines range markers.
C. Provides blanking and unblanking signals for the CRT.
D. All of the above.

The timer circuit determines pulse repetition rate and range markers, and provides blanking and unblanking signals for the cathode-ray tube in a raster scan display. **ANSWER D**

8-17B5 Pulse RADARs require precise timing for their operation. Which type circuit below might best be used to provide these accurate timing pulses?

A. Single-swing blocking oscillator.
B. AFC controlled sinewave oscillator.
C. Non-symmetrical astable multivibrator.
D. Triggered flip-flop type multivibrator.

The blocking oscillator operates for a period of time, followed by a period of time where no oscillation occurs. The circuit then oscillates again. This oscillator is often used to form the pulse in a RADAR transmitter. **ANSWER A**

8-17B6 Unblanking pulses are produced by the timer circuit. Where are they sent?

A. IF amplifiers.
B. Mixer.
C. CRT.
D. Discriminator.

The timer circuit determines pulse repetition rate and range markers, and provides blanking and unblanking signals for the cathode-ray tube in a raster scan display. **ANSWER C**

Key Topic 18 – Power Supplies

8-18B1 An advantage of resonant charging is that it:

A. Eliminates the need for a reverse current diode.
B. Guarantees perfectly square output pulses.
C. Reduces the high-voltage power supply requirements.
D. Maintains a constant magnetron output frequency.

Because a resonant charging circuit can develop extremely high voltages above the normal B+ voltage, it reduces the high-voltage power supply requirements. **ANSWER C**

8-18B2 The characteristics of a field-effect transistor (FET) used in a modern RADAR switching power supply can be compared as follows:

A. "On" state compares to a bipolar transistor. "Off" state compares to a 1-Megohm resistor.
B. "On" state compares to a pure resistor. "Off" state compares to a mechanical relay.
C. "On" state compares to an low resistance inductor. "Off" state compares to a 10-Megohm resistor.
D. "On" state compares to a resistor. "Off" state compares to a capacitor.

The field-effect transistor (FET) is the popular choice in switching power supplies where it is desirable for the ON state to be low resistance and the OFF state to offer infinite resistance similar to an open mechanical relay. **ANSWER B**

8-18B3 A pulse-width modulator in a switching power supply is used to:

A. Provide the reference voltage for the regulator.
B. Vary the frequency of the switching regulator to control the output voltage.
C. Vary the duty cycle of the regulator switch to control the output voltage.
D. Compare the reference voltage with the output voltage sample and produce an error voltage.

As the duty cycle (on time ÷ period of cycle) of the power supply increases, the voltage increases. Therefore, you can use the duty cycle to regulate the output voltage. **ANSWER C**

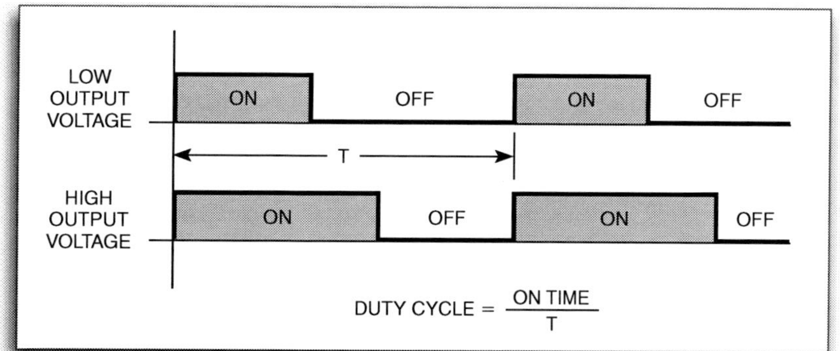

Duty Cycle Varies to Control Output Voltage in a Switching Power Supply

---

8-18B4 In a fixed-frequency switching power supply, the pulse width of the switching circuit will increase when:

A. The load impedance decreases.
B. The load current decreases.
C. The output voltage increases.
D. The input voltage increases.

A switching-mode power supply drops and regulates the output voltage by chopping the dc input to a square wave, and then recovering the voltage with a low-pass filter. As the input voltage or the load impedance changes, the duty cycle of the on/off cycle changes and keeps the voltage constant. **ANSWER A**

8-18B5 A major consideration for the use of a switching regulator power supply over a linear regulator is:

A. The switching regulator has better regulation.
B. The linear regulator does not require a transformer to step down AC line voltages to a usable level.
C. The switching regulator can be used in nearly all applications requiring regulated voltage.
D. The overall efficiency of a switching regulator is much higher than a linear power supply.

Switching power supplies offer incredible efficiency. Typical efficiency is 75 percent. For a linear power supply using a heavy transformer, efficiency is not great – only about 30 percent. **ANSWER D**

8-18B6 Which of the following characteristics are true of a power MOSFET used in a RADAR switching supply?

A. Low input impedance; failure mode can be gate punch-through.
B. High input impedance; failure mode can be gate punch-through.
C. High input impedance; failure mode can be thermal runaway.
D. Low input impedance; failure mode can be gate breakdown.

The MOSFET power transistor commonly found in RADAR switching power supplies has a high input impedance, but when the device fails, the failure is in the gate "punch-through" current (breakdown at gate insulation). **ANSWER B**

Subelement C – Receiving Systems
10 Key Topics, 10 Exam Questions

Key Topic 19 – Receiving Systems

8-19C1 Which of the following statements is true?
A. The front end of the receiver does not provide any amplification to the RADAR signal.
B. The mixer provides a gain of at least 6 db.
C. The I.F. amplifier is always a high gain, narrow bandwidth amplifier.
D. None of the above.

On microwave frequencies, a point-contact crystal diode is used instead of solid state or tube amplifiers, as these circuits would generate excessive noise. The Schottky barrier diode offers an improvement in signal-to-noise ratio over point contact diodes. There is no front end amplification of the signal. **ANSWER A**

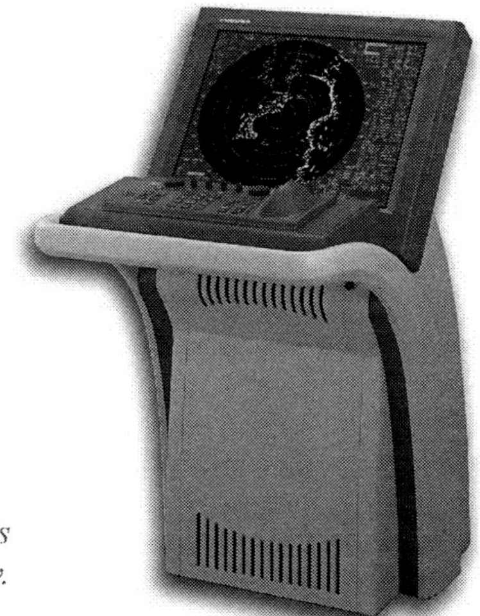

The video amplifier drives this large RADAR screen display.

Photo courtesy of Furuno Electric Co., Ltd.

8-19C2 Logarithmic receivers:
A. Can't be damaged.
B. Can't be saturated.
C. Should not be used in RADAR systems.
D. Have low sensitivity.

The logarithmic receiver has no gain control circuits with the linear log amplifier (lin-log) working for high-amplitude signals, without saturation. Exceptionally strong echoes will not saturate the amplifier, producing a small echo that might otherwise be undetected with a normal receiver going into saturation. **ANSWER B**

8-19C3 RADAR receivers are similar to:
A. FM receivers.
B. HF receivers.
C. T.V. receivers.
D. Microwave receivers.

The characteristics of a microwave receiver apply precisely to RADAR equipment – the demand for an extremely low noise floor, mechanical shielding between different receiver stages, receiver protection against strong transmitter overloads, and microsecond switching time between transmit and receive operations. **ANSWER D**

8-19C4 What section of the receiving system sends signals to the display system?
A. Video amplifier.
B. Audio amplifier.
C. I.F. Amplifier.
D. Resolver.

The word "display" gives it away – the section of the receiver system sending signals to the display system is indeed the video amplifier. **ANSWER A**

8-19C5 What is the main difference between an analog and a digital receiver?
A. Special amplification circuitry.
B. The presence of decision circuitry to distinguish between "on" and "off" signal levels.
C. The two cannot be compared.
D. Digital receivers produce no distortion.

The digital RADAR receiver has a much lower noise floor, thanks to digital signal processing. Digital circuitry distinguishes only on-and-off signal levels. **ANSWER B**

8-19C6 In a RADAR receiver, the RF power amplifier:
A. Is high gain.
B. Is low gain.
C. Does not exist.
D. Requires wide bandwidth.

In RADAR receivers, RF power amplifiers really don't exist. They are used with transmitters, not receivers. **ANSWER C**

Key Topic 20 – Mixers

8-20C1 The diagram in Fig. 8C9 shows a simplified RADAR mixer circuit using a crystal diode as the first detector. What is the output of the circuit when no echoes are being received?

A. 60 MHz CW.
B. 4095 MHz CW.
C. 4155 MHz CW.
D. No output is developed.

When no echoes are received, the output from the circuit is zero because there is no output from the first detector. **ANSWER D**

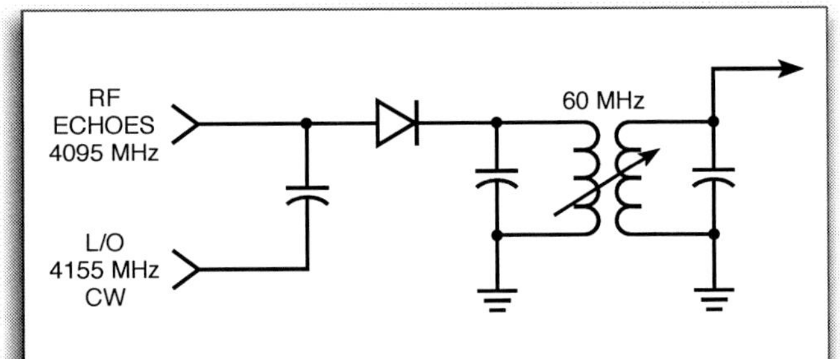

Figure 8C9

8-20C2 In the receive mode, frequency conversion is generally accomplished by a:

A. Tunable wave-guide section.
B. Pentagrid converter.
C. Crystal diode.
D. Ferrite device.

Frequency conversion is usually accomplished by the crystal diode. **ANSWER C**

8-20C3 An RF mixer has what purpose in a RADAR system?

A. Mixes the CW transmitter output to form pulsed waves.
B. Converts a low-level signal to a different frequency.
C. Prevents microwave oscillations from reaching the antenna.
D. Combines audio tones with RF to produce the RADAR signal.

To obtain the required gain and bandwidth of a signal, a crystal mixer is employed to lower the frequency of the received signal to the 30-60 MHz band. **ANSWER B**

8-20C4 In a RADAR unit, the mixer uses a:

A. Pentagrid converter tube.
B. Field-effect transistor.
C. Silicon crystal or PIN diode.
D. Microwave transistor.

The PIN diode is a point-contact, silicon-crystal diode with a fine tungsten, bronze-phosphor, or beryllium-copper "cat whisker" between two N- and P-type silicon crystals. These diodes must be treated carefully to avoid static electricity burnout. They normally plug into the T/R unit, and many times require replacement when another powerful RADAR operating nearby causes the diode to overload and immediately burnout. **ANSWER C**

8-20C5 What component of a RADAR receiver is represented by block 49 in Fig. 8A1?

A. Discriminator.
B. IF amplifier.
C. Klystron.
D. Crystal detector (the mixer).

This is the crystal detector section of the RADAR receiver. **ANSWER D**

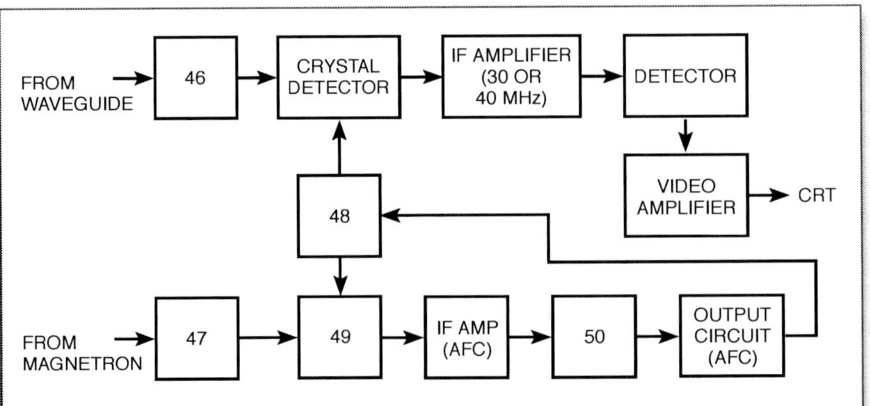

Figure 8A1

8-20C6 In a RADAR unit, the mixer uses:

A. PIN diodes and silicon crystals.
B. PIN diodes.
C. Boettcher crystals.
D. Silicon crystals.

A RADAR mixer almost always uses a PIN diode and silicon crystals. **ANSWER A**

Key Topic 21 – Local Oscillators

8-21C1 The error voltage from the discriminator is applied to the:

A. Repeller (reflector) of the klystron.
B. Grids of the IF amplifier.
C. Grids of the RF amplifiers.
D. Magnetron.

The reflex klystron and the local oscillator control circuit provides repeller plate voltage, causing the klystron to control the local oscillator frequency. The local oscillator operates 30 MHz below the transmit frequency on most RADARs. **ANSWER A**

8-21C2 In a RADAR unit, the local oscillator is a:

A. Hydrogen Thyratron.
B. Klystron.
C. Pentagrid converter tube.
D. Reactance tube modulator.

The local oscillator is the klystron (LOK). **ANSWER B**

8-21C3 What component of a RADAR receiver is represented by block 48 in Fig. 8A1?

A. Klystron (local oscillator).
B. Discriminator.
C. IF amplifier.
D. Crystal detector.

Block 48 in Fig. 8A1 is the klystron.
ANSWER A

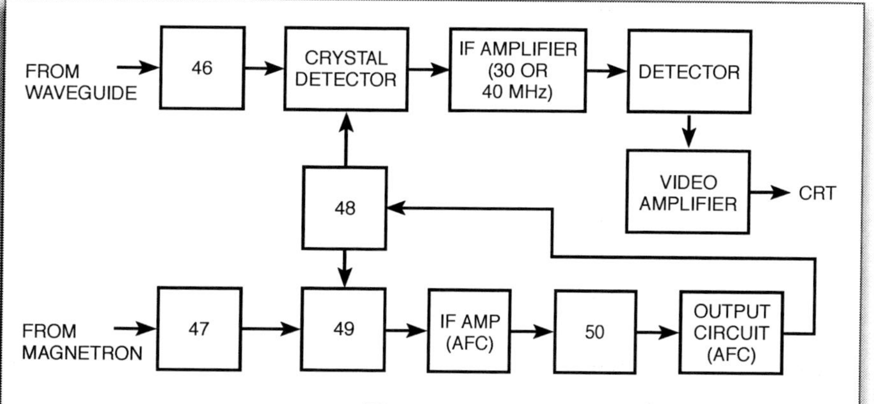

Figure 8A1

8-21C4 What device(s) could be used as the local oscillator in a RADAR receiver?

A. Thyratron.
B. Klystron.
C. Klystron and a Gunn diode.
D. Gunn diode.

Klystrons and Gunn diodes are used as oscillators in RADAR receivers. **ANSWER C**

8-21C5 The klystron local oscillator is constantly kept on frequency by:

A. Constant manual adjustments.
B. The Automatic Frequency Control circuit.
C. A feedback loop from the crystal detector.
D. A feedback loop from the TR box.

Because the magnetron's frequency can drift, the automatic frequency control circuit causes the klystron to track any changes in magnetron frequency and uses the difference between the two frequencies to maintain a constant intermediate frequency. **ANSWER B**

8-21C6 How may the frequency of the klystron be varied?

A. Small changes can be made by adjusting the anode voltage.
B. Large changes can be made by adjusting the frequency.
C. By changing the phasing of the buncher grids.
D. Small changes can be made by adjusting the repeller voltage and large changes can be made by adjusting the size of the resonant cavity.

The klystron frequency may be slightly varied by adjusting the repeller voltage. For large changes in the klystron frequency, ultra-careful adjustments of the size of the resonant cavity may also be employed.
ANSWER D

Klystron

Key Topic 22 – Amplifiers

8-22C1 Overcoupling in a RADAR receiver will cause?

A. Improved target returns.
B. Increase the range of the IAGC.
C. Decrease noise.
D. Oscillations.

Most gain in a receiver is developed within the intermediate frequency amplifier stages. Bandwidth is also determined by the bandwidth of the IF stages, but with overcoupling within these stages, oscillations will many times obscure weak targets. IF gain must be variable to provide a consistent voltage output for input signals of different amplitudes. **ANSWER D**

8-22C2 The usual intermediate frequency of a shipboard RADAR unit is:

A. 455 kHz.
B. 10.7 MHz.
C. 30 or 60 MHz.
D. 120 MHz.

The common intermediate frequency of shipboard RADAR units is 30 MHz or 60 MHz between the oscillator and the transmitted frequency. The RADAR may employ automatic frequency control (AFC), or a manual tune control, to keep the output of the local oscillator in a closed loop circuit. The output of the discriminator is a DC error voltage, and indicates the degree of mistuning between the transmitter and the local oscillator. **ANSWER C**

8-22C3 The I.F. Amplifier bandwidth is:

A. Wide for short ranges and narrow for long ranges.
B. Wide for long ranges and narrow for short ranges.
C. Constant for all ranges.
D. Adjustable from the control panel.

Intermediate frequency amplifier bandwidth is automatically selected by the range control adjustments at the RADAR display. On short range with a faster pulse repetition rate, amplifier bandwidth is wider, and for long range echoes, IF bandwidth becomes more narrow, which assists in mitigating receiver noise. **ANSWER A**

8-22C4 A logarithmic IF amplifier is preferable to a linear IF amplifier in a RADAR receiver because it:

A. Has higher gain.
B. Is more easily aligned.
C. Has a lower noise figure.
D. Has a greater dynamic range.

The characteristic of a log amplifier provides an output voltage that is proportional to the log of the input signal. This characteristic allows greater dynamic range for the receiver. **ANSWER D**

8-22C5 The high-gain IF amplifiers in a RADAR receiver may amplify a 2 microvolt input signal to an output level of 2 volts. This amount of amplification represents a gain of:

A. 60 db.
B. 100 db.
C. 120 db.
D. 1,000 db.

Voltage ratio equals 2 volts ÷ 2 microvolts = 1×10^6. Voltage gain in dB is = 20 log 1,000,000. The $\log_{10} 10^6 = 6$. Thus db = 20 × 6 = 120 dB. Remember 2 microvolts is 2.0×10^{-6}. IMPORTANT: When using VOLTAGE to calculate dB calculations (either loss or gain) it is ABSOLUTELY NECESSARY to assure that both the input and output impedances are the same, an assumption that is made in this problem. All dB calculations ultimately are POWER ratios. If input and output impedances are either different or unknown, we cannot use the "voltage dB" formula, $20 \log (V_{out}/V_{in})$. **ANSWER C**

8-22C6 In a RADAR receiver AGC and IAGC can vary between:

A. 10 and 15 DB.
B. 20 and 40 DB.
C. 30 and 60 DB.
D. 5 and 30 DB.

IAGC is *instantaneous automatic gain control* which works on individual pulses rather than an average level. Conventional AGC has the disadvantage of reducing the entire system gain in accordance with the strongest signal. This will cause weak signals to be masked by stronger signals. IAGC has a recovery time which is shorter than the PRF, allowing full sensitivity to be achieved between pulses. In either case, a typical AGC-controlled I.F. stage can undergo a change of gain of about 20 dB. Controlling two stages can give a control range of about 40 dB. **ANSWER B**

Key Topic 23 – Detectors - Video Amplifiers

8-23C1 Which of the following statements is correct?

A. The video amplifier is located between the mixer and the I.F. amplifier.
B. The video amplifier operates between 60 MHz and 120 Mhz.
C. The video amplifier is located between the I.F. amplifier and the display system.
D. The video amplifier is located between the local oscillator and the mixer.

The video amplifier receives pulses from the detector and amplifies the pulses to the input of the display circuit. This puts the video amp between the IF amplifier and the display system. The video amplifier is thus the final stage that will then send the recovered signal to the RADAR display screen. The video amp is an RC-coupled amplifier with high-gain transistors operating over a wide frequency response. Frequency compensation networks within the video amplifier mitigate common problems of component lead inductance and inter-electrode capacitance, each of which reduces the overall high- and low-frequency responses of the amplifier. **ANSWER C**

8-23C2 Video amplifiers in pulse RADAR receivers must have a broad bandwidth because:

A. Weak pulses must be amplified.
B. High frequency sine waves must be amplified.
C. The RADARs operate at PRFs above 100.
D. The pulses produced are normally too wide for video amplification.

One of the fundamental laws of physics that the RADAR technician has to deal with is the relationship between time and frequency. The narrower (or shorter) a pulse is in the time domain the wider it is in the frequency domain. We must never be misled by the temptation of looking at a short RADAR pulse and assuming it's a "narrow" radio signal. Quite the opposite is true. A short radio pulse distributes its energy over a wide range of frequencies surrounding its nominal carrier frequency, in the form of many sidebands. Unless our receiver can accommodate not only the carrier frequency but the sidebands as well, most of that energy will be lost. The receiver I.F. bandwidth as well as any subsequent amplifiers (such as the video amplifier in this example) MUST be as wide as the pulse spectrum or we just won't see the return pulse. This is true even if the pulse is a very strong one – the bandwidth principle is independent of power. A too-narrow bandwidth in either the receiver I.F. or video amplifier will make the RADAR signals appear weak even when they aren't, just because we aren't letting the sidebands "through the door." This is why the the weak pulses must be amplified. **ANSWER A**

8-23C3 In video amplifiers, compensation for the input and output stage capacitances must be accomplished to prevent distorting the video pulses. This compensation is normally accomplished by connecting:

A. Inductors in parallel with both the input and output capacitances.
B. Resistances in parallel with both the input and output capacitances.
C. An inductor in parallel with the input capacitance and an inductor in series with the output capacitance.
D. An inductor in series with the input capacitance and an inductor in parallel with the output capacitance.

A video amplifier must amplify a wide band of frequencies with minimal frequency and phase distortion. Parasitic capacitance at the input and output of any amplifier can limit the bandwidth and cause undesirable phase shift, which can distort the shape or size of any detected RADAR pulse. These parasitic capacitances can be compensated for by the use of proper complementary inductances. The most troublesome parasitic capacitances of any active device are input series capacitance and output parallel capacitance. The input series capacitance can be compensated for by an input series inductance, and the output parallel capacitance can be compensated for with a parallel inductance. **ANSWER D**

8-23C4 Which of the following signals is not usually an input to the video amplifier?

A. Resolver.
B. Range.
C. Brilliance.
D. Contrast.

The video amplifier will take selected inputs and apply them to the indicating device. The video amplifier must be capable of relatively wide frequency response, taking inputs of range, brilliance, and contrast. A frequency compensation network will help adjust for different levels. **ANSWER A**

8-23C5 Which of the following signals are usually an input to the video amplifier?

A. Range.
B. Brilliance.
C. Contrast.
D. All of the above.

All three of these – range, brilliance, and contrast – are inputs to the video amplifier. **ANSWER D**

8-23C6 The video (second) detector in a pulse modulated RADAR system would most likely use a/an:

A. Discriminator detector.
B. Diode detector.
C. Ratio detector.
D. Infinite impedance detector.

A crystal detector diode is most often used to extract the video modulation from the carrier. **ANSWER B**

Key Topic 24 – Automatic Frequency Control - AFC

8-24C1 The AFC system is used to:

A. Control the frequency of the magnetron.
B. Control the frequency of the klystron.
C. Control the receiver gain.
D. Control the frequency of the incoming pulses.

The automatic frequency control (AFC) system controls the frequency of the klystron. **ANSWER B**

8-24C2 A circuit used to develop AFC voltage in a RADAR receiver is called the:

A. Peak detector.
B. Crystal mixer.
C. Second detector.
D. Discriminator.

The discriminator is tuned to the IF frequency. When operating properly, the difference in magnetron and klystron frequency is the IF frequency. As long as the difference remains the same as the IF frequency, no error voltage is produced. If there is a difference, an error voltage is produced and is applied to the klystron to change the frequency difference back to the IF frequency. This corrective action is called automatic frequency control (AFC). **ANSWER D**

8-24C3 In the AFC system, the discriminator compares the frequencies of the:

A. Magnetron and klystron.
B. PRR generator and magnetron.
C. Magnetron and crystal detector.
D. Magnetron and video amplifier.

In the automatic frequency control system, the discriminator compares the frequencies of the magnetron and the klystron. **ANSWER A**

8-24C4 An AFC system keeps the receiver tuned to the transmitted signal by varying the frequency of the:

A. Magnetron.
B. IF amplifier stage.
C. Local oscillator.
D. Cavity duplexer.

The automatic frequency control function takes place in the local oscillator. **ANSWER C**

8-24C5 A RADAR transmitter is operating on 3.0 GHz and the reflex klystron local oscillator, operating at 3.060 GHz, develops a 60 MHz IF. If the magnetron drifts higher in frequency, the AFC system must cause the klystron repeller plate to become:

A. More positive.
B. More negative.
C. Less positive.
D. Less negative.

The reflector is also called a repeller plate. It is negatively charged to repel the electrons. The more negatively charged the repeller plate, the lower the klystron frequency. The drift is high so the repeller must be more negative. **ANSWER B**

8-24C6 What component is block 50 in Fig. 8A1?

A. IF amplifier.
B. AFC amplifier.
C. Discriminator.
D. Crystal detector.

Block 50 in Fig. 8A1 is the discriminator stage and is fed by the intermediate frequency amplifier and its output is the output AFC circuit. **ANSWER C**

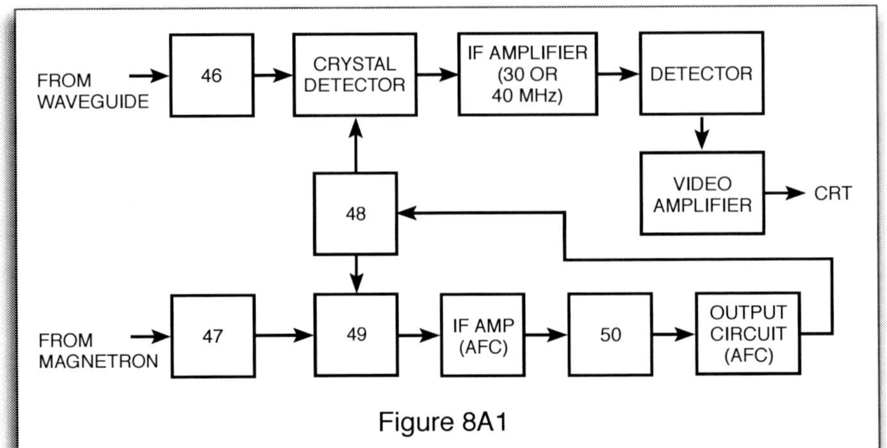

Figure 8A1

Key Topic 25 – Sea Clutter - STC

8-25C1 The STC circuit is used to:

A. Increase receiver stability.
B. Increase receiver sensitivity.
C. Increase receiver selectivity.
D. Decrease sea return on a RADAR receiver.

STC stands for Sensitivity Time Control, a circuit to adjust amplifier gain with TIME, during a single pulse repetition period. A bias voltage, that varies with time, is applied to the IF amplifier. Close-in targets are decreased in strength, with more distant targets maintained at their normal level. This will help mitigate false echoes from wind waves, dubbed "popcorn," close-in to your RADAR station. **ANSWER D**

8-25C2 The STC circuit:

A. Increases the sensitivity of the receiver for close targets.
B. Decreases sea return on the PPI scope.
C. Helps to increase the bearing resolution of targets.
D. Increases sea return on the PPI scope.

The STC (sensitivity time control) decreases return echoes from the sea in the direction of the incoming wind. It is the front side of the ocean waves that creates the greatest number of false-echo sea returns. **ANSWER B**

8-25C3 Sea return is:

A. Sea water that gets into the antenna system.
B. The return echo from a target at sea.
C. The reflection of RADAR signals from nearby waves.
D. None of the above.

Sea return is echoes reflected off the face of nearby waves. You can even see the wake of certain ships as sea return. **ANSWER C**

The STC circuit helps minimize sea clutter and improves resolution from multiple nearby targets.

Photo courtesy of Furuno Electric Co., Ltd.

8-25C4 Sea clutter on the RADAR scope cannot be effectively reduced using front panel controls. What circuit would you suspect is faulty?

A. Sensitivity Time Control (STC) circuit.
B. False Target Eliminator (FTE) circuit.
C. Fast Time Constant (FTC) circuit.
D. Intermediate Frequency (IF) circuit.

Sea clutter is actually a positive indication that a marine RADAR is operating properly in the short-range setting. However, in heavy weather, sea clutter may sometimes cover up close-in targets, such as small boats and buoys. Switching in the sensitivity time control (STC) circuit attenuates close-in echoes so you can better identify real close-in targets. Think of "STC" as short timing control. **ANSWER A**

8-25C5 What circuit controls the suppression of sea clutter?

A. EBL circuit.
B. STC circuit.
C. Local oscillator.
D. Audio amplifier.

Sea clutter – called "popcorn" – is always a problem on windy days, and the STC adjustable on/off circuit will help minimize ocean wave return. **ANSWER B**

8-25C6 The sensitivity time control (STC) circuit:

A. Decreases the sensitivity of the receiver for close objects.
B. Increases the sensitivity of the receiver for close objects.
C. Increases the sensitivity of the receiver for distant objects.
D. Decreases the sensitivity of the transmitter for close objects.

The sensitivity time control is used to minimize false echoes from sea return during heavy weather. It minimizes the pick up of echoes from nearby targets. **ANSWER A**

Key Topic 26 – Power Supplies

8-26C1 Prior to making "power-on" measurements on a switching power supply, you should be familiar with the supply because of the following:

A. You need to know where the filter capacitors are so they can be discharged.
B. If it does not use a line isolation transformer you may destroy the supply with grounded test equipment.
C. It is not possible to cause a component failure by using ungrounded test equipment.
D. So that measurements can be made without referring to the schematic.

When working on a switching power supply that is turned on, keep in mind that it may not use an isolation transformer and the internal resistance of your test equipment could destroy the transistors, unless your test probe is specifically designed for use with switching power supplies. **ANSWER B**

8-26C2 A constant frequency switching power supply regulator with an input voltage of 165 volts DC, and a switching frequency of 20 kHz, has an "ON" time of 27 microseconds when supplying 1 ampere to its load. What is the output voltage across the load?

A. It cannot be determined with the information given.
B. 305.55 volts DC.
C. 89.1 volts DC.
D. 165 volts DC.

The formula for the output voltage of a switching power supply is:

$V_O = V_{IN} \times D$ Where: V_O = Output voltage in volts
V_{IN} = Input voltage in volts
D = duty cycle = $t_{ON} \times f$

Since $D = 27 \times 10^{-6} \times 20 \times 10^{+3} = 0.540 \times 10^{-3}$

Then $V_O = 165 \times 0.54 = 89.1$ volts **ANSWER C**

8-26C3 The circuit shown in Fig. 8C10 is the output of a switching power supply. Measuring from the junction of CR6, CR7 and L1 to ground with an oscilloscope, what waveform would you expect to see?

A. Filtered DC.
B. Pulsating DC at line frequency.
C. AC at line frequency.
D. Pulsating DC much higher than line frequency.

Switching power supplies have eliminated the need for bulky transformers in RADAR equipment. Transistor "chopper" circuits create the pulsating DC, and what you see with an oscilloscope at the junction of CR6, CR7 and L1 to GND is this pulsating DC at a much higher frequency than the line input. The down side of switching power supply circuits is the broad-band noise that is sometimes radiated to other onboard marine electronic navigation receivers. **ANSWER D**

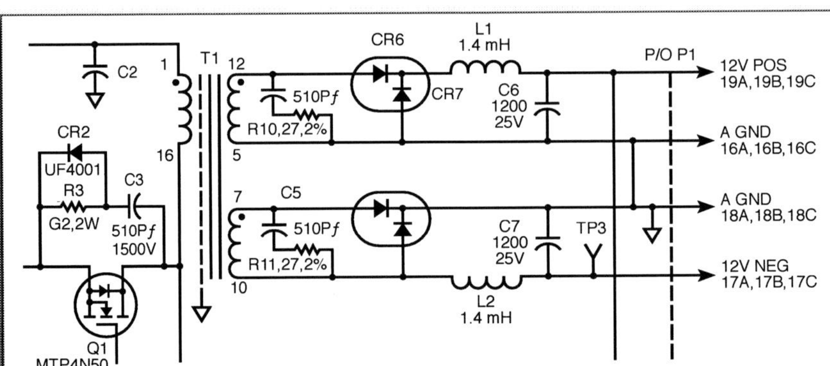

Figure 8C10

8-26C4 With regard to the comparator shown in Fig. 8C11, the input is a sinusoid. Nominal high level output of the comparator is 4.5 volts. Choose the most correct statement regarding the input and output.

A. The leading edge of the output waveform occurs 180 degrees after positive zero crossing of the input waveform.
B. The rising edge of the output waveform trails the positive zero crossing of the input waveform by 45 degrees.
C. The rising edge of the output waveform trails the negative zero crossing of the input waveform by 45 degrees.
D. The rising edge of the output waveform trails the positive peak of the input waveform by 45 degrees.

The sinusoidal input is connected to the negative terminal. When the input sine wave goes positive, the output goes negative and is a square wave. In comparison, the two waveforms are 180 degrees out of phase. **ANSWER A**

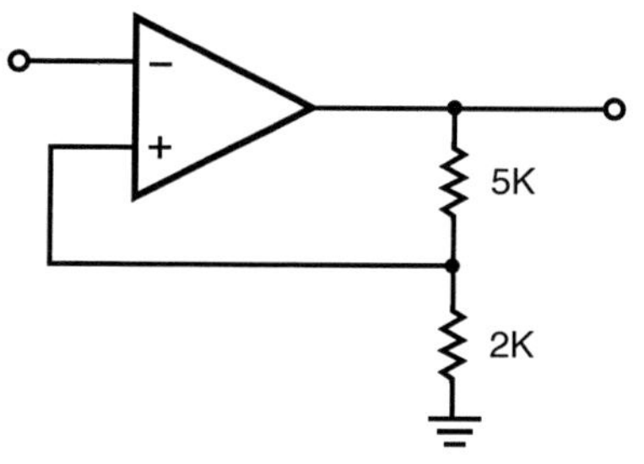

Figure 8C11

8-26C5 When monitoring the gate voltage of a power MOSFET in the switching power supply of a modern RADAR, you would expect to see the gate voltage change from "low" to "high" by how much?

A. 1 volt to 2 volts.
B. 300 microvolts to 700 microvolts.
C. Greater than 2 volts.
D. 1.0 volt to 20.0 volts.

The MOSFET is the preferred transistor in switching power supplies, and the MOSFET gate voltage change is always greater than 2.0 volts. **ANSWER C**

8-26C6 The nominal output high of the comparator shown in Fig. 8C11 is 4.5 volts. Choose the most correct statement which describes the trip points.

A. Upper trip point is 4.5 volts. Lower trip point is approximately 0 volts.
B. Upper trip point is 2.5 volts. Lower trip point is approximately 2.0 volts.
C. Upper trip point is 900 microvolts. Lower trip point is approximately 0 volts.
D. Upper trip point is +1.285 volts. Lower trip point is -1.285 volts.

In a comparator shown in Fig. 8C11, above, we have assigned R1 to the upper resistor and R2 to the lower resistor. The comparator operates from a 15V and 25V supply. The output swings from 14.5V to 24.5V. You calculate the trip points as follows:

$$V_{UT} = \frac{R_2}{R_1 + R_2} V_{O(MAX)} \qquad V_{LT} = \frac{R_2}{R_1 + R_2} (-V_{O(MAX)})$$

$$V_{UT} = \frac{2{,}000}{5000 + 2000} \times 4.5 = 1.285V \qquad V_{LT} = \frac{2{,}000}{5000 + 2000} \times (-4.5) = -1.285V$$

The trip points are +1.285V and −1.285V. **ANSWER D**

Key Topic 27 – Interference Issues

8-27C1 One of the best methods of reducing noise in a RADAR receiver is?
  A. Changing the frequency.
  B. Isolation.
  C. Replacing the resonant cavity.
  D. Changing the IF strip.

All microwave receivers must incorporate copper shielding, preventing stray noise from leaking into the circuit. This will also help isolate one stage from another. At microwave wavelengths, it takes a total copper enclosure to provide adequate isolation. **ANSWER B**

8-27C2 The primary cause of noise in a RADAR receiver can be attributed to:
  A. Electrical causes.
  B. Atmospheric changes.
  C. Poor grounding.
  D. Thermal noise caused by RADAR receiver components.

Noise in a RADAR receiver can be attributed to thermal noise caused by sensitive RADAR receiver components. Erratic video "blotches" with sharp changes in intensity on the screen may illustrate noise, and the imminent failure of a sensitive receiver component. **ANSWER D**

8-27C3 Noise can appear on the LCD as:
  A. Erratic video and sharp changes in intensity.
  B. Black spots on the screen.
  C. Changes in bearings.
  D. None of the above.

Anytime a RADAR operator spots erratic video and big changes in display intensity, they can assume that a component within the receiver "front end" may be failing, creating noise. **ANSWER A**

8-27C4 RADAR interference on a communications receiver appears as:
  A. A varying tone.
  B. Static.
  C. A hissing tone.
  D. A steady tone.

Interference from an energized RADAR heard over a communications receiver is a steady tone at the pulse repetition rate frequency. It sounds like a musical tone. **ANSWER D**

8-27C5 In a RADAR receiver the most common types of interference are?
  A. Weather and sea return.
  B. Sea return and thermal.
  C. Weather and electrical.
  D. Jamming and electrical.

RADAR operating at peak efficiency aboard a ship will many times display distant thunderstorms, close-in sea return, and flying birds several miles away. **ANSWER A**

8-27C6 Noise can:
  A. Mask larger targets.
  B. Change bearings.
  C. Mask small targets.
  D. Increase RADAR transmitter interference.

Internal circuits creating noise will often mask distant faint echo returns. This will cause the operator to miss small targets that do not exhibit sufficient echo strength to be resolved by the receiver, and appear on the display.
ANSWER C

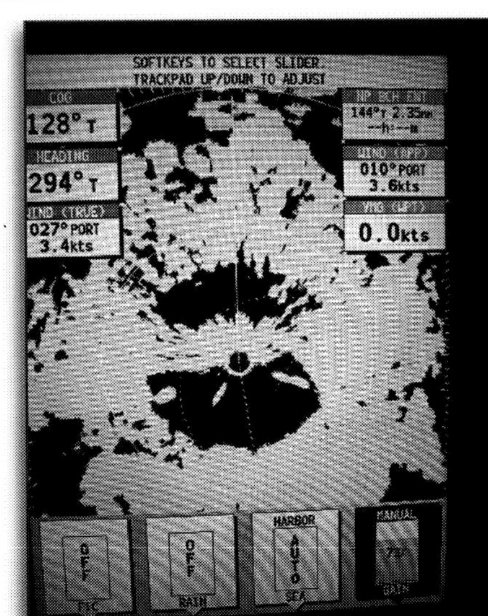

Too much gain can create noise that could mask close-in targets.

Key Topic 28 – Miscellaneous

8-28C1 The purpose of the discriminator circuit in a RADAR set is to:

A. Discriminate against nearby objects.
B. Discriminate against two objects with very similar bearings.
C. Generate a corrective voltage for controlling the frequency of the klystron local oscillator.
D. Demodulate or remove the intelligence from the FM signal.

The discriminator circuit in a RADAR is part of the automatic frequency control circuitry. It detects any change of the magnetron or klystron frequency so that their difference in frequency maintains the intermediate frequency. **ANSWER C**

8-28C2 The MTI circuit:

A. Acts as a mixer in a RADAR receiver.
B. Is a filter, which blocks out stationary targets, allowing only moving targets to be detected.
C. Is used to monitor transmitter interference.
D. Will pick up targets, which are not in motion.

MTI stands for Moving Target Indicator system. With an MTI circuit, only targets that are moving will show up on the display. This is a complex, advanced system that must distinguish between stationery targets, such as a breakwater, and moving targets, such as a small boat ahead. Phase detection and pulse-to-pulse comparison help separate the moving targets from stationery objects. **ANSWER B**

8-28C3 Where is a RF attenuator used in a RADAR unit?

A. Between the antenna and the receiver.
B. Between the magnetron and the antenna.
C. Between the magnetron and the AFC section of the receiver.
D. Between the AFC section and the klystron.

An RF attenuator is in circuit between the magnetron and the automatic frequency control circuit of the receiver. **ANSWER C**

8-28C4 The condition known as "glint" refers to a shifting of clutter with each RADAR pulse and can be caused by a:

A. Improperly functioning MTI filter.
B. Memory failure.
C. Low AFC voltage.
D. Interference from electrical equipment.

The term "glint" refers to patches of close-in sea clutter randomly appearing on the indicator. This is a problem in the MTI circuit, sometimes caused by misadjustment, that can cause random sea returns to look like moving targets! **ANSWER A**

8-28C5 An ion discharge (TR) cell is used to:

A. Protect the transmitter from high SWRs.
B. Lower the noise figure of the receiver.
C. Tune the local oscillator of the RADAR receiver.
D. Protect the receiver mixer during the transmit pulse.

The TR cell is similar to that a TR switch on a transceiver. The TR cell protects the receiver input during the transmit pulse output. **ANSWER D**

8-28C6 When the receiver employs an MTI circuit:

A. The receiver gain increases with time.
B. Only moving targets will be displayed.
C. The receiver AGC circuits are disabled.
D. Ground clutter will be free of "rabbits".

In the "moving target" mode, the MTI circuit eliminates stationary echoes and only displays those from vessels that are underway. **ANSWER B**

Subelement D – Display & Control Systems

10 Key Topics, 10 Exam Questions

Key Topic 29 – Displays

8-29D1 Modern liquid crystal displays have a pixel count of

A. Greater than 200 pixels per inch.
B. Greater than 50 pixels per inch.
C. Can have no more than 125 pixels per inch.
D. Can implement 1000 pixels per inch.

Newer RADARs have upgraded three generations of displays – early PPI indicators incorporated a slow resolving phosphor material, and then came the monochrome Raster scan display with various level of quantization. Now we have color liquid crystal displays with a crystal count greater than 200 pixels per inch. Conductive film of nematic crystal substance will change polarization to interrupt or reflect light in back of it or light coming in from the front. Without voltage, this crystal substance becomes clear. **ANSWER A**

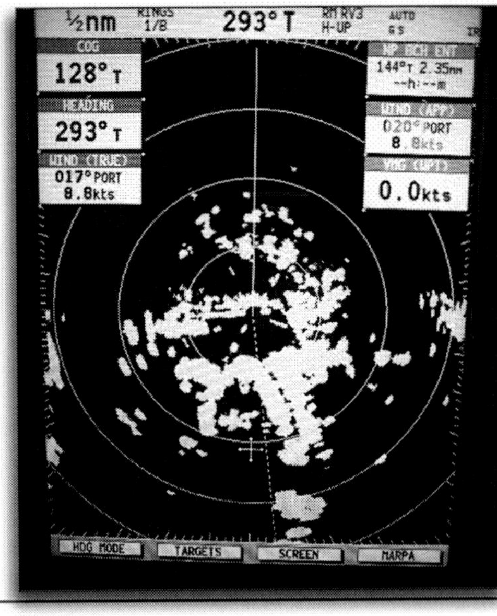

Color raster scan RADAR displays allow near-unlimited screen graphics.

8-29D2 Voltages used in CRT anode circuits are in what range of value?

A. 0.5-10 mV.
B. 10-50 kV.
C. 20-50 mV.
D. 200-1000 V.

The 10,000 to 50,000 volts used in CRTs can produce lethal results if a technician touches the anode of a RADAR cathode-ray tube. **ANSWER B**

8-29D3 The purpose of the aquadag coating on the CRT is:

A. To protect the electrons from strong electric fields.
B. To act as a second anode.
C. To attract secondary emissions from the CRT screen.
D. All of the above.

The aquadag coating on a CRT acts as a second anode and attracts secondary emissions from the CRT. It also provides shielding from strong electric fields that may be present near the CRT tube. **ANSWER D**

8-29D4 LCD patterns are formed when:

A. Current passes through the crystal causing them to align.
B. When voltage is reduced to the raster scan display.
C. When the deflection coils are resonant.
D. When the ships antenna's bearing is true North.

Liquid crystal display patterns may cause the nematic fluid to turn black as the crystals align. **ANSWER A**

8-29D5 In a raster-type display, the electron beam is scanned:

A. From the center of the display to the outer edges.
B. Horizontally and vertically across the CRT face.
C. In a rotating pattern which follows the antenna position.
D. From one specified X-Y coordinate to the next.

In a raster scan display, the electron beam scans horizontally and vertically across the CRT face, similar to the scanning in the picture tube of a standard TV set. **ANSWER B**

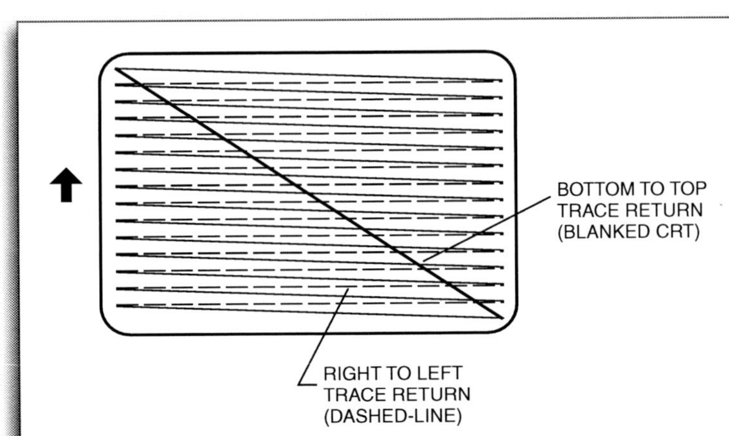

Raster Scan in RADAR Similar to TV Scan.

8-29D6 Select the statement, which is most correct regarding a raster scan display.

A. Raster displays are the same as conventional T.V. receivers.
B. The scan rate for a RADAR system is 30 frames per second.
C. Raster scanning is controlled by clock pulses and requires an address bus.
D. Raster scanning is not used in RADAR systems.

Raster scanning on a RADAR indicator requires an address bus and clock pulses from the video amplifier. **ANSWER C**

Key Topic 30 – Video Amplifiers and Sweep Circuits

8-30D1 What are the usual input signals to the video amplifier?

A. Low level video.
B. Fixed range rings.
C. Variable range rings.
D. All of the above.

The video amplifier will display low level video targets, fixed range rings depending on range selection, and variable range ring markers. **ANSWER D**

8-30D2 Which of the following would not normally be an input to the video amplifier?

A. Fixed range rings.
B. Variable range rings.
C. Resolver signal.
D. Low level video.

A resolver is an electro-mechanical transducer that relays a RADAR antenna position to a PPI console. A resolver is a form of SYNCHRO or SELSYN. This signal is not applied to the PPI video display, but controls the rotation of the CRT deflection yoke. It has no input to the video amplifier. **ANSWER C**

8-30D3 The purpose of the sweep amplifier is to:

A. Increase the power of the video amplifier.
B. Drive the CRT deflection coils.
C. Drive the resolver coils.
D. All of the above.

On older cathode ray tube indicators the sweep amplifier drives the CRT deflection coils to display changing imagery on the RADAR display scope. **ANSWER B**

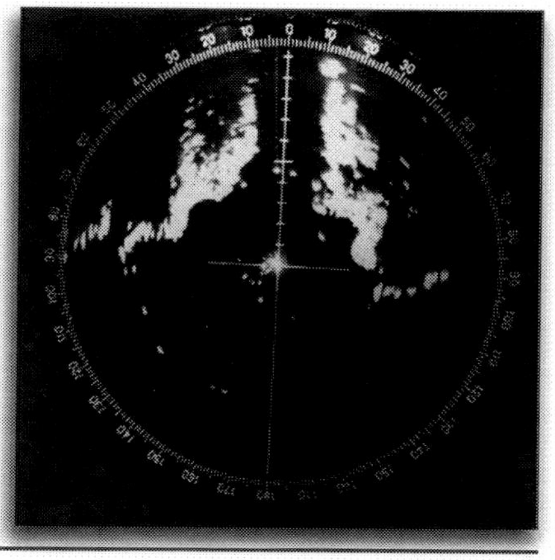

Older phosphor tube displays still have an important part of RADAR navigation.

8-30D4 How many deflection coils are driven by the sweep amplifier?

A. 4.
B. 3.
C. 2.
D. 1.

There are usually 2 deflection coils driven by the sweep amplifier. These two coils, seen as twin copper windings, are positioned on the neck of the cathode ray tube and are driven by sweep circuit current to produce the required beam deflection. The simple way of synchronizing antenna sweep rotation to CRT rotation is to electronically rotate the magnetic fields, as opposed to older methods of physically rotating the coils. Antenna azimuth information is converted into electrical signals, and sweep currents are applied to the coils. **ANSWER C**

8-30D5 The main purpose of the sweep generator is to provide:

A. Antenna information.
B. Range rings.
C. Composite video to the cathode of the CRT.
D. The drive signal to the sweep amplifier.

The sweep generator circuit develops currents applied to the sweep amplifier to deflect an electron beam across the face of the display. Gate voltage determines sweep rate, and effective distance covered by each sweep. The sweep generator is synchronized by an input from the preceding gate circuit. **ANSWER D**

8-30D6 The main purpose of the video amplifier is to provide:

A. Composite video to the cathode of the CRT.
B. Resolver signals.
C. Antenna X and Y signals.
D. Provide the drive signal to the sweep amplifier.

The video amplifier develops the video signal from the receiver and applies it to the control grid of the cathode ray tube for video intensity. **ANSWER A**

Key Topic 31 – Timing Circuits

8-31D1 Timing circuits are used to provide what function?
A. Develop synchronizing pulses for the transmitter system.
B. Synchronize the antenna and display system.
C. Adjust the sea return.
D. Control the North Up presentation.
The RADAR timing circuit synchronizer develops a time base critical for both transmit and receive RADAR pulses and echoes. It also times the interval between the transmitted pulse, and the interval of the pulse length. The timing pulse also insures a synchronized operation in the pulse repetition frequency. **ANSWER A**

8-31D2 The circuit that develops timing signals is called the:
A. Resolver.
B. Synchronizer.
C. Pulse forming network.
D. Video amplifier.
The synchronizer sends timing pulses to the pulse repetition frequency stage, and the pulse repetition frequency stage is usually managed by a sine wave oscillator, multivibrator, or blocking oscillator. **ANSWER B**

8-31D3 Which of the following functions is not affected by the timing circuit?
A. Resolver output.
B. Pulse repetition frequency.
C. Sweep drive.
D. Modulation.
Modulation, sweep, sweep drive, and pulse repetition frequency are all determined by the timing circuit, but resolver output is independent. **ANSWER A**

8-31D4 The synchronizer primarily affects the following circuit or function:
A. Mixer.
B. Receiver.
C. Modulator.
D. I.F. Amplifier.
The timing synchronizer will always drive the modulator. **ANSWER C**

8-31D5 The output from the synchronizer usually consists of a:
A. Sine wave.
B. Pulse or square wave.
C. Triangle wave.
D. None of the above.
The timer synchronizer output is applied to pulse shaping circuits to produce pulses or square waves. **ANSWER B**

8-31D6 The sweep drive is initiated by what circuit?
A. Resolver.
B. Sweep amplifier.
C. Video amplifier.
D. Synchronizer.
The sweep drive is initiated by the synchronizer, too. **ANSWER D**

Key Topic 32 – Fixed Range Markers

8-32D1 Accurate range markers must be developed using very narrow pulses. A circuit that could be used to provide these high-quality pulses for the CRT is a:
A. Ringing oscillator.
B. Monostable multivibrator.
C. Triggered bi-stable multivibrator.
D. Blocking oscillator.
A blocking oscillator, like all oscillators, makes use of feedback. That is, some of the output of the circuit is sent back to the input. But, as its name suggests, the blocking oscillator "blocks out" most of the input and feedback, and allows only a very sharp, narrow pulse at its output. Of course this is perfect for generating the thin, unobtrusive range rings that allow you to see the distances to targets on a RADAR scope without "blocking out" the targets. **ANSWER D**

8-32D2 Range markers are determined by:
A. The CRT.
B. The magnetron.
C. The timer.
D. The video amplifier.
It's the timer circuit that determines range rings. **ANSWER C**

8-32D3 A gated LC oscillator, operating at 27 kHz, is being used to develop range markers. If each cycle is converted to a range mark, the range between markers will be:

A. 3 nautical miles.
B. 6 nautical miles.
C. 8 nautical miles.
D. 12 nautical miles.

Once in a while, just memorizing some numbers is a great help. Working with nautical miles is one such case. The speed of light in nautical miles/second (knots) is 162,000. 162,000 ÷ 27,000 = 6 (nautical miles). But wait! The cycle is the round trip time. We need to calculate the ONE-WAY trip time. So 6 ÷ 2 = 3 nautical miles. **ANSWER A**

Two range rings show up on the 1/4 nm scale.

8-32D4 What would be the frequency of a range ring marker oscillator generating range rings at 10 nautical miles intervals?

A. 24 kHz.
B. 16 kHz.
C. 12 kHz.
D. 8 kHz.

Radio waves travel one nautical mile in six microseconds. How long do they take to travel 10 nautical miles? 10 × 6 = 60 microseconds. Now, it's always less confusing to do these calculations in the basic units, like seconds and Hertz, instead of "prefixed" units like microseconds and kilohertz. We can convert back to the desired units after we're done with the number crunching. So let's convert our time interval to seconds, which gives us 0.000060 seconds. But we need to double this number to get the ROUND TRIP time. That give us 0.000060 × 2 = 0.00012 seconds. To get our final answer in frequency, we just need to take the reciprocal of our time, or 1 ÷ 0.00012 sec. = 8333 Hz, or 8.333 kHz. Now, remember, the FCC wants the best answer, even if it's not a perfect answer. 8 kHz is the closest answer to 8.333 kHz, isn't it? **ANSWER D**

8-32D5 What is the distance between range markers if the controlling oscillator is operating at 20 kHz?

A. 1 nautical miles.
B. 2 nautical miles.
C. 4 nautical miles.
D. 8 nautical miles.

First, convert oscillator frequency into time, or period. This is easy enough – simply take the reciprocal. 1 ÷ 20,000 = 0.000050 seconds (50 microseconds). Our old friend "seasix" shows up again; 6 microseconds per nautical mile. Divide our 50 microseconds by 6 microseconds-per-mile, which gives us 8.333 nautical miles. We can round that off to 8 miles because the "seasix" 6 we use is also rounded off a bit. But wait! We have to cut that distance in half because our RADAR signal has to make a round trip. 8 ÷ 2 = 4. Ah, that's better! **ANSWER C**

8-32D6 What would be the frequency of a range ring marker oscillator generating range rings at intervals of 0.25 nautical miles?

A. 161 kHz.
B. 322 kHz.
C. 644 kHz.
D. 1288 kHz.

What's with all of these range marker questions, anyway? Well, this is one of the most important functions you'll do as a RADAR operator – being able to quickly detect the range of an object at sea. A few days out at sea and this will become second nature. You'll want to be able to think on your feet, in a blinding storm, in the dark! This is by far the "closest in" calculation that we've done, a mere quarter of a nautical mile. And, as you guessed, old "seasix" just won't go away. The difference here is that we're going to be LESS than 6 microseconds. We can either multiply our 6 by 0.25, or divide by 4; either way we get the same answer, 1.5 microseconds (0.0000015 sec). This is our period. But remember, we have to double our period to account for our round trip – there and back. So we have 2 x 0.0000015 sec = 0.000003 sec. Now, to get our frequency, we take the reciprocal of our time interval. 1 ÷ 0.000003 = 333,333 Hz. Or 333kHz, if you prefer. The closest answer provided is 322 kHz. Close enough for government work, or FCC exams. **ANSWER B**

Key Topic 33 – Variable Range Markers

8-33D1 The variable range marker signal is normally fed to the input of the:

A. Sweep amplifier.
B. Low voltage power supply regulator.
C. Video amplifier.
D. Range ring oscillator.

The variable range marker is an important display feature. It allows the operator to calculate the distance to a target. Non-variable range marks are equally spaced, and are produced only for the duration of the range marker gate pulse. The variable range marker signal is an additional range ring, fed to the input of the video amplifier, and ultimately displayed on the RADAR indicator. **ANSWER C**

8-33D2 The purpose of the variable range marker is to:

A. Provide an accurate means of determining the range of a moving target.
B. Provide a bearing line between own ship and a moving target.
C. Indicate the distance between two different targets.
D. Provide a means of calibrating the fixed range rings.

The variable range marker allows you to determine the range of any target on the display. **ANSWER A**

Variable range rings help determine range to a distant target.

8-33D3 How is the variable range marker usually adjusted for accuracy?

A. Adjusting the frequency of the VRM oscillator at the maximum range.
B. Adjusting the frequency of the VRM oscillator at the minimum range.
C. Adjusting the readout to match at the median range ring.
D. The minimum and maximum ranges are aligned with the matching fixed range ring.

The adjustment of variable range marker readouts is accomplished by first calibrating minimum range and maximum range readouts with the previously calculated fixed range rings. A technician will usually do this on several different range ring scales to insure that the variable range ring always agrees with minimum and maximum fixed range ring readouts. **ANSWER D**

8-33D4 The panel control for the variable range marker is normally a:

A. Variable resistor.
B. Variable inductance.
C. Variable capacitance.
D. Variable resolver.

The actual panel control for a variable range ring marker is a potentiometer – a variable resistor. **ANSWER A**

8-33D5 An important component of the VRM system is the:

A. Resolver.
B. Interference rejection.
C. STC sensitivity control.
D. Shift register.

The variable range marker system is fed timing triggers by a master oscillator, and then fed to dividing circuits, usually consisting of flip-flops, where pulses may be variably divided to longer range readouts. The shift register faithfully tracks any changes dialed-in by the operator adjusting the variable range marker. **ANSWER D**

8-33D6 Which of the following statements about the Variable Range Marker system is correct?

A. The VRM is an auxiliary output of the fixed range marker oscillator.
B. The VRM system develops a single adjustable range ring.
C. The VRM system is calibrated using a frequency counter.
D. The VRM system is controlled by a crystal oscillator.

A variable range marker indicator is usually a single adjustable range ring. **ANSWER B**

Key Topic 34 – EBL, Azimuth and True Bearing

8-34D1 The purpose of the Electronic Bearing Line is to:
A. Indicate your own vessel's heading.
B. Measure the bearing of a specific target.
C. Indicate True North.
D. Display the range of a specific target.
It is likely that a RADAR indicator may also include a variable electronic bearing line. This allows you to move the cursor so that the EBL intersects a specific echo. The EBL is a BEARING to that echo. **ANSWER B**

8-34D2 The Electronic Bearing Line is:
A. The ships heading line.
B. A line indicating True North.
C. Used to mark a target to obtain the distance.
D. A line from your own vessel to a specific target.
The electronic bearing line does not show distance – that is the job of the variable range ring. The EBL is a bearing from your own vessel to a specific target echo. **ANSWER D**

The electronic bearing line helps determine direction to a distant target.

8-34D3 Which of the following inputs is required to indicate azimuth?
A. Gyro signals.
B. Synchronizer.
C. Resolver.
D. Range rings.
In order for the RADAR to indicate azimuth, it will need an input from a gyro compass or the modern fluxgate compass readout. **ANSWER A**

8-34D4 Bearing information from the gyro is used to provide the following:
A. The heading of the nearest target.
B. Range and bearing to the nearest target.
C. Vessel's own heading.
D. The range of a selected target.
The gyro provides the vessel's true heading to a readout at the helm station, plus redundant readouts throughout the bridge and to the automatic pilot. The gyro also feeds the RADAR, which displays the vessel's true heading as well as true North.. **ANSWER C**

8-34D5 Which of the following statements about "true bearing" is correct?
A. The ship's heading flasher is at the top of the screen.
B. True North is at the top of the screen and the heading flasher indicates the vessel's course.
C. The true bearing of the nearest target is indicated.
D. The relative bearing of the nearest target is indicated.
RADARs fed by a gyro or fluxgate compass may incorporate an indicator outside the compass rose that will illustrate true north. This will be at the top of the screen, and the heading flasher indicates the vessel's course. **ANSWER B**

8-34D6 A true bearing presentation appears as follows:
A. The bow of the vessel always points up.
B. The course of the five closest targets is displayed.
C. North is at the top of the display and the ship's heading flasher indicates the vessel's course.
D. The course and distance of the closest target is displayed.
A true bearing presentation will show North at the top of the display, and the ship's heading flasher indicating the course. **ANSWER C**

Key Topic 35 – Memory Systems

8-35D1 In a digitized RADAR, the 360 degree sweep is divided into how many digitized segments?

A. 16 segments.
B. 64 segments.
C. 255 segments.
D. 4096 segments.

In a digital RADAR system, return pulse data is collected in a series of spokes. These are narrow regions of space radiating away from the rotating antenna, like the spokes of a wheel. There are 4096 radials or spokes of data for each revolution, a little over 11 spokes per degree. The return data from each of these spokes is collected in a data bin for further processing.

Since a typical radar beam can be several degrees wide, one might wonder why one would need so many spokes. For any position of the beam you could have perhaps 40 to 50 return spokes of overlapping data. This is a good question with an interesting answer.

The multiple spokes allow various bin processing methods, such as data averaging, to be used to greatly increase the effective signal-to-noise ratio of the system. Alternatively, by detecting subtle changes in return time between several spokes within a single beam heading, you can derive Doppler shift (motion) of targets, even at great distances.

The sample rate of a digital RADAR analog-to-digital converter determines the range resolution of a digital RADAR. A typical RADAR A/D converter takes 50,000 samples per second, which gives a range resolution of about 3.75 statute miles.

After the data is digitized, it's a simple matter to process it for useful and detailed display on a modern computer video monitor, such as using color to display return strength. Generally, strong returns will be displayed as "hot" colors, while weaker returns are displayed as "cool" colors. **ANSWER D**

8-35D2 While troubleshooting a memory problem in a raster scan RADAR, you discover that the "REFRESH" cycle is not operating correctly. What type of memory circuit are you working on?

A. SRAM.
B. DRAM.
C. ROM.
D. PROM.

Dynamic random access memory (DRAM) requires a refresh cycle. SRAM does not because SRAM memory is changed when another bit replaces a stored bit. ROM and PROM memory are permanently stored. **ANSWER B**

8-35D3 The term DRAM stands for

A. Digital refresh access memory.
B. Digital recording access memory.
C. Dynamic random access memory.
D. Digital response area motion.

DRAM stands for Dynamic Random Access Memory, where each RADAR sweep response is digitally compared to the previous response. Only signals which re-appear are then displayed on the indictor unit. **ANSWER C**

8-35D4 How does the dual memory function reduce sea clutter?

A. Successive sweeps are digitized and compared. Only signals appearing in both sweeps are displayed.
B. The dual memory system makes the desired targets larger.
C. It reduces receiver gain for closer signals.
D. It increases receiver gain for real targets.

The modern RADAR with DRAM may offer multiple levels of quantization, requiring a real target echo to continuously re-appear in each sweep before it is verified and passed on to the display unit. **ANSWER A**

8-35D5 How many sequential memory cells with target returns are required to display the target?

A. 1.
B. 2.
C. 4.
D. 8.

In the dynamic refresh access memory (DRAM) it usually takes two sequential memory cell target returns to cause the display to show the echo. **ANSWER B**

8-35D6 What is the primary purpose of display system memory?

A. Eliminate fluctuating targets such as sea return.
B. Display stationary targets.
C. Display the last available targets prior to a power dropout.
D. Store target bearings.

Newer RADAR DRAM circuitry helps minimize unwanted sea return, small birds, and other momentary false echoes. **ANSWER A**

Key Topic 36 – ARPA - CAS

8-36D1 The ship's speed indication on the ARPA display can be set manually, but does not change with changes in the vessel's speed. What other indication would point to a related equipment failure?

A. "GYRO OUT" is displayed on the ARPA indicator.
B. "LOG OUT" is displayed on the ARPA indicator.
C. "TARGET LOST" is displayed on the ARPA indicator.
D. "NORTH UP" is displayed on the ARPA indicator.

On an ARPA (Automatic RADAR Plotting Aid) RADAR display, a "Log Out" error message indicates the unit is not receiving information from an external speed source such as GPS, digital paddlewheel information, or Loran. "Gyro Out" indicates a failure of heading information, not speed. **ANSWER B**

8-36D2 What does the term ARPA/CAS refer to?

A. The basic RADAR system in operation.
B. The device which displays the optional U.S.C.G. Acquisition and Search RADAR information on a CRT display.
C. The device which acquires and tracks targets that are displayed on the RADAR indicator's CRT.
D. The device which allows the ship to automatically steer around potential hazards.

Big-ship RADAR systems feature ARPA/CAS, which allows automatic tracking of any target and leaves an electronic "wake" that is easily monitored on the RADAR's cathode-ray tube. **ANSWER C**

8-36D3 Which of the following would not be considered an input to the computer of a collision avoidance system?

A. Own ship's exact position from navigation satellite receiver.
B. Own ship's gyrocompass heading.
C. Own ship's speed from Doppler log.
D. Own ship's wind velocity from an anemometer.

Wind velocity is not considered a normal input to the computer collision avoidance system. **ANSWER D**

The modern mini-ARPA RADAR warns of target intrusion.

8-36D4 Which answer best describes a line on the display which indicates a target's position. The speed is shown by the length of the line and the course by the direction of the line.

A. Vector.
B. Electronic Bearing Line.
C. Range Marker.
D. Heading Marker.

The modern ARPA RADAR will compute surrounding RADAR vectors. These are electronic lines showing the target course and speed, an important indication of safe or dangerous encounters. The marine radio VHF Automatic Identification System will also superimpose traffic details on the RADAR to further alert watch personnel to details about all the other nearby vessels. **ANSWER A**

8-36D5 What is the purpose or function of the "Trial Mode" used in most ARPA equipment?

A. It selects trial dots for targets' recent past positions.
B. It is used to display target position and your own ship's data such as TCPA, CPA, etc.
C. It is used to allow results of proposed maneuvers to be assessed.
D. None of these.

Using the trial mode, you can execute intended maneuvers that cause the screen to display an intended course change. This allows you to judge the results of a proposed maneuver before you actually engage the steering system and engine speed for that maneuver. **ANSWER C**

8-36D6 The ARPA term CPA refers to:

A. The furthest point a ship or target will get to your own ship's bow.
B. Direction of target relative to your own ship's direction.
C. The combined detection and processing of targets.
D. The closest point a ship or target will approach your own ship.

CPA stands for Closest Point of Approach. (NOTE: The acronym ARPA stands for Automatic RADAR Plotting Aid.) **ANSWER D**

Big ship APRA RADARs provide the safest approach into a harbor.
Photo courtesy of Furuno Electric Co., Ltd.

Key Topic 37 – Display System Power Supplies

8-37D1 The display power supply provides the following:

A. +18 volts DC for the pulse forming network.
B. 5 volts DC for logic circuits and ± 12 volts DC for analog and sweep circuits.
C. 80 volts AC for the antenna resolver circuits.
D. All of the above.

Most logic circuits operate at 5 volts DC, and the analog and sweep circuits will operate at ± 12 volts DC. **ANSWER B**

8-37D2 The display power supply provides the following:

A. 5 volts DC for logic circuits.
B. ± 12 volts DC for analog and sweep circuits.
C. 17kV DC for the CRT HV anode.
D. All of the above.

The power supply in a cathode ray tube display will include 17 kV DC for the display tube high voltage anode, 12 volts for analog and sweep circuits, and 5 volts DC for logic circuits. **ANSWER D**

8-37D3 In a display system power supply what is the purpose of the chopper?

A. It acts as an electronic switch between the raw DC output and the inverter.
B. It interrupts the AC supply line at a varying rate depending on the load demands.
C. It regulates the 5 volt DC output.
D. It pre-regulates the AC input.

In a RADAR power supply, the purpose of the "chopper" is to switch the input DC current to a pulsating direct current that may then be fed to the inverter's power transformer. High speed switching transistors called "switchers" may handle large amounts of current. **ANSWER A**

8-37D4 In a display system power supply, what is the purpose of the inverter?

A. Inverts the polarity of the DC voltage applied to the voltage regulators.
B. Provides the dual polarity 12 volt DC supply.
C. Acts as the voltage regulator for the 5 volt DC supply.
D. Produces the pulsed DC input voltage to the power transformer.

The switching "chopper" circuit – the heart of the power supply inverter – develops a sawtooth DC input voltage that will step up the pulse DC voltage to a quasi alternating current (AC) voltage. Electrolytic capacitors will then help smooth the resulting AC waveform. **ANSWER D**

8-37D5 What would be a common switching frequency for a display system power supply?

A. 18 kHz.
B. 120 Hz.
C. 60 kHz.
D. 120 kHz.

Most switcher display power systems operate at 18 kHz. **ANSWER A**

8-37D6 What display system power supply output would use a tripler circuit?

A. The logic circuit supply.
B. The sweep circuit supply.
C. The HV supply for the CRT anode.
D. The resolver drive.

The tripler is found in the display system HV (high voltage) power supply for the 17 kV DC for the CRT anode. **ANSWER C**

Key Topic 38 – Miscellaneous

8-38D1 The heading flash is a momentary intensification of the sweep line on the PPI presentation. Its function is to:

A. Alert the operator when a target is within range.
B. Alert the operator when shallow water is near.
C. Inform the operator of the dead-ahead position on the PPI scope.
D. Inform the operator when the antenna is pointed to the rear of the ship.

The heading flash reveals the dead-ahead position of the bow of the vessel. A microswitch and a cam trigger send the heading flash to the scope. The heading flash should be adjusted about once a year. **ANSWER C**

8-38D2 The major advantage of digitally processing a RADAR signal is:

A. Digital readouts appear on the RADAR display.
B. Enhancement of weak target returns.
C. An improved operator interface.
D. Rectangular display geometry is far easier to read on the CRT.

Digital signal processing, abbreviated DSP, allows enhancement of return echoes by the use of sampling and Fast Fourier analysis. This allows reduction of interference that may come in with the return echo. A DSP circuit strips out the interference, and accentuates the real echo. **ANSWER B**

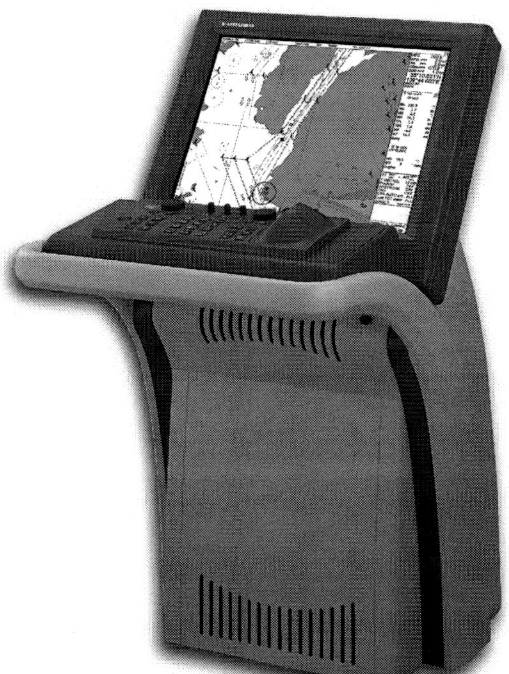

Modern RADARs overlay marine charts and show depth contours on the display.

Photo courtesy of Furuno Electric Co., Ltd.

8-38D3 In order to ensure that a practical filter is able to remove undesired components from the output of an analog-to-digital converter, the sampling frequency should be:

A. The same as the lowest component of the analog frequency.
B. Two times the highest component of the analog frequency.
C. Greater than two times the highest component of the sampled frequency.
D. The same as the highest component of the sampled frequency.

According to the Nyquist Theorem, to correctly represent the analog signal, the sampling frequency must be greater than two times the highest frequency of the analog signal being sampled. **ANSWER C**

8-38D4 Bearing resolution is:

A. The ability to distinguish two adjacent targets of equal distance.
B. The ability to distinguish two targets of different distances.
C. The ability to distinguish two targets of different elevations.
D. The ability to distinguish two targets of different size.

Bearing resolution is the ability to differentiate between two targets that are separated horizontally. If the RADAR antenna is relatively small, a wider beamwidth illustrates two individual targets as one echo. **ANSWER A**

8-38D5 The output of an RC integrator, when driven by a square wave with a period of much less than one time constant is a:

A. Sawtooth wave.
B. Sine wave.
C. Series of narrow spikes.
D. Triangle wave.

As the square wave goes positive, the capacitor in the RC circuit charges. With the square wave period much less than one time constant, the square wave returns to zero before the capacitor can charge or discharge, generating a triangular wave output. The amplitude of the wave is less than the square wave amplitude. **ANSWER D**

8-38D6 How do you eliminate stationary objects such as trees, buildings, bridges, etc., from the PPI presentation?

A. Remove the discriminator from the unit.
B. Use a discriminator as a second detector.
C. Calibrate the IF circuit.
D. Calibrate the local oscillator.

For RADAR installations used on land for the vessel traffic system (VTS), using a discriminator as a second detector is one way to minimize return echoes from stationary objects. **ANSWER B**

Subelement E – Antenna Systems

5 Key Topics, 5 Exam Questions

Key Topic 39 – Antenna Systems

8-39E1 Slotted waveguide arrays, when fed from one end exhibit:

A. Frequency scan.
B. High VSWR.
C. Poor performance in rain.
D. A narrow elevation beam.

The slotted waveguide antenna is a popular one for small shipboard RADAR antenna systems. The slots are in-phase radiating openings, similar to a stacked array for narrow horizontal beamwidth. The slotted array also has low SWR, good performance in rain, and a regular vertical beamwidth, and the slotted waveguide array is said to be frequency scan. **ANSWER A**

8-39E2 A typical shipboard RADAR antenna is a:

A. Rotary parabolic transducer.
B. Slotted waveguide array.
C. Phased planar array.
D. Dipole.

The slotted waveguide array is the most popular type of marine RADAR antenna. Although we occasionally will see a parabolic reflector or a printed circuit board antenna system, the slotted waveguide is preferred because of its low loss and predictable signal dispersion. **ANSWER B**

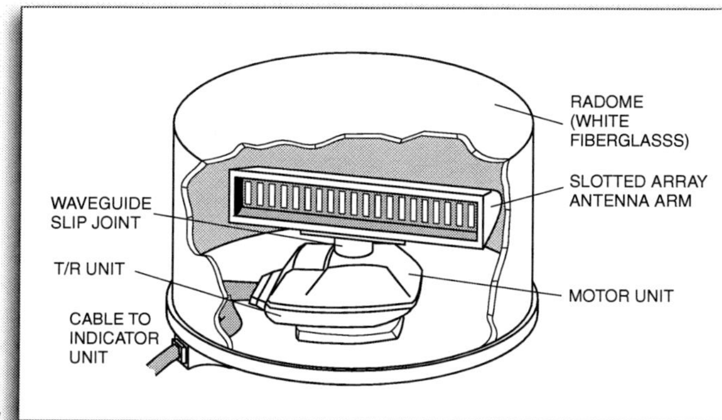

Slotted Array Antenna

8-39E3 Good bearing resolution largely depends upon:

A. A high transmitter output reading.
B. A high duty cycle.
C. A narrow antenna beam in the vertical plane.
D. A narrow antenna beam in the horizontal plane.

The resolution of the RADAR beam is proportional to the length of the rotating antenna. The narrower the beam in the horizontal plane, the better the ability to differentiate multiple targets. **ANSWER D**

8-39E4 The center of the transmitted lobe from a slotted waveguide array is:

A. Several degrees offset from a line perpendicular to the antenna.
B. Perpendicular to the antenna.
C. Maximum at the right hand end.
D. Maximum at the left hand end.

The slotted waveguide array eliminates the requirement of a large reflector parabola to concentrate the microwave energy in a tight horizontal pattern. While the slotted waveguide array may have a pair of "wings," these are usually relatively small, and will further increase the f-gain of the array as they provide a ground plane for the individual slots. Depending on how the slotted waveguide array is fed, the actual main lobe may be several degrees offset from a line perpendicular to the antenna. **ANSWER A**

8-39E5 How does antenna length affect the horizontal beamwidth of the transmitted signal?

A. The longer the antenna the wider the horizontal beamwidth.
B. The longer the antenna the narrower the horizontal beamwidth.
C. The horizontal beamwidth is not affected by the antenna length.
D. None of the above.

It is desirable to sharpen the horizontal beamwidth by providing the longest possible antenna array, improving resolution to individual targets. **ANSWER B**

PCB phased array recreational boat RADAR antenna.

8-39E6 What is the most common type of RADAR antenna used aboard commercial maritime vessels?

A. Parabolic.
B. Truncated parabolic.
C. Slotted waveguide array.
D. Multi-element Yagi array.

The slotted array is commonly used on many small boat X-band RADARs. It has a number of in-phase radiating slots that act as a stacked array, narrowing the horizontal beamwidth for increased resolution of individual targets. Slotted waveguide antennas are always encased in a fiberglass radome. **ANSWER C**

Key Topic 40 – Transmission Lines

8-40E1 The VSWR of a microwave transmission line device might be measured using:

A. A dual directional coupler and a power meter.
B. A network analyzer.
C. A spectrum analyzer.
D. A dual directional coupler, a power meter, and a network analyzer.

Instruments used to determine VSWR of a microwave transmission line or antenna system include a dual-directional coupler, a power meter, and a network analyzer that would also allow you to determine the cause of high VSWR. Now that most antenna units are integral to the T/R units, waveguide runs are either relatively short or non-existent. A network analyzer would NEVER be used on a live system, while the directional coupler and power meter can be used "live". **ANSWER D**

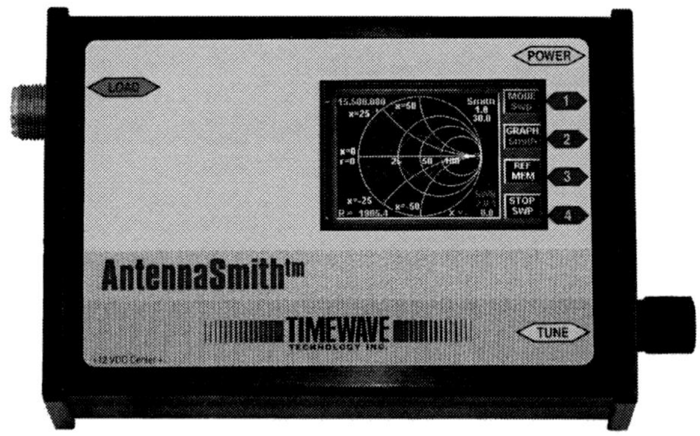

The AntennaSmith is an automatic antenna impedance analyzer. Built-in software allows you to instantly see the effect of an adjustment or a change made to your antenna system. No computer connection is required to see all the graphs – even the Smith chart – and the unit can run on battery power for use in the field.

8-40E2 The impedance total (Z_o) of a transmission line can be calculated by $Z_o = \sqrt{L/C}$ when L and C are known. When a section of transmission line contains 250 microhenries of L and 1000 picofarads of C, its impedance total (Z_o) will be:

A. 50 ohms.
B. 250 ohms.
C. 500 ohms.
D. 1,000 ohms.

$Z_O = \sqrt{(250 \times 10^{-6})/1000 \times 10^{-12}} = \sqrt{(0.25 \times 10^{6})} = \sqrt{250{,}000} = 500$ ohms. **ANSWER C**

8-40E3 If long-length transmission lines are not properly shielded and terminated:

A. The silicon crystals can be damaged.
B. Communications receiver interference might result.
C. Overmodulation might result.
D. Minimal RF loss can result.

If long RADAR cables are not properly shielded or terminated, they radiate signals that can interfere with onboard communication receivers. This could also effect Loran reception, and quite possibly GPS reception. **ANSWER B**

8-40E4 A certain length of transmission line has a characteristic impedance of 72 ohms. If the line is cut at its center, each half of the transmission line will have a Z_o of:

A. 36 ohms.
B. 144 ohms.
C. 72 ohms.
D. The exact length must be known to determine Z_O.

Changing the length of a transmission line will not affect its characteristic impedance. **ANSWER C**

8-40E5 Standing waves on a transmission line may be an indication that:

A. All energy is being delivered to the load.
B. Source and surge impedances are equal to Z_O and Z_L.
C. The line is terminated in impedance equal to Z_O.
D. Some of the energy is not absorbed by the load.

A transmission line with high SWR will sometimes feel warm to the touch. This is because some of the energy is not absorbed by the load (the antenna system) and points along the transmission line will exhibit hot spots. **ANSWER D**

8-40E6 What precautions should be taken with horizontal waveguide runs?

A. They should be sloped slightly downwards at the elbow and a small drain hole drilled in the elbow.
B. They should be absolutely level.
C. They should not exceed 10 feet in length.
D. None of the above.

While newer RADAR transmitters and receivers might be built directly into the bottom side of the rotating antenna assembly, there are still many RADAR installations with the TR equipment many feet away from the rotating antenna. This requires waveguide to interconnect the TR to the antenna array, and long horizontal runs should be sloped slightly downward with a small drain hole drilled in the elbow to allow condensation either to escape or dry out. **ANSWER A**

Key Topic 41 – Antenna to Display Interface

8-41E1 The position of the PPI scope sweep must indicate the position of the antenna. The sweep and antenna positions are frequently kept in synchronization by the use of:

A. Synchro systems.
B. Servo systems.
C. DC positioning motors.
D. Differential amplifiers.

It is the job of the synchro system to keep the RADAR sweep and antenna in alignment. After several years of operation, mechanical slippage could require you to get into the antenna unit and readjust the synchro settings and heading marker microswitch. **ANSWER A**

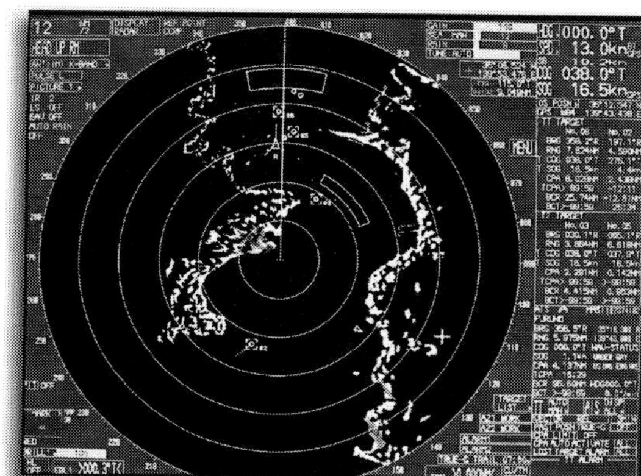

Big screen radars offer plenty of room for graphic texts, too.

Photo courtesy of Furuno Electric Co., Ltd.

8-41E2 On a basic synchro system, the angular information is carried on the:

A. DC feedback signal.
B. Stator lines.
C. Deflection coils.
D. Rotor lines.

Usually, 3 stator lines carry the angular information from the antenna to the indicator. **ANSWER B**

8-41E3 What is the most common type of antenna position indicating device used in modern RADARs?

A. Resolvers.
B. Servo systems.
C. Synchro transmitters.
D. Step motors.

Early RADARs used synchro systems that were geared at a ratio of around 10:1 and required a cam and microswitch system at both the antenna and the indicator to insure that the antenna was in step with the indicator. Today, modern RADARs rely on a resolver circuit, which constantly keeps antenna alignment in sync with the indicator readout. **ANSWER A**

8-41E4 Which of the following statements about antenna resolvers is correct?

A. Most resolvers contain a rotor winding and a delta stator winding.
B. Resolvers consist of a two rotor windings and two stator windings that are 90 degrees apart.
C. The basic resolver contains a rotor winding and two stator windings that are 90 degrees apart.
D. Resolvers consist of a "Y" connected rotor winding and a delta connected stator winding.

The resolver is a rotor-winding and two stator windings that are ninety degrees apart, continuously yielding information on the travels of the rotating antenna. **ANSWER C**

8-41E5 An antenna synchro transmitter is composed of the following:

A. Three rotor and two stator windings.
B. Two rotor and three stator windings.
C. Three rotor and three stator windings.
D. A single rotor and 3 stator windings.

Another type of antenna position monitoring device is the antenna synchro transmitter, with TWO rotor and THREE stator windings. The rotor winding is usually split into two series parts, for balance, with just one external connection, and for this question and answer, remember TWO rotor windings, three stator windings! **ANSWER B**

8-41E6 RADAR antenna direction must be sent to the display in all ARPAs or RADAR systems. How is this accomplished?

A. 3-phase synchros.
B. 2-phase resolvers.
C. Optical encoders.
D. Any of the above.

The modern ARPA RADAR system might employ synchros, resolvers, and optical encoders to reference the physical RADAR antenna direction to the scanned display. Any of these are found in ARPA systems. **ANSWER D**

Key Topic 42 – Waveguides-1

8-42E1 Waveguides can be constructed from:

A. Brass.
B. Aluminum.
C. Copper.
D. All of the above.

Waveguide construction requires tighter tolerances, and must be protected from any forces that could damage the integrity of the waveguide run. Sturdy brass may be good for long runs, but aluminum and copper will keep weight down aloft. **ANSWER D**

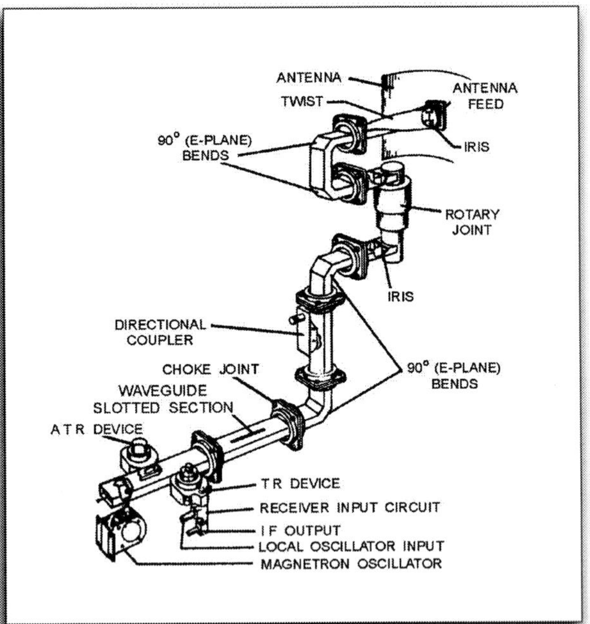

Waveguide "plumbing."

8-42E2 A microwave transmission line constructed of a center conductor suspended between parallel conductive ground planes is called:

A. Microstrip.
B. Coax.
C. Stripline.
D. Waveguide.

Stripline circuitry is used at microwave frequencies in both transmitting and receiving circuits. In fact, small boat marine RADARs utilize stripline circuitry as the actual radiating antenna elements. **ANSWER C**

8-42E3 Waveguide theory is based upon:

A. The movement of an electromagnetic field.
B. Current flow through conductive wires.
C. Inductance.
D. Resonant charging.

Waveguides allow the movement of microwave electromagnetic fields within the inner walls of the conductor. RADAR technicians aptly name waveguides as "microwave plumbing". **ANSWER A**

8-42E4 A waveguide is used at RADAR microwave frequencies because:

A. It is easier to install than other feedline types.
B. It is more rugged than other feedline types.
C. It is less expensive than other feedline types.
D. It has lower transmission losses than other feedline types.

Waveguide is used exclusively at RADAR microwave frequencies because losses are lower than coaxial feedline. Waveguide analysis is similar to open line; that is, two parallel lines. At microwave frequencies it is not practical to have open line because it must be insulated and high-quality insulators are hard to make for microwave frequencies. **ANSWER D**

8-42E5 Waveguide theory is based on the principals of:

A. Ohm's Law.
B. High standing waves.
C. Skin effect and use of ¼ wave stubs.
D. None of the above.

It is important to never allow moisture to collect on the inside wall of waveguides. This would reduce the desired skin effect of the waveguide conductive material. **ANSWER C**

8-42E6 How is the signal removed from a waveguide or magnetron?

A. With a thin wire called a T-hook.
B. With a thin wire called a J-Hook.
C. With a coaxial connector.
D. With a waveguide flange joint.

It is called the "J-Hook" – that section in the waveguide that actually removes the signal. **ANSWER B**

Key Topic 43 – Waveguides-2

8-43E1 A rotary joint is used to:

A. Couple two waveguides together at right angles.
B. Act as a switch between two waveguide runs.
C. Connect a stationary waveguide to the antenna array.
D. Maintain pressurization at the end of the waveguide.

The rotary joint connects the piece of fixed waveguide from the transceiver to the antenna that is rotating. This joint should be inspected often to insure proper alignment and that no moisture has leaked in. **ANSWER C**

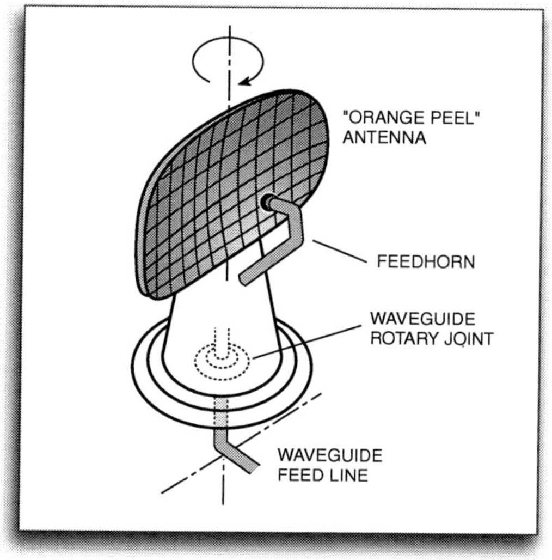

Typical Surface Search Radar Antenna.

8-43E2 Resistive losses in a waveguide are very small because:

A. The inner surface of the waveguide is large.
B. The inner surface of the waveguide is small.
C. The waveguide does not require a ground connection.
D. The heat remains in the waveguide and cannot dissipate.

A waveguide provides a boundary which allows the wave to be propagated freely. Only the boundary's inner surface, which is relatively large in comparison with the wave, confines the energy. Waveguides with proper dimensions must be used for the particular band on which the RADAR is operating. **ANSWER A**

8-43E3 A right-angle bend in an X-band waveguide must have a radius greater than:

A. Three inches.
B. Six inches.
C. One inch.
D. Two inches.

Sharp bends in an X-band waveguide run will cause signal transmission problems. A bend with a radius of 2 inches or greater will provide a smooth path with minimal "bumps" along the transmission line. Since most waveguide runs are short and confined to the antenna unit, the factory that has manufactured the equipment has already taken this into account. **ANSWER D**

8-43E4 To insert RF energy into or extract RF energy from a waveguide, which of the following would not be used?

A. Coupling capacitance.
B. Current loop.
C. Aperture window.
D. Voltage probe.

For this question, they want to know what would NOT be used to insert RF or extract energy to a waveguide. Coupling capacitance would not be used for this purpose. All of the other answers would be appropriate methods for inserting RF energy into, or extracting RF energy from, the waveguide. **ANSWER A**

8-43E5 The following is true concerning waveguides:

A. Conduction is accomplished by the polarization of electromagnetic and electrostatic fields.
B. Ancillary deflection is employed.
C. The magnetic field is strongest at the center of the waveguide.
D. The magnetic field is strongest at the edges of the waveguide.

The magnetic field is strongest along the copper edges of the waveguide, and the electric field is strongest at the center of the waveguide. It's important that the inside of a waveguide be as clean as possible. Since new ship installations don't require long waveguide runs, this waveguide question really applies to older (very old!) installations. **ANSWER D**

8-43E6 At microwave frequencies, waveguides are used instead of conventional coaxial transmission lines because:

A. They are smaller and easier to handle.
B. They have considerably less loss.
C. They are lighter since they have hollow centers.
D. Moisture is never a problem with them.

At microwave frequencies, waveguides are the preferred method of coupling microwave energy from the transmitter to the antenna because they have considerably less loss than coaxial cable. **ANSWER B**

Subelement F – Installation, Maintenance & Repair

7 Key Topics, 7 Exam Questions

Key Topic 44 – Equipment Faults-1

8-44F1 When you examine the RADAR you notice that there is no target video in the center of the CRT. The blank spot gets smaller in diameter as you increase the range scale. What operator front panel control could be misadjusted?

A. TUNE.
B. Sensitivity Time Control (STC).
C. Anti-Clutter Rain (ACR).
D. False Target Elimination (FTE).

One way of minimizing sea echo return is to switch on the STC circuit. If the control is turned too high, you won't receive any echoes in that small spot in the center of the screen. Limit the use of STC when cruising in calm waters. This way, you won't miss small, close-in targets. **ANSWER B**

8-44F2 Range rings on the PPI indicator are oval in shape. Which circuit would you suspect is faulty?

A. Timing circuit.
B. Video amplifier circuit.
C. Range marker circuit.
D. Sweep generation circuit.

If the range rings on the scope are more oval than round, you are going to need to make a slight adjustment to the sweep generation circuit to bring the rings back into a round shape. This is a yearly calibration process that should be checked every time you are inside the RADAR lower unit. **ANSWER D**

8-44F3 What would be the most likely defective area when there is no target video in the center of the CRT and the blank spot gets smaller in diameter as your range scale is increased?

A. The TR (TRL) Cell.
B. The local oscillator is misadjusted.
C. Video amplifier circuit.
D. The IF amplifier circuit.

If you still have a blank spot in the middle of the tube, you might suspect misadjustment of the internal TR cell timing circuitry. If not properly timed, the TR cell desensitizes the receiver; this affects close-in echoes most. **ANSWER A**

8-44F4 While the vessel is docked the presentation of the pier is distorted near the center of the PPI with the pier appearing to bend in a concave fashion. This is a primary indication of what?

A. The deflection coils need adjusting.
B. The centering magnets at the CRT neck need adjusting.
C. The waveguide compensation delay line needs adjusting.
D. The CRT filaments are weakening.

This is a common problem when RADAR cables have been lengthened or shortened since the last timing adjustment. In the old days of waveguides, they called this "compensation delay line" adjusting. Now it's simply the "timing" circuit to compensate for different lengths of cable between the indicator unit and the antenna/TR unit. Aboard small boats where different cable lengths may be ordered, it is essential that you time the RADAR properly by observing a straight pier or breakwater, and adjust the timing control so the echo indeed looks straight – not concave or convex. **ANSWER C**

8-44F5 In a RADAR using digital video processing, a bright, wide ring appears at a fixed distance from the center of the display on all digital ranges. The transmitter is operating normally. What receiver circuit would you suspect is causing the problem?

A. VRM circuit.
B. Video storage RAM or shift register.
C. Range ring generator.
D. EBL circuit.

In diagnosing this symptom of a wide ring that appears on all digital ranges at the same distance from the center of the display, you should suspect a problem in the video random access memory (VRAM). It should generate proper range rings that change in proportion to the RADAR range you select. **ANSWER B**

8-44F6 The raster scan RADAR display has missing video in a rectangular block on the screen. Where is the most likely problem area?

A. Horizontal sweep circuit.
B. Power supply.
C. Memory area failure.
D. Vertical blanking pulse.

When a modern raster scan RADAR all of a sudden develops a blank video block on the screen, part of the digitally-stored signal is missing. Because it is only a small rectangular block on the screen, the most likely failure area is in the RAM or memory area. The other problems create failures of the entire video display, not a small block. **ANSWER C**

Key Topic 45 – Equipment Faults-2

8-45F1 A circuit card in a RADAR system has just been replaced with a spare card. You notice the voltage level at point E in Fig. 8F12 is negative 4.75 volts when the inputs are all at 5 volts. The problem is:

A. The 25 K resistor is open.
B. The 100 K resistor has been mistakenly replaced with a 50 K resistor.
C. The op amp is at the rail voltage.
D. The 50 K resistor has been mistakenly replaced with a 25 K resistor.

The circuit is a summing amplifier that uses an inverting op amp operating from a power supply of 15V and 25V. In the summing amplifiers of 8F12, the output voltage is:

$$V_O = -R_F \; \frac{V_A}{200\text{ k}\Omega} + \frac{V_B}{100\text{ k}\Omega} + \frac{V_C}{50\text{ k}\Omega} + \frac{V_D}{25\text{ k}\Omega}$$

when $V_A = V_B = V_C = V_D = 5\text{ V}$

$$V_O = -10\text{ k}\Omega \; \frac{5}{200} \times 10^{-3} + \frac{5}{100} \times 10^{-3} + \frac{5}{50} \times 10^{-3} + \frac{5}{25} \times 10^{-3}$$

$$V_o \quad = -10 \times 10^3 \,(0.025 + 0.05 + 0.1 + 0.2) \times 10^{-3}$$

$$= -10 \times 0.375 = -3.75\text{ V}$$

If the output $V_O = 4.75$, the current has increased by 0.1 mA, caused by the mistaken replacement of a 25-kΩ resistor for a 50-kΩ input resistor. **ANSWER D**

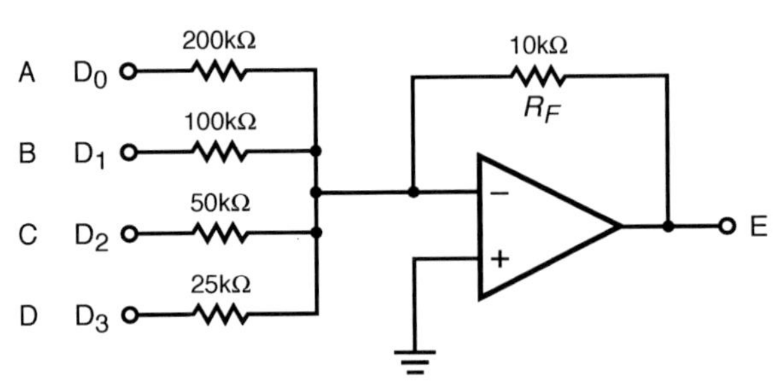

Figure 8F12

8-45F2 A defective crystal in the AFC section will cause:

A. No serious problems.
B. Bright flashing pie sections on the PPI.
C. Spiking on the PPI.
D. Vertical spikes that constantly move across the screen.

When a crystal goes out in the AFC section of a phosphorous PPI scope, bright flashing pie sections appear on the screen. **ANSWER B**

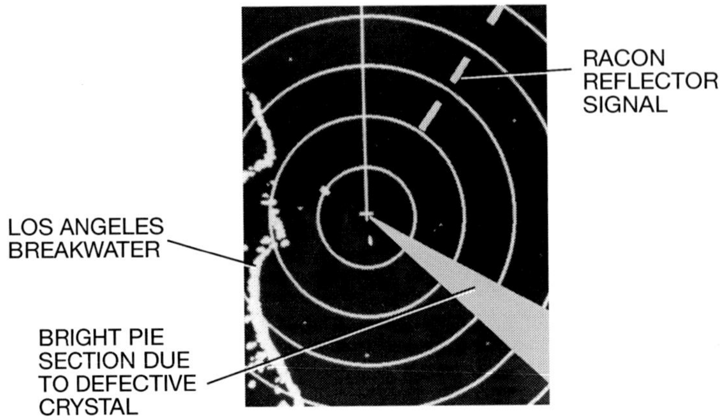

Effect of defective crystal on RADAR screen.

8-45F3 The RADAR display has sectors of solid video (spoking). What would be the first thing to check?

A. Antenna information circuits failure.
B. Frequency of raster scan.
C. For interference from nearby ships.
D. Constant velocity of antenna rotation.

Spoking is common in big harbors where other RADARs are turned on and operating in the vicinity. Spoking is caused by interference from nearby ships operating RADAR within the same band that your RADAR is tuned to. Turning on the interference rejection circuits helps minimize spoking, but the IR circuitry might cause you to miss on-purpose "interference" echoes from a nearby RACON. **ANSWER C**

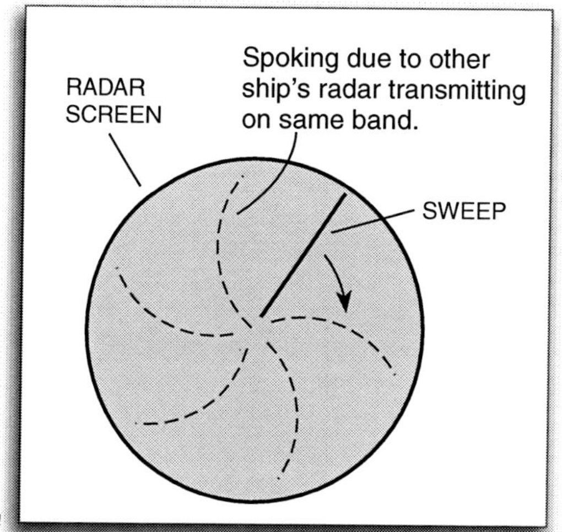

Spoking on RADAR screen

8-45F4 In the circuit contained in Fig. 8F12, there are 5 volts present at points B and C, and there are zero volts present at points A and D. What is the voltage at point E?

A. −1.5 Volts.
B. 3.75 Volts.
C. 23.75 Volts.
D. 4.5 Volts.

Using the formula developed for question 8-45F1, the output voltage is:

$$V_o = -10\ \text{k}\Omega\ (0 + 0.05 + 0.1 + 0) \times 10^{-3} = -10 \times 0.15 = -1.5\ \text{V}$$

Only inputs B and C generate input current, not A and D. **ANSWER A**

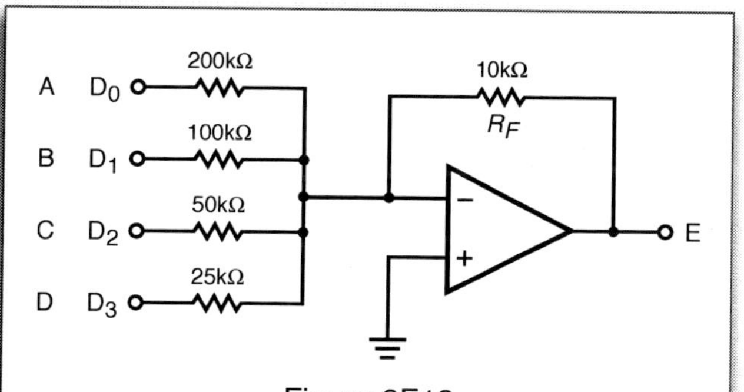

Figure 8F12

8-45F5 If the TR tube malfunctions:

A. The transmitter might be damaged.
B. The receiver might be damaged.
C. The klystron might be damaged.
D. Magnetron current will increase.

If the TR tube malfunctions in the RADAR, it could cause RADAR energy to leak into the receiver section, permanently damaging the receiver. **ANSWER B**

8-45F6 The indicated distance from your own vessel to a lighthouse is found to be in error. What circuit would you suspect?

A. Range ring oscillator.
B. Video amplifier.
C. STC circuit.
D. FTC circuit.

When troubleshooting a RADAR, a lighthouse calculated to be two miles away is found to be three miles away! The most likely circuit that will require re-tuning is the range ring oscillator. **ANSWER A**

Key Topic 46 – Equipment Faults-3

8-46F1 Silicon crystals are used in RADAR mixer and detector stages. Using an ohmmeter, how might a crystal be checked to determine if it is functional?

A. Its resistance should be the same in both directions.
B. Its resistance should be low in one direction and high in the opposite direction.
C. Its resistance cannot be checked with a dc ohmmeter because the crystal acts as a rectifier.
D. It would be more appropriate to use a VTVM and measure the voltage drop across the crystal.

Sometimes you're out there with nothing more than a VOM to run a basic check of a RADAR system that may not be operating properly. A silicon crystal looks similar to a diode if the crystal is functional; that is, resistance is low in one direction and high in the other. **ANSWER B**

8-46F2 In a RADAR unit, if the crystal mixer becomes defective, replace the:

A. Crystal only.
B. The crystal and the ATR tube.
C. The crystal and the TR tube.
D. The crystal and the klystron.

The point-contact crystals in the mixer can be damaged by any current exceeding 1 mA. The TR switch (tube) helps to protect the crystal from damage. If the crystal goes out, it may have gone bad because of a defective TR tube. Check out the tube before replacing the crystal. **ANSWER C**

8-46F3 An increase in magnetron current that coincides with a decrease in power output is an indication of what?

A. The pulse length decreasing.
B. A high SWR.
C. A high magnetron heater voltage.
D. The external magnet weakening.

When a magnetron begins to grow weak, the problem is usually weakening of the permanent magnet, which results in an increase in magnetron current, a slight change in transmit frequency, and reduced power output (or no power output at all). **ANSWER D**

8-46F4 It is reported that the RADAR is not receiving small targets. The most likely causes are:

A. Magnetron, IF amplifier, or receiver tuning.
B. PFN, crystals, or processor memory.
C. Crystals, local oscillator tuning, or power supply.
D. Fuse blown, IF amp, or video processor.

The magnetron is the main transmitting tube used for transmitting RADAR microwaves. If the tube gets weak, power output is reduced, and so are the echoes. Large targets still show up, but small targets with weak echoes won't have enough return signal strength because of the weak tube. A weak IF amp in the receiver or improperly tuned outside receiver can also cause small targets to not return a strong enough echo to show up on the display. All of the other answers are incorrect because any failure of those other components would lead to absolutely no reception of any targets. **ANSWER A**

8-46F5 A high magnetron current indicates a/an:

A. Defective AFC crystal.
B. Increase in duty cycle.
C. Defective external magnetic field.
D. High standing wave ratio (SWR).

If there is high magnetron current, chances are the permanent magnet inside the magnetron is getting weak. **ANSWER C**

8-46F6 Low or no mixer current could be caused by:

A. Local oscillator frequency misadjustment.
B. TR cell failure.
C. Mixer diode degradation.
D. All of the above.

Low mixer current could be caused by the inputs of the local oscillator/TR cell failing, or by the mixer diode itself. Local oscillator frequency misadjustment can also cause low or no mixer current. **ANSWER D**

Key Topic 47 – Equipment Faults-4

8-47F1 If the magnetron is allowed to operate without the magnetic field in place:
A. Its output will be somewhat distorted.
B. It will quickly destroy itself from excessive current flow.
C. Its frequency will change slightly.
D. Nothing serious will happen.
If the magnetic field in the magnetron gets weak, the current flow to the anode is unrestrained, and the tube overheats and destroys itself. **ANSWER B**

8-47F2 Targets displayed on the RADAR display are not on the same bearing as their visual bearing. What should you first suspect?
A. A bad reed relay in the antenna pedestal.
B. A sweep length misadjustment.
C. One phase of the yoke assembly is open.
D. Incorrect antenna position information.
Discrepancies between antenna and display position are usually the result of either the antenna or the display device becoming loose in its mounting and rotating. In the case of a geared synchro system with large discrepancies, the synchro microswitch in the antenna or display should be checked. **ANSWER D**

8-47F3 Loss of distant targets during and immediately after wet weather indicates:
A. A leak in waveguide or rotary joint.
B. High atmospheric absorption.
C. Dirt or soot on the rotary joint.
D. High humidity in the transmitter causing power supply loading.
Moisture can accumulate and leak into the rotary waveguide joint feeding the antenna. Moisture inside a waveguide can severely attenuate the RADAR signals. **ANSWER A**

8-47F4 In a marine RADAR set, a high VSWR is indicated at the magnetron output. The waveguide and rotary joint appear to be functioning properly. What component may be malfunctioning?
A. The magnetron.
B. The waveform generator.
C. The STC circuit.
D. The waveguide array termination.
High VSWR at the magnetron output indicates a problem beyond the output to the antenna system. The incorrect answers are all components within the RADAR itself, not at the antenna. You would check the waveguide array termination for corrosion. **ANSWER D**

8-47F5 On a vessel with two RADARs, one has a different range indication on a specific target than the other. How would you determine which RADAR is incorrect?
A. Check the sweep and timing circuits of both indicators for correct readings.
B. Triangulate target using the GPS and visual bearings.
C. Check antenna parallax.
D. Use the average of the two indications and adjust both for that amount.
The timing circuit, or sychronizer, performs a number of functions, including control of range markers and pulse repetition frequency. If a RADAR cable between the antenna unit and the indicator unit is replaced, lengthened, or shortened, you need to recalibrate the sweep and timing circuits for correct readings. You know you have a timing circuit problem if a straight breakwater looks concave or convex on the screen. Consult the technical manual on how to recalibrate and synchronize the timing circuits. **ANSWER A**

8-47F6 An increase in the deflection on the magnetron current meter could likely be caused by:
A. Insufficient pulse amplitude from the modulator.
B. Too high a B1 level on the magnetron.
C. A decrease of the magnetic field strength.
D. A lower duty cycle, as from 0.0003 to 0.0002.
In checking current at the magnetron, an increase in the deflection of the magnetron current could be caused by an older magnetron losing its permanent magnetic field. **ANSWER C**

Key Topic 48 – Maintenance

8-48F1 A thick layer of rust and corrosion on the surface of the parabolic dish will have what effect?

A. No noticeable effect.
B. Scatter and absorption of RADAR waves.
C. Decrease in performance, especially for weak targets.
D. Slightly out of focus PPI scope.

Older RADARs employ a waveguide and a horn feed which concentrates the energy into a parabola. Rust and corrosion on the surface of the parabolic dish will decrease the performance to detect distant target echoes. **ANSWER C**

8-48F2 The echo box is used for:

A. Testing the wavelength of the incoming echo signal.
B. Testing and tuning of the RADAR unit by providing artificial targets.
C. Amplification of the echo signal.
D. Detection of the echo pulses.

The echo box simulates artificial targets so you can check RADAR system performance against a known base line. **ANSWER B**

8-48F3 What should be done to the interior surface of a waveguide in order to minimize signal loss?

A. Fill it with nitrogen gas.
B. Paint it with nonconductive paint to prevent rust.
C. Keep it as clean as possible.
D. Fill it with a high-grade electrical oil.

The inside surface of waveguide must be kept absolutely clean from dirt, moisture, or filings as the RADAR antenna unit is matched up to the TR section. Always cover the waveguide with a cap to insure it stays absolutely clean. Most of the problems occur up in the antenna unit. **ANSWER C**

8-48F4 Which of the following is the most useful instrument for RADAR servicing?

A. Oscilloscope.
B. Frequency Counter.
C. R. F. Wattmeter.
D. Audio generator.

A portable LCD display oscilloscope will be your best instrument for servicing RADAR, allowing you to carefully observe waveforms when compared to those correct waveforms indicated in the service manual. **ANSWER A**

8-48F5 A non-magnetic screwdriver should always be used when replacing what component?

A. TR tube.
B. Mixer.
C. Video amplifier.
D. Magnetron.

Since the magnetron is surrounded by a massive permanent magnet, never use a metal screwdriver that could be attracted to the magnet, causing damage to the magnet, and potential personal injury. “Non-magnetic” and “demagnetized” are two very different things! Most techs are careful about demagnetizing ferrous tools around video and computer equipment, but might be clueless about the huge magnets in magnetrons! **ANSWER D**

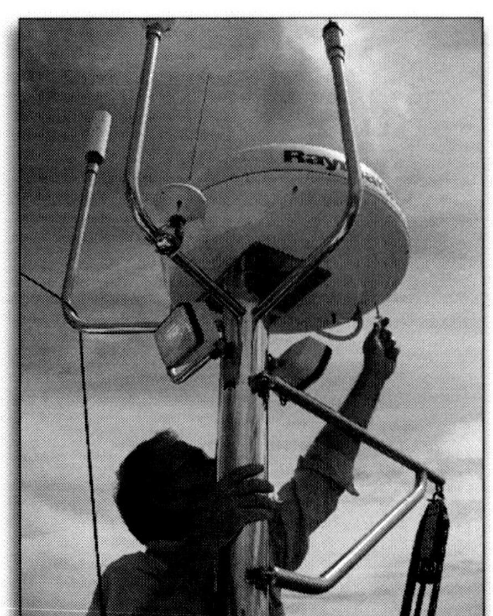

Only non-magnetic tools should be used around RADAR electronics, including the antenna unit.

8-48F6 What kind of display would indicate water in the waveguide?
A. Spoking.
B. Large circular rings near the center.
C. Loss of range rings.
D. Wider than normal targets.
Water in the horizontal run of a waveguide will show up on the indicator unit as large circular rings, coming out from the center. **ANSWER B**

Key Topic 49 – Installation

8-49F1 Why is coaxial cable often used for S-band installations instead of a waveguide?
A. Losses can be kept reasonable at S-band frequencies and the installation cost is lower.
B. A waveguide will not support the power density required for modern S-band RADAR transmitters.
C. S-band waveguide flanges show too much leakage and are unsafe for use near personnel.
D. Dimensions for S-band waveguide do not permit a rugged enough installation for use by ships at sea.
Since S-band RADARs operate below 3,000 MHz, hard-line coax cable is sometimes employed instead of waveguide because installation is easier and cost is lower. Newer installations eliminate the need for long runs of waveguides because the transmitter and receiver units now are up inside the antenna – not down at the display. **ANSWER A**

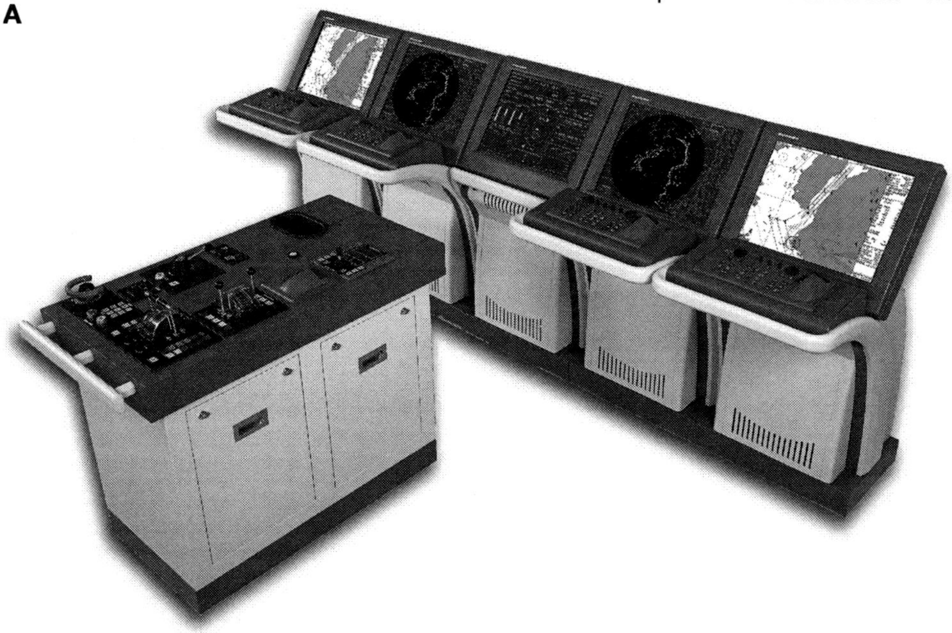

Photo courtesy of Furuno Electric Co., Ltd.
GMDSS bridge electronics are tied into the automatic pilot for track control.

8-49F2 RADAR interference to a communications receiver is eliminated by:
A. Not operating other devices when RADAR is in use.
B. Properly grounding, bonding, and shielding all units.
C. Using a high pass filter on the power line.
D. Using a link coupling.
Grounding, bonding, and shielding all RADAR units keeps noise from being radiated out of the chassis. **ANSWER B**

8-49F3 Why should long horizontal runs of waveguide be avoided?
A. They must be insulated to prevent electric shock.
B. To prevent damage from shipboard personnel.
C. To minimize reception of horizontally polarized returns.
D. To prevent accumulation of condensation.
Long horizontal runs of waveguide should be avoided because condensation may settle on the bottom of the waveguide, creating signal losses and the deterioration of the copper walls of the waveguide. After heavy weather, you notice that your RADAR is losing distant target response. Chances are moisture has leaked into the waveguide. Check all joints to insure their gaskets are supple with no evidence of moisture inside the joint. Most of the problem occurs up in the antenna unit. **ANSWER D**

8-49F4 Long horizontal sections of waveguides are not desirable because:
A. Moisture can accumulate in the waveguide.
B. The waveguide can sag, causing loss of signal.
C. Excessive standing waves can occur.
D. The polarization of the signal might shift.

This question dates back to the old days of RADAR when the TR unit was down below. Back then, long horizontal runs of waveguide were avoided because moisture would accumulate. **ANSWER A**

8-49F5 In a RADAR system, waveguides should be installed:
A. Slightly bent for maximum gain.
B. As straight as possible to reduce distortion.
C. At 90 degree angles to improve resonance.
D. As long as possible for system flexibility.

Keep waveguide runs straight, and minimize the number of elbows that may slightly attenuate the propagation of the microwaves within the walls of the tuned waveguide sections. **ANSWER B**

8-49F6 What is the most important factor to consider in locating the antenna?
A. Allow the shortest cable/waveguide run.
B. Maximum height for best long range operation.
C. The antenna is in a location that is not shadowed by other structures.
D. Easy access for maintenance.

When assessing the optimum location for a RADAR antenna, look for an area that is relatively high and absolutely unobstructed by other structures aboard the craft. Also make sure there is no walkway in line with the antenna, as RADAR emissions could be harmful to the human body. **ANSWER C**

Key Topic 50 – Safety

8-50F1 Choose the most correct statement with respect to component damage from electrostatic discharge:
A. ESD damage occurs primarily in passive components which are easily identified and replaced.
B. ESD damage occurs primarily in active components which are easily identified and replaced.
C. The technician will feel a small static shock and recognize that ESD damage has occurred to the circuit.
D. ESD damage may cause immediate circuit failures, but may also cause failures much later at times when the RADAR set is critically needed.

The sensitive microwave receivers in aeronautical RADAR are constantly subjected to electrostatic discharge (ESD), either from the aircraft flying through dry air and building up a charge-to-discharge state, or from the aircraft flying through heavily charged clouds within a severe weather system. Sometimes ESD damage is subtle, and you won't realize you have an impending failure until a critical system ceases to function. **ANSWER D**

8-50F2 Before testing a RADAR transmitter, it would be a good idea to:
A. Make sure no one is on the deck.
B. Make sure the magnetron's magnetic field is far away from the magnetron.
C. Make sure there are no explosives or flammable cargo being loaded.
D. Make sure the Coast Guard has been notified.

Never run a RADAR at the dock when the vessel is loading explosive or flammable cargo. Never energize a RADAR with anyone standing next to the rotating antenna unit. **ANSWER C**

8-50F3 While making repairs or adjustments to RADAR units:
A. Wear fire-retardant clothing.
B. Discharge all high-voltage capacitors to ground.
C. Maintain the filament voltage.
D. Reduce the magnetron voltage.

Before working on any RADAR, be sure that all high-voltage capacitors are safely discharged. Also be sure that no one can accidentally turn on the equipment while you are working on it. **ANSWER B**

8-50F4 While removing a CRT from its operating casing, it is a good idea to:
A. Discharge the first anode.
B. Test the second anode with your fingertip.
C. Wear gloves and goggles.
D. Set it down on a hard surface.

If ever you are required to pull out a defective cathode-ray tube and install a new one, wear protective gloves and goggles for safety. **ANSWER C**

8-50F5 If a CRT is dropped:

A. Most likely nothing will happen because they are built with durability in mind.
B. It might go out of calibration.
C. The phosphor might break loose.
D. It might implode, causing damage to workers and equipment.

If a CRT is dropped, it could implode, causing glass to fly in all directions. Always handle a new or used RADAR CRT as if it were an expensive television picture tube. It is! **ANSWER D**

8-50F6 Prior to removing, servicing or making measurements on any solid state circuit boards from the RADAR set, the operator should ensure that:

A. The proper work surfaces and ESD grounding straps are in place to prevent damage to the boards from electrostatic discharge.
B. The waveguide is detached from the antenna to prevent radiation.
C. The magnetic field is present to prevent over-current damage or overheating from occurring in the magnetron.
D. Only non-conductive tools and devices are used.

Before working on any RADAR equipment, be sure the system is turned off and cannot be accidentally turned on while you are working on it. Also double check that all work surfaces are well grounded to prevent damage to circuit boards and components from electrostatic discharge. **ANSWER A**

New RADARs Use Pulse Compression Techniques

The effectiveness of any RADAR system is a function of the total *received* radio frequency energy. Traditionally, the high energy levels required for effective RADAR operation involved the use of very high peak power levels and short pulses. These high-energy shorter pulses result in better range resolution, but the power levels impose severe demands on RADAR transmitters, antennas, and transmission lines.

By using less peak power and longer pulses, the same effective energy can be created, but at a cost of *range resolution.* The range resolution problem of a long pulse can be overcome if pulse compression is used, which involves modification of both the transmitting and receiving methods. If the frequency of a long pulse is swept downward during the pulse transmission, or "chirped," the use of a *matched filter* on the receiving end can greatly shorten the pulse while taking advantage of the received signal's high energy content. The *matched filter* at the receiver has a *propagation time* that is inversely proportional to the frequency; higher frequencies take longer to pass through the filter than lower frequencies. This property causes the long chirped pulse to "bunch up," or be compressed, as the lower frequency part of the pulse catches up with the higher frequency part, resulting in a short, high amplitude detected pulse. This pulse compression process is illustrated with an animation you can view at the following link:

http://www.radartutorial.eu/08.transmitters/pic/pulscompression.big.gif

Ideally, a perfectly matched FM chirp detector would have an infinite number of different time delay filters, but practical limitations usually place the maximum number of "taps" at between five and ten, as shown in the animation. This gives a significant improvement of performance over an uncompressed pulse, resulting in improved range resolution.

FM chirp is the simplest and oldest method of pulse compression. Much more sophisticated pulse compression methods, such as *digital phase encoding*, are now commonplace in most modern RADAR systems.

Appendix

COLEMS

Commercial Operator License Examination Managers

A COLEM is a non-government organization that has entered into a written agreement with the Federal Communications Commission to provide Commercial Radio Operator license examinations and tests to the public. There are nine such organizations, and each charge an examination fee. Applicants for the various commercial radio operator licenses, permits and endorsement should apply directly to one of these organizations to schedule their exams and tests.

COLEM	Examination Availability
National Radio Examiners P.O. Box 200065 Arlington, TX 76006-0065 UPS or Fed Ex: 2000 E. Randol Mill Rd. Suite 608-A Arlington, TX 76011-8208 Contact: Larry Pollock PHONE: 800-669-9594 FAX: 817-548-9594 Website: www.nationalradioexaminers.com E-mail: colem@radioexaminers.com	$50.00 for the first 2 Elements and $25.00 for each additional Element per sitting. (minimum sitting fee $50.00)
BFT Training Unlimited, Inc. P.O. Box 2677 Santa Rosa, CA 95405 Contact: J. David Byrd PHONE: 800-821-0906 or 707-792-5678 FAX: 707-792-5677 Website: www.elkinstraining.com E-mail: info@elkinstraining.com	$75.00 per 1-2 exam elements, with a minimum fee of $75.00 per exam session.
Computer Assisted Testing Services, Inc./Comira 777 Mariners Island Blvd. Suite 200 San Mateo, CA 94404 Contact: Tina Silva PHONE: 650-532-1917 FAX: 650-692-9307 Website: www.catstest.com or www.comiratesting.com E-mail: Tina.Silva@catstest.com	$60.00 for the first element. $30.00 for each additional element taken on the same day.
Electronic Technicians Association International, Inc. 5 Depot Street Greencastle, IN 46135 Contact: Teresa Maher PHONE: 765-653-4301 or 1-800-288-3824 FAX: 765-653-4287 Website: www.etainternational.org/ E-mail: eta@eta-i.org	$50.00 per 1 or 2 exam elements.

Elkins International/Elkins Institute, Inc.
P.O. Box 797666
Dallas, TX 75379
Contact: Laura Elkins
PHONE: 888-621-8876
Website: www.elkinsinternational-fcc.com
E-mail: elkins.fcc@gmail.com

$25.00 per element, with a minimum of $50.00 per exam session.

Exemplar Global, Inc.
600 N Plankinton Ave.
PO Box 602
Milwaukee, WI 53201
Contact: Elouise Baylor, FCC Exam Administrator
PHONE: 1-888-722-2440 or 1-414-272-3937 Ext 7823
FAX: 1-414-765-8661
Website: http://www.inarte.org/what-we-offer/federal-communications-commission-fcc-licensing-examinations
E-mail: fcc@exemplarglobal.org

$65.00 for up to 2 exam Elements.

International Society of Certified Electronics Technicians (ISCET)
3608 Pershing Avenue
Fort Worth, TX 76107-4527
Contact: Mack Blakeley
PHONE: 817-921-9101
FAX: 817-921-3741
Website: www.iscet.org
E-mail: testing@iscet.org

Fee: $50.00 - 90.00

Kennedy Point Maritime School
320 N. Atlantic Ave 6B
Cocoa Beach, FL 32931
Contact: John Beasley
PHONE: 321-543-3241
FAX: 321-252-2144
Website: www.kpmaritime.org
E-mail: johnbeasley1@msn.com or johnbeasley1000@att.com

$60 for the first element. $15 for any additional element.

LaserGrade Computer Testing, Inc.
16821 SE McGillivray
Suite 201
Vancouver, WA 98683
Contact: Linda Small
PHONE: 1-800-211-2753, ext. 5203 or 360-859-1277
FAX: 360-891-0958
Website: www.lasergrade.com
E-mail: Linda.small@lasergrade.com

$60.00 for the first element and $30.00 for up to two additional elements per day. Computer based testing is offered on a daily basis throughout the year.

Marine Technical Institute
1915 S. Andrews Avenue
Fort Lauderdale, FL 33316
Contact: Jack Krantz
PHONE: 954-525-1014
FAX: 954-764-0431
Website: www.mptusa.com
E-mail: info@MPTUSA.com

Fee: $50.00 for the first 2 Elements (minimum sitting fee $50.00) and $25.00 for each additional Element per sitting.

National Marine Electronics Association
692 Ritchie Highway
Suite 104
Severna Park, MD 21146
Contact: Cindy Love
PHONE: 1-410-975-9425
FAX: 1-410-975-9450
Website: www.nmea.org
E-mail: clove@nmea.org

$50 for the first two elements (minimum sitting fee $50) and $25 for each additional element per sitting.

Sea School
8440 4th Street North
St. Petersburg, FL 33702
Contact: Mary Foster
PHONE: 1-800-237-8663
FAX: 727-522-3155
Website: www.seaschool.com
E-mail: hqstaff@seaschool.com

$25.00 - $55.00

US Captain's Training
8256 S. Watkins Lane
Traverse City, MI 49684
Contact: Sue Smith
PHONE: 1-866-293-9308
FAX: 1-231-946-5136
Website: www.uscaptain.us
E-mail: info@uscaptain.us

$75.00 per 1-2 exam elements, with a minimum fee of 75.00 per exam session. $50.00 for any additional element.

FEDERAL COMMUNICATIONS COMMISSION

The Commercial Operator License Examination (COLE) program is regulated by the FCC's Wireless Telecommunications Bureau (WTB.)
Federal Communications Commission
Public Safety & Critical Infrastructure Division
445 12th Street, SW
Washington, DC 20554
Tel. (202)418-0680

All Commercial Radio Operator licenses are issued by the FCC's licensing facility located at:
Federal Communications Commission
1270 Fairfield Road
Gettysburg, PA 17325-7245
Tel. (717)338-2888
For information on the status of your Commercial Radio Operator license, permit or endorsement, call: FCC Consumer Assistance Hot Line Toll Free: (888) CALL FCC (225-5322)

FCC FIELD ORGANIZATION

The various FCC Field Offices may be contacted by mail or through the FCC's Consumer Assistance Hotline at (toll free) 888-CALL-FCC (225-5322).

FCC Anchorage Resident Agent Office
PO Box 231949
Anchorage, AK 99523-1923
(907) 271-6342

FCC Atlanta Field Office
3575 Koger Blvd., Suite 320
Duluth, GA 30096-4958
770-935-3370

FCC Boston Field Office
1 Batterymarch Park
Quincy, MA 02169-7495
617-786-7746

FCC Chicago Field Office
Park Ridge Office Center, Suite 306
1550 Northwest Hwy.
Park Ridge, IL 60068-1460
847-813-4672

FCC Columbia Field Office
9200 Farm House Ln.
Columbia, MD 21046
301-725-1996

FCC Dallas Field Office
9330 LBJ Freeway, Suite 1170
Dallas, TX 75243-3429
214-575-6361

FCC Denver Field Office
165 South Union Blvd., Suite 860
Lakewood, CO 80228-2213
303-231-5212

FCC Honolulu Field Office
P.O.B. 971030
Waipahu, Hawaii 96797-1030
808-675-7021

FCC Los Angeles Field Office
Cerritos Corporate Tower
18000 Studebaker Rd., Suite 660
Cerritos, CA 90701-3684
562-860-7474

FCC Miami Field Office
Rochester Building, Suite 301
8390 N.W. 3rd Street
Miami, FL 33166
305-526-7420

FCC New Orleans Field Office
2424 Edenborn Ave. Suite 460
Metarie, LA 70001
504-219-8999

FCC New York Field Office
201 Varick St.
New York, NY 10014-4870
212-337-1865

FCC Portland Office
1220 SW Third Ave
Portland, OR 97204
503-325-4114

San Francisco Field Office
3777 Depot Rd., Suite 420
Hayward, CA 94545-2756
925-416-9717

Metric Conversions

International System of Units (SI) – Metric Units

Prefix	Symbol			Multiplication Factor
exa	E	10^{18}	=	1,000,000,000,000,000,000
peta	P	10^{15}	=	1,000,000,000,000,000
tera	T	10^{12}	=	1,000,000,000,000
giga	G	10^{9}	=	1,000,000,000
mega	M	10^{6}	=	1,000,000
kilo	k	10^{3}	=	1,000
hecto	h	10^{2}	=	100
deca	da	10^{1}	=	10
(unit)		10^{0}	=	1

Prefix	Symbol			Multiplication Factor
deci	d	10^{-1}	=	0.1
centi	c	10^{-2}	=	0.001
milli	m	10^{-3}	=	0.000001
micro	u	10^{-6}	=	0.000000001
nano	n	10^{-9}	=	0.000000000000
pico	p	10^{-12}	=	0.000000000001
femto	f	10^{-15}	=	0.000000000000001
atto	a	10^{-18}	=	0.000000000000000001

Power

	Btu/h	ft – lb/s	hp	kW	Watt
1 British thermal unit per hour =	1	0.2161	3.929×10^{-4}	2.930×10^{-4}	0.2930
1 foot-pound per second =	4.628	1	1.818×10^{-3}	1.356×10^{-3}	1.356
1 horsepower =	2545	550	1	0.7457	745.7
1 kilowatt =	3413	737.6	1.341	1	1000
1 Watt =	3.413	0.7376	1.341×10^{-3}	0.001	1

Schematic Symbols

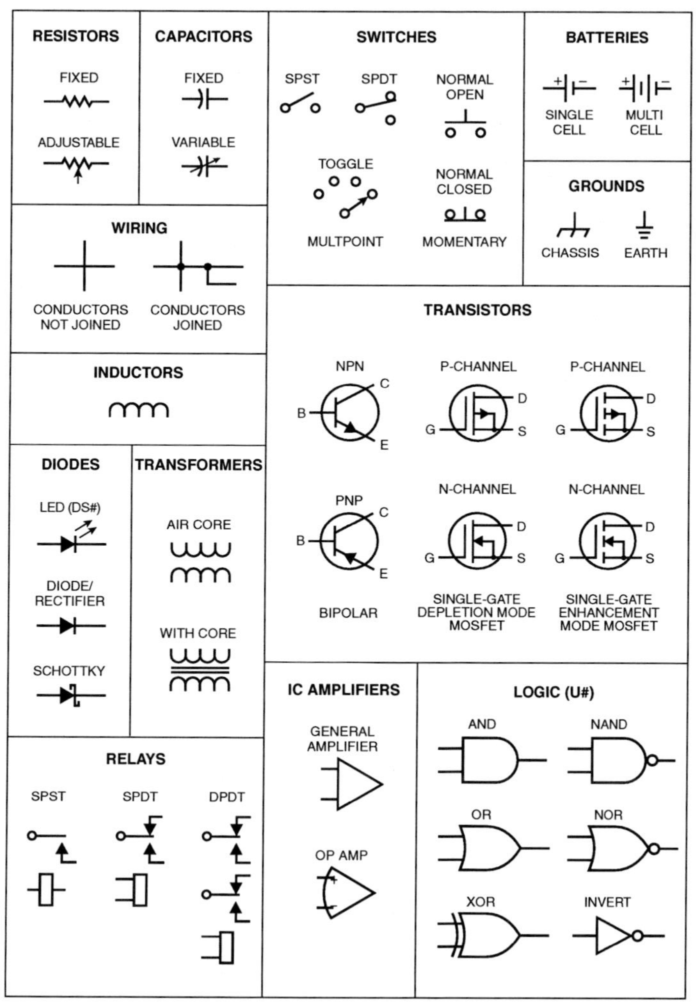

Glossary

A

AFC (Automatic Frequency Control): An arrangement whereby the frequency of an oscillator or the tuning of a circuit is automatically maintained within specified limits with respect to a reference frequency. A magnetron drifts in frequency over a period of time. The AFC of a RADAR makes the local oscillator shift by an equal amount so the IF frequency will remain constant.

AGC (Automatic Gain Control): A method for automatically obtaining an essentially constant receiver output amplitude. The amplitude of the received signal in the range gate determines the AGC bias: A DC voltage that controls the receiver gain so as to maintain a nearly constant output even though the amplitude of the input signal changes.

Amplifier: An electronic device used to increase signal magnitude or power. See also GaAs FET Amplifier, Klystron Amplifier, Traveling-Wave Tube Amplifier.

Amplitude Modulation (AM): A method of impressing a message upon a carrier signal by causing the carrier amplitude to vary proportionally to the message waveform.

Amplitude Shift Keying (ASK): A method of impressing a digital signal upon a carrier signal by causing the carrier amplitude to take different values corresponding to the different values of the digital signal.

Antenna Beamwidth: The angle, in degrees, between the half-power points (-3 dB) of an antenna beam. This angle is also nearly that between the center of the main lobe and the first null. The angle is given for both horizontal and vertical planes unless the beam is circular. When so indicated, the term may refer to the angular width of the main lobe between first nulls [beamwidth between first nulls (BWFN)]. See also Antenna Pattern.

Antenna Cross Talk: A measure of undesired power transfer through space from one antenna to another. Ratio of power received by one antenna to power transmitted by the other, usually expressed in decibels.

Antenna Isolation: The ratio of the power input to one antenna to the power received by the other. It can also be viewed as the insertion loss from transmit antenna input to receive antenna output to circuitry.

Antenna Loging: Two lobes are created that overlap and intercept at -1 to -3dB. The difference between the two lobes produces much greater spatial selectivity than provided by either lobe alone. See also "Lobe, Antenna".

Antenna Pattern: A cross section of the radiating pattern (representing antenna gain or loss) in any plane that includes the origin (source reference point) of the pattern. Both horizontal and vertical polar plots are normally used to describe the pattern. Also, termed "polar diagram" and "radiation pattern."

Antenna, Pencil-Beam: A highly directional antenna designed so that cross sections of the major lobe are approximately circular, with a narrow beamwidth.

Anti-Clutter Circuits (in RADAR): Circuits which attenuate undesired reflections to permit detection of targets otherwise obscured by such reflections.

Aperture: In an antenna, that portion of the plane surface area near the antenna perpendicular to the direction of maximum radiation through which the major portion of the radiation passes. The effective and/or scattering aperture area can be computed for wire antennas which have no obvious physical area.

A-Scope: A cathode-ray oscilloscope used in RADAR systems to display vertically the signal amplitude as a function of time (range) or range rate. Sometimes referred to as Range (R)-Scope.

Attenuation: Decrease in magnitude of current, voltage, or power of a signal in transmission between two points. May be expressed in decibels.

Authorized Bandwidth: The maximum permitted band of frequencies as specified in the FCC authorization to be occupied by an emission. Includes a total of the frequency departure above and below the carrier frequency.

Authorized Frequency: The radio frequency assigned to a station by the FCC and specified in the station license or other instrument of authorization.

Authorized Power: Unless transmitting distress calls, the minimum amount of output power necessary to carry on the telecommunications for which the station is licensed. (See § 80.63(a))

Avionics: The electronics including the radio system aboard an aircraft which must be maintained by a General Radiotelephone Operator.

Azimuth: In marine RADAR, the bearing in degrees measured in a clockwise direction.

B

Balanced Mixers: The two most frequently encountered mixer types are single-balanced and double-balanced. In a double-balanced mixer, four Schottky diodes and two wideband transformers are employed to provide isolation of all three ports.

Bandpass Filter: A type of frequency discrimination network designed to pass a band or range of frequencies and produce attenuation to all other frequencies outside of the pass region. The upper and lower frequencies are usually specified to be the half power (-3dB) or half voltage points (-6dB).

Bandwidth: The amount of frequency spectrum space taken up by 99% of a radiated signal. An expression used to define the actual operational frequency range of a receiver when it is tuned to a certain frequency. For a RADAR receiver, it is the difference between the two frequencies at which the receiver response is reduced to some fraction of its maximum response (such as 3 dB, 6 dB, or some other specified level). The frequencies between which "satisfactory" performance is achieved. See also Receiver Bandwidth and Spectrum Width.

Base Station: A fixed land station used to communicate with mobile radio stations installed in motor vehicles.

Beamwidth: See Antenna Beamwidth.

Beat Frequency Oscillator (BFO): Any oscillator whose output is intended to be mixed with another signal to produce a sum or difference beat frequency. Used particularly in reception of CW transmissions.

Blanking: The process of making a channel or device non-effective for a certain interval. Used for retrace sweeps on CRTs or to mask unwanted signals, such as blanking one's own RADAR from the onboard RWR.

Break: Phrase spoken just before a brief pause in radiotelephone conversation to allow the other station to acknowledge your transmission.

Bridge-to-Bridge Station: A VHF radio station located on a ship's navigational bridge used for navigational purposes.

C

Calling Frequency: Agreed-upon frequency which stations use to initially call one another. Upon contact, both stations switch over to another "working" frequency so others may use the calling frequency.

Call Sign: A unique identifier issued to a radio station by the FCC as an aid to enforcement of radio regulations. A call sign identifies the country of origin and individual station.

Cargo Vessel: Any ship not licensed to carry more than 12 passengers.

Carrier Frequency: The basic radio frequency of the wave upon which modulations are impressed. Also called "Carrier" or fc.

Cathode-Ray Tube (CRT): An indicator device which consists of an electron gun which focuses an electron beam on a fluorescent screen. The screen is painted when the beam is diverted by four deflection plates or a magnetic field deflection system.

Cavity: A space enclosed by a conducting surface used as a resonant circuit at microwave frequencies. Cavity space geometry determines the resonant frequency. A storage area for oscillating electromagnetic energy.

Center Frequency: The tuned or operating frequency. Also referred to as center operating frequency. In frequency diversity systems, the midband frequency of the operating range. See also Carrier Frequency.

CHIRP: A pulse compression technique which uses frequency modulation (usually linear) on pulse transmission. CHIRP RADAR - See PC.

Clutter: In RADAR, undesired returns or echoes resulting from man-made or natural objects including chaff, sea, ground, and rain, which interfere with normal RADAR system observations.

Coast Station: Land-based radio station for the maritime services. Class-I stations provide long distance communications to ships at sea. Class-II coast stations provide regional service. A public coast station is open to public correspondence; a limited coast station may not transmit telecommunications for the public.

Coaxial Cable: A two-conductor transmission line consisting of a single center wire surrounded by a braided metal shield. It is called "hard line" when the outer conductor is solid metal. The insulator between the two conductors is called the dielectric. The outer non-conductive rubber sheath protects the cable from the weather.

Co-Channel: This term is used to indicate that two (or more) equipments are operating on the same frequency.

Coherent: Two signals that have a set (usually fixed) phase relationship.

Coincidence Detector: This RADAR video process requires more than one hit in a range cell before a target is displayed. This prevents video interference from pulses coming from another RADAR, because such interference is unlikely to occur twice in the same range cell.

Commercial Operator Licensing Examination Manager (COLEM): Entity approved by the FCC to provide commercial radio license exams.

Communications Act of 1934: The basic document for controlling telecommunications in the United States.

Communications Priority: The order of priority of communications in the mobile service is: (1) Distress messages, (2) Communications preceded by the urgency signal, (3) Communications preceded by the safety signal, (4) Radio direction finding communications, (5) Navigation and safe movement of aircraft, (6) Navigation, movement of ships, and weather observations, (7) Government radio-telegrams, (8) Government communications for which priority has been requested, (8) Communications relating to previously exchanged and; (9) All other communications.

Compass Bearing: The azimuth (compass direction) based on the north magnetic pole. True North is azimuth zero degrees, east is 90°, south 180° and west 270°.

Compulsory Ship: Ship required by international law to carry radio equipment, licensed radio operators and to keep logs.

COSPAS-SARSAT System: An international satellite-based search and rescue system established by Canada, France, the USA and Russia. Used to detect and locate land, sea and airborne radio beacons.

Cross Modulation: Intermodulation caused by modulation of the carrier by an undesired signal wave.

Crystal: Piezoelectric material (such as quartz) that mechanically vibrates at specific frequencies when an ac current is applied. Crystals are very stable frequency oscillators.

CW (Continuous Wave): In RADAR and EW systems this term means that the transmitter is on constantly; i.e., not pulsed (100% duty cycle). These systems may frequency or phase modulate the transmitter output. A CW RADAR has the ability to distinguish moving targets against a stationary background while conserving spectrum bandwidth compared to pulsed RADAR requirements. A CW RADAR extracts accurate target range-rate data but cannot determine target range.

D

Decibel (dB): A dimensionless unit for expressing the ratio of two values of power, current, or voltage. Normally, used for expressing transmission gains, losses, levels, and similar quantities.

Demodulator: A device employed to separate the modulation signal from its associated carrier, also called Second Detector. See also Detection.

Detection: Usually refers to the technique of recovering the amplitude modulation signal (envelope) superimposed on a carrier.

Deviation: In FM, the maximum amount that a frequency-modulated signal changes from the center frequency.

Digital Selective Calling (DSC): An automatic calling system which allows a specific station to be contacted. An "all ships" call can also be made for distress alerting and navigation safety communications.

Diode: An electronic device which restricts current flow chiefly to one direction. See also Gunn diode, IMP ATT diode, PIN diode, point contact diode, Schottky barrier diode, step recovery diode, tunnel diode, varactor diode.

Diode Switch: PIN-diode switches provide state-of-the-art switching in most present-day microwave receivers. These switches are either reflective or nonreflective in that the former reflect incident power back to the source when in the isolated state. While both types of switches can provide high isolation and short transition times, the reflective switch offers multi octave bandwidth in the all shunt diode configuration, while the non-reflective switch offers an octave bandwidth.

Directional Coupler: A 4-port transmission coupling device used to sample the power traveling in one direction through the main line of the device. There is considerable isolation (typically 20 dB) to signals traveling in the reverse direction. Because they are reciprocal, the directional coupler can also be used to directively combine signals with good reverse isolation. The directional coupler is implemented in waveguide and coaxial configurations.

Directivity: For antennas, directivity is the maximum value of gain in a particular direction. (Isotropic point source has directivity =1). For directional couplers, directivity is a measure (in dB) of how good the directional coupling is and is equal to the isolation minus the coupling.

Dish: A microwave reflector used as part of a RADAR antenna system. The surface is concave and is usually parabolic shaped. Also called a parabolic reflector.

Distress: Requiring immediate assistance. Distress traffic in radio communications receives the highest priority because distress calls indicate imminent disaster. MAYDAY is the radiotelephone distress signal; SOS in radiotelegraphy.

Distress Frequency: An internationally recognized frequency set aside for distress traffic such as 2182 kHz (single sideband), 156.8 MHz (FM) and 500 kHz (telegraph.) 121.5 MHz (AM double sideband-full carrier) is the aircraft distress channel.

Doppler Effect: The apparent change in frequency of an electromagnetic wave caused by a change in distance between the transmitter and the receiver during transmission/reception. The Doppler increase is realized by comparing the signal received from a target approaching the RADAR site to the transmitted reference signal. An apparent frequency decrease is noted for targets departing the RADAR location. Differences can be calibrated to provide target range-rate data.

Downlink: A one-way wide angle radio beam from a communications satellite in Earth orbit to a station on the surface of the Earth. See uplink, transponder

Ducting: The increase in range that an electromagnetic wave will travel due to a temperature inversion of the atmosphere. The temperature inversion forms a channel or waveguide (duct) for the waves to travel in, and they can be trapped, not attenuating as would be expected from the RADAR equation.

Dummy Load: A dissipative but essentially nonradiating substitute device having impedance characteristics simulating those of the antenna. This allows power to be applied to the RADAR transmitter without radiating into free space. Dummy loads are commonly used during EMCON conditions or when troubleshooting a transmitter at a workbench away from it's normal environment. A artificial antenna device used to prevent an antenna from radiating an interfering signal into space. Its resistance is similar to the impedance of an actual antenna.

Duplex: Two-way communications with both stations transmitting and receiving on different frequencies.

Duplexer: A switching device used in RADAR to permit alternate use of the same antenna for both transmitting and receiving.

Duty Cycle: The ratio of average power to peak power, or ratio of pulse length to interpulse period for pulsed transmitter systems. Interpulse period is equal to the reciprocal of the pulse repetition rate.

E

Earth Station: A station in the Earth-space service located on the surface of the Earth or on a ship or aircraft.

Echo: In RADAR, a return radio-frequency pulse reflected from a remote object which is detected a few milliseconds or microseconds after transmission. The round trip time can accurately be converted to distance.

Effective Radiated Power (ERP): Input power to the antenna in watts times the gain ratio of the antenna. When expressed in dB, ERP is the transmitter power (PT), in dBm (or dB W) plus the antenna gain (GT) in dB. The term EIRP is used sometimes and reiterates that the gain is relative to an isotropic radiator.

Electromagnetic Coupling: The transfer of electromagnetic energy from one circuit or system to another circuit or system. An undesired transfer is termed EMI (electromagnetic interference).

EMC (Electromagnetic Compatibility): That condition in which electrical/electronic systems can perform their intended function without experiencing degradation from, or causing degradation to other electrical/electronic systems. More simply stated, EMC is that condition which exists in the absence of EMI. See also Intersystem and Intrasystem EMC tests.

EME (Electromagnetic Environment): The total electromagnetic energy in the RF spectrum that exists at any given location.

EMI (Electromagnetic Interference): Any induced, radiated, or conducted electrical emission, disturbance, or transient that causes undesirable responses, degradation in performance, or malfunctions of any electrical or electronic equipment, device, or system. Also synonymously referred to as RFI (Radio Frequency Interference).

Emergency Locator Transmitter (ELT): An airborne distress alerting beacon that is detected by the COPAS-SARSAT polar-orbiting satellites. The ELT is the aviation counterpart of the maritime EPIRB. It is battery operated and transmits automatically when activated.

Emergency Position Indicating Radio Beacon (EPIRB): A battery-operated radio transmitter in the maritime mobile service that activates upon the sinking of a ship to facilitate search and rescue operations.

Emission Designator: A series of three alphanumeric characters used to identify radio signal properties. For example: A1A, manual radiotelegraphy. (A = Double-sideband amplitude modulation, full carrier; I = One channel of digital modulation; A = Morse code for manual reception.)

Emitter: Any device or apparatus which emits electromagnetic energy.

EMP (Electromagnetic PULSE): The generation and radiation of a very narrow and very high-amplitude pulse of electromagnetic noise. It is associated with the high level pulse as a result of a nuclear detonation and with intentionally generated narrow, high-amplitude pulse for ECM applications. In the case of nuclear detonations, EMP consists of a continuous spectrum with most of its energy distributed through the low frequency band of 3 KHz to 1 MHz.

Enhanced Group Calling (EGG): A feature of the INMARSAT SafetyNET System that permits the addressing of messages to a group (or all vessels) in specific geographical areas.

Error Signal: In servomechanisms, the signal applied to the control circuit that indicates the degree of misalignment between the controlling and the controlled members.

In tracking RADAR systems, a voltage dependent upon the signal received from a target whose polarity and magnitude depend on the angle between the target and the center axis of the scanning beam.

F

Facsimile: The transmission and reception of fixed images by converting scanned lines to digital signals. The lines are redrawn on paper at the receiving site. An important maritime use is the transmission of weather maps for satellites.

FCC - Federal Communications Commission: The official telecommunications agency in the United States. Among its duties is the allocation and regulation of radio frequency assignments within a framework of international agreements.

Fiber Optics: The conveying of information by transmitting light waves down a thin thread of glass or plastic. A signal modulates a light emitting diode (LED) which is fed into an optical fiber. Fiber optic cables are inexpensive, light in weight, immune to electromagnetic interference and can carry more signals than coaxial cables.

Field Strength: The magnitude of a magnetic or electric field at any point, usually expressed in terms of ampere turns per meter or volts per meter. Sometimes called field intensity and is expressed in volts/meter or dBv/meter. Above 100 MHz, power density terminology is used more often.

First Harmonic: The fundamental (original) frequency.

Frequency Agility: A RADAR's ability to change frequency within its operating band, usually on a pulse-to-pulse basis.

Frequency Allocation: A radio frequency or band of frequencies internationally or nationally assigned by an authorized body.

Frequency Deviation: VHF-FM transmitters in the maritime service are determined to be operating properly (100% modulation) when the maximum amount by which the carrier frequency changes either side of center frequency is plus-or-minus (±) 5 kHz.

FM - Frequency Modulation: The process of varying a radio signal to convey intelligence by changing the transmitting frequency.

Frequency Range: (1) A specifically designated portion of the frequency spectrum; (2) of a device, the band of frequencies over which the device may be considered useful with various circuit and operating conditions; (3) of a transmission system, the frequency band in which the system is able to transmit power without attenuating or distorting it more than a specified amount.

Frequency Shift Keying (FSK): A form of FM where the carrier is shifted between two frequencies in accordance with a predetermined code. In multiple FSK, the carrier is shifted to more than two frequencies. FSK is used principally with teletype communications.

Full Duplex: Simultaneous two-way communication in both directions on separate frequencies. Normal two-way conversation, similar to the telephone, is possible.

Fundamental Frequency: Used synonymously for tuned frequency, carrier frequency, center frequency, output frequency, or operating frequency.

G

GaAs FET Amplifier: Because of their low noise, field-effect transistors are often used as the input stage of wideband amplifiers. Their high input resistance makes this device particularly useful in a variety of applications. Since the FET does not employ minority current carriers, carrier storage effects are eliminated giving the device faster operating characteristics and improved radiation resistant qualities.

Gain: For antennas, the value of power gain in a given direction relative to an isotropic point source radiating equally in all directions. Frequently expressed in dB (gain of an isotropic source = 0 dB).

Gate (Range): A signal used to select RADAR echoes corresponding to a very short range increment. Range is computed by moving the range gate or marker to the target echo; an arrangement which permits RADAR signals to be received in a small selected fraction of the time period between RADAR transmitter pulses.

General Radiotelephone Operator License (GROL): License issued by the FCC to individuals qualified to service, maintain, repair and operate radiotelephony communications equipment.

Geostationary Satellite: An orbiting satellite positioned 22,285 miles above the equator has exactly a 24-hour orbital period. Since this is the same as the rotation of the Earth, the satellite appears to hang motionless in space.

Giga: A prefix meaning 10^9 (times a billion). For example, gigahertz (GHz).

GLINT (In RADAR): The random component of target location error caused by variations in the phase front of the target signal (as contrasted with Scintillation Error). Glint may affect angle, range of Doppler measurements, and may have peak values corresponding to locations beyond the true target extent in the measured coordinate.

Global Maritime Distress and Safety System (GMDSS): An automated ship-to-shore distress alerting system that uses satellites and advanced terrestrial communications systems. It is coordinated throughout the world by the IMO, the International Maritime Organization. It picks up radio distress messages and relays them to the proper authorities.

Global Positioning System (GPS): Also called Navstar, the GPS uses multiple satellites to provide world-wide positional fix capability. This is accomplished by measuring the propagation time of satellite signals at the GPS receiver.

Great Lakes Radio Agreement: A rule applying to all ships in the Great Lakes region. Technical requirements are stated for radio equipment and operators.

Ground: A connection with the Earth to establish ground potential. A common connection in an electrical or electronic circuit. The area directly below an antenna. With Marconi antennas, the ground becomes one-half of the antenna.

Guardband: A frequency band to which no other emitters are assigned as a precaution against interference to equipments susceptible to EMI in that band.

Gunn Diode: The Gunn diode is a transferred electron device which, because of its negative resistance, can be used in microwave oscillators or amplifiers. When the applied voltage exceeds a certain critical value, periodic fluctuations in current occur. The frequency of oscillation depends primarily upon the drift velocity of electrons through the effective length of the device. This frequency may be varied over a small range by means of mechanical tuning.

H

Half Duplex: Two-way communications over two separate channels or frequencies but not at the same time.

Harmful Interference: Any emission or radiation which interrupts or degrades a radio communications service operating in accordance with the rules. Operators must never deliberately interfere with any radio signal.

Harmonic: A sinusoidal component of a periodic wave or quantity having a frequency that is an integral multiple of the fundamental frequency. For example, a component which is twice the fundamental frequency is called the second harmonic, (the fundamental is the first harmonic, which is frequently misunderstood).

Hertz: The unit of frequency equal to one cycle per second.

Hertz Antenna: A one-half wave dipole antenna that is usually fed at its center and horizontally polarized.

HF - High Frequency: The radio frequency band that occupies 3 MHz to 30 MHz.

Horn Antenna: A flared, open-ended section of waveguide used to radiate the energy from a waveguide into space. Also termed “horn” or “horn radiator.” Usually linearly polarized, it will be vertically polarized when the feed probe is vertical, or horizontally polarized if the feed is horizontal. Circular polarization can be obtained by feeding a square horn at a 45° angle and phase shifting the vertical or horizontal excitation by 90°.

Hyperabrupt Varactor Oscillator: Due to a non-uniform concentration of N-type material (excess electrons) in the depletion region, this varactor produces a greater capacitance change in tuning voltage and a far more linear voltage-vs.-frequency tuning curve. As a result, this device has an improved tuning linearity and low tuning voltage.

I

Image Frequency: That frequency to which a given superheterodyne receiver is inherently susceptible, thereby rendering such a receiver extremely vulnerable to EMI at that frequency. The image frequency is located at the same frequency difference to one side of the local oscillator as the tuned (desired) frequency is to the other side. An undesired signal received at the image frequency by a superheterodyne receiver not having preselection would, therefore, mix with the oscillator, produce the proper receiver IF, and be processed in the same manner as a signal at the desired frequency. See also receiver selectivity.

Insertion Loss: The loss incurred by inserting an element, device, or apparatus in an electrical/electronic circuit. Normally expressed in decibels determined as 10 log of the ratio of power measured at the point of insertion prior to inserting the device to the power measured after inserting the device.

ILS - Instrument Landing System: An electronic aircraft guidance system using a radio beam to direct a pilot along a glide path.

INMARSAT: A four geostationary satellite network operated by the International Mobile Satellite Organization. It provides telex, telephone, data, fax and SafetyNET (maritime safety information, weather and navigational warning) service for ships at sea. INMARSAT is responsible for the space segment of GMDSS.

Integration Effect: Pulse RADARs usually obtain several echoes from a target. If these echoes are added to each other, they enhance the S/N ratio, making a weak target easier to detect. The noise and interference do not directly add from pulse to pulse, so the ratio of target strength to undesired signal strength improves making the target more detectable. Random noise increases by the square root of the number of integrations, whereas the signal totally correlates and increases directly by the number of integrations, therefore the S/N enhancement is equal to the square root of the number of integrations.

Interference: The presence of unwanted atmospheric noise or man-made signals that obstructs or inhibits the reception of radio communications.

Interferometer: When two widely spaced antennas are arrayed together, they form an interferometer. The radiation pattern consists of many lobes, each having a narrow beamwidth. This antenna can provide good spatial selectivity if the lobe-to-lobe ambiguity can be solved, such as using amplitude comparison between the two elements.

IF (Intermediate Frequency): The difference frequency resulting from mixing the received signal in a superheterodyne receiver with the signal from the local oscillator. The difference frequency product provides the advantages inherent to the processing (amplification, detection, filtering, and such) of low frequency signals. The receiver local oscillator may operate either below or above the receiver tuned frequency. A single receiver may incorporate multiple IF detection.

Intermodulation: The production, in a nonlinear element (such as a receiver mixer), of frequencies corresponding to the sums and differences of the fundamentals and harmonics of two or more frequencies which are transmitted through the element; or, the modulation of the components of a complex wave by each other, producing frequencies equal to the sums and differences of integral multiples of the component frequencies of the complex wave.

International Fixed Public Radio Service: A point-to-point radio communications service open to public correspondence.

International Maritime Organization (IMO): A United Nations agency headquartered in London specializing in safety of shipping and preventing ships from polluting the seas.

International Phonetic Alphabet: Worldwide method of substituting words for individual letters to increase understanding.

International Telecommunication Union - ITU: The worldwide governing body controlling wire and radio telecommunications.

Inverse Gain: Amplification, inverse modulation, and re-radiation of a RADAR's pulse train at the rotation rate of the RADAR scan. Deceives a conical scanning RADAR in angle.

Ionosphere: An electrically charged frequency sensitive portion of the upper atmosphere that has the ability to reflect radio waves back to Earth.

Isotropic Antenna: A theoretical antenna in free space that radiates an RF signal equally well in all directions. Usually used as a reference point in the measurement of antennas. The gain of a half-wave antenna over an isotropic radiator is 2.15 dB.

I/S Ratio (Interference-to-Signal Ratio) (ISR): The ratio of electromagnetic interference level to desired signal level that exists at a specified point in a receiving system. The ratio, normally expressed in dB, is employed as a tool in prediction of electronic receiving system performance degradation for a wide range of interference receiver input levels. Performance evaluations compare actual I/S ratios to minimum acceptable criteria.

K

Kilo: A prefix meaning 10^3 (times one thousand). For example, kilohertz, kHz.

Klystron Amplifier: An electron beam device which achieves amplification by the conversion of periodic velocity variations into conduction-current modulation in a field-free drift region. Velocity variations are launched by the interaction of an RF signal in an input resonant cavity and are coupled out through an RF output cavity. Several variations including reflex and multi cavity klystrons are used.

Klystron Multicavity: An electron tube which employs velocity modulation to generate or amplify electromagnetic energy in the microwave region. Since velocity modulation employs transit time of the electron to aid in the bunching of electrons, transient time is not a deterrent to high frequency operations as is the case in conventional electron tubes. See also Velocity Modulation.

Klystron Reflex: A klystron which employs a reflector (repeller) electrode in place of a second resonant cavity to redirect the velocity-modulated electrons through the resonant cavity. The repeller causes one resonant circuit to serve as both input and output, which simplifies the tuning operation. This type of klystron is well adapted for use as an oscillator because the frequency is easily controlled by varying the position of the repeller. See also Velocity Modulation.

L

Land-Mobile Service: A mobile communications service between land-based movable and permanently located base stations - or between land-based movable stations.

Leakage: Undesired radiation or conduction of RF energy through the shielding of an enclosed area or of an electronic device.

Lens RADAR (Microwave): The purpose of any such lens is to refract (focus) the diverging beam from an RF feed into a parallel beam (transmitting) or vice versa (receiving). The polarization is feed dependent.

LF - Low Frequency: This band occupies 30 kHz to 300 kHz.

Licensee: The entity to which a radio station is licensed by the Federal Communications Commission.

License Term: (effective March 25, 2008) First, Second, and Third Class Radiotelegraph Operator's Certificates are normally valid for a term of five (5) years from the date of issuance. General Radiotelephone Operator Licenses, Restricted Radiotelephone Operator Permits, Restricted Radiotelephone Operator Permits-Limited Use, GMDSS Radio Operator's Licenses, Restricted GMDSS Radio Operator's Licenses, GMDSS Radio Maintainer's Licenses, GMDSS Operator/Maintainer Licenses, and Marine Radio Operator Permits are normally valid for the lifetime of the holder.

Light Amplification by Stimulated Emission of Radiation (LASER): A process of generating coherent light. The process utilizes a natural molecular (and atomic) phenomenon whereby molecules absorb incident electromagnetic energy at specific frequencies, store this energy for short but usable periods, and then release the stored energy in the form of light at particular frequencies in an extremely narrow frequency-band.

Limiting: A term to describe that an amplifier has reached its point of saturation or maximum output voltage swing. Deliberate limiting of the signal is used in FM demodulation so that AM will not also be demodulated.

Line of Sight: The unobstructed distance to the horizon. Communication above the VHF level is usually by direct "line of sight" radio-wave propagation. Range depends upon antenna height and terrain.

Lobe, Antenna: Various parts of the antenna's radiation pattern are referred to as lobes, which may be subclassified into major and minor lobes. The major lobe is the lobe of greatest gain and is also referred to as the main lobe or main beam. The minor lobes are further subclassified into side and back lobes as indicated in the figure to the right. The numbering of the side lobes are from the main lobe to the back lobe.

Local Oscillator Frequency: An internally generated frequency in a superheterodyne receiver. This frequency differs from the receiver operating frequency by an amount equal to the IF of the receiver. The local oscillator frequency may be designed to be either above or below the incoming signal frequency.

Log: Diary of radio communications kept by the station operator. It must contain frequencies used, any technical problems encountered, what action has been taken to correct technical problems, and if any distress traffic has been intercepted. The log is the written report of the station's performance and activities.

Long Pulse Mode: Many pulsed RADARs are capable of transmitting either long or short pulses of RF energy. When the long pulses of RF energy are selected manually (or sometimes automatically), the RADAR is said to be operating in the long pulse mode. In general, "long pulse mode" is used to obtain high average power for long-range search or tracking, and "short pulse mode" gives low average power for short-range, high-definition, tracking or search.

Loran-C: Acronym for LOng RAnge Navigation. Loran-C is a system of radio transmitters broadcasting low-frequency pulses to allow ships to determine their positions. The difference in time it takes for pulses from different transmitters to reach a ship allows Loran-C to determine the ship's location very accurately.

M

Magnetron: A magnetron is a thermionic vacuum tube which is constructed with a permanent magnet forming a part of the tube and which generates microwave power. These devices are commonly used as the power output stage of RADAR transmitters operating in the frequency range above 1000 MHz and are used less commonly down to about 400 MHz. A magnetron has two concentric cylindrical electrodes. On a conventional magnetron, the inner one is the cathode and the outer one is the anode. The opposite is true for a coaxial magnetron.

Magnetron Oscillator: A high-vacuum tube in which the interaction of an electronic space charge and a resonant system converts direct current power into ac power, usually at microwave frequencies. The magnetron has good efficiency, is capable of high power outputs, and is stable.

Marconi Antenna: A one-quarter wavelength antenna fed at one end and operated against a good RF grounding system.

Marine Radio Operator Permit (MROP): A permit earned by passing a 24 question examination on regulations, operating techniques and practices in the maritime services. The MROP is granted by passing Element 1 Radio Law. MROP holders may not make internal adjustments to radio transmitting equipment.

Maritime: Relating to navigation or commerce on the seas.

Maritime Mobile Radio Service: A two-way mobile communications service between ships, or ships and coast stations.

Maritime Mobile Repeater Station: Aland station at a fixed location established for the automatic retransmission of signals to extend the range of communication of ship and coast stations.

Maritime Safety Information (MSI): Important navigational and meteorological information.

Master: A person licensed to command a merchant ship.

Matched Filter: This describes the bandwidth of an IF amplifier that maximizes the signal-to-noise ratio in the receiver output. This bandwidth is a function of the pulse-width of the signal.

Mayday: A word spoken three times spoken during radiotelephone distress messages.

MDS (Minimum Detectable / Discernible Signal): The receiver input power level that is just sufficient to produce a detectable/discernible signal in the receiver output.

Mega: A prefix meaning 10^6 (times one million). For example megahertz (MHz).

MF - Medium Frequency: This radio frequency band occupies 300 kHz to 3,000kHz (or 0.3 MHz to 3 MHz.)

Microvolt Per Meter: A commonly used unit of field strength at a given point. The field strength is measured by locating a standard receiving antenna at that point, and the "microvolts per meter" value is then the ratio of the antenna voltage in microvolts to the effective antenna length in meters. Usually used below 100 MHz. Above 100 MHz, power density terminology is normally used.

Microwaves: Radio waves generally beginning at 1,000 MHz or 1 GHz. Most microwave activity is in the 1 to 50 GHz range.

Microwave Amplification by Stimulated Emission of Radiation (MASER): A low-noise radio-frequency amplifier. The emission of energy stored in a molecular or atomic system by a microwave power supply is stimulated by the input signal.

Mission Control Center (MCC): Land-based rescue authorities who collect, store and exchange distress and alerting information.

Mixers: See Balanced and Schottky Diode Mixers.

Modem: A modulator/demodulator that converts digital signals to audio tones for transmission over wirelines or via radio wave. The process is reversed at the receiver.

Modulation: The process whereby some characteristic of one wave is varied in accordance with some characteristic of another wave. The basic types of modulation are angle modulation (including the special cases of phase and frequency modulation) and amplitude modulation.

Modulation, Amplitude (AM): This type of modulation changes the amplitude of a carrier wave in responses to the amplitude of a modulating wave. This modulation is used in RADAR and EW only as a switch to turn on or turn off the carrier wave; i.e., pulse is a special form of amplitude modulation.

Modulation, Frequency (FM): The frequency of the modulated carrier wave is varied in proportion to the amplitude of the modulating wave and therefore, the phase of the carrier varies with the integral of the modulating wave. See also Modulation.

Modulation, Phase (PM): The phase of the modulated carrier is varied in proportion to the amplitude of the modulating wave. See also Modulation.

Monopulse: A type of tracking RADAR that permits the extracting of tracking error information from each received pulse and offers a reduction in tracking errors as compared to a conical-scan system of similar power and size. Multiple (commonly four) receiving antennas or feeds are placed symmetrically about the center axis and operate simultaneously to receive each RF pulse reflected from the target. A comparison of the output signal amplitude or phase among the four antennas indicates the location of the target with respect to the RADAR beam center line. The output of the comparison circuit controls a servo system that reduces the tracking error to zero and thereby causes the antenna to track the target.

MTI (Moving Target Indicator): This RADAR signal process shows only targets that are in motion. Signals from stationary targets are subtracted out of the return signal by a memory circuit.

Multipath: The process by which a transmitted signal arrives at the receiver by at least two different paths. These paths are usually the main direct path, and at least one reflected path. The signals combine either constructively or destructively depending upon phase, and the resultant signal may be either stronger or weaker than the value computed for free space.

Multiplex: Simultaneous transmission of two or more signals on a common carrier wave. The three types of multiplex are called time division, frequency division, and phase division.

Multiband RADAR: A type of RADAR which uses simultaneous operation on more than one frequency band through a common antenna. This technique allows for many sophisticated forms of video processing.

N

Nautical Mile: The fundamental unit of distance used in navigation. One nautical mile = 1.15 statute miles (or 6,080 feet). One knot is one nautical mile per hour.

Navigational Communications: Safety communications exchanged between ships and/or coast stations concerning the maneuvering of vessels.

NAVTEX: Navigational Telex, an international, automated system for instantly distributing maritime navigational warnings, weather forecasts, search and rescue notices and similar information to ships on 518 kHz worldwide.

Noise Figure, Receiver: A figure of merit (NF or F) of a system given by the ratio of the signal-to-noise ratio at the input, Si/Ni, divided by the signal-to-noise ratio at the output, So/No. It essentially expresses the ratio of output noise power of a given receiver to that of a theoretically perfect receiver which adds no noise. Noise figure is usually expressed in dB and given for an impedance matched condition. Impedance mismatch will increase the noise figure by the amount of mismatch loss.

Notch: The portion of the RADAR velocity display where a target disappears due to being notched out by the zero Doppler filter. If not filtered (notched), ground clutter would also appear on the display. A notch filter is a narrow band-reject filter. A "notch maneuver" is used to place a tracking RADAR on the beam of the aircraft so it will be excluded.

Null, Antenna Pattern: The directions of minimum transmission (or reception) of a directional antenna. See also Lobe, Antenna.

O

Omega: A radio navigation system relying on eight land transmitters throughout the world. Ships carry special receivers to listen to the transmitters and determine from the information carried exactly where the ship is at all times. Omega relies on phase differences between received signals.

Omni-Directional: Performing equally well in all directions.

Oscillators: Devices which generate a frequency. See also Backward Wave, Dielectrically Stabilized Oscillator, Hyperabrupt Varactor Oscillator, Magnetron Oscillator, Varactor Tuned Oscillator, and YIG tuned oscillator.

Oscillator, Local: See Local Oscillator Frequency.

Over: Word spoken in radiotelephone conversation to indicate that it is the other station's turn to speak.

Over-the-Horizon (OTH) RADAR: A method of tracking remote objects by refracting high-frequency (HF) radio signals off of the ionosphere down towards the Earth and analyzing the return echoes.

Overmodulation: Driving a transmitter over its designed parameters causes adjacent frequency interference. Can be caused by shouting into a microphone. Peak modulation should not exceed 100%.

P

PAN: The internationally recognized radiotelephone urgency signal. The words "PAN PAN" are spoken three times in succession to indicate that an urgent message will follow.

Parabolic Antenna: A receiving or transmitting reflector (dish) antenna that focuses the RF energy into a narrow beam to insure maximum gain.

Parasitic Oscillations: Unwanted spurious signals at frequencies removed from the operating frequency. Parasitics can cause distortion, loss of power and possible interference to others. They are usually eliminated by shielding, RF chokes and neutralization.

Part 13: The rules issued by the FCC governing commercial radio operators.

Part 23: The group of rules issued by the FCC governing stations in the international fixed public radio communication services.

Part 73: The group of rules issued by the FCC governing radio broadcasting services.

Part 80: The group of rules issued by the FCC governing stations in the maritime services.

Part 87: The group of rules issued by the FCC governing stations in the aviation services.

Passenger Ship: Any ship carrying more than twelve paying passengers. (Six passengers when used in reference to the Great Lakes Radio Agreement.)

PC (Pulse Compression): The process used in search and tracking pulse RADARs whereby the transmitted pulse is long, so as to obtain high average transmitter output power, and the reflected pulse is processed in the RADAR receiver to compress it to a fraction of the duration of the transmitted pulse to obtain high definition and signal strength enhancement. Pulse compression may be accomplished by sweeping the transmitted frequency (carrier) during the pulse. The returned signal is then passed through a frequency-dependent delay line. The leading edge of the pulse is therefore delayed so that the trailing edge catches up to the leading edge to produce effectively a shorter received pulse than that transmitted. Pulse compression RADARs are also referred to as CHIRP RADARs. Other more sophisticated pulse compression techniques are also possible and are becoming more popular.

Peak Envelope Power (PEP): Method of measuring the output power of a single sideband signal since no carrier is transmitted.

Pencil Beam: A narrow circular RADAR beam from a highly directional antenna (such as a parabolic reflector).

Personal Locator Beacon (PLB): An land-based satellite beacon that is detected by the CO-PAS-SARSAT polar-orbiting satellites.

Phased Array RADAR: RADAR using many antenna elements which are combined in a controlled phase relationship. The direction of the beam can be changed as rapidly as the phase relationships (usually less than 20 microseconds). Thus, the antenna typically remains stationary while the beam is electronically scanned. The use of many antenna elements allows for very rapid and high directivity of the beam(s) with a large peak and/or average power. There is also a potential for greater reliability over a conventional RADAR since the array will fail gracefully, one element at a time.

Phonetic Alphabet: A system of substituting easily understood words for corresponding letters.

Pin Diode: A diode with a large intrinsic (I) region sandwiched between the P- and N- doped semiconducting regions. The most important property of the PIN diode is the fact that it appears as an almost pure resistance at RF. The value of this resistance can be varied over a range of approximately one-10,000 ohms by direct or low frequency current control. When the control current is varied continuously, the PIN diode is useful for attenuating, leveling and amplitude modulation of an RF signal. When the control current is switched on and off or in discrete steps, the device is useful in switching, pulse modulating, and phase shifting an RF signal.

Plan Position Indicator (PPI): A cathode ray tube display map used to highlight target objects illuminated by a scanning RADAR antenna.

Point Contact Diode: This was one of the earliest semiconductor device to be used at microwave frequencies. Consisting of a spring-loaded metal contact on a semiconducting surface, this diode can be considered an early version of the Schottky barrier diode. Generally used as a detector or mixer, the device is somewhat fragile and limited to low powers.

Polarization: The direction of the electric field (E-field) vector of an electromagnetic (EM) wave. See Section 3-2. The most general case is elliptical polarization with all others being special cases. The E-field of an EM wave radiating from a vertically mounted dipole antenna will be vertical and the wave is said to be vertically polarized. In like manner, a horizontally mounted dipole will produce a horizontal electric field and is horizontally polarized. Equal vertical and horizontal E-field components produce circular polarization.

Power (Average) for Pulsed RADARS: Average power for a pulse RADAR is the average power transmitted between the start of one pulse and the start of the next pulse (because the time between pulses is many times greater than the pulse duration time, the average power will be a small fraction of peak power).

Power-Driven Vessel: Any ship propelled by machinery.

Power Output: Power output of a transmitter or transmitting antenna is commonly expressed in dBW or dBm.

Power (Peak) for Pulsed RADARS: Peak power for a pulsed RADAR is the power radiated during the actual pulse transmission (with zero power transmitted between pulses).

Power for CW RADARS: Since the power output of CW transmitters (such as illuminator transmitters) usually have a duty cycle of one (100%), the peak and average power are the same.

Power Density: The density of power in space expressed in Watts/meter2 , dBW/m2, etc. Generally used in measurements above 100 MHz. At lower frequencies, field intensity measurements are taken.

PPI-Scope: A RADAR display yielding range and azimuth (bearing) information via an intensity modulated display and a circular sweep of a radial line. The RADAR is located at the center of the display.

Preselector: A device placed ahead of the mixer in a receiver, which has bandpass characteristics such that the desired (tuned) RF signal, the target return, is allowed to pass, and other undesired signals (including the image frequency) are attenuated.

Priority of Communications: Maritime service rules require that priority be given to: (1) distress calls concerning grave and imminent danger, (2) urgent messages concerning safety of life and property, and (3) traffic concerning navigational safety and weather warnings. (See §80.91)

Propagation: In electrical practice, the travel of waves through or along a medium. The path traveled by the wave in getting from one point to another is known as the propagation path (such as the path through the atmosphere in getting from a transmitting antenna to a receiving antenna, or the path through the waveguides and other microwave devices in getting from an antenna to a receiver).

Propagation, Radio Wave: The method of radio wave travel which may be along the Earth's surface, directly through space or reflected from the upper atmosphere.

Public Correspondence: Any third-party telecommunication (message, image or voice traffic) that must be accepted for transmission.

Pulse: In RADAR, a repetitious burst of radio-frequency energy of very short duration which is directed at a remote object.

Pulsed Doppler (PD): A type of RADAR that combines the features of pulsed RADARs and CW Doppler RADARs. It transmits pulses (instead of CW) which permits accurate range measurement. This is an inherent advantage of pulsed RADARs. Also, it detects the Doppler frequency shift produced by target range rate which enables it to discriminate between targets of only slightly different range rate and also enables it to greatly reduce clutter from stationary targets.

Pulse Length: Same meaning as Pulse-width.

Pulse Modulation: A special case of amplitude modulation wherein the carrier wave is varied at a pulsed rate. The modulation of a carrier by a series of pulses generally for the purpose of transmitting data. The result is a short, powerful burst of electromagnetic radiation which can be used for measuring the distance from a RADAR set to a target.

Pulse Repetition Frequency (PRF): The rate of occurrence of a series of pulses, such as 100 pulses per second. It is equal to the reciprocal of the pulse spacing (T) or PRT. (PRF = 1/T = 1/PRI). Sometimes the term pulse repetition rate (PRR) is used.

Pulse Repetition Frequency (PRF) Stagger: The technique of switching PRF (or PRI) to different values on a pulse-to-pulse basis such that the various intervals follow a regular pattern. This is useful in compensating for blind speeds in pulsed MTI RADARs. Interpulse intervals which differ but follow a regular pattern.

Pulse Repetition Interval: Time between the beginning of one pulse and the beginning of the next.

Pulse Spacing: The interval of time between the leading edge of one pulse and the leading edge of the next pulse in a train of regularly recurring pulses. See also Pulse Repetition Frequency. Also called "the interpulse period."

Pulsewidth: The interval of time between the leading edge of a pulse and the trailing edge of a pulse (measured in microseconds for the short pulses used in RADAR). Usually measured at the 3 dB midpoint (50-percent power or 70% voltage level) of the pulse, but may be specified to be measured at any level.

Q

Quantize: The process of restricting a variable to a number of discrete values. For example, to limit varying antenna gains to three levels.

R

RADAR: Acronym for RAdio Detection And Ranging. A method of tracking objects by analyzing reflected microwave radio signals or echoes. Doppler RADAR is used to measure speed.

RADAR Cross Section: A measure of the RADAR reflection characteristics of a target. It is equal to the power reflected back to the RADAR divided by power density of the wave striking the target. For most targets, the RADAR cross section is the area of the cross section of the sphere that would reflect the same energy back to the RADAR if the sphere were substituted. RCS of sphere is independent of frequency if operating in the far field region.

RADAR Mile: The time required for a radio frequency signal to travel from the transmitter to a target one nautical mile away and back. A RADAR mile is considered to be 12.3 microseconds.

RADAR Range Equation: The RADAR range equation is a basic relationship which permits the calculation of received echo signal strength, if certain parameters of the RADAR transmitter, antenna, propagation path, and target are known.

Radio Direction Finding (RDF): The art of determining the direction of a radio signal using a radio receiver with a signal strength indicator and a rotatable directional antenna. The location of a radio signal requires bearings to be taken by two radio stations.

Radio-Frequency Wave: An electromagnetic wave that travels at the speed of light; 186,282 statute miles or 162,000 nautical miles per second. Its frequency ranges from 10 kHz to 3000 GHz.

Radio Operator: The FCC licensed operator in charge of the station who is responsible for its proper use and operation.

Radioprinter: A means of exchanging alphanumeric codes by direct printing.

Radio Services: Radio operations are classified into services according to the nature and purpose of the transmission.

Radiotelephony: Method of transmitting voice over radio waves.

Relative Bearing: The direction when the reference point is the ship's heading. The number of degrees port or starboard of the bow.

RF (Radio Frequency): A term indicating high frequency electromagnetic energy.

RFI (Radio Frequency Interference): Any induced, radiated, or conducted electrical disturbance or transient that causes undesirable responses or malfunctioning in any electrical or electronic equipment, device, or system. Same as EMI.

Receiver Bandwidth: The difference between the limiting frequencies within which receiver performance in respect to some characteristic falls within specified limits. (In most receivers this will be the difference between the two frequencies where the intermediate frequency (IF) amplifier gain falls off 3 dB from the gain at the center IF frequency.)

Receiver Selectivity: The degree to which a receiver is capable of differentiating between the desired signal and signals or interference at other frequencies. (The narrower the receiver bandwidth, the greater the selectivity.)

Reflection: The turning back (or to the side) of a radio wave as a result of impinging on any conducting surface which is at least comparable in dimension to the wavelength of the radio wave.

Repeater: A receiver/transmitter installation that receives signals on one channel and instantly retransmits them on another frequency usually at higher power from tall antennas. A duplexer keeps the transmitter from overloading the receiver. The advantage of a repeater is increased radio range. A repeater on a satellite is called a transponder.

Rescue Coordination Center (RCC): In GMDSS, the unit responsible for organizing and conducting search and rescue operations within a specific geographical area.

Resolution: In RADAR, the capability to separate two objects that are close to each other in range or bearing and to show them as two distinct echoes on the RADARscope.

Restricted Radiotelephone Operator Permit: Permit allowing certain radio privileges in the aviation, broadcast and maritime services. No examination is required.

Roger: A word in radiotelephone conversation to indicate that you have received and understood all of the other station's transmission.

S

Safety Communications: A radio transmission indicating that a station is about to transmit an important navigation or weather warning.

Safety Convention: International agreement which spells out certain safety requirements for on-board radio equipment and operators.

Scan: To transverse or sweep a sector or volume of airspace with a recurring pattern, by means of a controlled directional beam from a RADAR antenna.

Schottky Barrier Diode: The Schottky barrier diode is a simple metal-semiconductor boundary with no P-N junction. A depletion region between the metal contact and the doped semiconductor region offers little capacitance at microwave frequencies. This diode finds use as detectors, mixers, and switches.

Schottky Diode Mixer: The mixer is a critical component in modern RF systems. Any nonlinear element can perform the mixing function, but parameters determining optimal mixing are noise figure, input admittance, and IF noise and impedance. The Schottky diode is particularly effective because of its low noise figure and nearly square law characteristics.

Schottky Diode Switch: Standard P-N diodes are limited in switching ability at high frequencies because of capacitance provided by the minority carriers. The Schottky diode overcomes this problem by use of the metal-semiconductor junction with inherently low carrier lifetimes, typically less than 100 picoseconds.

Sea Area: Radio equipment on ships is considered in terms of its range and the areas in which the vessel will travel. There are four watchkeeping areas. Sea area Al is within VHF communications range (20-30 miles); A2 within the coverage of a shore-based MF coast station (about 100 miles); A3 within coverage of an INMARSAT geostationary satellite and A4 is the remaining polar areas of the world.

Search and Rescue RADAR Transponders (SARTs): These are portable devices which are taken into a survival craft when abandoning ship. When switched on, they transmit a series of dots on a rescuing ship's 9-GHz RADAR display. SARTs can also notify persons in distress that a rescue ship or aircraft is within range.

Search RADAR: A RADAR whose prime function is to scan (search) a specified volume of space and indicate the presence of any targets on some type of visual display.

Security: The word "SECURITY" is spoken three times prior to the transmission of a safety message.

Secrecy of Communications: Other than broadcasts to the general public, persons may not divulge the content of telecommunications nor use the information obtained to benefit anyone other than the intended recipient.

Selective Calling: A coded transmission to a particular radio station. Other stations do not hear it.

Selectivity: The ability of a radio receiver to separate the desired signal from other signals.

Sensitivity: The sensitivity of a receiver is taken as the minimum signal level required to produce an output signal having a specified signal-to-noise ratio.

Separation: A method of minimizing mutual interference by spacing stations using the same frequency at required distance intervals.

Sensitivity: The ability of a radio receiver to respond to weak input signals.

Shielding: The physical arrangement of shields for a particular component, equipment, or system, (A shield is a housing, screen, or other material, usually conducting, that substantially reduces the effect of electric or magnetic fields on one side of the shield upon devices or circuits on the other side.) Examples are tube shields, a shielded enclosure or cabinet for a RADAR receiver, and the screen around a screen room.

Ship Earth Station: A mobile satellite station located on board a vessel.

Short Pulse Mode: See Long Pulse Mode.

Sideband: A signal either above or below the carrier frequency, produced by the modulation of the carrier wave by some other wave.

Signal Strength: The magnitude of a signal at a particular location. Units are volts per meter or dBV/m.

Signature: The set of parameters which describe the characteristics of a RADAR target or an RF emitter and distinguish one emitter from another. Signature parameters include the radio frequency of the carrier, the modulation characteristics (typically the pulse modulation code), and the scan pattern.

Silent Period: The three-minute duration of time during a continuous watch on a distress frequency when a maritime radio operator must not transmit.

Silicon Controlled Switch: A P-N-P-N device able to operate at sub-microsecond switching speeds by the application of gate signals. Because it is a four layer device, this switch is also known as a tetrode thyristor.

Simplex: Two-way communications with both stations transmitting and receiving on the same frequency. Only one station may transmit at a time.

Single Sideband (SSB): Method of transmitting radiotelephony where one sideband is filtered out and the carrier suppressed or reduced. SSB is more efficient than double sideband signals since it takes up less radio spectrum. Emission: J3E

SITOR: Acronym for SIplex Teleprinter On Radio. Error-free teleprinting of news and weather over the high-frequency radio bands.

SOS: The radiotelegraphy distress signal sent three times followed by DE ("this is") and the call sign of the station in danger.

Spectrum: The distribution of power versus frequency in an electromagnetic wave.

Spectrum Analyzer: An electronic device for automatically displaying the spectrum of the electromagnetic radiation from one or more devices. A cathode ray tube display is commonly used to display this power-versus frequency spectrum.

Spoking (RADAR): Periodic flashes of the rotating radial display. Sometimes caused by mutual interference.

Spurious Emission: Electromagnetic radiation transmitted on a frequency outside the bandwidth required for satisfactory transmission of the required waveform. Spurious emissions include harmonics, parasitic emissions, and intermodulation products, but exclude necessary modulation sidebands of the fundamental carrier frequency.

Squint Angle: The angular difference between the axis of the antenna main lobe and the geometric axis of the antenna reflector, such as the constant angle maintained during conical scan as the main lobe rotates around the geometric axis of the reflector.

Staggered PRF: Staggered PRF allows an increase in MTI blind speeds such that no zeros exist in the velocity response at lower velocities. In a two-period mode, the usual "blind speed" or occurrence of a zero in the velocity response is multiplied by a factor which is a function of the ratio of the two repetition periods.

Standing-Wave Ratio (SWR): The ratio of the current values at the maximum and minimum points on a transmission line. There are no standing waves when the load impedance matches the line impedance. SWR of 3:1 and less is generally considered satisfactory. Standing waves are measured with an SWR meter or reflectometer.

Station Authorization: Any construction permit, license or special temporary authorization issued by the FCC for activating a radio station.

Statute Mile: 5,280 feet. Unit of distance commonly used on land in the United States. One statute mile equals 0.8684 nautical miles.

STC (Sensitivity Time Control): Control that reduces the RADAR receiver gain for nearby targets as compared to more distant targets. STC prevents receiver saturation from close-in targets.

Sunspot Cycle: The height, thickness and intensity of the ionosphere from which radio waves are reflected vary according to a cycle of approximately 11 years.

Superheterodyne Receiver: A receiver that mixes the incoming signal with a locally generated signal (local oscillator) to produce a fixed, low intermediate frequency (IF) signal for amplification in the IF amplifiers.

Suppression: Elimination or reduction of any component of an emission, such as suppression of a harmonic of a transmitter frequency by band rejection filter.

Survival Craft Station: A mobile station on a lifeboat, life raft or other survival equipment aboard a ship or aircraft intended for emergency purposes

Sweep: In RADAR, the line across the cathode-ray tube that is synchronized with the rotation of the scanning antenna.

Synchrodyne: A klystron mixer amplifier stage in a transmitter, where two signal frequencies are applied as inputs and a single amplified signal is taken out.

T

Target: In RADAR, a term frequently used to denote a boat, buoy, island or other object that is RADAR conspicuous and produces an echo on the RADARscope.

Target Size: A measure of the ability of a RADAR target to reflect energy to the RADAR receiving antenna. The parameter used to describe this ability is the "RADAR cross section" of the target. The size (or RADAR cross section) of a target, such as an aircraft, will vary considerably as the target maneuvers and presents different views to the RADAR. A side view will normally result in a much larger RADAR cross section than a head-on view. See also RADAR Cross Section.

Telecommunication: The transfer of sound, images or other intelligence by electromagnetic means.

Telegraphy: The process of sending and receiving information through the use of Morse code.

Telephony: The process of exchanging information through the use of speech transmissions.

Teleprinter: A mechanical typewriter-like device that prints text sent over wire or radio circuits. In a radioteleprinter, a modem converts the audio output of a receiver into electrical impulses to drive the individual keys. See modem.

Terminal Impedance: The equivalent impedance as seen by the transmitter/receiver.

Thermistor: A resistor whose resistance varies with temperature in a defined manner. The word is formed from the two words "thermal" and "resistor."

Traffic: Radio messages exchanged between stations.

Transducer: A device that converts one form of energy into another. In depth sounder applications, electronic impulses are converted into sound impulses and vice versa.

Transient: A phenomenon (such as a surge of voltage or current) caused in a system by a sudden change in conditions, and which may persist for a relatively short time after the change (sometimes called ringing).

Translator, Broadcast: A relay station used to improve the reception of weak television and FM broadcast signals in remote locations. Translator equipment rebroadcast the input signal on another frequency or channel.

Transmission Line: The conduit by which radio frequency energy is transferred from the transmitter to the antenna.

Transponder: A transmitter-receiver capable of accepting the electronic challenge of an interrogator and automatically transmitting an appropriate reply. There are four modes of operation currently in use for military aircraft. Mode 1 is a no secure low cost method used by ships to track aircraft and other ships. Mode 2 is used by aircraft to make carrier controlled approaches to ships during inclement weather. Mode 3 is the standard system used by commercial aircraft to relay their position to ground controllers throughout the world. Mode 4 is IFF.

Traveling-Wave Tube (TWT): A vacuum tube used to amplify UHF and microwave frequencies.

Travelling-Wave Tube Amplifier: The TWT is a microwave amplifier capable of operation over very wide bandwidths. In operation, an electron beam interacts with a microwave signal which is traveling on a slow wave helical structure. The near synchronism of the beam and RF wave velocities results in amplification. Bandwidths of 3:1 are possible. Operation at high powers or at millimeter wavelengths is possible at reduced bandwidths.

Tunnel Diode: A heavily doped P-N junction diode that displays a negative resistance over a portion of its voltage-current characteristic curve. In the tunneling process, electrons from the p-side valence bands are able to cross the energy barrier into empty states in the N-side conduction band when a small reverse bias is applied. This diode is used as a microwave amplifier or oscillator.

Type Acceptance: Radio equipment that has met FCC specifications. All transmitters in the Fixed, Aviation and Maritime Services must be "type accepted." Type acceptance is based on data submitted by the manufacturer. "Type Approval" is granted after tests are made by FCC technical personnel.

U

UHF - Ultra High Frequency: This radio frequency band occupies 300 MHz to 3,000 MHz (or 0.3 GHz to 3 GHz).

Universal Coordinated Time - (UTC): Sometimes called Greenwich Mean Time (GMT), the time appearing at the zero meridian near Greenwich, England. UTC is the standard for time throughout the world.

Universal Licensing System (ULS): FCC licensing system using one integrated database to electronically issue and maintain radio operator licenses.

Uplink: The ground-based frequency on which an orbiting satellite receives its radio signals from Earth. See downlink, transponder.

Upper Sideband: The information carrying portion of the signal just above the amplitude modulated (AM) carrier frequency, which is reduced or eliminated before transmission.

Urgency Communication: Urgent message concerning the safety of a ship, aircraft, other vehicle or person. Urgency traffic has slightly lower priority than distress traffic.

V

Varactor Diode: A P-N junction employing an external bias to create a depletion layer containing very few charge carriers. The diode effectively acts as a variable capacitor.

Varactor Tuned Oscillator: A varactor diode serves as a voltage-controlled capacitor in a tuned circuit to control the frequency of a negative resistance oscillator. The major feature of this oscillator is its extremely fast tuning speed. A limiting factor is the ability of the external voltage driver circuit to change the voltage across the varactor diode, which is primarily controlled by the driver impedance and the bypass capacitors in the tuning circuit.

Vertical Polarization: Standard method of orienting maritime antennas operating at frequencies above 30 MHz: perpendicular to the ground or water. Polarization is determined by the direction of the electric component of the electromagnetic field. Vertically oriented antennas produce vertically polarized waves.

Vessel Traffic Service (VTS): Traffic management service operated by the U.S. Coast Guard in certain water areas to prevent ship collisions, groundings and environmental harm.

VHF - Very High Frequency: This radio frequency band occupies 30 MHz to 300 MHz. (or 3,000 kHz to 30,000 kHz).

Video: Receiver RF signals that have been converted (post detection) into a pulse envelope that can be seen when applied to some type of visual display, such as a RADAR; also used to describe the actual display itself (such as the video on an A-scope).

Violation Notices: Notification from the FCC of a rule infraction. A written response must be made within 10 days containing a full explanation and action taken to prevent reoccurrence.

VLF - Very Low Frequency: This band occupies 3 kHz to 30 kHz.

Voluntary Ship: A ship not required to carry radio equipment, licensed radio operators or keep logs. When radio equipment is carried, however, appropriate listening watches must be maintained on 2182 kHz and Channel 16.

W X Y Z

WWV: The precise standard frequency and time transmissions of the National Bureau of Standards.

Watch: The act of keeping close observation on distress frequencies for any distress messages.

Waveguide: A low-loss circular or rectangular hollow pipe-like feedline used at UHF and microwave frequencies. The waves are carried inside the pipe.

Wavemeter: An instrument for measuring the frequency of a radio wave. The wavemeter is a mechanically tunable resonant circuit. It must be part of a reflection of transmission measurement system to measure the maximum response of a signal. Below 20 GHz, the wavemeter has been replaced by the frequency counter with much greater accuracy and ease of use.

Wilco: Phrase spoken in radiotelephony to acknowledge that a message has been received and that the receiving station will comply. "Wilco" is short for "will comply."

Working Frequency: A frequency establishing for conducting communications after first being established on a Calling Frequency.

YIG Tuned Oscillator: A YIG (yttrium iron garnet) sphere, when installed in the proper magnetic environment with suitable coupling will behave like a tunable microwave cavity with Q on the order of 1,000 to 8,000. Since spectral purity is related to Q, the device has excellent AM and FM noise characteristics.

Zener Diode: A diode that exhibits in the avalanche-breakdown region a large change in reverse current over a very narrow range in reverse voltage. This characteristic permits a highly stable reference voltage to be maintained across the diode despite a wide range of current.

Index

M

N

O

P

Q

R

S

T

U

V

W

Z